Lecture Notes in Mathematics

Edited by A. Dold and B. Eckmann

1104

Computation and Proof Theory

Proceedings of the Logic Colloquium
held in Aachen, July 18–23, 1983
Part II

Edited by M. M. Richter, E. Börger,
W. Oberschelp, B. Schinzel and W. Thomas

Springer-Verlag
Berlin Heidelberg New York Tokyo 1984

Editors

Egon Börger
Lehrstuhl für Informatik II, Universität Dortmund
Postfach 500500, 4600 Dortmund, Federal Republic of Germany

Walter Oberschelp
Lehrstuhl für Angewandte Mathematik, insbesondere Informatik, RWTH Aachen
Templergraben 57, 5100 Aachen, Federal Republic of Germany

Michael M. Richter
Lehrgebiet mathematische Grundlagen der Informatik, RWTH Aachen
Templergraben 64, 5100 Aachen, Federal Republic of Germany

Brigitta Schinzel
Lehrgebiet Theoretische Informatik, RWTH Aachen
Büchel 29–31, 5100 Aachen, Federal Republic of Germany

Wolfgang Thomas
Lehrstuhl für Informatik II, RWTH Aachen
Büchel 29–31, 5100 Aachen, Federal Republic of Germany

AMS Subject Classification (1980): 03Dxx, 03Fxx, 68Cxx

ISBN 3-540-13901-X Springer-Verlag Berlin Heidelberg New York Tokyo
ISBN 0-387-13901-X Springer-Verlag New York Heidelberg Berlin Tokyo

Printing and binding: Beltz Offsetdruck, Hemsbach/Bergstr.
2146/3140-543210

VORWORT

Dieser Band enthält einen Teil der Proceedings des Logic Colloquium '83, welches vom 18. - 23. Juli 1983 in Aachen stattfand; es war dies gleichzeitig der Europäische Sommerkongreß der Association for Symbolic Logic. Ein weiterer Band der Proceedings erscheint unter dem Titel "Models and Sets" ebenfalls in den Lecture Notes in Mathematics des Springer-Verlages.

Insgesamt hatte das Logic Colloquium '83 189 Teilnehmer aus 26 Ländern. Zusätzlich zu den eingeladenen Hauptvorträgen wurden siebzig angemeldete Vorträge gehalten. Ein Teil davon fand in "Special Sessions" statt: Boole'sche Algebren (organisiert von S. Koppelberg), Topologische Modelltheorie (J. Flum), Nonstandard Analysis (K.-H.Diener), Logic versus Computer Science (E. Börger). Abstracts aller angemeldeten Vorträge sowie eine vollständige Liste aller eingeladenen Vorträge werden im Bericht der Veranstalter im Journal of Symbolic Logic veröffentlicht.

Das Logic Colloquium '83 wurde ermöglicht durch großzügige finanzielle Unterstützung der Deutschen Forschungsgemeinschaft, des Landes Nordrhein-Westfalen, der Division of Logic, Methodology and Philosophy of Science, dem Deutschen Akademischen Austauschdienst, der Deutschen Stiftung für Internationale Entwicklung, der Maas-Rhein-Euregio, der Stadt Aachen, der RWTH Aachen und nicht zuletzt der deutschen Industrie. Ihnen allen sei herzlich gedankt!

Die alte Kaiserstadt Aachen gab einen würdigen Rahmen ab. Oberstadtdirektor Dr. H. Berger eröffnete den Kongreß als Schirmherr im Namen der Maas-Rhein-Euregio. Die erste Bürgermeisterin Frau Prof. Dr. W. Kruse lud ein zu einem Empfang im Krönungssaal des Rathauses, der für die Teilnehmer zu einem bleibenden Erlebnis wurde. Auch hierfür ein herzliches "Danke"!

Kurz vor Drucklegung erreichte uns die traurige Nachricht vom Ableben von Prof. Dr. D. Rödding. Er war ein eingeladener Sprecher des Kongresses; sein Beitrag ist in diesem Buche abgedruckt.

Die Herausgeber.

TABLE OF CONTENTS

† Professor Dr. D. Rödding died on June 4, 1984

* An asterisk indicates a contributed paper.

CONTENTS - PART I

(published in LNM vol. 1103)

* Invited Lecture

CONTIGUOUS R.E. DEGREES.

Klaus Ambos-Spies
Lehrstuhl für Informatik II
Universität Dortmund
D-4600 Dortmund 50

Ladner and Sasso [13] observed that strong reducibility notions can be used to obtain results about the structure $\mathop{R}\limits_{\sim}$ of the r.e. (Turing) degrees. Namely they showed that certain results about weak truth table (wtt) degrees can be transferred to ones about Turing degrees. Weak truth table reducibility is very similar to Turing reducibility: A wtt (sometimes also called bounded Turing) reduction is a Turing reduction whose use function is recursively bounded. Despite this similarity, technically weak truth table degrees can be handled much more easily than Turing degrees. For instance density of the r.e. wtt degrees can be proved by a finite injury priority argument whereas the analogous result for r.e. Turing degrees requires an infinite injury proof. Moreover, the structure of r.e. wtt degrees seems to be much more well behaved than that of Turing degrees. So Ladner and Sasso showed that a combined density and splitting theorem holds for the r.e. wtt degrees whereas by a famous result of Lachlan in general density and splitting do not combine for r.e. Turing degrees. The probably most significant difference between both structures is that the r.e. wtt degrees, but not the r.e. Turing degrees, form a *distributive* upper semi-lattice (Lachlan [10]).

The key fact used in transferring results from wtt to Turing degrees is the existence of socalled *contiguous* degrees, i.e. r.e. (Turing) degrees which contain only one r.e. wtt degree. Since, for r.e. sets A, B, C such that degC is contiguous, $deg_{wtt}A \leq deg_{wtt}C$ iff $deg_T A \leq deg_T C$ and $deg_{wtt}A \cup deg_{wtt}B = deg_{wtt}C$ iff $deg_T A \cup deg_T B = deg_T C$, the structures of the r.e. wtt respective Turing degrees which cup to a contiguous degree show certain similarities.

Unfortunately contiguous degrees are rather scarce. Cohen [7] and others have shown that contiguous degrees are low_2, and Ambos-Spies and Fejer [4] proved that the class of contiguous degrees is nowhere dense in the set of low degrees. Still it seems that the method of transferring results from wtt to Turing degrees using contiguity is quite powerful. Though, by the lack of density of the contiguous degrees, this method cannot be used to prove any homogeneity results for the r.e. Turing degrees, in various cases it has been applied to show that certain phenomena occur in the structure $\mathop{R}\limits_{\sim}$. In particular this method has been used to study cupping properties of $\mathop{R}\limits_{\sim}$ (Ladner and Sasso [13], Stob [17], Ambos-Spies [2]) and degree theoretical splitting properties of r.e. sets (Ambos-Spies [1], Ambos-Spies and Fejer [4]). Also parts of the proof for the existence of infinitely many 1-types of r.e. degrees in Ambos-Spies and Soare [6] use this method. Most of the quoted applications of Ladner and Sasso's transfer technique exploit the fact that contiguity implies local distributivity.

The purpose of this paper is twofold. We first prove a series of existence results for contiguous degrees which will be used for applications of the Ladner-Sasso method in [1,2,4,6], and which we hope will also be useful for further applications of this method. We then apply the transfer method to answer some questions about finite distributive sublattices of initial segments of the r.e. Turing degrees.

After some preliminaries in section 1, in section 2 we describe the basic construction of a contiguous degree, and we analyze which properties of the construction suffice to guarantee contiguity. In section 3 we prove some results on the distribution of the contiguous degrees among the r.e. degrees. E.g. we extend Ladner and Sasso's result that every nonzero r.e degree bounds a nonzero r.e. contiguous degree and we show that noncappable contiguous degrees exist and that, for any low degrees $\underset{\sim}{a} < \underset{\sim}{b}$, there is a contiguous degree which is below $\underset{\sim}{b}$ but not below $\underset{\sim}{a}$. In section 4 we study sublattices of initial segments of $\underset{\sim}{R}$. We prove a theorem on lattice embeddings in the contiguous degrees. This result will imply that every finite (infact countable) distributive lattice can be embedded in any initial segment $\underset{\sim}{R}(\leq\underset{\sim}{a})$, $\underset{\sim}{0} < \underset{\sim}{a}$, of $\underset{\sim}{R}$. By Lachlan's non-bounding theorem [11] these embeddings in general do not preserve the least element. For embeddings which preserve the least element we obtain the following criterion: A finite distributive lattice L is embeddable in an initial segment $\underset{\sim}{R}(\leq\underset{\sim}{a})$ by a map which preserves the least element if and only if the maximum number of elements in L which are pairwise minimal pairs (= the number of minimal elements of L) is not greater than the maximum number of r.e. degrees below $\underset{\sim}{a}$ which are pairwise minimal pairs of r.e. degrees. Finally, in section 5 we list some limiting results on the existence of contiguous degrees.

§1. PRELIMINARIES.

Our notation is standard as in Soare [15], with a few modifications. All sets - denoted by A, B, C, ... - and all (Turing) degrees - denoted by $\underset{\sim}{a}$, $\underset{\sim}{b}$, $\underset{\sim}{c}$, ... - are recursively enumerable. $\deg_T A$, or shortly degA, is the degree of A. We identify a set and its characteristic function; so $x \in A$ iff $A(x) = 1$ and $x \notin A$ iff $A(x) = 0$. For a (partial) function $f : \omega \to \omega$, $f \upharpoonright x$ denotes the restriction of f to arguments less than x. In particular $A \upharpoonright x = \{y : y \in A \;\&\; y < x\}$. A *string* is an element of $\{0,1\}^{<\omega}$, i.e. a finite sequence of 0s and 1s. Strings are denoted by σ, τ, ... We write $\sigma \subseteq \tau$ if σ is an initial segment of τ, and $\sigma \subset \tau$ if $\sigma \subseteq \tau$ and $\sigma \neq \tau$. $\sigma * \tau$ denotes the concatenation of σ with τ. $|\sigma|$ is the length of σ. For $i < |\sigma|$, $\sigma(i)$ denotes the (i+1)st component of σ, i.e. $\langle x_0,\dots,x_{n-1}\rangle(i) = x_i$, and $\sigma \upharpoonright i$ denotes the initial segment of σ of length i. We say σ is *to the left of* τ, $\sigma <_L \tau$, if $\sigma \upharpoonright i = \tau \upharpoonright i$ and $\sigma(i) = 0$ and $\tau(i) = 1$ for some $i < |\sigma|,|\tau|$. Moreover, a total ordering of $\{0,1\}^{<\omega}$ is defined by $\sigma < \tau$ iff $\sigma <_L \tau$ or $\sigma \subset \tau$. We identify $A \upharpoonright x$ with the string σ of length x where $\forall i<x\ (A(i) = \sigma(i))$ and we write $\sigma \subset A$ ($\sigma <_L A$, $\sigma < A$) iff $\sigma \subset A \upharpoonright |\sigma| + 1$ ($\sigma <_L A \upharpoonright |\sigma| + 1$, $\sigma < A \upharpoonright |\sigma| + 1$).

For $\Phi_e(A;x)$, the e-th Turing reduction with oracle A applied to x, we write $\{e\}^A(x)$ or $\{e\}(A;x)$. Similarly, for $\Phi_{e,s}(A;x)$ we write $\{e\}_s^A(x)$ or $\{e\}_s(A;x)$. The use functions of $\{e\}(A;x)$ and $\{e\}_s(A;x)$ are denoted by $u(e,A,x)$ and $u(e,A,x,s)$, respectively. By convention, for any e, A, s, x,

(1.1) $\{e\}_s(A;x)\downarrow \ \rightarrow\ e,x,u(e,A,x,s) < s.$

For $n \geq 2$, let $\langle\ \rangle : \omega^n \to \omega$ be a recursive bijection which is nondecreasing in each argument. $A^{(x)}$ denotes the set $\{\langle x,y\rangle : y \in \omega \text{ and } \langle x,y\rangle \in A\}$.

Recall that a set A is *weak truth table (wtt) reducible to* a set B, $A \leq_{wtt} B$, if, for some e, $A = \{e\}^B$ and the use function $u(e,B,x)$ is dominated by a recursive function. For $e = \langle e_0,e_1\rangle$ we let

$$[e](x) = \{e_1\}(x), \qquad [e]_s(x) = \{e_1\}_s(x),$$

$$[e]^A(x) = [e](A;x) = \begin{cases} \{e_0\}(A;x) & \text{if } \{e_0\}(A;x)\downarrow \text{ and } u(e_0,A,x) \leq [e](x)\downarrow \\ \uparrow & \text{otherwise,} \end{cases}$$ and

$$[e]_s^A(x) = [e]_s(A;x) = \begin{cases} \{e_0\}_s(A;x) & \text{if } \{e_0\}_s(A;x)\downarrow \text{ and } u(e_0,A,x,s) \leq [e]_s(x)\downarrow \\ \uparrow & \text{otherwise.} \end{cases}$$

Obviously, $A \leq_{wtt} B$ iff $A = [e]^B$ for some e. $\deg_{wtt} A$ denotes the wtt degree of A; $\underline{a}$, $\underline{b}$, $\underline{c}$, ... stand for r.e. wtt degrees.

<u>1.1. Definition</u> (Ladner and Sasso [13]). An r.e. (Turing) degree $\underset{\sim}{a}$ is *contiguous* if it contains only one r.e. wtt degree; i.e., if for any r.e. sets A_0, A_1 in $\underset{\sim}{a}$, $A_0 \equiv_{wtt} A_1$.

Note that the degree $\underset{\sim}{0}$ of recursive sets is contiguous. The usefulness of contiguous degrees for transferring results about wtt degrees to ones about Turing degrees stems from the following fact.

<u>1.2. Proposition.</u> Let A, B, C be r.e sets such that degA is contiguous. Then $B \leq_T A$ iff $B \leq_{wtt} A$ and $B \oplus C \equiv_T A$ iff $B \oplus C \equiv_{wtt} A$. In other words, $\deg_T B \leq \deg_T A$ iff $\deg_{wtt} B \leq \deg_{wtt} A$ and $\deg_T B \cup \deg_T C = \deg_T A$ iff $\deg_{wtt} B \cup \deg_{wtt} C = \deg_{wtt} A$.

<u>Proof.</u> Immediate. □

The following result on wtt reducibility is exploited in many applications of contiguous degree results.

1.3. Theorem (Lachlan). Let $A, B_o, \ldots, B_n$ $(n \geq 1)$ be r.e. sets such that $A \leq_{wtt} B_o \oplus \ldots \oplus B_n$. Then there is a partition of A into r.e. sets $A_o, \ldots, A_n$ such that $A_i \leq_{wtt} B_i$ for $i \leq n$.

A proof for the case $n = 1$ can be found in Stob [17]. The general case follows by induction on n.

Note that, for $A_o, \ldots, A_n$ as in Theorem 1.3, $A =_{wtt} A_o \oplus \ldots \oplus A_n$. So Theorem 1.3 implies that the upper semi-lattice of r.e. wtt degrees is distributive:

1.4. Corollary (Lachlan). Let $\underline{a}, \underline{b}_o, \ldots, \underline{b}_n$ be r.e. wtt degrees such that $\underline{a} \leq \underline{b}_o \cup \ldots \cup \underline{b}_n$. Then there are r.e. wtt degrees $\underline{a}_o, \ldots, \underline{a}_n$ such that $\underline{a} = \underline{a}_o \cup \ldots \cup \underline{a}_n$ and $\forall i \leq n\ (\underline{a}_i \leq \underline{b}_i)$. □

Theorem 1.3 and Corollary 1.4 in general fail for Turing reducibility in place of wtt reducibility (Lachlan [10]). By Proposition 1.2, however, these results can be partially transferred.

1.5. Corollary. (a) Let $A, B_o, \ldots, B_n$ be r.e. sets such that $\deg(B_o \oplus \ldots \oplus B_n)$ is contiguous and $A \leq_T B_o \oplus \ldots \oplus B_n$. Then there is a partition of A into r.e. sets $A_o, \ldots, A_n$ such that $A_i \leq_{wtt} B_i$ for $i \leq n$.

(b) Let $\underset{\sim}{a}, \underset{\sim}{b}_o, \ldots, \underset{\sim}{b}_n$ be r.e. degrees such that $\underset{\sim}{b}_o \cup \ldots \cup \underset{\sim}{b}_n$ is contiguous and $\underset{\sim}{a} \leq \underset{\sim}{b}_o \cup \ldots \cup \underset{\sim}{b}_n$. Then there are r.e. degrees $\underset{\sim}{a}_o, \ldots, \underset{\sim}{a}_n$ such that $\underset{\sim}{a} = \underset{\sim}{a}_o \cup \ldots \cup \underset{\sim}{a}_n$ and $\forall i \leq n\ (\underset{\sim}{a}_i \leq \underset{\sim}{b}_i)$. □

We conclude this section with one more fact on contiguous degrees which we will need later.

1.6. Proposition. Let $\underset{\sim}{b}_o, \ldots, \underset{\sim}{b}_n$ $(n \geq 1)$ be r.e. degrees such that $\underset{\sim}{b}_o \cup \ldots \cup \underset{\sim}{b}_n$ is contiguous and $\forall i,j \leq n\ (i \neq j \rightarrow \underset{\sim}{b}_i \cap \underset{\sim}{b}_j = \underset{\sim}{0})$. Then the degrees $\underset{\sim}{b}_o, \ldots, \underset{\sim}{b}_n$ are contiguous too.

Proof. Let A_i, B_i be any r.e. sets with $\deg A_i = \deg B_i = \underset{\sim}{b}_i$ $(i \leq n)$. It suffices to show $A_i \leq_{wtt} B_i$.

Since $A_i \leq_{wtt} B_o \oplus \ldots \oplus B_n$, by Corollary 1.5 there are r.e. sets A_i^j, $j \leq n$, such that $A_i^j \leq_{wtt} B_j$ and $A_i =_{wtt} A_i^o \oplus \ldots \oplus A_i^n$. Now, for $j \neq i$, $A_i^j \leq_{wtt} B_j$ and $A_i^j \leq_{wtt} A_i \leq_T B_i$. So, since $\underset{\sim}{b}_i \cap \underset{\sim}{b}_j = \underset{\sim}{0}$, A_i^j is recursive. It follows that $A_i =_{wtt} A_i^i \leq_{wtt} B_i$. □

§2. THE BASIC CONSTRUCTION OF A CONTIGUOUS DEGREE.

Contiguous degree constructions can be found in Ladner and Sasso [13] and Stob [17]. In both cases stronger results are proved. The plain construction of a nonzero contiguous degree does not appear in the literature. In this section we therefore present the basic construction of a contiguous degree and analyze which features of the construction are sufficient to guarantee contiguity. Based on this analysis, in the following section we will reduce the construction of special contiguous degrees to the basic construction given here. Our construction which uses a tree argument is very similar to the construction one can extract from Stob's construction of a minimal pair with contiguous supremum ([17, Theorem 4.1]). Also some of our terminology is taken from [17].

__2.1. Theorem.__ There is a nonzero contiguous r.e. degree.

__Proof.__ By a priority argument we construct a nonrecursive r.e. set A such that degA is contiguous. Nonrecursiveness of A is ensured by meeting the requirements

$$R_e \; : \; A \neq \{e\} \qquad (e \in \omega).$$

To make degA contiguous we have to satisfy the condidtion

$$\forall e \,(A \equiv_T W_e \; \rightarrow \; A \equiv_{wtt} W_e).$$

This condition can be broken down in an infinite list of requirements

$$N_{<e,i,j>} \; : \; A = \{i\}^{W_e} \;\&\; W_e = \{j\}^A \; \rightarrow \; A \equiv_{wtt} W_e \qquad (e,i,j \in \omega).$$

Each requirement $N_{<e,i,j>}$ can be further split into the subrequirements

$$N^1_{<e,i,j>} \; : \; A = \{i\}^{W_e} \;\&\; W_e = \{j\}^A \; \rightarrow \; A \leq_{wtt} W_e \quad \text{and}$$

$$N^2_{<e,i,j>} \; : \; A = \{i\}^{W_e} \;\&\; W_e = \{j\}^A \; \rightarrow \; W_e \leq_{wtt} A.$$

Obviously the N^1 (N^2) requirements imply that $\deg_{wtt} A$ is the least (greatest) r.e. wtt degree in $\deg_T A$.

The priority ordering of the requirements is given by $R_0 > N_0 > R_1 > N_1 > \ldots$.

In the course of the construction the requirements R_e will have certain *followers* $x \in \omega$ for which we try to ensure $A(x) \neq \{e\}(x)$. I.e. if $\{e\}(x) = 0$ then we will restrain x from A; otherwise we attempt to enumerate x in A. It will be crucial for our strategies for meeting the N requirements that the followers of the R requirements obey certain rules.

Therefore we will next define what it means for a set A to be constructed by a system of followers. This definition will be independent of the context of the proof for Theorem 2.1.

Let A be effectively constructed in stages, say $A = \cup \{A_s : s \in \omega\}$, where A_s denotes the part of A enumerated by the end of stage s. If $\{A_s : s \in \omega\}$ is a construction by a system of followers then certain requirements which we attempt to meet in the course of the construction will have followers. We will call the requirements, which are eligible for having followers, the R requirements.

A *follower* is a number. Once a follower is *appointed* it can be eventually *cancelled* or it can exist for ever. In the latter case we say the follower is *permanent*. We say x is a *follower at stage* s if x has been appointed at a stage $\leq s$ and has not been cancelled by the end of stage s. A follower is called *relevant at stage* s if x is a follower at stage s and $x \notin A_s$. Finally, followers can be *active* at certain stages. We say an R requirement is *active* at some stage if some of its followers is.

We now list some rules for followers which each system of followers will obey.

(F1) For each number x which is a follower there is a unique R requirement which x follows. Moreover, there is a total priority ordering defined on the set of R requirements.

(F2) If $x \in A_{s+1}-A_s$ then x is a follower at both stage s and stage $s+1$. $A_0=\emptyset$.

(F3) If x is appointed to be a follower at stage s then $s > 0$ and $x = s - 1$.

(F4) At every stage at most one follower is active.

(F5) Every follower is active at most finitely often.

(F6) If x is appointed at stage s then x is active at stage s.

(F7) If x is the least follower in $A_s - A_{s-1}$ then x is active at stage s.

Note that, by (F4) and (F6), at most one follower is appointed at every stage. So, by (F1), we can define a total *priority ordering* on the set of followers by

$x \propto y$ ("x has higher priority than y") iff there are requirements R_1 and R_2 such that x and y follow R_1 and R_2, respectively, and either R_1 has higher priority than R_2 or $R_1 = R_2$ and x has been appointed earlier than y.

This priority ordering is used in the formulation of the following two *cancellation rules*.

(F8) If y is a follower at stage s-1, $x \propto y$ and x is active at stage s but not enumerated in A at stage s, then y is cancelled at stage s (i.e. $A(y) = A_{s-1}(y)$ by (F2)).

(F9) If y is a follower at stage s-1, $x \propto y$ and x is active at stage s and $x \in A_s - A_{s-1}$ then y is either cancelled or enumerated in A at stage s (i.e. y is not relevant at stage s).

We now discuss the intention behind the above rules. (F2) ensures that only followers enter A. So, by cancellation, we can restrain certain numbers from A. (F3) ensures that the set of followers is recursive and that a follower appointed at stage $s+1$ has no impact on the A-correctness of any existing computation at stage s. I.e. if $\{e\}_s(A_s;y)\downarrow$ then, by (1.1), $u(e,A_s,y,s)<s$; so, by (F3) and (F2), the A-correctness of this computation only depends on the values $A(x)$ for followers x appointed prior to stage s. Also by (F2) and (F3), $A_s \subseteq \{0,\dots,s-2\}$. So, for any follower x appointed at stage $s+1$, $x \notin A_s$. Moreover, (F3) ensures that every follower is appointed only once. (F6) and (F7) are special instances of activities of a follower. Other activities will be specified in the respective constructions. Note that by (F5) a requirement which has only finitely many followers is active only finitely often. The cancellation rules ensure that the priority ordering of the followers at any stage is just the natural ordering.

2.1.1. Proposition. (Assuming (F1) - (F9)). If x,y are followers at stage s then $x \propto y$ iff $x<y$.

Proof. Let x,y be followers at stage s such that $x<y$. Then, by (F3), $x+1 < y+1 \le s$ and x (y) is appointed at stage $x+1$ ($y+1$). So x is a follower at stage y and, by (F6) and (F2), y is active at stage $y+1$ but it is not enumerated in A at this stage. Since x is not cancelled at stage $y+1$, by (F8) this implies that y does not have higher priority than x. Hence $x \propto y$. □

The following fact is a direct consequence of the cancellation rules.

2.1.2. Proposition. (Assuming (F1) - (F9)). (a) If x is active at stage $s+1$ then $\forall y\,(x<y\le s \rightarrow A(y)=A_{s+1}(y))$.

(b) If x is active at stage $s+1$ but x is not enumerated in A at this stage then $\forall y\,(x<y\le s \rightarrow A(y)=A_s(y))$.

Proof. For $y<s$ this follows from Proposition 2.1.1 and (F2,F3,F8,F9); for $y=s$, by (F3,F4,F6), y is not a follower and thus, by (F2), $A(y)=0$. □

We now return to the proof of Theorem 2.1 and describe the strategy for meeting the N^1 subrequirements in a construction using a follower concept as just described. Let

$$u(\langle e,i,j\rangle,x,s)=\max\{u(i,W_{e,s},y,s) : y\le x\}.$$

The strategy is based on the following observation.

2.1.3. Proposition. Assume that

(2.0) $A=\{i\}^{W_e}$ and $W_e=\{j\}^A$

and that x and s satisfy

(2.1) $A_s \upharpoonright x+1 = \{i\}_s^{W_{e,s}} \upharpoonright x+1$ &

$W_{e,s} \upharpoonright u(<e,i,j>,x,s) = \{j\}_s^{A_s} \upharpoonright u(<e,i,j>,x,s)$

and

(2.2) $\forall y\ (\ x < y < s \ \rightarrow\ A_s(y) = A(y)\ )$.

Then

(2.3) $A_s \upharpoonright x+1 = A \upharpoonright x+1$ iff $W_{e,s} \upharpoonright u(<e,i,j>,x,s) = W_e \upharpoonright u(<e,i,j>,x,s)$.

Proof. The direction "$\Leftarrow$" in (2.3) is immediate by (2.1) and $A = \{i\}^{W_e}$. For the other direction assume $A_s \upharpoonright x+1 = A \upharpoonright x+1$, and let $u = u(<e,i,j>,x,s)$. Then, by (2.2), $A_s \upharpoonright s = A \upharpoonright s$ and thus (since by (2.1) $\forall z<u\ (\ \{j\}_s^{A_s}(z)\downarrow\ )$) $\{j\}_s^{A_s} \upharpoonright u = \{j\}^A \upharpoonright u$. Since $W_e = \{j\}^A$, this and (2.1) imply $W_{e,s} \upharpoonright u = W_e \upharpoonright u$. □

Now the idea for meeting a requirement $N^1_{<e,i,j>}$ is as follows. W.l.o.g. assume that (2.0) holds. Then we have to define a recursive function $u(x)$ such that, for (almost) all x, $A(x)$ can be computed from $W_e \upharpoonright u(x)$ uniformly in x. So fix x. Since only followers can enter A and since we can decide if a number is a follower, w.l.o.g. assume that x is a follower. Note that, by (2.0), (2.1) holds for all sufficiently large s. So we can effectively compute s_x where s_x is the least $s > x+1$, i.e. by (F3) the least stage s after the stage x was appointed, such that (2.1) holds. Now if x is still relevant at stage s_x and no higher priority follower wants to act at stage s_x+1, then we let x become active at stage s_x+1 and say x is *confirmed at* <e,i,j> (at stage s_x+1). Now let $u(x) = u(<e,i,j>,x,s_x)$. We show that $A(x)$ can be computed from $W_e \upharpoonright u(x)$. Namely first check if x is relevant at stage s_x+1. If not then $A(x) = A_{s_x+1}(x)$. If so then, by (F8) and (F9), no higher priority follower acts at stage s_x+1 and x is relevant at stage s_x. So x is confirmed at <e,i,j> at stage s_x+1 (and $x \notin A_{s_x+1}$), whence, by Proposition 2.1.2(b), (2.2) holds for $s = s_x$. Therefore, by Proposition 2.1.3, $A_{s_x} \upharpoonright x+1 = A \upharpoonright x+1$ iff $W_{e,s_x} \upharpoonright u(x) = W_e \upharpoonright u(x)$. Now using $W_e \upharpoonright u(x)$ as oracle decide whether $W_{e,s_x} \upharpoonright u(x) = W_e \upharpoonright u(x)$. If so then $A(x) = A_{s_x}(x)$. Otherwise, $A \upharpoonright x+1 \neq A_{s_x} \upharpoonright x+1$ and we compute $t > s_x$ minimal such that $A_{s_x} \upharpoonright x+1 \neq A_t \upharpoonright x+1$. By (F7), some follower $y \leq x$ which is enumerated in A_t is active at stage t. Hence, by Proposition 2.1.2, $A(x) = A_t(x)$. This completes the reduction procedure.

Note that by (F5) a follower cannot be confirmed at each <e,i,j>, but only at finitely many ones. So we will confirm a follower at <e,i,j> only if $N_{<e,i,j>}$ has higher priority than the requirement x follows. If there are only finitely many R requirements having higher priority than $N_{<e,i,j>}$ and if each of these requirements acts only finitely often, then the above strategy will still succeed since almost all followers will be considered for confirmation at <e,i,j>.

Also note that the above argument doesn't require that x is confirmed at the least possible stage but only that there is a recursive function f such that for almost all followers x,

(2.4) $x \in A_{f(x)+1}$ or x is cancelled by the end of stage $f(x)+1$ or (2.1) holds for $s = f(x)$ and x is active at stage $f(x)+1$ (i.e. x is confirmed at $\langle e,i,j\rangle$).

So we proved the following lemma.

2.1.4. Lemma. Assume that $\{A_s : s \in \omega\}$ is a construction using a follower concept satisfying (F1) - (F9) and the additional condition

(F10) For every $\langle e,i,j\rangle$ such that (2.0) holds there is a recursive function f such that for almost all followers x (2.4) holds.

Then A meets all requirements $N^1_{\langle e,i,j\rangle}$, $\langle e,i,j\rangle \in \omega$; i.e. $\deg_{wtt} A$ is the least r.e. wtt degree in $\deg_T A$. □

To give an example of a construction satisfying (F1) to (F10) we describe the finite injury construction of an r.e. set A satisfying the requirements R_n and N^1_n from above. The priority ordering of the requirements is $R_0 > N^1_0 > R_1 > N^1_1 > \ldots$. A follower x of R_n is called *realized* at stage s if $\{n\}_s(x) = 0$. A_s denotes the part of A enumerated by the end of stage s.

2.1.5. Construction.

Stage 0. Do nothing.

Stage s+1. Requirement R_n, $n \leq s$, *requires attention* if there is no follower y of R_n at stage s such that $y \in A_s$ and one of the following holds:

(i) R_n has no follower at stage s.

(ii) There is a realized follower x of R_n at stage s.

(iii) There is a follower x of A at stage s such that for some $\langle e,i,j\rangle < n$, (2.1) holds and x is not yet confirmed at $\langle e,i,j\rangle$.

Choose n minimal such that R_n requires attention.

If (i) holds *appoint* $x = s$ to be follower of R_n. Otherwise choose x minimal such that (ii) or (iii) holds for x. If (ii) holds enumerate x in A; otherwise say x is *confirmed* at $\langle e,i,j\rangle$ for each $\langle e,i,j\rangle < n$ such that (2.1) holds.

In all cases say x is *active* and cancel all followers of requirements $R_{n'}$ with $n' > n$.

Note that every requirement R_n has at most one follower at every stage. Properties (F1) to (F9) are easily verified. E.g. for a proof of (F5), note that a follower x of requirement R_n acts at most $n+2$ times: namely once appointed, once enumerated in A and n-times confirmed at numbers $< n$.

By a straightforward induction on n, each requirement R_n acts only finitely often, eventually has a permanent follower x, and for this x $A(x) \neq \{n\}(x)$. So R_n is met.

Finally, for $<e,i,j>$ such that (2.0) holds and x a follower of a requirement R_n with $n > <e,i,j>$, at the least stage $s+1 > x+1$ such that (2.1) holds x is not relevant at stage s+1 or x is confirmed at $<e,i,j>$ at stage s+1. Since, as shown above, the requirements R_n with $n \leq <e,i,j>$ have only finitely many followers, this implies that (F10) holds. So, by Lemma 2.1.4, the N^1 requirements are met.

We now turn to the N^2 requirements. We first give the "basic module" for meeting a single requirement.

Fix $<e,i,j>$ and assume that the premise (2.0) of $N^2_{<e,i,j>}$ is satisfied. Let $l'(e,j,s) = \max\{x : W_{e,s} \upharpoonright x = \{j\}^{A_s}_s \upharpoonright x\}$. Note that by (2.0) $\lim_s l'(e,j,s) = \omega$. Now the basic idea for meeting requirement $N^2_{<e,i,j>}$ is to confine the construction of A to the successor stages of an infinite set of stages S at which the length $l'(e,j,s)$ of agreement between W_e and $\{j\}^A$ is (essentially) increasing. To be more precise, we will cancel all followers appointed at a stage s+1 with $s \notin S$ at the next S stage.

2.1.6. Lemma. Assume that $\{A_s : s \in \omega\}$ is a construction by followers satisfying (F1) to (F9). Moreover assume that (2.0) holds for $<e,i,j>$, that S is an infinite recursive set and that l is a recursive function such that

(2.5) $\forall s\ (\ l(s) \leq l'(e,j,s)\ )\ \&\ \forall s,t \in S(\ s < t \rightarrow l(s) < l(t)\ ).$

Finally assume

(F11) For almost all followers x the following holds: If x is appointed at stage s+1, $s \notin S$, and $t > s$ is a stage in S then $A(x) = A_t(x)$.

Then requirement $N^2_{<e,i,j>}$ is met.

Proof. We have to show that there is a recursive function u such that $W_e(x)$ can be computed from $A \upharpoonright u(x)$ uniformly in x. Fix x_o, s_o such that (F.11) applies to all followers $x \geq x_o$ and $A_{s_o} \upharpoonright x_o = A \upharpoonright x_o$, and let

$u(x) = \mu s \in S$ ($s > s_o$ & $l(s) > x$).

We claim that for every x,

(2.6) $W_e(x) = W_{e,t(x)}(x)$ where $t(x) = \mu t \in S$ ($t \geq u(x)$ & $A_t \upharpoonright u(x) = A \upharpoonright u(x)$).

Obviously this will give the desired wtt reduction of W_e to A. To verify (2.6), for a contradiction assume that (2.6) fails for x, i.e. $W_e(x) \neq W_{e,t(x)}(x)$. Since, by definition of t(x), $l'(e,j,t(x)) > x$, i.e. $W_{e,t(x)}(x) = \{j\}^{A_{t(x)}}_{t(x)}(x)$, this and (2.0) imply

$A_{t(x)} \upharpoonright u(j,A_{t(x)},x,t(x)) \neq A \upharpoonright u(j,A_{t(x)},x,t(x))$.

So for y, the least element in $A - A_{t(x)}$, $u(x) \leq y < u(j,A_{t(x)},x,t(x)) < t(x)$. It follows that y is a follower appointed at stage y+1 with $u(x) < y+1 < t(x)$. Therefore, by (F.11), $y \in S$ and, by (F.7), Proposition 2.1.2 and definition of t(x), $A_y \upharpoonright u(x) = A_{t(x)} \upharpoonright u(x) = A \upharpoonright u(x)$. So t(x) is not the least t such that $t \in S$, $t \geq u(x)$ and $A_t \upharpoonright u(x) = A \upharpoonright u(x)$. Contradiction. □

Note that for <e,i,j> such that (2.0) holds, S and l as in Lemma 2.1.6 exist. Namely we can let $l(s) = l(\langle e,i,j\rangle,s)$, where

(2.7) $l(\langle e,i,j\rangle,s) = \max \{x : A_s \upharpoonright x = \{i\}^{W_{e,s}}_s \upharpoonright x \ \& \ W_{e,s} \upharpoonright u(x) = \{j\}^{A_s}_s \upharpoonright u(x)\}$,

where $u(x) = \max \{u(j,A_s,y,s) : y < x\} \cup \{x\}$(Our choice of l is motivated in part by our strategy for meeting the N^1 requirements; note that (2.1) holds if $l(\langle e,i,j\rangle,s) > x$.), and we let S be the set of <e,i,j>-*expansionary stages*

$\{s : \forall t < s \ (\ l(\langle e,i,j\rangle,t) < l(\langle e,i,j\rangle,s) \ )\}$.

When we try to apply this strategy to all N^2 requirements then two problems arise: The first problem is that for a N^2 requirement with false hypothesis S and l as in Lemma 2.1.6 in general will not exist. Moreover, we cannot decide whether the hypothesis (2.0) of an N^2 requirement is correct or not. The second problem is that if N_n and $N_{n'}$ are requirements with correct premises and we let S and S' be the n- respective n'-expansionary stages then it might happen that $S \cap S' = \emptyset$. So, by (F.11), almost all followers are eventually cancelled. In general this will imply that we fail to meet the R requirements.

The solution to these problems is to use a tree argument, i.e. a tree of R requirements where each R requirement is provided with a guess at the correctness of premises of higher priority N requirements, and to use a *nested* version of the expansionary stages for the sets S (as e.g. in the minimal pair construction in [15]). For a general presentation and treatment of the tree method we refer the reader to Soare [16].

We now describe the priority tree required for the proof of Theorem 2.1 and give the tree construction of a set A satisfying the theorem. Thereafter we will state a lemma which lists properties of a tree construction using a follower concept which suffice to guarantee that the degree of the constructed set is contiguous.

Instead of the requirements R_n we will now have requirements R_σ, $\sigma \in \{0,1\}^{<\omega}$. Requirement R_σ has higher priority than R_τ iff $\sigma < \tau$. N_n has higher priority than R_σ if $n < |\sigma|$. For σ with $|\sigma| = n$, R_σ is a copy of requirement R_n, i.e.

R_σ : $A \neq \{n\}$.

In addition R_σ is provided with a guess about the correctness of the hypotheses of the N requirements of higher priority. Namely, for $n < |\sigma|$, R_σ believes that the premise of N_n is correct iff $\sigma(n) = 0$. Consequently, R_σ will ignore N_n requirements with $\sigma(n) = 1$. Activities of R_σ are (essentially) limited to stages at which the guess of R_σ seems to be correct.

<u>2.1.7. Definition.</u> The set of *σ-stages* is defined by induction on $|\sigma|$ as follows.

(i) Every stage s is a $\langle\rangle$-stage.

(ii) If s is a σ-stage, $|\sigma| = n$, and

$$\forall t < s \; (t \text{ is a } \sigma\text{-stage} \rightarrow l(n,t) < l(n,s))$$

(for l as in (2.7)) then s is a $\sigma * \langle 0\rangle$ - stage. Otherwise s is a $\sigma * \langle 1\rangle$ - stage.

Note that for each s and n there is a unique σ such that $|\sigma| = n$ and s is σ-stage.

<u>2.1.8. Lemma.</u> There is a unique T such that for all $n \in \omega$,

(i) there are infinitely many $T\restriction n$-stages and

(ii) there are only finitely many stages s such that s is σ-stage and $\sigma <_L T\restriction n$.

<u>Proof</u>. Straightforward. □

We call T the *left-most path* or *true path*.

<u>2.1.9. Lemma.</u> For all $\langle e,i,j\rangle$ such that $A = \{i\}^{W_e}$ and $W_e = \{j\}^A$,

(i) $T(\langle e,i,j\rangle) = 0$ and

(ii) $\forall s,t \; (s,t \text{ are } T\restriction\langle e,i,j\rangle+1\text{-stages} \;\&\; s<t \rightarrow l(\langle e,i,j\rangle,s) < l(\langle e,i,j\rangle,t))$.

<u>Proof.</u> By definition of σ-stages and T. □

So the requirements R_σ with $\sigma \subset T$ have the correct guess in the sense that if the hypothesis of an N requirement is satisfied then R_σ guesses so. Moreover the lemma shows that if we essentially restrict the construction to T-stages then, by Lemmata 2.1.6 and 2.1.8, the N^2 requirements are met. To achieve this we make requirement R_σ only accessible at stages s+1 such that s is a σ-stage. Furthermore if s is a σ-stage,

$|\sigma| = s$, then all followers of requirements R_τ, $\sigma <_L \tau$, are cancelled. So, for $\sigma \subset T$, R_σ is accessible infinitely often and from some stage on followers of R_σ can be cancelled only by activities of requirements R_τ with $\tau \subseteq \sigma$. Since the R requirements are finitary, this will suffice to meet those copies of the requirements R_n which are on the true path.

We now describe the construction of a set A witnessing Theorem 2.1. Here - as in all subsequent constructions - A_s will denote the part of A enumerated by the end of stage s.

2.1.10. Construction.

Stage 0. Do nothing.

Stage s+1. Step 1. Fix the unique σ_s such that $|\sigma_s| = s$ and s is σ_s-stage. Cancel all followers of requirements R_τ with $\sigma_s <_L \tau$.

Step 2. Requirement R_σ, $\sigma \leq \sigma_s$, *requires attention* if at the end of stage s there is no follower y of R_σ in A_s and one of the following holds

(i) $\sigma \subseteq \sigma_s$ and R_σ has no follower.

(ii) There is a follower x of R_σ which is *realized*, i.e. $\{|\sigma|\}_s(x) \simeq 0$.

(iii) There is a follower x of R_σ such that for some $\langle e,i,j\rangle < |\sigma|$ (2.1) holds and x is not yet confirmed at $\langle e,i,j\rangle$.

Choose σ minimal (w.r.t. <) such that R_σ requires attention.

If (i) holds, *appoint* x = s to be follower of R_σ. Otherwise choose x minimal such that (ii) or (iii) holds for x. If (ii) holds, enumerate x in A; otherwise say x is *confirmed at* each $\langle e,i,j\rangle < |\sigma|$ such that (2.1) holds.

In all cases say x is *active* and cancel all lower priority followers.

Note that R_{σ_s} has no follower at the end of stage s. So at any stage at least one requirement requires attention. Also note that at every stage every requirement has at most one follower. So in step 2 of stage s+1 just the followers of requirements R_τ with $\sigma < \tau$ are cancelled.

To prove the construction correct we verify the following claims.

Claim 1. Conditions (F1) - (F9) are satisfied.

Proof. This is shown as for construction 2.1.5. □

Claim 2. For each n, there is a stage t_n such that there is no stage $s \geq t_n$

and no $\tau <_L T \restriction n$ such that s is a τ-stage or R_τ is active at stage s.

Proof. Since followers for a requirement R_τ are only appointed at stage s+1 where s is τ-stage, this follows from definition of T and (F.5). □

Claim 3. Let $R_n = R_{T\restriction n}$. Then R_n acts only finitely often, has a permanent follower and is met.

Proof (by induction on n). Fix n and, by inductive hypothesis, choose s_o such that no requirement $R_{n'}$, $n' < n$, acts after stage s_o, and let $s_1 = \max\{s_o, t_n\}$ for t_n as in Claim 2. Then no R_n follower will be cancelled after stage s_1 and R_n becomes active whenever it requires attention. So if s_2 is the least $T\restriction n$-stage greater than s_1 (note that by definition of T there are infinitely many $T\restriction n$-stages) then either R_n has some follower x at stage s_2 or $x = s_2$ is appointed R_n follower at stage s_2+1. Furthermore, either x is never realized and $A(x) = 0$ or x is eventually realized and $A(x) = 1$. So R_n is met and x is the last follower of R_n whence, by (F.5), R_n eventually stops acting. □

Claim 4. If $T \restriction |\tau| <_L \tau$ then every follower of R_τ is eventually cancelled.

Proof. By step 1 of the construction since there are infinitely many $T\restriction|\tau|$-stages.

Claim 5. (F.10) holds.

Proof. Fix <e,i,j> such that (2.0) holds, i.e. $A = \{i\}^{W_e}$ and $W_e = \{j\}^A$. It suffices to show that almost all followers are eventually enumerated in A, cancelled, or confirmed at <e,i,j>. Moreover, by Claims 2, 3 and 4, it suffices to show this for followers of requirements R_σ with $T \restriction \langle e,i,j\rangle \subset \sigma$. So fix a follower x of such a requirement and assume that x is neither cancelled nor enumerated in A. Note that, by (2.0), for almost all s, (2.1) holds . Since x is the only follower of R_σ at all stages $\geq$ x, at the least stage $s \geq x$ with (2.1), R_σ requires attention via (iii), and, since x is never cancelled, x is confirmed at <e,i,j> at stage s+1. □

Claim 6. Let <e,i,j> be given such that (2.0) holds and let

$S = \{s : s \text{ is } T\restriction\langle e,i,j\rangle+1\text{-stage}\}$

and

$l(s) = l(\langle e,i,j\rangle, s)$.

Then the hypothesis of Lemma 2.1.6 is satisfied.

Proof. By Lemma 2.1.9, S is infinite and (2.5) holds. For a proof of (F.11), by Claims 2 and 3, it suffices to consider followers of requirements R_σ with

$T \restriction \langle e,i,j\rangle < \sigma$. Now if $T \restriction \langle e,i,j\rangle \subset \sigma$ then R_σ followers are appointed only at stages $s+1$ with $s \in S$, and if $T \restriction \langle e,i,j\rangle <_L \sigma$, then by step 1 of stage $s+1$ any follower x of R_σ is cancelled at the beginning of the least stage $s+1 > x+1$ such that $s \in S$, and thus $A_s(x) = A(x)$. □

The above claims show that degA has the desired properties. By Claim 3, the R requirements are met and thus A is nonrecursive. Claims 1 and 5 and Lemma 2.1.4 show that A meets the N^1 requirements. Finally, Claim 6 and Lemma 2.1.6 imply that the N^2 requirements are met. So degA is contiguous.

This completes the proof of Theorem 2.1. □

2.2. Remarks. (1) Note that in the above construction activities of requirement R_σ are not strictly confined to σ-stages but only the appointment of new followers. By (F.5) this suffices to prevent that requirements to the left of the true path act infinitely often. This extra accessibility of the R requirements will be crucial when the construction is combined with the permitting method (see Theorem 3.1 below).

(2) Also note that for a follower x of a requirement R_σ it suffices to confirm x at such $\langle e,i,j\rangle$ such that $\sigma(\langle e,i,j\rangle) = 0$. Namely assume that (2.0) holds for $\langle e,i,j\rangle$. Then $T(\langle e,i,j\rangle) = 0$. So every follower of a requirement R_σ with $T \restriction \langle e,i,j\rangle+1 \subseteq \sigma$ will be considered for confirmation at $\langle e,i,j\rangle$ as before, while (by Claims 2 and 3) there are only finitely many followers of requirements R_σ with $\sigma < T \restriction \langle e,i,j\rangle+1$ and (by step 1 of the construction) followers of requirements R_σ with $T \restriction \langle e,i,j\rangle+1 <_L \sigma$ are eventually cancelled. So (F.10) will still hold.

On the other hand we can require a follower x of a requirement R_σ to be *completely confirmed*, i.e. to be confirmed at each $n < |\sigma|$ for which $\sigma(n) = 0$, before x is enumerated in A. Namely, for any n such that $T(n) = 0$, $\lim_s \{l(n,s) : s \text{ is } T\restriction n\text{-stage}\} = \omega$. So, for a follower x of a requirement R_σ with $\sigma \subset T$ and for any $n = \langle e,i,j\rangle$ with $\sigma(n)=0$, (2.1) holds for all sufficiently large σ-stages. Hence for such requirements the permanent followers (cf. Claim 3) will eventually be completely confirmed and thus we can argue as before that these requirements are met.

So in construction 2.1.10 in the definition of requiring attention clauses (ii) and (iii) may be replaced by

(ii') There is a realized and completely confirmed follower x of R_σ.

(iii') There is a follower x of R_σ such that for some $\langle e,i,j\rangle < |\sigma|$, $\sigma(\langle e,i,j\rangle) = 0$, (2.1) holds and x is not yet confirmed at $\langle e,i,j\rangle$.

This variant of the construction will be a more convenient basis for extensions of Theorem 2.1 than the original construction.

We conclude this section with a lemma summarizing properties of a construction which suffice to make the degree of the constructed set contiguous.

We say that an effective construction $\{A_s : s \in \omega\}$ by a follower concept is a *tree construction* if the following holds.

1) The set of R followers is ordered by a priority tree; i.e. the R requirements are indexed by strings and R_σ has higher priority than R_τ iff $\sigma < \tau$.

2) There is an (effective) notion of σ-stage such that for every s and n there is a unique string σ with $|\sigma| = n$ and s is σ-stage. Furthermore, if s is σ-stage then s is $\sigma \upharpoonright n$-stage for each $n < |\sigma|$.

Note that Lemma 2.1.8 holds for any notion of σ-stage as above. So for every tree construction the *true path* T can be defined by 2.1.8.

<u>2.3. Lemma.</u> Assume that $\{A_s : s \in \omega\}$ is an effective tree construction using a follower concept satisfying (F1) to (F9) such that the following hold.

(2. 8) If a follower of R_σ is appointed at stage s+1 then $|\sigma| \leq s$ and s is a σ-stage.

(2. 9) If s is a σ-stage, $|\sigma| = s$, then all followers x of requirements R_τ with $\sigma < \tau$ are cancelled at the beginning of stage s+1 and $A_s(x) = A(x)$.

(2.10) If $\sigma \subset T$, T the true path of the construction, then R_σ acts only finitely often.

(2.11) For every <e,i,j> such that (2.0) holds there is some n such that for any follower x of a requirement R_σ with $T \upharpoonright n \subseteq \sigma$ there is a stage $s > x$ such that

(2.11.1) x is not relevant at stage s+1 or (2.1) holds and x is active at stage s+1 (i.e. x is confirmed at <e,i,j>).

(2.12) For every <e.i.j> such that (2.0) holds there is some n such that for $T \upharpoonright n$-stages s and t,

$$s < t \rightarrow l(\langle e,i,j\rangle,s) < l(\langle e,i,j\rangle,t)$$

(where l is defined as in (2.7)).

Then degA is contiguous.

<u>Proof.</u> We have to show that the N^1 and N^2 requirements are satisfied. Fix <e,i,j> and w.l.o.g. assume that (2.0) holds. Now, for a proof that $N^1_{\langle e,i,j\rangle}$ is met, fix n as in (2.11). By Lemma 2.1.4, it suffices to show that for almost all followers x there is a stage s > x satisfying (2.11.1). Note that by definition of T there are

only finitely many stages s such that s is a σ-stage for some $\sigma <_L T \upharpoonright n$, whence, by (2.8), there are only finitely many followers of requirements to the left of $T \upharpoonright n$, and, by (2.10), there are only finitely many followers of requirements R_σ with $\sigma \subseteq T \upharpoonright n$. So it suffices to consider followers of requirements R_σ where $T \upharpoonright n \subset \sigma$ or $T \upharpoonright n <_L \sigma$. For the former (2.11.1) holds by (2.11), while the latter are eventually cancelled by definition of T and (2.9).

To show that $N^2_{<e,i,j>}$ is met, fix n as in (2.12), let S be the set of $T\upharpoonright n$-stages and set $l(s) = l(<e,i,j>,s)$. Then, by (2.12) and Lemma 2.1.6, it suffices to show that (F.11) holds. As above it suffices to consider followers of requirements R_σ where $T\upharpoonright n < \sigma$. So let x be follower of such a requirement R_σ. Then, for $T \upharpoonright n \subset \sigma$, x is appointed at some stage s+1 with $s \in S$ by (2.8); and, for $T \upharpoonright n <_L \sigma$, x satisfies (F.11) by (2.9). □

§3. SOME EXISTENCE RESULTS FOR CONTIGUOUS DEGREES.

By combining the construction of a contiguous degree in the preceding section with other techniques we now prove some positive results on the existence of contiguous degrees. We first reprove Ladner and Sasso's result that every nonzero r.e. degree has a nonzero contiguous predecessor.

<u>3.1. Theorem</u> (Ladner and Sasso [13]). Let B be any nonrecursive r.e. set. Then there is a nonrecursive r.e. set A such that $A \leq_{wtt} B$ and degA is contiguous.

<u>3.2. Corollary</u> (Ladner and Sasso [13]).
$\forall \underset{\sim}{b} > \underset{\sim}{0} \; \exists \underset{\sim}{a} \; (\underset{\sim}{0} < \underset{\sim}{a} \leq \underset{\sim}{b} \;\&\; \underset{\sim}{a}$ is contiguous). □

<u>Proof of Theorem 3.1.</u> We combine the proof of Theorem 2.1 with the permitting method. Let $\{B_s : s \in \omega\}$ be a recursive enumeration of B. We will ensure

(3.1) $\forall x,s \, (x \in A_{s+1} - A_s \rightarrow B_{s+1} \upharpoonright x \neq B_s \upharpoonright x)$.

Obviously this will imply $A \leq_{wtt} B$.

The construction of A is essentially construction 2.1.10 (modified as in Remark 2.2). Only the definition of requiring attention at stage s+1 has to be slightly modified. We now have

(3.2) Requirement R_σ, $\sigma \leq \sigma_s$, requires attention if at the end of stage s there is no follower y of R_σ in A_s and one of the following holds:

(i) $\sigma \subseteq \sigma_s$ and R_σ either has no follower or all followers x of R_σ are *completely confirmed* and *realized*, i.e. x is confirmed at each n with $\sigma(n)=0$ and $\{|\sigma|\}_s(x)=0$, and $B_{s+1} \upharpoonright x = B_s \upharpoonright x$.

(ii) There is a completely confirmed and realized follower x of R_σ and $B_{s+1} \upharpoonright x \neq B_s \upharpoonright x$.

(iii) There is a follower x of R_σ and an $\langle e,i,j\rangle < |\sigma|$ such that (2.1) holds, $\sigma(\langle e,i,j\rangle) = 0$ and x is not yet confirmed at $\langle e,i,j\rangle$.

Note that the clause $B_{s+1} \restriction x \neq B_s \restriction x$ in (ii) ensures that (3.1) is satisfied. Due to this clause requirement R_σ will need more than one follower at a time. Whenever all followers of R_σ are ready to be enumerated in A -i.e. completely confirmed and realized- and only wait for being permitted by B then a new follower will be appointed for R_σ. The standard permitting argument will show that one of the followers will eventually be permitted (or never ready to enter A).

We now show that the construction is correct, i.e. that the R and N requirements are met. Obviously the construction satisfies (F1) to (F9), (2.8) and (2.9). Moreover, by Lemmata 2.1.8 and 2.1.9, (2.11) and (2.12) hold for $n = \langle e,i,j\rangle+1$. So, by Lemma 2.3, it suffices to prove the following claim.

<u>Claim 1</u>. Let $R_n = R_{T\restriction n}$. Then R_n acts only finitely often and is met.

<u>Proof</u>. The proof is by induction. Fix n and assume the claim correct for $n' < n$. Then, as in the proof of Claim 3 in the proof of Theorem 2.1, there is a stage s_1 such that no requirement R_σ, $\sigma < T\restriction n$, is active after stage s_1 and no stage $s > s_1$ is σ-stage for any $\sigma <_L T\restriction n$. So after stage s_1 R_n receives attention whenever it requires attention and an R_n follower can be cancelled only by activity of a smaller follower of R_n (cf. Proposition 2.1.1). Moreover, if at some stage $s > s_1$ a follower y of R_n is in A_s, then $\{n\}(y) = 0$, whence R_n is met, and y is permanent, whence R_n stops acting at stage s. Similarly, if some permanent follower of R_n is never realized then R_n is met and eventually stops acting. So in the following we may assume no follower of R_n is in A after stage s_1 and that all permanent followers are eventually realized. It follows, as outlined in Remark 2.2, that all permanent followers of R_n are eventually completely confirmed.

Now for a contradiction assume that R_n acts infinitely often. Then, by (F5), R_n has infinitely many followers. Moreover the least follower x_1 of R_n after stage s_1 is never cancelled. Similarly the first follower x_2 appointed after x_1 stopped acting is permanent, and so on. So we can effectively enumerate permanent followers $x_1 < x_2 < \ldots < x_n < \ldots$ of R_n and stages $t_1 < t_2 < \ldots < t_n < \ldots$ such that x_n is completely confirmed and realized at stage t_n. Moreover, x_n is prevented from entering A only by the clause "$B_{s+1} \restriction x_n \neq B_s \restriction x_n$" in (ii) of the definition (3.2). So for all n, $B_{t_n} \restriction x_n = B \restriction x_n$. It follows that B is recursive contrary to our assumption.

So R_n acts only finitely often. Furthermore R_n is met, since otherwise R_n will act infinitely often.

This completes the proof of the claim and that of Theorem 3.1. □

3.3. Theorem. Let $B_o, \ldots, B_n$ $(n \geq 0)$ be nonrecursive r.e. sets. Then there are nonrecursive r.e. sets $A_i \leq_{wtt} B_i$, $i \leq n$, such that $\deg A_o, \ldots, \deg A_n$ and $\deg(A_o \oplus \ldots \oplus A_n)$ are contiguous.

3.4. Corollary. Let $\mathbf{b}_o, \ldots, \mathbf{b}_n$ $(n \geq 0)$ be nonzero r.e. degrees. Then there are nonzero r.e. degrees $\mathbf{a}_i \leq \mathbf{b}_i$, $i \leq n$, such that $\mathbf{a}_o, \ldots, \mathbf{a}_n$ and $\mathbf{a}_o \cup \ldots \cup \mathbf{a}_n$ are contiguous. □

Proof of Theorem 3.3. By induction on n. For $n = 0$ Theorem 3.3 is just Theorem 3.1. So fix $n > 0$. By Theorem 3.1, we may assume that $\deg B_o, \ldots, \deg B_n$ are contiguous. Moreover, w.l.o.g $\deg B_i \cap \deg B_j = \mathbf{0}$ for $i \neq j \leq n$. Otherwise w.l.o.g. $\deg B_{n-1} \cap \deg B_n \neq \mathbf{0}$, i.e. by contiguity of $\deg B_{n-1}$ and $\deg B_n$ there is a nonrecursive $B_n' \leq_{wtt} B_{n-1}, B_n$, and the claim follows by inductive hypothesis.

So, by Proposition 1.6, it suffices to construct pairwise recursively separable nonrecursive r.e. sets $A_i \leq_{wtt} B_i$, $i \leq n$, such that, for $A = A_o \cup \ldots \cup A_n$, $\deg A$ is contiguous. To construct such sets $A_o, \ldots, A_n, A$, we essentially use the construction of Theorem 3.1. The R requirements are now

$$R_\sigma : A_i \neq \{e\} \qquad \text{for } \sigma \text{ with } |\sigma| = (n+1)e+i,\ i \leq n.$$

The only changes in the construction are the following ones in the definition (3.2) of requiring attention: If $|\sigma| = (n+1)e+i$ for some $e < \omega$ and $i \leq n$, then in (i) and (ii) replace the clause "$B_{s+1} \restriction x =(\neq) B_s \restriction x$" by "$B_{i,s+1} \restriction x =(\neq) B_{i,s} \restriction x$", where $\{B_{i,s} : s \in \omega\}$ is a fixed recursive enumeration of B_i $(i \leq n)$, and in (i) now say that x is *realized* if $\{e\}_s(x) = 0$.

We then let A_i be the intersection of A with the set of followers of requirements R_σ where $|\sigma| = (n+1)e+i$ for some $e < \omega$. We obtain a proof for the correctness of the construction, by making the above changes throughout the proof of correctness for Theorem 3.1. □

For applications of the Ladner-Sasso transfer technique Theorem 3.3 is considerably more powerful than Theorem 3.1. The (technical) main theorem of Stob [17] stating the existence of a minimal pair with contiguous supremum directly follows from Corollary 3.4 and the existence of minimal pairs (see [9]).

3.5. Corollary (Stob [17]). There is a minimal pair of r.e. degrees with contiguous supremum.

Proof. Apply Corollary 3.4 to degrees $\mathbf{b}_o, \mathbf{b}_1$ which form a minimal pair. □

We now turn to another extension of Theorem 3.1.

<u>3.6. Theorem.</u> Let $\mathbf{b}$ and $\mathbf{c}$ be r.e. degrees such that $\mathbf{b} \not\leq \mathbf{c}$ and $\mathbf{c}$ is low. Then there is a contiguous degree $\mathbf{a}$ such that $\mathbf{a} \leq \mathbf{b}$ but $\mathbf{a} \not\leq \mathbf{c}$.

The proof of the theorem uses a technique due to Robinson [14] for testing the C-correctness of computations $\{e\}_s(C_s;x)$ for a low oracle set C and $\{C_s : s \in \omega\}$ a recursive enumeration of C. The variant of the Robinson technique we use here is due to Soare. For a more detailed explanation we refer the reader to Fejer [8,§4], where also a proof for Theorem 3.6.1 below is given.

Note that a convergent computation $\{e\}_s(C_s;x)$ is C-correct iff
$\{z : z < u(e,C_s,x,s)\ \&\ z \notin C_s\} \subseteq \bar{C}$.
So it suffices to have a procedure answering questions of the form "$D_x \subseteq \bar{C}$?", where D_x is the finite set with canonical index x. The existence of such a procedure follows from the following result.

<u>3.6.1. Theorem (Soare).</u> Let C be a low r.e. set. Then there is a recursive function f such that for all e,

(3.3) $W_e \cap \{x : D_x \subseteq \bar{C}\} = W_{f(e)} \cap \{x : D_x \subseteq \bar{C}\}$ and

(3.4) $W_e \cap \{x : D_x \subseteq \bar{C}\}$ finite $\rightarrow$ $W_{f(e)}$ finite.

Now the test procedure is as follows. In the course of the construction we will uniformly build r.e. *test sets* V_e, $e \in \omega$. By the recursion theorem, we may assume to know the r.e. indices of the sets V_e in advance, say $V_e = W_{g(e)}$, g a recursive function.

A *test of the C-correctness of a computation* $\{i\}_s(C_s;y)\downarrow$ *using test set* V_e goes as follows: Let D_x be the set of numbers used in the computation but not in C_s. Put x in $V_e = W_{g(e)}$, and simultaneously enumerate C and $W_{f(g(e))}$ until either x shows up in $W_{f(g(e))}$ or an element of D_x shows up in C (By (3.3), one of them must happen!). In the former case we say the test *outcome* is *positive*, in the latter case *negative*. If the test is negative then we <u>know</u> that the tested computation is not C-correct. If the test is positive then we <u>guess</u> that the computation is C-correct. A positive outcome may *turn out to be incorrect* at a later stage t when an element of D_x shows up in C_t. If from some point on, however, we use V_e for a new test after a test with positive outcome only if the outcome turned out to be incorrect, then, by (3.4), the positive outcomes of V_e-tests (if there are any) will eventually be correct.

In the construction below we will use the test procedure quite informally using the above introduced terminology.

<u>Proof of Theorem 3.6.</u> Fix r.e. sets $B \in \mathbf{b}$ and $C \in \mathbf{c}$, and let $\{B_s : s \in \omega\}$ and $\{C_s : s \in \omega\}$ be recursive enumerations of B and C respectively. By a tree argument using a follower concept we construct an r.e. set A meeting the contiguity re-

quirements N_n and the requirements

$$R_\sigma \quad : \quad A \neq \{|\sigma|\}^C.$$

As in the proof of Theorem 3.1, $A \leq_T B$ is ensured by satisfying (3.1). For handling the R requirements we will use the above described Robinson test procedure. We will have a test set V_σ for each requirement R_σ and in addition a test set V_x for each follower x.

Basically the construction of A is the one given in the proof of Theorem 3.1. The construction here is more involved, however, by the now necessary guesses at the correctness of computations $\{|\sigma|\}_s(C_s;x) = 0$, x a follower of R_σ.

<u>Construction.</u>

<u>Stage 0.</u> Do nothing.

<u>Stage s+1.</u> The stage consists of four steps.

<u>Step 1.</u> Fix the unique σ_s such that $|\sigma_s| = s$ and s is a σ_s-stage. Cancel all followers of requirements R_τ with $\sigma_s <_L \tau$.

<u>Step 2.</u> For each requirement R_σ, $\sigma \leq \sigma_s$, such that there is no follower of R_σ in A_s and for each follower x of R_σ do the following:

(i) If x is completely confirmed and $\{|\sigma|\}_s(C_s;x) = 0$ then test the C-correctness of this computation using test set V_x. If the outcome is positive then say x is *x-realized*.

(ii) If x is the least follower of R_σ such that x is completely confirmed, $\{|\sigma|\}_s(C_s;x) = 0$ and $B_{s+1} \restriction x \neq B_s \restriction x$ then test the correctness of the computation $\{|\sigma|\}_s(C_s;x)$ using test set V_σ. If the test outcome is positive then say x is *σ-realized*.

<u>Step 3.</u> Requirement R_σ, $\sigma \leq \sigma_s$, *requires attention* if at the end of stage s no follower of R_σ is in A_s and one of the following holds

(i) $\sigma \subseteq \sigma_s$ and either R_σ has no follower or each follower x of R_σ is x- but not σ-realized.

(ii) There is a σ-realized follower x of R_σ.

(iii) There is a follower x of R_σ such that for some $\langle e,i,j\rangle < |\sigma|$, (2.1) holds, $\sigma(\langle e,i,j\rangle) = 0$ and x is not yet confirmed at $\langle e,i,j\rangle$.

Choose σ minimal such that R_σ requires attention.

If (i) holds *appoint* x = s to be follower of R_σ. Otherwise choose x minimal such that (ii) or (iii) holds for x. If (ii) holds, enumerate x in A; otherwise say x is *confirmed at* each $\langle e,i,j\rangle$ such that (2.1) holds and $\sigma(\langle e,i,j\rangle)=0$

(and if x is now confirmed at each n such that $\sigma(n) = 0$ then say x is *completely confirmed*).

In all cases say x is *active* and cancel all followers of requirements R_τ, $\sigma < \tau$, and all followers of R_σ later appointed than x.

Step 4. For all σ and x, if x is an R_σ follower in A_s and $C_s \upharpoonright u \neq C_{s+1} \upharpoonright u$ for $u = u(|\sigma|, C_s, x, s)$ then cancel x.

This completes the construction.

The V_x tests ensure that for a permanent relevant follower x of R_σ with $\{|\sigma|\}(C;x) \neq 0$ this fact eventually becomes apparent, and appointment of R_σ followers is stopped. The V_σ tests ensure that for R_σ on the true path T only finitely many followers x with $\{|\sigma|\}(C;x) \neq 0$ enter A (and step 4 ensures that such followers are eventually cancelled).

As in the proof of Theorem 3.1 the proof of correctness of the construction can be reduced to the proof of the following claim.

Claim. Let $R_n = R_{T\upharpoonright n}$. Then, for all n, R_n acts only finitely often and is met.

Proof. Fix n. As in the proof of the analogous claim in the proof of Theorem 3.1, there is a stage s_1 such that after stage s_1 R_n receives attention whenever it requires attention and a follower of R_n can be cancelled only by activity of a smaller, i.e. earlier appointed, R_n follower or in step 4 of some stage.

Now first assume that there is a permanent follower of R_n in A, say y. Then R_n acts only finitely often. Moreover, if s is the least stage such that $y \in A_{s+1}$ then $\{n\}_s(C_s;y) = 0$, and, since y is never cancelled in step 4 of a stage, this computation is C-correct. Hence R_n is met.

So in the following we may assume that there is no permanent follower of R_n in A. We will show next that only finitely many R_n followers enter A. For a contradiction assume that there are infinitely many followers $x_1, \ldots, x_n, \ldots$ of R_n entering A after stage s_1, say at stages $t_1 < \ldots < t_n < \ldots$, respectively. Note that after stage s_1, any R_n follower which is σ-realized at stage s+1 is enumerated in A at this stage. So the only tests using V_σ and having positive outcome are tests of the computations $\{n\}_{t_m-1}(C_{t_m-1};x_m)$, $m < \omega$. Since no x_m is permanent, i.e. by choice of s_1 is cancelled in step 4 at some stage, none of these computations is correct. But, by (3.4), this is impossible. (All for $\sigma = T\upharpoonright n$).

So we can fix $s_2 > s_1$ such that no follower of R_n is in A after stage s_2 and no R_n follower is $T\upharpoonright n$-realized after stage s_2.

We now show that R_n has only finitely many followers and thus acts only finitely often. Namely, for a contradiction assume that R_n has infinitely many followers. Then -as in the proof of Theorem 3.1- there are infinitely many permanent followers of R_n and every permanent follower of R_n is eventually completely confirmed . Moreover, since a completely confirmed follower of R_n cannot become active after stage s_2, we can effectively list the permanent followers $x_1 < \dots < x_m < \dots$ of R_n and stages $s_2 < t_1 < \dots < t_m < \dots$ such that x_m is completely confirmed by the end of stage t_m. Now fix such a follower x_m. Since R_n has infinitely many followers, x_m is x_m-realized at infinitely many stages. Note that V_{x_m} is only used to test the computations $\{n\}_s(C_s;x_m)$, $s \in \omega$. So after a positive outcomeof a test, we don't make a new test (but repeat the old test) before the outcome turns out to be incorrect. Hence (3.4) impliesthat from some point on either all outcomes are negative or after some positive test no new test is made. Since x_m is x_m-realized infinitely often, the second alternative must hold and we can conclude that $\{n\}(C;x_m) = 0$. So, using C as oracle, we can compute a stage $u_m > t_m$ such that $\{n\}_{u_m}(C_{u_m};x) = 0$ via a C-correct computation. Since x_m is not $T\lceil n$-realized after stage u_m this implies $B_{u_m} \lceil x_m = B \lceil x_m$. It follows $B \leq_T C$ contrary to assumption.

So R_n acts only finitely often. It remains to show that R_n is met. For a contradiction assume that $A = \{n\}^C$, and choose $s_3 > s_2$ such that R_n doesn't act after stage s_3. Note that there are infinitely many $T\lceil n$-stages. So R_n has certain followers at stage s_3, say $x_1, \dots, x_m$, since otherwise at the least stage $s+1 > s_3$ where s is $T\lceil n$-stage R_n obtains a new follower. By definition of s_3, the followers $x_1, \dots, x_m$ are permanent and thus, as pointed out above, completely confirmed at stage s_3. Furthermore, since $x_i \notin A$ and R_n is not met, $\{n\}(C;x_i) = 0$, for $i = 1,\dots,m$. So we may fix $s_4 > s_3$ such that for each R_n follower x_i, $\{n\}_{s_4}(C_{s_4};x_i) = 0$ via a C-correct computation. So follower x_i is x_i-realized at every stage $s > s_4$ and contrary to choice of s_3 requirement R_n receives a new follower at the least stage $s+1 > s_4$, s a $T\lceil n$-stage.

This completes the proof. □

Theorem 3.6 implies that the contiguous degrees form an automorphism basis (see [3]). It also shows that for any nonzero low degree $\underset{\sim}{c}$ there is a contiguous degree incomparable with $\underset{\sim}{c}$.

<u>3.7. Corollary</u>. $\forall \underset{\sim}{c} > \underset{\sim}{0}\ (\underset{\sim}{c}$ low $\rightarrow \exists \underset{\sim}{a}\ (\underset{\sim}{a}$ contiguous $\&\ \underset{\sim}{a} \mid \underset{\sim}{c}))$.

<u>Proof</u>. For $\underset{\sim}{c} \neq \underset{\sim}{0}$ and $\underset{\sim}{c}$ low, by Sacks splitting there is an r.e. degree $\underset{\sim}{b}$ incomparable with $\underset{\sim}{c}$. Theorem 3.6 applied to $\underset{\sim}{b}$ and $\underset{\sim}{c}$ yields a contiguous degree $\underset{\sim}{a} \mid \underset{\sim}{c}$.□

The following variant of Theorem 3.6 is used for the study of splitting properties of r.e. sets in [4].

3.8. Theorem. Let B and C be r.e. sets such that $B \not\leq_{wtt} C$ and C is low. Then there is an r.e. set A such that degA is contiguous, $A \leq_{wtt} B$, but $A \not\leq_{wtt} C$.

Proof. In the proof of Theorem 3.6 replace the R requirements by

R_σ : $A \neq [|\sigma|]^C$

and, consequently, throughout the construction replace $\{|\sigma|\}_s(C_s;x)$ by $[|\sigma|]_s(C_s;x)$.

For a proof that the thus modified construction works, first note that (3.1) implies $A \leq_{wtt} B$. So it suffices to show that the claim stated in the proof of Theorem 3.6 still holds. The there given proof will still work with the following modification. Since we may replace $\{n\}^C s$ by $[n]^C s$ throughout the proof, in order to compute the least stage u_m such that $[n]_{u_m}(C_{u_m};x_m) = 0$ via a C-correct computation we only have to know $C \restriction [n](x_m)$. So the assumption that R_n has infinitely followers will now imply $B \leq_{wtt} C$ (contrary to assumption) since now $B \restriction x_m$ is computable from $C \restriction [n](x_m)$ for the recursively given sequence $x_1 < \ldots < x_m < \ldots$ of the previous proof.

This completes the proof. □

By Theorem 3.1, nonzero cappable contiguous degrees exist. We now show that there are also noncappable contiguous degrees. This has been claimed without proof in [5].

3.9. Theorem (Ambos-Spies, Jockusch, Shore and Soare [5]). There is a noncappable contiguous degree $\underset{\sim}{a}$, i.e. a contiguous degree $\underset{\sim}{a}$ such that for every $\underset{\sim}{b} > \underset{\sim}{0}$ there is an r.e. degree $\underset{\sim}{c}$ such that $\underset{\sim}{0} < \underset{\sim}{c} \leq \underset{\sim}{a},\underset{\sim}{b}$.

Proof. We simultaneously construct r.e. sets A and B such that degA is contiguous, $B \leq_T A$ and degB is noncappable. Then degA will be both contiguous and noncappable since the class of noncappable degrees is closed upwards.

To make degB noncappable we use the standard requirements

$R_{<e_0,e_1>}$: W_{e_0} nonrecursive $\rightarrow B^{(e_0)} \neq \{e_1\}$

and ensure

(3.5) $B^{(e)} \leq_T W_e$

for all e_0, e_1, e.

The basic strategy for meeting requirement $R_{<e_0,e_1>}$ is to wait for the least stage s such that there is some z of the form $<e_0,<e_1,y>>$, $y \in \omega$, such that $\{e_1\}_s(z) = 0$ and $W_{e_0,s+1} \restriction z \neq W_{e_0,s} \restriction z$. Then we enumerate the least such z in $B^{(e_0)}$ at stage s+1. Now if W_{e_0} is nonrecursive and $\{e_1\}(<e_0,<e_1,y>>) = 0$ for (almost) all y then the standard permitting argument shows that s and z as above exist and thus $R_{<e_0,e_1>}$ is met; and otherwise $R_{<e_0,e_1>}$ is met trivially if we ensure that no z of the above form enters B.

$B \leq_T A$ is ensured by permitting. Namely requirement R_e will have certains followers. If x is a follower of R_e, $e = \langle e_0, e_1 \rangle$, then we consider only such z in the basic strategy for meeting R_e which are greater than x, and when z is enumerated in B then simultaneously x is enumerated in A.

To make degA contiguous we use the usual construction. So we will replace the R requirements above by requirements R_σ where each R_σ is a copy of $R_{|\sigma|}$.

The actual construction of A and B is obtained by modifying construction 2.1.10 as follows. In the definition of requiring attention at stage s+1 replace (ii) by

(ii) There are e_0, e_1, x, y, z such that $|\sigma| = \langle e_0, e_1 \rangle$, $z = \langle e_0, \langle e_1, y \rangle\rangle$, x is a follower of R_σ, $x < z$, $\{e_1\}_s(z) = 0$ and $W_{e_0,s+1} \upharpoonright z \neq W_{e_0,s} \upharpoonright z$.

Then proceed as in 2.1.10. Only in the case where (ii) holds for (the minimal) x, in addition enumerate the least z as in (ii) above in B.

Then, by permitting, $B \leq_T A$ and (3.5) is satisfied for each e. Furthermore, the construction obviously satisfies (F1) to (F9), (2.8), (2.9) and, by Lemmata 2.1.8 and 2.1.9, (2.11) and (2.12) for $n = \langle e,i,j \rangle + 1$. So, by Lemma 2.3, the correctness of the construction will follow from

Claim. For each n, $R_{T\upharpoonright n}$ acts only finitely often, has a permanent follower and is met.

The proof of the first two parts of the claim parallels the proof of Claim 3 in 2.1. That $R_{T\upharpoonright n}$ is met then follows by the standard argument. Namely once $R_{T\upharpoonright n}$ has a permanent follower x then at all subsequent stages the only restraint on the basic strategy for meeting $R_{T\upharpoonright n}$ is that only numbers $z > x$ may be considered.

This completes the proof. □

Note that noncappable degrees are cuppable (indeed low cuppable; cf. [5]). So Theorem 3.9 implies the existence of cuppable contiguous degrees. This is contrasted by Theorem 5.3 below which asserts that, for any contiguous degree $\underset{\sim}{a}$, the r.e. wtt degree contained in $\underset{\sim}{a}$ is noncuppable (in the structure $\underline{R}_{wtt}$ of r.e. wtt degrees).

§4. DISTRIBUTIVE SUBLATTICES OF INITIAL SEGMENTS OF THE R.E. DEGREES.

In this section we use a new result on contiguous degrees to prove two theorems on lattice embeddings in initial segments of the r.e. degrees. Thomason [18] and, independently, Lerman have shown that every finite (in fact countable) distributive lattice can be (lattice-)embedded in the r.e. degrees by a map which preserves the least element. Here we look at questions about embeddings of finite distributive lattices in initial segments of the r.e. degrees. Lachlan [11] has shown that there is a nonzero r.e. degree $\mathbf{a}$ such that the two-atom Boolean algebra cannot be embedded in $\mathbf{R}(\leq\mathbf{a})$, the class of r.e. degrees below $\mathbf{a}$, by a map which preserves the least element. So in general Lerman and Thomason's result doesn't carry over to initial segments of $\mathbf{R}$. We first show, however, that if the requirement of preserving the least element is dropped then we still can embed any finite (in fact countable) distributive lattice in any initial segment $\mathbf{R}(\leq\mathbf{a})$, $\mathbf{a} > \mathbf{0}$. We then give a criterion for 0-preserving embeddings. Obviously if f is a 0-preserving embedding of a finite lattice L in an initial segment $\mathbf{R}(\leq\mathbf{a})$ of $\mathbf{R}$, then f maps every minimal pair of L to a minimal pair of r.e. degrees below $\mathbf{a}$. So the maximum number of elements in L which are pairwise minimal pairs cannot be greater than the maximum number of degrees in $\mathbf{R}(\leq\mathbf{a})$ which are pairwise minimal pairs. We will show that for distributive lattices this condition is also sufficient to guarantee 0-preserving embeddability. Since for a finite distributive lattice L the maximum number of elements forming pairwise minimal pairs is just the number of minimal elements of L, we will have: A finite distributive lattice L can be embedded in the initial segment $\mathbf{R}(\leq\mathbf{a})$ by a map which preserves the least element if and only if the number of minimal elements of L is not greater than the maximum number of degrees below $\mathbf{a}$ which are pairwise minimal pairs.

We first introduce some additional notation. Let $\langle L_1,\cup,\cap,<\rangle$ be a lattice and $\langle L_2,\cup,\cap,<\rangle$ be an (upper semi-)lattice. A one-to-one map f from L_1 to L_2 is a (lattice-) embedding of L_1 in L_2 if for all $a,b \in L_1$, $f(a\cup b) = f(a) \cup f(b)$, $f(a) \cap f(b)\downarrow$ and $f(a\cap b) = f(a) \cap f(b)$. f preserves the least element if L_1 doesn't have a least element or the least element of L_1 is mapped to the least element of L_2. We write $L_1 \overset{}{\underset{f}{===>}} L_2$ ($L_1 \overset{0}{\underset{f}{===>}} L_2$) if f embeds L_1 in L_2 (preserving the least element), and $L_1 ===> L_2$ ($L_1 \overset{0}{===>} L_2$) if such an f exists. L_1 is a sublattice of L_2 if $L_1 \subseteq L_2$ and $L_1 \underset{id}{===>} L_2$, id the inclusion mapping.

Let F be the class of finite subsets of ω and let $\alpha, \beta, \gamma, \ldots$ denote elements of F. It is well known that F together with the set theoretical operations $\cup$ (union) and $\cap$ (intersection) forms a distributive lattice. Moreover, any finite distributive lattice can be embedded in F by a map which preserves the least element. So we may identify the finite distributive lattices with the finite sublattices of F with least element $\emptyset$.

Finally, we need the following definition.

<u>4.1. Definition.</u> Let $\langle L,\cup,\cap,<\rangle$ be an upper semi-lattice with least element 0. Then the *mp-rank* of L is defined by

$$\text{mp-rank}(L) = \sup\{n : \exists\, a_0,\ldots,a_{n-1} \in L\ \forall\, i,j < n\ (a_i > 0\ \&\ (i\neq j \rightarrow a_i \cap a_j = 0))\}.$$

Note that, for L with least element and $|L| \geq 2$, mp-rank(L) ≥ 1. If mp-rank(L)=1 then L has no minimal pairs; if mp-rank(L) ≥ 2 then it is the maximum number of elements of L which are pairwise minimal pairs. One can easily show that for a finite distributive lattice L the mp-rank of L is just the number of minimal elements of L; in particular for a finite Boolean algebra the mp-rank is the number of atoms.

Instead of mp-rank($\underset{\sim}{R}(\leq \underset{\sim}{a})$) we also write mp-rank($\underset{\sim}{a}$). Note that mp-rank($\underset{\sim}{O}$) = 0 and, by the above quoted result of Lachlan, there is an r.e. degree $\underset{\sim}{a}$ with mp-rank($\underset{\sim}{a}$) = 1. It follows from the Thomason-Lerman embedding theorem that all noncappable (and some cappable) degrees have infinite mp-rank. Ambos-Spies and Soare [6] show that, for each $n < \omega$, there is an r.e. degree with mp-rank n.

4.2. Theorem. Let $B_o, \ldots, B_n$ $(n \geq 0)$ be nonrecursive r.e. sets. Then there are r.e. sets $A_{i,\alpha}$ $(i \leq n, \alpha \in F)$ such that

(4.1) $A_{i,\alpha} \leq_{wtt} B_i$ $(i \leq n, \alpha \in F)$

(4.2) For $k \leq n$, $i_o, \ldots, i_k \leq n$ and $\alpha_o, \ldots, \alpha_k \in F$, $\deg(A_{i_o,\alpha_o} \oplus \ldots \oplus A_{i_k,\alpha_k})$ is contiguous.

(4.3) For $f_i : F \to \underset{\sim}{R}$ defined by $f_i(\alpha) = \deg A_{i,\alpha}$, $F \underset{f_i}{\Longrightarrow} \underset{\sim}{R}$ $(i \leq n)$.

Proof. We construct an r.e. set A and distinguish pairwise recursively separable subsets A_i and $A_{i,j}$, $i \leq n$, $j \in \omega$, such that

$$A = \bigcup \{A_{i,j} : i \leq n, j \in \omega\} \cup \bigcup \{A_i : i \leq n\}.$$

For $i \leq n$ and $\alpha \in F$, we let

$$A_{i,\alpha} = \bigcup \{A_{i,j} : j \in \alpha\} \cup A_i.$$

Then

(4.4) $\alpha \subseteq \beta \rightarrow A_{i,\alpha} \leq_{wtt} A_{i,\beta}$ and

(4.5) $\deg A_{i,\alpha} \cup \deg A_{i,\beta} = \deg A_{i,\alpha\cup\beta}$.

Let F^n denote the set of all tuples $((i_o,\alpha_o),\ldots,(i_k,\alpha_k))$ with $k \leq n$, $i_o,\ldots,i_k \leq n$, $\alpha_o,\ldots,\alpha_k \in F$, and denote elements of F^n by $\bar{\alpha}, \bar{\beta}, \ldots$. For $\bar{\alpha} = ((i_o,\alpha_o),\ldots,(i_k,\alpha_k))$, we let

$$A_{\bar{\alpha}} = A_{i_o,\alpha_o} \cup \ldots \cup A_{i_k,\alpha_k}.$$

Note that $A_{\bar{\alpha}} \equiv_{wtt} A_{i_o,\alpha_o} \oplus \ldots \oplus A_{i_k,\alpha_k}$.

Now to ensure (4.1) it suffices to make $A_i \leq_{wtt} B_i$ and $A_{i,j} \leq_{wtt} B_i$, which will be done by the permitting method. Let $\{B_{i,s} : s \in \omega\}$ be a recursive enumeration of B_i, $i \leq n$, and let A^s $(A_i^s, A_{i,j}^s, A_{i,\alpha}^s, A_{\bar{\alpha}}^s)$ be the part of A (A_i, ...) enumerated by the end of stage s of the construction of A below. Then it suffices to satisfy for $i \leq n$ and $j \in \omega$

(4.6) $\forall x,s\ (\ x \in (A_i^{s+1} \cup A_{i,j}^{s+1}) - (A_i^s \cup A_{i,j}^s) \ \rightarrow\ B_{i,s+1} \restriction x \neq B_{i,s} \restriction x\ ).$

To ensure (4.2), for each $\bar{\alpha} \in F^n$ we have to guarantee

(4.7) $\deg A_{\bar{\alpha}}$ is contiguous.

To do so, we have to meet the requirements

$$N_{(\bar\alpha,e,i,j)} \quad : \quad A_{\bar\alpha} = \{i\}^{W_e} \ \& \ W_e = \{j\}^{A_{\bar\alpha}} \quad \rightarrow \quad A_{\bar\alpha} =_{wtt} W_e$$

for $\bar\alpha \in F^n$, $e,i,j \in \omega$.

Finally, one can easily show that, by (4.4) and (4.5), (4.3) will follow if we satisfy for $i \leq n$, $j \in \omega$ and $\alpha, \beta \in F$

(4.8) $\quad A_{i,j} \not\leq_T A_i$ $\qquad$ and

(4.9) $\quad$ For any r.e. set C, if $C \leq_T A_{i,\alpha}$ and $C \leq_T A_{i,\beta}$ then $C \leq_T A_{i,\alpha\cap\beta}$.

Condition (4.9) is ensured by meeting minimal pair type requirements. Note that, by (4.7), in (4.9) it suffices to consider sets C with $C \leq_{wtt} A_{i,\alpha}, A_{i,\beta}$. So it is enough to meet the requirements

$$N'_{(\alpha,\beta,i,j,k)} \quad : \quad [j]^{A_{i,\alpha}} = [k]^{A_{i,\beta}} \text{ total} \quad \rightarrow \quad [j]^{A_{i,\alpha}} \leq_T A_{i,\alpha\cap\beta}.$$

(4.8) is broken down in the requirements

$$R_{(n+1)<e,j>+i} \quad : \quad A_{i,j} \neq \{e\}^{A_i} \qquad (e,j \in \omega,\ i \leq n).$$

Since the N requirements require a tree construction we will actually have requirements R_σ, $\sigma \in \{0,1\}^{<\omega}$, where each R_σ is a copy of requirement $R_{|\sigma|}$ above. Requirement R_σ, $|\sigma| = (n+1)<e,j>+i$, will have two types of followers: Primary followers and trace followers. For a primary follower x we attempt to ensure $A_{i,j}(x) \neq \{e\}(A_i;x)$ by enumerating x in $A_{i,j}$ or restraining x from $A_{i,j}$. A trace follower x' is trace for a primary follower x. Each primary follower x has at most one trace x', and x will enter $A_{i,j}$ iff x' will enter A_i. So the sets A_i and $A_{i,j}$ are defined by

$$A_{i,j} = A \cap \{x : x \text{ is primary follower of a requirement } R_\sigma \text{ with } |\sigma| = (n+1)<e,j>+i,\ e < \omega\}$$

$$A_i = A \cap \{x : x \text{ is trace follower of a requirement } R_\sigma \text{ with } |\sigma| = (n+1)<e,j>+i,\ e,j < \omega\}.$$

The traces are required by our strategy for meeting the N′ requirements.

Now let N_e, $e \in \omega$, be an effective listing of the N and N′ requirements such that N (N′) requirements have even (odd) index.

As part of the construction of A we define the length function $l(e,s)$ of N_e by

- $l(e,s) = \max\{x : A^s_{\bar\alpha} \restriction x = \{i\}_s(W_{e,s}) \restriction x \ \& \ W_{e,s} \restriction u_x = \{j\}_s(A^s_{\bar\alpha}) \restriction u_x$, where $u_x = \max\{u(i,W_{e,s},y,s) : y < x\} \cup \{x\}$, if N_e is requirement $N_{(\bar\alpha,e,i,j)}$
- $l(e,s) = \max\{x : [j]_s(A^s_{i,\alpha}) \restriction x \downarrow = [k]_s(A^s_{i,\beta}) \restriction x\}$ if N_e is requirement $N'_{(\alpha,\beta,i,j,k)}$.

Based on this definition of l, define σ-stages as in Definition 2.1.7. Then there is a true path T of the construction satisfying Lemma 2.1.8, and as one can easily show the following holds.

<u>4.2.1. Lemma</u>. (i) If the hypothesis of N_e holds then $T(e) = 0$.
(ii) If $T(e) = 0$ then $\forall s,t$ (s,t $T\restriction e+1$-stages & $s<t \rightarrow l(s) < l(t)$).

As in the previous constructions a follower x of R_σ will be confirmed at such $2n < |\sigma|$ such that $\sigma(2n) = 0$ (There is no need to confirm odd indexed requirements of the minimal pair type). Again we say that x is completely confirmed if it is confirmed at each 2n such that $\sigma(2n) = 0$. A primary follower x of R_σ, $|\sigma|=(n+1)<e,j>+i$, is *realized* at stage s if $\{e\}_s(A_i^s;x) = 0$; it is *completely realized* if it has a completely confirmed trace.

We now describe the stages of the construction.

<u>Stage 0.</u> Do nothing.

<u>Stage s+1.</u> <u>Step 1.</u> Fix the unique σ_s such that $|\sigma_s| = s$ and s is a σ_s-stage. Cancel all followers of requirements R_τ with $\sigma_s <_L \tau$.

<u>Step 2.</u> Requirement R_σ, $\sigma \leq \sigma_s$, $|\sigma| = (n+1)<e,j>+i$, $i \leq n$, *requires attention* if at the end of stage s there is no follower y of R_σ in A^s and one of the following holds.

(i) $\sigma \subseteq \sigma_s$ and either R_σ has no follower or each primary follower x of R_σ is completely realized and $B_{i,s+1} \restriction x = B_{i,s} \restriction x$.

(ii) There is a completely realized primary follower x of R_σ such that $B_{i,s+1} \restriction x \neq B_{i,s} \restriction x$.

(iii) $\sigma \subseteq \sigma_s$ and there is a realized and completely confirmed follower x of R_σ which not yet has a trace.

(iv) There is a follower x of R_σ such that for some n with $\sigma(2n) = 0$ x is not yet confirmed at 2n and $l(2n,s) > x$.

Choose σ minmal such that R_σ requires attention.

If (i) holds appoint s to be a primary follower of R_σ. Otherwise fix x minimal such that (ii), (iii) or (iv) holds. If (ii) holds enumerate x and its trace x′ in A. If (iii) holds appoint s to be trace follower for x. Finally, if (iv) holds then say x is confirmed at each 2n with $\sigma(2n) = 0$ and $l(2n,s) > x$.

In cases (i) and (iii) we say s is active, in cases (ii) and (iv) x is active. In any case we cancel all followers of requirements R_τ with $\sigma < \tau$. In case (ii) we also cancel all followers of R_σ appointed after x with the exception of the trace x′ of x. In case (iii) all R_σ followers appointed after x with the exception of s are cancelled. Finally, in case (iv) all followers of R_σ appointed after x are cancelled.

This completes the construction.

Proof of correctness. We start with some observations. Obviously conditions (F1) to (F9) for followers are satisfied, and the construction is a tree construction in the sense of Lemma 2.3. Moreover, if x is appointed we can decide if x is primary or trace follower. So, by definition of A_i and $A_{i,j}$, these sets are mutually recursively separable.

If an R_σ follower x, $|\sigma| = (n+1)<e,j>+ i$, is appointed a trace x′ at stage s+1 then $x' = s$ and x′ -not x- is active at stages s+1. The cancellation at stage s+1, however, is like the cancellation of an activity of x. This ensures that after s+1 there is no follower y such that $x < y < x'$ (i.e. $x \propto y \propto x'$). Furthermore, once x has a trace x′ the only activity of x is via (ii) in which case x and x′ are both enumerated in A (i.e. x in $A_{i,j}$ and x′ in A_i). Also note that a primary or trace follower cannot enter A alone. So either x,x′ are both eventually cancelled, or both eventually enter A, or both are permanently relevant. It follows that a primary follower has at most one trace. Also note that the trace x′ of x is greater than x. So the clause "$B_{i,s+1} \restriction x \neq B_{i,s} \restriction x$" in (ii) of the definition of requiring attention implies that (4.6) is satisfied.

It remains to show that the R, N and N' requirements are met.

Claim 1. Let $R_m = R_{T\restriction m}$. Then R_m acts only finitely often and is met.

Proof. Fix m, say $m = (n+1)<e,j>+i$.
Note that for a permanent primary follower x of R_m the following hold. By Remark 2.2, x is eventually completely confirmed. If x never has a trace then $\{e\}(A_i;x) = 0$ and $x \notin A_{i,j}$, whence R_m is met. On the other hand, if x has a trace, say x′ appointed at stage s+1, then $\{e\}_s(A_i^s;x) = 0$, $x' = s > u(e,A_i^s,x,s)$ and, by cancellation at stage s+1 and the fact that x is permanent, $A_i^s \restriction s = A_i \restriction s$, i.e. the computation $\{e\}_s(A_i^s;x) = 0$ is A_i-correct. Moreover, x′ will be eventually completely confirmed, say at stage t. It follows that either x enters A and R_m is met, or $x \notin A$ and $B_i \restriction x = B_{i,t} \restriction x$.

Using these facts on permanent followers, Claim 1 can be proved in a way similar to the proof the analogous claim in Theorem 3.1. We leave the details to the reader.□

Claim 2. For each $\bar{\alpha} \in F^n$, $\deg A_{\bar{\alpha}}$ is contiguous.

Proof. Fix $\bar{\alpha} = ((i_o,\alpha_o),\ldots,(i_k,\alpha_k))$. It suffices to show that for $A_{\bar{\alpha}}$ and $\{A_{\bar{\alpha}}^s : s \in \omega\}$ in place of A and $\{A_s : s \in \omega\}$ the premises of Lemma 2.3 are satisfied. As noted already above, (F1) to (F9) are satisfied. (2.8) and (2.9) are immediate by construction, and (2.10) holds by claim 1. Finally, given e,i,j such that $A_{\bar{\alpha}} = \{i\}^{W_e}$ and $W_e = \{j\}^{A_{\bar{\alpha}}}$, choose m such that N_{2m} is the requirement $N_{(\bar{\alpha},e,i,j)}$. Then, by definition of T and Lemma 4.2.1, (2.11) and (2.12) hold for $n = 2m+1$. □

Claim 3. For every m, requirement N_{2m+1} is met.

Proof. Fix m and say N_{2m+1} is the requirement $N'_{(\alpha,\beta,i,j,k)}$. W.l.o.g. assume $[j](A_{i,\alpha}) = [k](A_{i,\beta}) = f$ is total.
We have to show $f \leq_T A_{i,\alpha\cap\beta}$.

We say requirement R_σ is of type i if $|\sigma| = (n+1)\langle e,j\rangle+i$ for some e,j, and $i \leq n$. Note that the sets $A_{i,\alpha}$ and $A_{i,\beta}$ contain only followers of requirements of type i. Furthermore, $A_i \subseteq A_{i,\alpha\cap\beta}$ contains just the trace followers of requirements of type i which enter A. Also note that, by definition of T and Lemma 4.2.1, $T(2m+1) = 0$, there are infinitely many $T\lceil 2m+2$-stages, and the length function $\lambda s.l(2m+1,s)$ of N_{2m+1} is strictly increasing on the set of $T\lceil 2m+2$-stages. Furthermore, we can fix s_1 such that no requirement R_σ with $\sigma < T\lceil 2m+2$ acts after stage s_1.

Now to compute f(x) from $A_{i,\alpha\cap\beta}$ for given x, find the least $T\lceil 2m+2$-stage $s > s_1$ such that

(4.10) $l(2m+1,s) > x$ and

(4.11) If y is primary follower of a requirement of type i and $y < u(x)$, where $u(x) = \max\{[j](x),[k](x)\}$, then $A^s_{i,\alpha\cap\beta}(y) = A_{i,\alpha\cap\beta}(y)$. Furthermore, if y has a trace y′ at stage s then $A^s_i(y') = A_i(y')$.

Obviously such a stage s exists and can be found recursively in $A_{i,\alpha\cap\beta}$. Note that, by (4.10), for some z, $[j]_s(A^s_{i,\alpha};x) = [k]_s(A^s_{i,\beta};x) = z$. We claim that $f(x) = z$. To prove this claim we have to show

$$\forall t > s\ (\ [j]_t(A^t_{i,\alpha};x) = z \ \text{ or } \ [k]_t(A^t_{i,\beta};x) = z\ ).$$

Since at $T\lceil 2m+2$-stages $t \geq s$, $[j]_t(A^t_{i,\alpha};x) = [k]_t(A^t_{i,\beta};x)$, it suffices to show that between two $T\lceil 2m+2$ stages one side of this equation remains unchanged, i.e. that

(4.12) $\forall k\ (\ A^{t_k}_{i,\alpha} \lceil u(x) = A^{t_{k+1}}_{i,\alpha} \lceil u(x) \ \text{ or } \ A^{t_k}_{i,\beta} \lceil u(x) = A^{t_{k+1}}_{i,\beta} \lceil u(x)\ )$,

where $s = t_0 < t_1 < \ldots < t_n < \ldots$ are the $T\lceil 2m+2$-stages $\geq s$.

Note that, by choice of s, a number $y < u(x)$ which enters $A_{i,\alpha}$ or $A_{i,\beta}$ after stage s is a follower at stage s. So y follows a requirement R_σ with $T\lceil 2m+2 \subseteq \sigma$. Furthemore R_σ is of type i. So, by (4.11), y is a primary follower which has no trace at stage s and which can enter only $A_{i,\alpha}$ or $A_{i,\beta}$ not both. Now before y enters A, it must be appointed a trace at a σ-stage, i.e. a $T\lceil 2m+2$-stage.

So if we assume that (4.12) fails, say for k, then there are followers $y_1 < y_2$ as just described, which enter A at stages v_p, $t_k < v_p \leq t_{k+1}$, $p = 1,2$. But then a trace for y_1 has been appointed at a stage w+1 such that w is $T\lceil 2m+2$-stage, i.e. at a stage w+1 with $s < w+1 \leq t_k+1$. It follows that y_2 is cancelled at stage w+1 and thus cannot enter A after stage t_k. Contradiction. □

This completes the proof of Theorem 4.2. □

4.3. Corollary. Let L be a finite distributive lattice and $\underset{\sim}{a}$ be a nonzero r.e. degree. Then $L \overset{}{\Longrightarrow} \underset{\sim}{R}(\leq \underset{\sim}{a})$. □

4.4. Remark. In Theorem 4.2 we may replace the lattice F by the countable atomless Boolean algebra $\mathcal{B}$ (The proof requires only some notational changes; cf. Lachlan [10]). Since every countable distributive lattice can be embedded in $\mathcal{B}$, this shows that in Corollary 4.3 we may replace "finite" by "countable".

4.5. Corollary. Let L be a finite distributive lattice and $\underset{\sim}{a}$ be an r.e. degree. Then

$L \overset{o}{\Longrightarrow} \underset{\sim}{R}(\leq \underset{\sim}{a})$ if and only if mp-rank(L) $\leq$ mp-rank($\underset{\sim}{a}$).

The proof of Corollary 4.5 requires the following lemma.

4.5.1. Lemma. Let U_n ($n \geq 2$) be the closure under finite unions of the set $\bigcup\{\{i, k\cdot n+i\} : i < n, k \in \omega\} \cup \{\emptyset\}$.

Then

(i) U_n is a sublattice of F with least element $\emptyset$.

(ii) mp-rank$(U_n) = n$.

(iii) For any finite distributive lattice L with mp-rank(L) $\leq n$, $L \overset{o}{\Longrightarrow} U_n$.

We omit the proof of Lemma 4.5.1, which is straightforward but somewhat tedious. Note that for any $\alpha \in U_m - \{\emptyset\}$ there are unique $k, i_o, \ldots, i_k < n$ and $\alpha_o, \ldots, \alpha_k \in F$ such that

(4.13) $\alpha = i_o\alpha_o \cup \ldots \cup i_k\alpha_k, \quad 0 \leq i_o < \ldots < i_k, \quad 0 \in \alpha_o \cap \ldots \cap \alpha_k,$

where $i\alpha$ is defined to be the set $\{k\cdot n+i : k \in \alpha\}$.

Proof of Corollary 4.5. For the nontrivial direction fix $\underset{\sim}{a}$ such that mp-rank of $\underset{\sim}{a}$ is greater than or equal to n. By Corollary 4.3, we may assume that $n \geq 2$. So, by Lemma 4.5.1, it suffices to show $U_n \overset{o}{\Longrightarrow} \underset{\sim}{R}(\leq \underset{\sim}{a})$.

Since mp-rank($\underset{\sim}{a}$) $\geq$ n, there are r.e. sets $B_o, \ldots, B_n$ such that

(4.14) $\forall i,j < n \; (\underset{\sim}{0} < \deg B_i \leq \underset{\sim}{a} \;\&\; (i \neq j \rightarrow \deg B_i \cap \deg B_j = \underset{\sim}{0}))$.

By Theorem 4.2, choose r.e. sets $A_{i,\alpha}$ ($i < n, \alpha \in F$) such that (4.1) to (4.3) hold with n-1 in place of n.

The embedding $f : U_n \rightarrow \underset{\sim}{R}(\leq \underset{\sim}{a})$ is defined by

$f(\emptyset) = \underset{\sim}{0}$

and, for $\alpha \in U_m - \{\emptyset\}$

$f(\alpha) = \deg(A_{i_o,\alpha_o} \oplus \ldots \oplus A_{i_k,\alpha_k})$

where $k, i_o, \ldots, i_k, \alpha_o, \ldots, \alpha_k$ are chosen according to (4.13).

Then, obviously, $f(\mathcal{U}_m) \subseteq \mathcal{R}(\leq a)$, f preserves the least element, and for $\alpha,\beta \in \mathcal{U}_m$,

(4.15) $\quad \alpha \subseteq \beta \quad \rightarrow \quad f(\alpha) \leq f(\beta)$

and

(4.16) $\quad f(\alpha\cup\beta) = f(\alpha) \cup f(\beta)$.

Also note that $f(\alpha)$ is contiguous for every $\alpha \in \mathcal{U}_m$. It remains to show that f is one-to-one, i.e. that for $\alpha,\beta \in \mathcal{U}_m$,

(4.16) $\quad \alpha \neq \beta \quad \rightarrow \quad f(\alpha) \neq f(\beta)$,

and that f preserves infima. By (4.15), the latter will follow from

(4.17) For any r.e. degree $\underset{\sim}{c}$, if $\underset{\sim}{c} \leq f(\alpha)$ and $\underset{\sim}{c} \leq f(\beta)$ then $\underset{\sim}{c} \leq f(\alpha\cap\beta)$.

For a proof of (4.16) fix $\alpha,\beta \in \mathcal{U}_m$ such that $\alpha \neq \beta$. W.l.o.g. $\beta \nsubseteq \alpha$, say $p\cdot n+i$ $(i < n)$ is not in α but in β. Then $\deg A_{i,\{p\}} \leq f(\beta)$. So it suffices to show that $\deg A_{i,\{p\}} \nleq f(\alpha)$. Note that, by (4.3), $\deg A_{i,\{p\}} > 0$. So w.l.o.g $\alpha \neq \emptyset$ and we may choose a representation $\alpha = i_o\alpha_o \cup \ldots \cup i_k\alpha_k$ satisfying (4.13). Then $f(\alpha) = \deg(A_{i_o,\alpha_o} \oplus \ldots \oplus A_{i_k,\alpha_k})$. Now for a contradiction assume that $\deg A_{i,\{p\}} \leq f(\alpha)$. Then, by contiguity of $f(\alpha)$ and Corollary 1.5, there are r.e. sets $C_o,\ldots,C_k$ such that $A_{i\{p\}} \equiv_T C_o \oplus \ldots \oplus C_k$ and $C_j \leq_T A_{i_j,\alpha_j}$, $j = 0,\ldots,k$. Now, by (4.1) and (4.14), for j such that $i \neq i_j$, C_j is recursive. So, by nonrecursiveness of $A_{i,\{p\}}$ there is a (unique) j such that $i = i_j$ and $A_{i,\{p\}} \equiv_T C_j \leq_T A_{i_j,\alpha_j}$. By (4.3), this implies $\{p\} \subseteq \alpha_j$ and thus $p\cdot n+i \in \alpha$ contrary to assumption. So $f(\alpha) \neq f(\beta)$.

(4.17) is proved by a similar application of Corollary 1.5. □

Note that the embeddings in the proofs of Corollaries 4.3 and 4.5 are by maps into the contiguous degrees. So, by the fact that in (4.1) of Theorem 4.2 we have wtt reductions, the above proofs carry over to wtt degrees.

<u>4.6. Corollary.</u> Let $\underset{\sim}{a}$ be a nonzero r.e. wtt degree and let L be a finite distributive lattice. Then

$L \Longrightarrow \underset{\sim}{R}_{wtt}(\leq \underset{\sim}{a})$.

Moreover, if mp-rank(L) $\leq$ mp-rank($\underset{\sim}{a}$) then $\quad L \overset{o}{\Longrightarrow} \underset{\sim}{R}_{wtt}(\leq \underset{\sim}{a})$. □

The relation between mp-rank($\deg_T A$) and mp-rank($\deg_{wtt} A$) is the following.

<u>4.7. Lemma.</u> For any r.e. set A, mp-rank($\deg_{wtt} A$) $\leq$ mp-rank($\deg_T A$). Moreover, if degA is contiguous then mp-rank($\deg_{wtt} A$) = mp-rank($\deg_T A$).

<u>Proof.</u> Fix A and assume that mp-rank($\deg_{wtt} A$) $\geq$ n. Then there are r.e. sets $A_o, \ldots, A_{n-1} \leq_{wtt} A$ such that $\deg_{wtt} A_o, \ldots, \deg_{wtt} A_{n-1}$ are pairwise minimal pairs. By Theorem 3.1 , w.l.o.g. $\deg A_o, \ldots, \deg A_{n-1}$ are contiguous, and thus are pairwise minimal pairs of r.e. Turing degrees too. So mp-rank($\deg_T A$) $\geq$ n.

The second part of the lemma follows from Proposition 1.2 and the fact that for r.e sets A_o and A_1 such that $\deg A_o$, $\deg A_1$ form a minimal pair, $\deg_{wtt} A_o$ and $\deg_{wtt} A_1$ form a minimal pair too. □

§5. LIMITING RESULTS ON THE EXISTENCE OF CONTIGUOUS DEGREES.

In this final section we list some negative results on the existence of contiguous degrees limiting the Ladner-Sasso transfer technique.

Theorem 5.1 (Cohen [7]). If $\underset{\sim}{a}$ is contiguous then $\underset{\sim}{a}$ is low_2 (i.e. $\underset{\sim}{a}'' = \underset{\sim}{0}''$).

A proof for Theorem 5.1 can be found in Stob [17]. Theorem 5.1 shows that contiguous degrees exist only in the lower part of $\underset{\sim}{R}$. Using a result of Ladner [12], Ambos-Spies and Fejer [4] have shown that there actually are contiguous degrees which are low_2 but not low. Ambos-Spies and Fejer also showed that the contiguous degrees are sparse in the low degrees.

Theorem 5.2 (Ambos-Spies and Fejer [4]). The contiguous degrees are nowhere dense in the low degrees. I.e. if $\underset{\sim}{a}$, $\underset{\sim}{b}$ are low degrees such that $\underset{\sim}{a} < \underset{\sim}{b}$ then there are r.e. degrees $\underset{\sim}{c}$ and $\underset{\sim}{d}$ such that $\underset{\sim}{a} \leq \underset{\sim}{c} < \underset{\sim}{d} \leq \underset{\sim}{b}$ and no r.e. degree in the interval $[\underset{\sim}{c},\underset{\sim}{d}]$ is contiguous.

A further limiting result has been claimed in [5], but no proof has been given there. Since the result has some interesting consequences, we will give a proof here.

Theorem 5.3 (Ambos-Spies, Jockusch, Shore and Soare [5]). Let A, B, K be r.e. sets such that K is wtt-complete, $K \leq_{wtt} A \oplus B$ but $K \not\leq_{wtt} B$. Then degA is not contiguous.

Proof. We shall show that there is an r.e. set C such that $C \equiv_T A$ but $A \not\leq_{wtt} C$. Let $\{A_s : s \in \omega\}$, $\{B_s : s \in \omega\}$ and $\{K_s : s \in \omega\}$ be recursive enumerations of A, B and K, respectively. We define a system of markers $\gamma(x,s)$ for $\{A_s : s \in \omega\}$ by

$$\gamma(x,0) = \langle x,0\rangle \quad \text{and} \tag{5.1}$$
$$\gamma(x,s+1) = \begin{cases} \langle x,s+1\rangle & \text{if } A_{s+1} \upharpoonright x+1 \neq A_s \upharpoonright x+1 \\ \gamma(x,s) & \text{otherwise,} \end{cases}$$

and we let $\gamma(x) = \sup_s \gamma(x,s)$. Obviously, $\gamma(x) = \lim_s \gamma(x,s) < \omega$ and $\lambda x.\gamma(x)$ is recursive in A. Conversely, $A \leq_T \lambda x.\gamma(x)$ since $x \in A$ iff $x \in A_{\gamma(x)}$.

Now let $C_s = \{\gamma(x,t) : t < s \;\&\; x \in A_{t+1} - A_t\}$ and set $C = \bigcup\{C_s : s \in \omega\}$. Then $C \leq_{wtt} A$ by permitting, i.e. $\forall x,s\,(A \upharpoonright x = A_s \upharpoonright x \rightarrow C \upharpoonright x = C_s \upharpoonright x)$, and $A \leq_T C$ since $\forall x,s\,(\gamma(x) = \gamma(x,s)$ iff $\gamma(x,s) \notin C)$ and thus $\lambda x.\gamma(x) \leq_T C$.

It remains to show that $A \not\leq_{wtt} C$, i.e. that for all e the requirement

$$R_e : \quad A \neq [e]^C$$

is met. For meeting requirement R_e it suffices to show that there is a recursive function u such that for all e,n,

$$R_{e,n} : \begin{cases} n \notin K & \text{or} \\ \exists s\ (n \in K_s \ \&\ B_s \upharpoonright u(e,n) \neq B \upharpoonright u(e,n)\ ) & \text{or} \\ A \upharpoonright u(e,n) \neq [e]^C \upharpoonright u(e,n). \end{cases}$$

Namely, if for each e there is at least one n such that the third clause in $R_{e,n}$ holds then R_e is met; otherwise, however, $K \leq_{wtt} B$ contrary to assumption.

In order to meet requirements $R_{e,n}$ we use a variant of the non-diamond technique developped in [5] for the proof of Theorem 1.24 there. We construct an r.e. set D. Since $A \oplus B$ is wtt-complete, by an application of the recursion theorem, we may assume that we know a wtt-reduction of D to $A \oplus B$ in advance, say

(5.2) $\quad D = [i]^{A \oplus B}$,

where w.l.o.g. the use function [i] is strictly increasing. So by enumerating a number x in D we can force A or B to change below [i](x). Now let

$$u(e,n) = [i](\langle e,n,[i](\langle e,n,0\rangle)\rangle)$$

and let D_s be the part of D enumerated by the end of stage s of the construction below. In order to meet requirement $R_{e,n}$ we will enumerate certain numbers $\langle e,n,x\rangle$, $x \leq [i](\langle e,n,0\rangle)+1$ into D. We say the attack on $R_{e,n}$ is in *state* k at stage s, if k is the least number such that $\langle e,n,k\rangle \notin D_s$.

Now if the attack on $R_{e,n}$ is in state k at stage s, $k \leq [i](\langle e,n,0\rangle)+1$ and the following conditions are satisfied

(5.3) $\quad n \in K_s$

(5.4) $\quad$ There is no $t \leq s$ such that $n \in K_t$ and $B_t \upharpoonright u(e,n) \neq B_s \upharpoonright u(e,n)$.

(5.5) $\quad D_s \upharpoonright \langle e,n,k\rangle+1 = [i]_s(A_s \oplus B_s) \upharpoonright \langle e,n,k\rangle+1$.

(5.6) $\quad A_s \upharpoonright u(e,n) = [e]_s(C_s) \upharpoonright u(e,n)$

then enumerate $\langle e,n,k\rangle$ in D at stage s+1.

This completes the construction of D. Note that at each stage the attack on $R_{e,n}$ is in a state $\leq [i](\langle e,n,0\rangle)+2$ and the state of the attack is nondecreasing. So the attack will have a *final* state from some point on.

Now, for a contradiction, assume that $R_{e,n}$ is not met. Then (5.4) holds for all all and (5.3) and (5.6) for sufficiently large s, whence, by (5.2), the attack on $R_{e,n}$ will have final state $[i](\langle e,n,0\rangle)+2$. So there are stages s_k, $k \leq [i](\langle e,n,0\rangle)+1$, such that (5.3) to (5.6) hold for $s = s_k$ and $\langle e,n,k\rangle$ is enumerated in D at stage s_k+1, and $s_0 < s_1 < \ldots < s_{[i](\langle e,n,0\rangle)+1}$. Hence, by (5.5) (for $s = s_k, s_{k+1}$),

$$A_{s_k} \oplus B_{s_k} \upharpoonright [i](\langle e,n,k\rangle) \neq A_{s_{k+1}} \oplus B_{s_{k+1}} \upharpoonright [i](\langle e,n,k\rangle)$$

for each $k \leq [i](\langle e,n,0\rangle)$. Since $[i](\langle e,n,k\rangle) \leq u(e,n)$, by (5.4) this implies

(5.7) $A_{s_k} \upharpoonright [i](\langle e,n,k\rangle) \neq A_{s_{k+1}} \upharpoonright [i](\langle e,n,k\rangle) \quad (k \leq [i](\langle e,n,0\rangle))$.

Furthermore, since (5.6) holds for all stages s_k,

(5.8) $[e](u(e,n)) < s_o$

and, by (5.7),

(5.9) $C_{s_k} \upharpoonright [e](u(e,n)) \neq C_{s_{k+1}} \upharpoonright [e](u(e,n)) \quad (k \leq [i](\langle e,n,0\rangle))$.

Now, by (5.7) for $k = 0$, (5.8) and by definition of γ,

$$\forall x \geq [i](\langle e,n,0\rangle)\ \forall s > s_o\ (\ \gamma(x,s) \geq \langle x,s_o\rangle \geq s_o > [e](u(e,n))\ ).$$

So (5.9) and the definition of C imply

$$\forall k \leq [i](\langle e,n,0\rangle)\ (\ A_{s_k} \upharpoonright [i](\langle e,n,0\rangle) \neq A_{s_{k+1}} \upharpoonright [i](\langle e,n,0\rangle)\ ).$$

It follows that A has $[i](\langle e,n,0\rangle)+1$ elements which are less than $[i](\langle e,n,0\rangle)$. But this is impossible.

This completes the proof of Theorem 5.3. □

Theorem 5.3 shows that the class of r.e. wtt degrees which are contained in contiguous degrees does not generate the class of all r.e. wtt degrees (under $\cup$ and $\cap$). We do not know, however, whether or not the contiguous degrees generate the set $\underset{\sim}{R}$ of r.e. (Turing) degrees.

As a further consequence of Theorem 5.3 there are low r.e. degrees with no contiguous successors .

<u>5.4. Corollary</u>. For any low r.e. degree $\underset{\sim}{a}$ there is a low r.e. degree $\underset{\sim}{b} \geq \underset{\sim}{a}$ such that $\forall \underset{\sim}{c} \geq \underset{\sim}{b}$ ($\underset{\sim}{c}$ is not contiguous).

<u>Proof</u>. Given a low r.e. degree $\underset{\sim}{a}$ fix any r.e. set $A \in \underset{\sim}{a}$, and let K be a wtt-complete set. By Robinson's splitting theorem [14] there are r.e. sets K_o and K_1 which split K and such that $K_o \oplus A$ and $K_1 \oplus A$ are low. Now let $\underset{\sim}{b} = \deg K_o \oplus A$. By Theorem 5.3, it suffices to show that for any $\underset{\sim}{c} \geq \underset{\sim}{b}$ there are r.e. sets C and D such that $C \in \underset{\sim}{c}$, $K \not\leq_{wtt} D$ but $K \leq_{wtt} C \oplus D$. As one can easily show, these conditions are satisfied for the sets $D = K_1 \oplus A$ and $C = K_o \oplus A \oplus C'$, C' any r.e. set in $\underset{\sim}{c}$. □

REFERENCES.

1. K.Ambos-Spies, Anti-mitotic recursively enumerable sets, Zeitschrift f. Math. Logik u. Grundlagen d. Mathematik, to appear.
2. K.Ambos-Spies, Cupping and noncapping in the r.e. wtt and Turing degrees, to appear.
3. K.Ambos-Spies, Automorphism bases for the r.e. degrees, to appear.
4. K.Ambos-Spies and P.A.Fejer, Degree theoretical splitting properties of recursively enumerable sets, to appear.
5. K.Ambos-Spies, C.G.Jockusch, Jr., R.A.Shore and R.I.Soare, An algebraic decomposition of the recursively enumerable degrees and the coincidence of several degree classes with the promptly simple degrees, Trans. A.M.S., to appear.
6. K.Ambos-Spies and R.I.Soare, One-types of the recursively enumerable degrees, in preparation.
7. P.F.Cohen, Weak truth table reducibility and the pointwise ordering of 1-1 recursive functions, Thesis, Univ. Illinois Urbana-Champaign (1975)
8. P.A.Fejer, Branching degrees above low degrees, Trans. A.M.S. 273 (1982) 157-180
9. A.H.Lachlan, Lower bounds for pairs of recursively enumerable degrees, Proc. London Math. Soc. 16 (1966) 537-569.
10. A.H.Lachlan, Embedding nondistributive lattices in the recursively enumerable degrees, Springer Lecture Notes Math. 255 (1972) 149-177 (Conference in Math. Logic, London 1970)
11. A.H.Lachlan, Bounding minimal pairs, J. Symbolic Logic 44 (1979) 626-642
12. R.E.Ladner, A completely mitotic nonrecursive recursively enumerable degree, Trans. A.M.S. 184 (1973) 479-507.
13. R.E.Ladner and L.P.Sasso, The weak truth table degrees of recursively enumerable sets, Ann. Math. Logic 4 (1975) 429-448.
14. R.W.Robinson, Interpolation and embedding in the recursively enumerable degrees, Ann. of Math. (2) 93 (1971) 285-314.
15. R.I.Soare, Fundamental methods for constructing recursively enumerable degrees, in Recursion Theory: its Generalisations and Aplications, F.R.Drake and S.S. Wainer (Ed.), Cambridge University Press, Lecture Notes 45 (1980) 1-51.
16. R.I.Soare, Tree arguments in recursion theory and the $0'''$-priority method, to appear.
17. M.Stob, wtt-degrees and T-degrees of recursively enumerable sets, J.Symbolic Logic, to appear.
18. S.K.Thomason, Sublattices of the recursively enumerable degrees, Zeitschrift f. Math. Logik u. Grundlagen d. Mathematik 17 (1971) 273-280.

Abstract Construction of Counterexamples in Recursive Graph Theory

Hans-Georg Carstens and Peter Päppinghaus
Department of Mathematics
University of Bielefeld
Universitätsstraße

4800 Bielefeld 1
Federal Republic of Germany
and

Institute of Mathematics
University of Hannover
Welfengarten 1

3000 Hannover 1
Federal Republic of Germany

§1 Introduction

Many theorems of graph theory take the form of stating the existence of a solution of a certain problem for a suitable class of graphs. We think of problems such as coloring, match-making, infinite paths, Hamilton or Euler paths, embedding etc., and of classes of graphs given e.g. by invariants such as maximal degree, genus, connectedness etc. Recursive graph theory is concerned with the recursive or degree-theoretic complexity of such solutions for classes of countable graphs.

In this paper we focus on negative results. In order to exclude trivialities and dirty tricks it is natural to impose certain restrictions on the graphs providing counter-examples. E.g. the non-existence of a recursive 3-coloring for a graph G will be of interest only provided G itself is recursive. Two less obvious restrictions are suggested by the following results. By Kleene's Basis Theorem [R, Corollary XLI(b), p. 419] there is a recursive tree having an infinite path, however no hyperarithmetical one. Such phenomena would take us too high up in complexity. To stay down to earth, we hence restrict attention to <u>locally finite</u> graphs throughout this paper. Furthermore Bean [Bl, Theorem 2, p. 470] has constructed a locally finite, recursive graph G, which is connected, planar, and 3-colorable, but has no recursive coloring using only finitely many colors. G as a graph is not very complicated, the mystery is that by a clever numbering of its points the function f_G with $f_G(n) = \{m|m$ is adjacent to n in G$\}$ is not recursive. To exclude this, one considers graphs for which the function f_G is recursive. These graphs are usually called highly recursive and can equivalently be defined as graphs for which the set of vertices, the adjacency relation, and the degree function are recursive.

We are interested in the construction of highly recursive graphs, for which there exists a solution of a certain problem, but no recursive one.

Perhaps the first result of this kind is an old theorem of Kleene [Kl], which he established as a counter-example in recursive mathematics to Brouwer's fan theorem (i.e. classically: König's Lemma). Kleene constructed a highly recursive binary tree, which has an infinite path, but no recursive one. On the other hand Kreisel's Basis Theorem shows that every infinite, highly recursive tree has an infinite path, which is recursive in 0'. Since virtually all problems for highly recursive graphs can be modelled in the infinite path problem for highly recursive trees, this gives an upper complexity bound in this area. The gap between the negative and the positive result was narrowed by Specker in an also old paper, which remained somewhat hidden [Sp]. Specker was interested in the complexity of complete and consistent extensions of number theory. As a tool he constructed an infinite highly recursive binary tree having no infinite path in the Boolean algebra generated by the r.e. sets. 17 years later the result was rediscovered by Jockusch [J] who was interested in the theory of Π^0_1-classes of sets.

With a growing interest in recursive combinatorics in the 1970's a lot of combinatorial constructions were investigated w.r.t. their recursive complexity, and positive as well as negative results were obtained for problems such as coloring, match-making, tiling, chains and anti-chains, Hamiltonian and Eulerian paths. The positive results are usually proved by explicitly describing a uniform algorithm. For the negative results one finds different methods of proof in the literature. In many cases a technique of diagonalization is applied as familiar from recursion theory, see. e.g. [BI], [CP1], [MR 1], [MR 2]. These proofs are completely similar. What is specific in the different examples is just the "geometry" which makes the diagonalization work and must be suitable for the problem under consideration.

The aim of this paper is to carry out this particular construction on an abstract level so that one obtains a sufficient condition for recursive unsolvability, which is of a finite combinatorial nature.

Let us explain this in more detail. The technique of diagonalization quoted above works as follows. With every algorithm one associates an infinite subgraph of the graph to be constructed and thinks of the e-th algorithm as working on the e-th subgraph. With different stages of the computations one associates growing finite subgraphs of these infinite subgraphs. Whenever one of the algorithms, say the e-th one, has computed sufficiently many values, a trap is built into the e-th subgraph obstructing these values as part of a solution of the problem for the whole graph.

What makes this construction work, is the possibility to define a family of finite graphs s.t. arbitrarily "far out" one can build in a finite number of traps simultaneously and independently of each other. In §3 we give an abstract formulation of the existence of such a family of finite graphs, called a <u>trap.</u>

The conditions to be formulated for such a trap pertain to the finite graphs of the trap, whereas the graph to be constructed from the trap is infinite and should not admit a recursive answer for a certain question, say Q^*. Therefore we first need an abstract description of an infinite problem P^* consisting of a question Q^* for a class I^* of countable graphs, of an associated finite problem P consisting of a question Q for a class I of finite graphs, and of a proper relation between P^* and P. In §2 we define these notions and thereby provide the abstract framework, in which our constructions are formulated.

In §3 we prove that the existence of a recursive trap for problem P implies the existence of a highly recursive graph admitting no recursive solution of question Q^*. This construction contains the recursion theoretic diagonalization in abstract terms and thereby reduces the construction of an infinite counter-example to that of an infinite family of finite graphs.

We have, however, not yet fully captured the uniformity encountered in the literature and not yet fully reduced to finite constructions. We observe that the existence of a trap is usually shown to be satisfied by glueing together in a suitable way certain finite graphs, which we call <u>chips.</u> There is a <u>start chip</u> S with a sequence of exit points carrying certain information about a solution of question Q for supergraphs of S. There is a <u>cable chip</u> C with a sequence of entry points and a sequence of exit points. The task of the cable chip is to transfer certain relevant information about a solution of question Q for supergraphs of C from the entries to the exits. Finally there are a number of <u>trap chips</u> T_i with a sequence of entry points. T_i has the property that for any solution of question Q for supergraphs of T_i the relevant information at the entries cannot be of the i-th type.

We give a simple example for the problem of 3-coloring. The start chip is  , and the relevant information is, which pair of points carries the same color. This information is obviously transferred from the entries to the exits by the following cable chip.

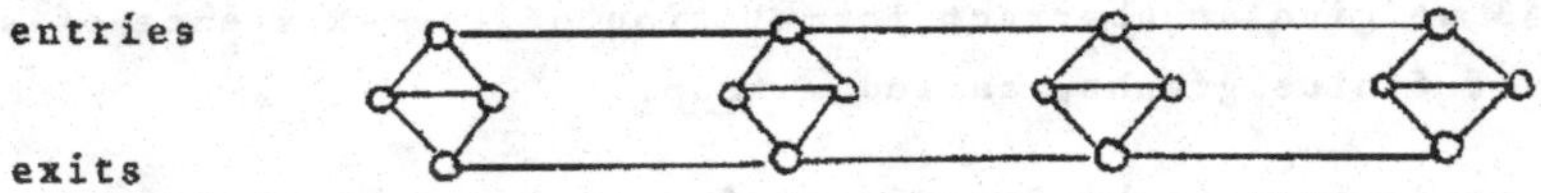

For the relevant information, say e.g. "the first entry point carries the same color as the third entry point", the trap chip $T_{(1,3)}$ is the following graph.

In §4 we define the notion of a <u>trap equipment</u> as consisting of these chips together with a classification of the "relevant information". We show how to construct a recursive trap from a trap equipment.

Put together we have succeeded to reduce the construction of a highly recursive graph $G \in I^*$ for which no recursive solution of question Q^* exists to the finite combinatorial task of constructing a trap equipment for the finite problem P.

In §5 we apply this to recapture several recursive unsolvability results from the literature as well as to add a new one, namely recursive unsolvability of the embedding problem. These results now fall out very simply: one just has to draw a few finite graphs and verify that they satisfy the conditions required for the chips of a trap equipment. Clearly these pictures of the chips capture the "geometrical" essence of the individual counter-example under consideration, free from the routine part of the method.

§2 Finite Approximation of Problems

In this section we discuss problems for countable graphs which can be approximated by problems for finite graphs.

We consider graphs G as structures having a domain $D(G)$ which is the set of points of the graph, and a symmetric and irreflexive relation L which is the adjacency relation. If necessary we expand this format by constants or relations. E.g. a rooted tree T is a structure $(D(T),L,\rho)$ where ρ is the root of T, and a bipartite graph G is a structure $(D(G),L,M,F)$ where M,F are the two classes of independent points in G.

To speak about problems for countable graphs, we use a class I^* of graphs with domain $\subseteq \mathbb{N}$, and a relation Q^* of G and r, thinking of $Q^*(G,r)$ as expressing "the partial function r answers question Q^* for the I^*-graph G". E.g. let I^* be the class of planar graphs with domain $\subseteq \mathbb{N}$, and let $Q^*(G,r)$ say: "$r(i)$ is the color of point i in a 4-coloring of G". We want to allow an answer r to question Q^* to contain redundant information. I.e. if $r \prec^* s :\Longleftrightarrow \forall i(r(i)\downarrow \Rightarrow r(i) \simeq s(i))$ and r is an answer to Q^* for G, then also s is an answer.

<u>Definition 1:</u> $P^* = (I^*,Q^*)$ is called a <u>problem for countable graphs</u> if the following is satisfied:

(i) I^* is a class of countable locally finite graphs with domain $\subseteq \mathbb{N}$

(ii) $Q^* \subseteq I^* \times \{r \mid r \text{ partial function } \mathbb{N} \to \mathbb{N}\}$

(iii) $r \prec^* s \wedge Q^*(G,r) \Rightarrow Q^*(G,s)$.

We represent a locally finite countable graph G with domain $\subseteq \mathbb{N}$ by an infinite sequence g s.t. $g_i = 0$ iff $i \notin D(G)$, and for $i \in D(G)$, g_i codes the neighbourhood of i in G. Note that a graph G is highly recursive iff its representing sequence g is recursive.

In the case of a problem for finite graphs we want an answer to a question to be a definite finite object. So we use finite sequences r of natural numbers and intend that $r_i = 0$ in the finite case corresponds to $r(i)\uparrow$ in the countable case. Hence we define
$r \prec s :\Leftrightarrow \forall i(r_i \neq 0 \Rightarrow r_i = s_i)$.

<u>Definition 2:</u> $P = (I,Q)$ is called a <u>problem for finite graphs</u> iff the following is satisfied:

(i) I is a class of finite graphs with domain $\subseteq \mathbb{N}$
(ii) $Q \subseteq I \times \{r \mid r$ finite sequence of natural numbers$\}$
(iii) $r \prec s \wedge Q(G,r) \Rightarrow Q(G,s)$.

Since we are discussing approximations of problems for countable graphs by problems for finite graphs, we will approximate countable graphs by ascending sequences of their finite subgraphs.

<u>Definition 3:</u> $(G_n)_{n\geq 0}$ is called an <u>ascending sequence in I</u> iff the following is satisfied:

$$\forall n: G_n \in I \wedge G_n \subseteq G_{n+1}.$$

Now we have to say what it means that a question Q^* for countable graphs is approximated by a question Q for finite graphs. We think of a partial function r as answering Q^* for G, if answers to Q for the finite subgraphs of G are contained in suitable finite initial segments of r. To make this precise, we define for a partial function $r:\mathbb{N} \to \mathbb{N}$ and $k \in \mathbb{N}$ the finite sequence $r \upharpoonright k$ as follows. For $i < k$: $(r\upharpoonright k)_i := r(i) + 1$, if $r(i)\downarrow$, and $(r\upharpoonright k)_i := 0$ otherwise.

Definition 4: Let $P = (I,Q)$ be a problem for finite graphs.

(i) $\lim I := \{G \mid G \text{ locally finite} \wedge \exists (H_n)_{n\geq 0} \text{ ascending sequence in } I \wedge G = \bigcup_n H_n\}$

(ii) $\lim Q \subseteq \lim I \times \{r \mid r \text{ partial function } \mathbb{N} \to \mathbb{N}\}$ is defined by $\lim Q(G,r) :\Longleftrightarrow \forall H' \in I (H' \subseteq G \Rightarrow \exists k \exists H \in I : H' \subseteq H \subseteq G \wedge Q(H, r \upharpoonright k))$

(iii) $\lim P := (\lim I, \lim Q)$.

In the way we have defined a problem $P^* = (I^*,Q^*)$ for countable graphs, I^* may contain finite graphs. Hence we can extract from P^* a problem for finite graphs.

Definition 5:

(i) $\text{fin } I^* := \{G \in I^* \mid G \text{ is finite}\}$

(ii) $\text{fin } Q^* \subseteq \text{fin } I^* \times \{r \mid r \text{ finite sequence of natural numbers}\}$ is defined by $\text{fin } Q^*(G,r) :\Longleftrightarrow Q^*(G,r^*)$, where $r^*(i) := r_i - 1$ if $i < \ell h(r) \wedge r_i \neq 0$, and $r^*(i)$ is undefined otherwise.

(iii) $\text{fin } P^* := (\text{fin } I^*, \text{fin } Q^*)$.

We are interested in problems P^* where we can reconstruct P^* as $\lim \text{fin } P^*$, for these problems can be viewed as natural infinite extensions of a finite graph problem.

Definition 6: A problem P^* for countable graphs is <u>finitely based</u> iff $\lim \text{fin } P^* = P^*$.

Now we present some examples for problems which are finitely based.

Example 1: We consider the problem to find in a binary rooted tree a path of maximal length starting at the root, i.e. we look at the problem

$$\text{PATHBINTREE} = (I^*,Q^*)$$

where $I^* := \{T \mid T \text{ is a binary rooted tree}\}$.
$Q^*(T,r) :\Longleftrightarrow r(0) = \rho(T) \wedge \forall n: (r(n)\downarrow \Rightarrow \forall i < n: r(i)\downarrow \wedge r(i) \neq r(n)) \wedge$
$\wedge \forall n: (r(n+1)\downarrow \Rightarrow \{r(n), r(n+1)\} \text{ is a line of } T) \wedge \forall a \in D(T) \forall k:$

(a has distance k to the root $\rho(T)$ in $T \Rightarrow r(k)\downarrow$) . fin PATHBINTREE is easily read off from PATHBINTREE.
Then obviously the following hold:
lim fin $I^* = I^*$ and lim fin $Q^* = Q^*$ (left to the reader).
Hence we see that the problem to find in a binary rooted tree a path of maximal length is finitely based.

Example 2: Here the problem is to find a k-coloring for a connected locally finite graph with chromatic number $\leq k$ ($k \geq 3$) .

$$\text{k-COLCON} = (I^*,Q^*)$$

where $I^* := \{G \mid G$ is a k-colorable, connected, locally finite graph$\}$,
and $Q^*(G,r) :\Longleftrightarrow \forall n \in D(G): (r(n)\downarrow \wedge r(n)<k) \wedge$
$\wedge\ \forall i,j \in D(G): (\{i,j\}$ line of $G \Rightarrow r(i) \neq r(j))$.
Again it is easy to verify that lim fin $I^* = I^*$ and lim fin $Q^* = Q^*$.
Hence we see that the problem to find a k-coloring for a k-colorable, connected, and locally finite graph is finitely based.

Example 3: The next problem is to find an embedding in an orientable surface of genus g for connected, locally finite graphs of genus g .

$$\text{g-EMBEDCON} = (I^*,Q^*)$$

where $I^* := \{G \mid G$ is a connected, locally finite graph of genus $g\}$.
$Q^*(G,r) :\Longleftrightarrow \forall n \in D(G): (r(n)\downarrow \wedge r(n)$ is a (code of a) sequence in which every neighbour of n occurs exactly once and taking for every $n \in D(G)$ the order of its neighbours in $r(n)$ as clockwise orientation, one obtains in total an embedding of G in an orientable surface of genus g . Again we omit the verification of lim fin $I^* = I^*$ and of lim fin $Q^* = Q^*$. So one sees that the problem to find an embedding in an orientable surface of genus g for locally finite connected graphs of genus g is finitely based.

Example 4: The last problem we consider here is to find a matching in a bipartite "almost k-regular" graph ($k \geq 2$) .

$$\text{MATCHING}(k) = (I^*,Q^*)$$

where $I^* := \{G|G$ is a locally finite, bipartite graph, $\forall f\in F(G)$: the degree of f in G is k, $\forall m\in M(G)$: the degree of m in G is $\leq k\}$. $Q^*(G,r)$: $\Longleftrightarrow \forall f\in F(G)$: $(r(f)\downarrow \wedge \{f,r(f)\}$ is a line of $G) \wedge$ $\wedge \forall f_1,f_2\in F(G)$: $(f_1 \neq f_2 \Rightarrow r(f_1)\neq r(f_2))$. As is easily verified the k-regularity for $f \in F(G)$ and the bound for the degrees of $m \in M(G)$ imply that G fulfills the Hall condition.

So $\lim \text{fin } I^* = I^*$ and $\lim \text{fin } Q^* = Q^*$, and we see that the problem to find a matching in a bipartite "almost k-regular" graph is finitely based.

§3 Traps

In this section we give a criterion implying recursive unsolvability of a problem for a particular highly recursive graph.

Let $P^* = (I^*,Q^*)$ be a finitely based problem, $P = (I,Q) = \text{fin } P^*$, and $P^* = \lim P$. We want to find a sufficient condition for the existence of a highly recursive graph $G \in I^*$ s.t. there is no recursive r with $Q^*(G,r)$. To obtain this criterion we analyze a particular construction of such a graph G. The idea is as follows. We construct an ascending sequence $(G_n)_{n\geq 0}$ s.t. $G = \bigcup_n G_n$ by stages. We want to make every algoritm a non-candidate for a solution to question Q^* for the graph G constructed eventually. Stage $n + 1$ corresponds to computations of length $n + 1$ for the algorithm no. 0 up to no. n. At this stage we want to take care of the first of these algorithms, which has finished its computations on all j in a certain significant interval $[\alpha,\alpha+t] \subseteq \mathbb{N}$ in at most $n + 1$ steps, and which is not yet taken care of. Let e be the no. of this algorithm, and collect the significant values $\{e\}(\alpha),\ldots,\{e\}(\alpha+t)$ in a sequence v. We must be able to obstruct these values as part of a solution r. In other words, r has to differ from the values collected in v at one of the arguments $\alpha,\ldots,\alpha + t$. t can be taken to be constant for the whole construction, α however has to depend on e. We fix α for e at stage e, immediately before algorithm no. e is first taken into consideration. In other words, α depends on e and G_e. At stage $n + 1$ then, when algorithm no. e is to be taken care of, G_n has to be extended to G_{n+1} in such a way that the values collected in v are obstructed as part of any solution to question Q for any graph H with $G_{n+1} \subseteq H \subseteq G_m$.

We denote finite sequences by $<\ldots>$. ℓh denotes the length function, $*$ the concatenation function, and $\upharpoonright$ the restriction function for finite sequences
(i.e. $\ell h(<x_0,\ldots,x_{n-1}>) = n$, $<x_0,\ldots,x_{n-1}>*<y_0,\ldots,y_{k-1}> =$
$= <x_0,\ldots,x_{n-1},y_0,\ldots,y_{k-1}>$, and $<x_0,\ldots,x_{n-1}> \upharpoonright k = <x_0,\ldots,x_{k-1}>$
for $k \leq n$). Furthermore, if r is a finite sequence of finite sequences, we denote by $r_{k,m}$ the m-th member of the sequence r_k. And for a finite sequence r and $k \leq n < \ell h(r)$ we put
$r[k,n] := <r_k, r_{k+1},\ldots,r_n>$. Where necessary to use recursion theory, we interpret $<x_0,\ldots,x_{n-1}>$ at the same time as its code in a canonical coding of finite sequences. The symbols ℓh, $*$, $\upharpoonright$, and $[\ ,\ ]$ are then taken to denote also the corresponding elementary recursive functions on the codes.

We call w a veto of arity t iff w is a finite sequence of finite sequences s.t. $\forall n < \ell h(w)$: $(w_n \neq 0 \Rightarrow \ell h(w_n) = t + 2 \wedge \forall i \leq t\colon w_{n,i} \neq 0 \wedge$
$\wedge\ w_{n,t+1} \leq n \wedge \forall j < n\colon w_{j,t+1} \neq w_{n,t+1})$.

The intention is that $v = w_n \upharpoonright t+1$ codes a veto coming from algorithm no. $e = w_{n,t+1}$.

Definition 1: T is a (recursive) trap for P iff for some $t \in \mathbb{N}$ $(T^w | w$ veto a arity $t)$ is a (recursive) family of I-graphs s.t. for some (recursive) total function $\alpha : \mathbb{N} \to \mathbb{N}$ the following is satisfied:

(T1) $D(T^w)$ is an initial segment $\wedge$
$(\ell h(w) = n+1 \Rightarrow T^{w\upharpoonright n} \subsetneq T^w) \wedge$

$(\ell h(w) = n+2 \Rightarrow T^{w\upharpoonright n+1} \underset{D(T^{w\upharpoonright n})}{=} T^w)$, where the relation $G \underset{M}{=} H$ is for $M \subseteq \mathbb{N}$ defined by

$G \underset{M}{=} H :\Longleftrightarrow \forall a \in M\colon ((a \in D(G) \Longleftrightarrow a \in D(H)) \wedge \forall b\ \mathbb{N}\colon (\{a,b\}$ line of $G \Longleftrightarrow$
$\Longleftrightarrow \{a,b\}$ line of $H)$.

A segment is an interval $[k,n] \subseteq \mathbb{N}$, and and initial segment is a segment $[0,n]$.

(T2) $e < \ell h(w) \wedge T^{w\upharpoonright e} \subseteq H \subseteq T^w \wedge Q(H,r) \Rightarrow$
$\forall i \leq t\colon \alpha(w\upharpoonright e) + i < \ell h(r) \wedge r_{\alpha(w\upharpoonright e)+i} \neq 0$.

(T3) $e \leq n < \ell h(w) \wedge w_n = v * <e> \wedge T^{w\upharpoonright n+1} \subseteq H \subseteq T^w \wedge Q(H,r) \Rightarrow$
$\Rightarrow r[\alpha(w\upharpoonright e), \alpha(w\upharpoonright e)+t] \neq v$.

<u>Definition 2:</u> G is called a <u>universal witness</u> for recursive unsolvability of P^* iff $G \in I^* \wedge \forall r(Q^*(G,r) \Rightarrow r$ is not recursive).

<u>Theorem 1:</u> If there is a recursive trap for $P = \mathrm{fin}\, P^*$, then there is a highly recursive universal witness for recursive unsolvability of P^*.

<u>Proof:</u> Let T be a recursive trap for P with associated arity t and recursive α. We inductively define a sequence $(w_n)_{n\geq 0}$ s.t. for every n: $\ell h(w_n) = n \wedge w_n$ is a veto of arity $t \wedge w_{n+1} \upharpoonright n = w_n$. $w_0 := \langle\ \rangle$. Assume that $w_0, \ldots, w_n$ are already defined, and in particular for all $e \leq n$: $w_n \upharpoonright e = w_e$.

We define $w_{n+1} := w_n * \langle\langle u_0, \ldots, u_t, e\rangle\rangle$, if $e \leq n$ is the smallest number satisfying condition (+). In this case $u_i := \{e\}(\alpha(w_e)+i) + 1$ is well-defined for $i \leq t$. If there is no e satisfying (+), we let $w_{n+1} := w_n * \langle 0\rangle$. The condition which determines e is the following:

(+) $e \leq n \wedge \forall i \leq t$: the computation for $\{e\}$ with input $\alpha(w_e) + i$ has stopped at step $n + 1 \wedge \forall j < n$: $(w_{n,j} \neq 0 \Rightarrow e \neq w_{n,j,t+1})$.

Clearly w_{n+1} is a veto of length $n + 1$ and arity t extending the veto w_n. An easy inspection of the definition also shows the sequence $(w_n)_{n\geq 0}$ to be recursive.

The conditions (T1) ensure that the sequence $(T^{w_n})_{n\geq 0}$ is an ascending I-sequence s.t. $D(\bigcup_n T^{w_n}) = \mathbb{N}$, and $G := \bigcup_n T^{w_n} \in I^*$. Since $(w_n)_{n\geq 0}$ and T are recursive, G has to be highly recursive. Suppose that there is a recursive partial function $r : \mathbb{N} \to \mathbb{N}$ s.t. $Q^*(G,r)$ and $r = \{e\}$.

T^{w_e} is an I-graph, and $T^{w_e} \subseteq G$. Hence there is an I-graph H, an $f > e$, and a $k \in \mathbb{N}$ s.t. $T^{w_e} \subsetneq H \subseteq T^{w_f} \wedge Q(H, r\upharpoonright k)$. So by (T2) we have $k > \alpha(w_e) + t$ and $\forall i \leq t$: $(r\upharpoonright k)_{\alpha(w_e)+i} \neq 0$, and hence $\forall i \leq t$: $\{e\}(\alpha(w_e)+i)\downarrow$. By the definition of $(w_j)_{j\geq 0}$ there has to be a uniquely determined $n \geq e$ s.t. $\forall i \leq t$: the computation for $\{e\}$ with input $\alpha(w_e) + i$ has stopped at step $n + 1 \wedge w_{n+1,n} =$
$= \langle\{e\}(\alpha(w_e))+1, \ldots, \{e\}(\alpha(w_e)+t)+1\rangle * \langle e\rangle =: v * \langle e\rangle$.
As above one sees that there is an I-graph H, an $m > n$, and a $k > \alpha(w_e) + t$ s.t. $T^{w_{n+1}} \subsetneq H \subseteq T^{w_m} \wedge Q(H, r\upharpoonright k)$. Now we conclude from

(T3) that $v = \langle \{e\}(\alpha(w_e))+1, \ldots, \{e\}(\alpha(w_e)+t)+1 \rangle = (\{e\}\restriction k)[\alpha(w_e), \alpha(w_e)+t]$ $= (r\restriction k)[\alpha(w_e), \alpha(w_e)+t] \neq v$, which is a contradiction. □

To give an example, we apply this theorem to PATHBINTREE. We obtain so an old result of Kleene.

<u>Theorem 2:</u> There is a highly recursive binary rooted tree, which is a universal witness for the recursive unsolvability of PATHBINTREE.

<u>Proof:</u> To show that there is a recursive trap for PATHBINTREE, we define $\alpha(w) := \ell h(w)$, $t := 0$. Now we define T. The graphs T^w will be binary trees with root 0, and will contain at least two points of level $\ell h(w) + 1$, but no point of level $\ell h(w) + 2$. By the level of a point in a rooted tree we mean its distance to the root. The points in T^w are numbered in such a way, that every point of level k has a smaller number than any of the points of level $k + 1$. Let T_n be a binary tree with root 0 having the following properties: T_n has no points of level $>n$. Points of level n have degree 1. Every point $\neq 0$ of level $<n$ has degree 3. The root has degree 2. Obviously T_n has exactly 2^i points of level i, $i \leq n$. T is now defined as follows:

$T^{\langle\,\rangle} :=$ 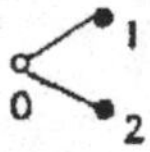 .

If w is a veto of arity 0 and length $n + 1$, then let T^w be a copy of a subgraph of T_{n+2} to be obtained from T_{n+2} as follows. Assume inductively $T^{w\restriction n}$ to be defined already. Let $e \leq m \leq n$ and a be such that $w_m = \langle a, e \rangle$ and a is of level $e + 1$ in $T^{w\restriction n}$. Then let every point b of level $n + 1$ in $T^{w\restriction n}$ with the property that a lies on the unique path from b to the root, be a leaf in T^w. All points of level $n + 1$ in $T^{w\restriction n}$ not declared leaves by this token, obtain two new neighbours of level $n + 2$ in T^w. Number the new points suitably s.t. $D(T^w)$ is an initial segment. By this rule, for each level $e + 1$, $e \leq n$, at most 2^{n+1-e} points of level $n + 2$ in T_{n+2} are omitted, this are in total at most $2^{n+2}-2$, so T^w has at least 2 points of level $n + 2$. It is easy to check (T1) – (T3). So the claim follows by Theorem 1.

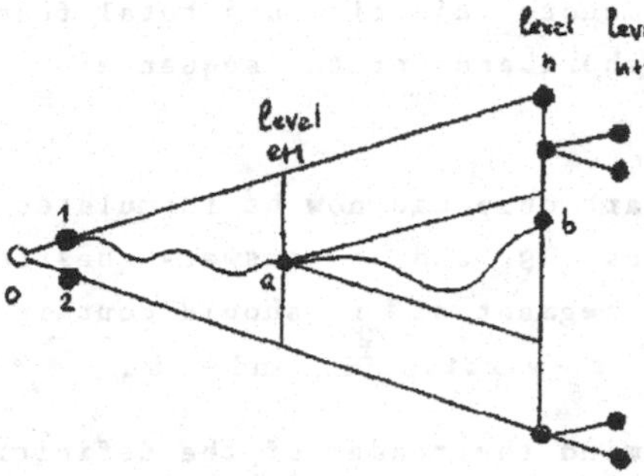

□

§4 Chips

In this section we present a machinery to construct from start, cable, and trap chips a highly recursive universal witness for the recursive unsolvability of a problem.

We assume to be given a finitely based problem $P^* = (I^*,Q^*)$ and $P = (I,Q) = \text{fin}\, P^*$. Furthermore we use a total function $\tau:\{G|G \text{ finite graph}\} \times \{r|r \text{ finite sequence of natural numbers}\} \times \mathbb{N}^{a+1} \rightarrow$ $\rightarrow \{M|M \subseteq \mathbb{N} \wedge M \text{ finite}\}$. Intuitively $i \in \tau(G,r,s_0,\ldots,s_a)$ says that an answer r to question Q for G is of type i w.r.t. relevant points $s_0,\ldots,s_a$. We require τ to have the following property: $r' \prec r \Rightarrow \tau(G,r',\vec{s}) \subseteq \tau(G,r,\vec{s})$, where $\vec{s} = s_0,\ldots,s_a$.

A start chip S is a finite graph, given together with a sequence $c_0,\ldots,c_\ell$ of core points, a sequence $s_0,\ldots,s_a$ of exit points, and merging subgraphs M_0,M_1 which are isomorphic. M_0 should be such that the union of any two I-graphs is an I-graph, provided their intersection is isomorphic to M_0. Typically M_0 will consist of just a single point or M_0 will be empty.

Our mental picture of a start chip is as follows:

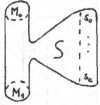

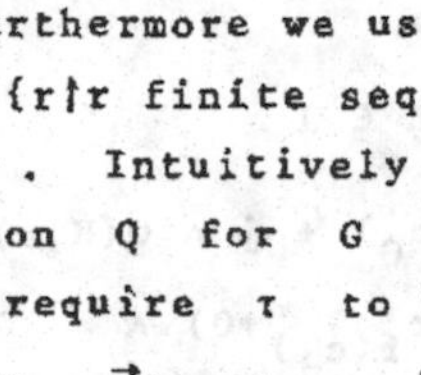

In our construction carried out later, we will use suitably renumbered copies of chips. We indicate a renumbering of a graph G by ${}^{f}G$, where

$f,g,h,\ldots$ are henceforth used to denote injective and total functions from $\mathbb{N}$ to $\mathbb{N}$. ${}^{f}\vec{s}$ is used as a shorthand for the sequence $f(s_0),\ldots,f(s_a)$.

The crucial requirement for the start chip can now be formulated as follows. If an I-graph H contains ${}^{f}S$ and r answers question Q for H, then a certain significant segment of r should contain enough information to determine a type of r w.r.t. ${}^{f}\vec{s}$ and ${}^{f}S$.

We define $0^m := \langle 0,\ldots,0\rangle$ and remind the reader of the definition of $r * r'$ and $r[k,n]$ given in §3.

<u>Definition 1:</u> S is called a <u>start chip</u> with <u>core</u> $c_0,\ldots,c_\ell$, <u>exits</u> $s_0,\ldots,s_a$, and <u>mergers</u> M_0, M_1 iff the following is satisfied:

(i) $M_0, M_1 \subseteq S \wedge M_0 \cong M_1 \wedge M_0 \cap M_1 = \emptyset \wedge$

$\wedge\ c_0,\ldots,c_\ell, s_0,\ldots,s_a \in S \setminus (M_0 \cup M_1) \wedge \#\{c_0,\ldots,c_\ell\} = \ell + 1 \wedge$

$\wedge\ \#\{s_0,\ldots,s_a\} = a + 1$.

(ii) $G, H \in I \wedge G \cap H \cong M_0 \Rightarrow G \cup H \in I$.

(iii) $H \supseteq {}^{f}S \wedge \forall j \leq \ell\colon f(c_j) = f(c_0) + j \wedge Q(H,r) \Rightarrow$

$\Rightarrow \forall j \leq \ell\colon (f(c_j) < \ell h(r) \wedge r_{f(c_j)} \neq 0) \wedge$

$\wedge\ \tau({}^{f}S, 0^{f(c_0)} * r[f(c_0), f(c_\ell)], {}^{f}\vec{s}) \neq \emptyset$.

A cable chip C is a finite graph, given together with a sequence $t_0,\ldots,t_a$ of entry points and a sequence $s_0,\ldots,s_a$ of exit points. C is used to transfer the types of a solution of Q from the entries to the exits. Such a transfer can be achieved by glueing a copy of C at its entries to some given graph. We picture a cable chip as follows:

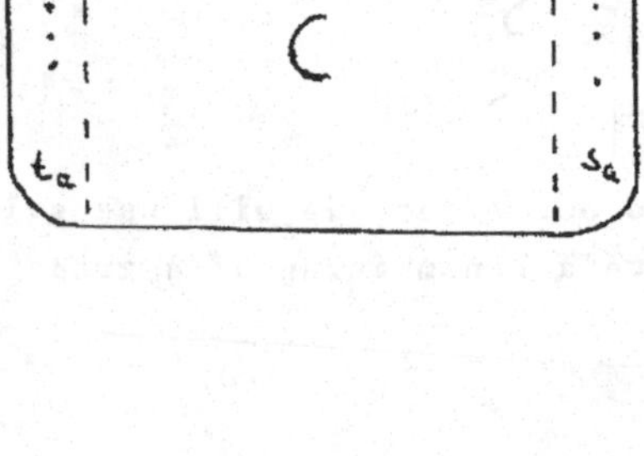

<u>Definition 2:</u> C is called a <u>cable chip</u> with <u>entries</u> $t_0,\dots,t_a$ and <u>exits</u> $s_0,\dots,s_a$ iff the following holds:

(i) $s_0,\dots,s_a,t_0,\dots,t_a \in C \wedge \#\{s_0,\dots,s_a,t_0,\dots,t_a\} = 2(a+1)$.

(ii) $D(G\cap {}^fC) = \{{}^f\vec{t}\} \wedge H \supseteq G \cup {}^fC \wedge Q(H,r) \Rightarrow \tau(G,r,{}^f\vec{t}) \subset \tau(G\cup {}^fC,r,{}^f\vec{s})$.

For every i in the range of τ we have a trap chip T_i, which is a finite graph, given together with a sequence $t_0,\dots,t_a$ of entry points. The trap chip T_i is capable of enforcing that no answer r to Q for an I-graph H containing ${}^fT_i$ can be of type i w.r.t. the entry points ${}^f\vec{t}$.

Our picture for a trap chip is as follows:

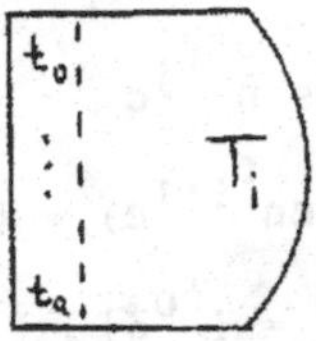

<u>Definition 3:</u> T_i is called a <u>trap chip of type i</u> with <u>entries</u> $t_0,\dots,t_a$ iff the following is satisfied:

(i) $t_0,\dots,t_a \in T_i \wedge \#\{t_0,\dots,t_a\} = a+1$.

(ii) $D(G\cap {}^fT_i) = \{{}^f\vec{t}\} \wedge H \supseteq G \cup {}^fT_i \wedge Q(H,r) \Rightarrow i \notin \tau(G,r,{}^f\vec{t})$.

To give an idea of the construction we want to carry out with these chips we give a rough picture.

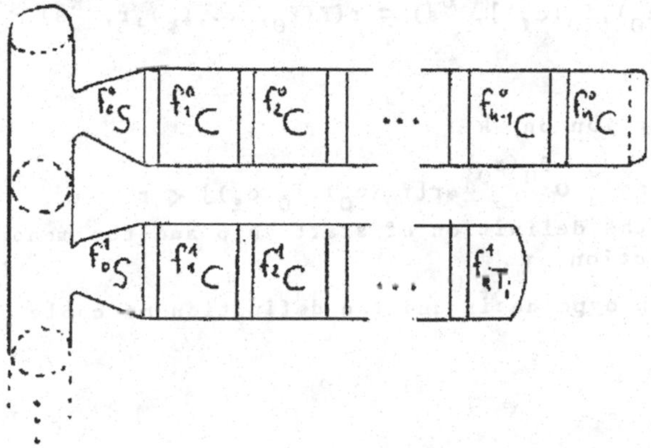

T_0^w: open cable

T_1^w: i-trapped cable

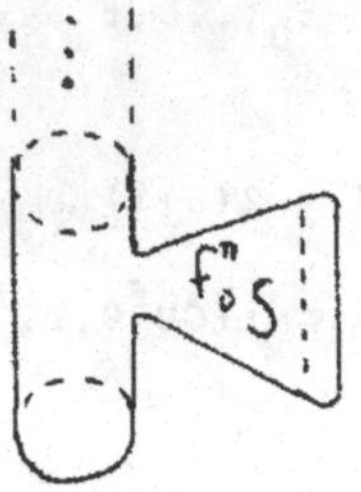

T^w_n: open cable

$n = \ell h(w)$

As indicated by the picture, we want to construct "cables" transmitting information about the type of potential answers of question Q.

<u>Definition 4:</u> $K(f_0,\dots,f_k)$ is called an <u>open cable</u> iff the following is satisfied for every $i,j \le k$:

(i) $i+1 < j \Rightarrow {}^{f_i}C \cap {}^{f_j}C = \emptyset = {}^{f_0}S \cap {}^{f_j}C$.

(ii) $0 < i < k \Rightarrow {}^{f_i}\vec{s} = {}^{f_{i+1}}\vec{t} \wedge D({}^{f_i}C \cap {}^{f_{i+1}}C) = \{{}^{f_i}\vec{s}\}$.

(iii) $k > 0 \Rightarrow {}^{f_0}\vec{s} = {}^{f_1}\vec{t} \wedge D({}^{f_0}S \cap {}^{f_1}C) = \{{}^{f_0}\vec{s}\}$.

(iv) $K(f_0,\dots,f_k) = {}^{f_0}S \cup {}^{f_1}C \cup \dots \cup {}^{f_k}C$.

<u>Cable Lemma:</u> Let $K(f_0,\dots,f_k)$ be an open cable, then the following is true:

$$H \supseteq K(f_0,\dots,f_k) \wedge \forall j \le \ell: f_0(c_j) = f_0(c_0) + j \wedge Q(H,r)$$
$$\Rightarrow \forall j \le \ell: (f_0(c_j) < \ell h(r) \wedge r_{f_0(c_j)} \neq 0) \wedge$$
$$\wedge\ \emptyset \neq \tau({}^{f_0}S, 0^{f_0(c_0)} * r[f_0(c_0), f_0(c_\ell)], {}^{f_0}\vec{s}) \subseteq \tau(K(f_0,\dots,f_k), r, {}^{f_k}\vec{s}).$$

<u>Proof:</u> We proceed by induction on k.

$k = 0$: Then $K(f_0) = {}^{f_0}S$, and $0^{f_0(c_0)} * r[f_0(c_0), f_0(c_\ell)] \prec r$.
So the claim follows from the definition of start chip and the monotonicity assumed for the function τ.

$k \Rightarrow k+1$: By the induction hypothesis and the definition of cable chip.

□

Next we need a cable which carries the information that a certain type is trapped.

<u>Definition 5:</u> $K_i(f_0,\dots,f_{k+1})$ is called an <u>i-trapped cable</u> iff the following is true:

(i) $K(f_0,\dots,f_k)$ is an open cable.

(ii) $D(K(f_0,\dots,f_k) \cap {}^{f_{k+1}}T_i) = \{{}^{f_k}\vec{s}\} \wedge {}^{f_k}\vec{s} = {}^{f_{k+1}}\vec{t}$.

(iii) $K_i(f_0,\dots,f_{k+1}) = K(f_0,\dots,f_k) \cup {}^{f_{k+1}}T_i$.

<u>Trap Lemma:</u> Let $K_i(f_0,\dots,f_{k+1})$ be an i-trapped cable, then the following holds:

$$H \supseteq K_i(f_0,\dots,f_{k+1}) \wedge \forall j \leq \ell\colon f_0(c_j) = f_0(c_0) + j \wedge Q(H,r)$$
$$\Rightarrow \forall j \leq \ell\colon (f_0(c_j) < \ell h(r) \wedge r_{f_0(c_j)} \neq 0)$$
$$\wedge\ i \in \tau({}^{f_0}S,0^{f_0(c_0)} * r[f_0(c_0),f_0(c_\ell)],\ {}^{f_0}\vec{s}) \neq \emptyset .$$

<u>Proof:</u> $K(f_0,\dots,f_k)$ is an open cable. So by the Cable Lemma we have that $\emptyset \neq \tau({}^{f_0}S,0^{f_0(c_0)} * r[f_0(c_0),f_0(c_\ell)],\ {}^{f_0}\vec{s}) \subseteq \tau(K(f_0,\dots,f_k),r,\ {}^{f_{k+1}}\vec{t})$. Using the definition of trap chip T_i we see that $i \in \tau(K(f_0,\dots,f_k),r,\ {}^{f_{k+1}}\vec{t})$.

<u>Definition 6:</u> $[S,C,\{T_i \mid i \in \bigcup \text{range } \tau\},\tau]$ is called a <u>trap equipment for P</u> iff the following is satisfied:

(i) With respect to τ, S is a start chip, C is a cable chip, and T_i is a trap chip of type i for $i \in \bigcup$ range τ.

(ii) $\tau\colon \{G \mid G \text{ finite graph}\} \times \{r \mid r \text{ finite sequence of natural numbers}\} \times \mathbb{N}^{a+1} \to \{M \mid M \subseteq \mathbb{N} \wedge M \text{ finite}\}$ is a recursive total function s.t. $\bigcup$ range τ is recursive and $r' \prec r \Rightarrow \tau(G,r',\vec{s}) \subseteq \tau(G,r,\vec{s})$.

(iii) Every open cable and every i-trapped cable for $i \in \bigcup$ range τ is an I-graph.

Now we are ready to prove our main theorem. It will be useful for the reader to look at the rough picture above when following the construction and the argument in the proof below.

<u>Theorem:</u> If there is a trap equipment for $P = \mathrm{fin}\ P^*$, then there is a highly recursive universal witness for the recursive unsolvability of P^*.

<u>Proof:</u> We show that there is a recursive trap for P, and use Theorem 1 of the preceding section.
Let $[S, C, \{T_i : i \in U \text{ range } \tau\}, \tau]$ be a trap equipment for P. Using this equipment we construct a recursive trap T of arity ℓ and with associated function α. We do this by induction on $\ell h(w)$ defining simultaneously T^w and $\alpha(w)$ for vetos w of arity ℓ.

$\ell h(w) = 0$: We put $\alpha(w) := 0$, and $T^w := T^w_0 := {}^{f^0_0}S \in I$ where $f^0_0(c_j) = j \wedge D({}^{f^0_0}S)$ is an initial segment.

$\ell h(w) = n + 1$: As an induction hypothesis we use the following: $T^{w\restriction n} \in I$ is of the form $T_0^{w\restriction n} \cup \ldots \cup T_n^{w\restriction n}$, where each $T_e^{w\restriction n}$ is either an open cable for the form $K(f_0^e, \ldots, f_{n-e}^e)$ or for some i an i-trapped cable of the form $K_i(f_0^e, \ldots, f_k^e)$ with $1 \le k \le n-e$ s.t.

1. $e + 1 < j \le n \Rightarrow T_e^{w\restriction n} \cap T_j^{w\restriction n} = \emptyset$.

2. $e < n \Rightarrow T_e^{w\restriction n} \cap T_{e+1}^{w\restriction n} = {}^{f_0^e}M_1 = {}^{f_0^{e+1}}M_0$.

3. $D(T^{w\restriction n})$ is an initial segment.

4. $e \le n \wedge j \le \ell \Rightarrow f_0^e(c_j) = \alpha(w\restriction e) + j$.

Now we define $\alpha(w) := \max(D(T^{w\restriction n})) + 1$. Furthermore we define for $e \le n$:

$T_e^w := T_e^{w\restriction n}$, if for some i $T_e^{w\restriction n}$ is an i-trapped cable.

$T_e^w := T_e^{w\restriction n} \cup {}^{f^e_{n+1-e}}T_i$, if $T_e^{w\restriction n}$ is an open cable and there is a v

s.t. $w_n = v*\langle e\rangle$ and i is the least element of

$\tau({}^{f_0^e}S,{}_0{}^{\alpha(w\restriction e)}*v,\ {}^{f_0^e}\vec{s})$.

$T_e^w :\equiv T_e^{w\restriction n} \cup {}^{f_{n+1-e}^e}C$ else.

$T_{n+1}^w := {}^{f_0^{n+1}}S$.

$T^w := T_0^w \cup \ldots \cup T_{n+1}^w$

The injections f_{n+1-e}^e are to be defined in such a way, that the properties used as induction hypothesis for $T^{w\restriction n}$ carry over from $T^{w\restriction n}$ to T^w. In particular we define $f_0^{n+1}(c_j) := \alpha(w) + j$ for $j \leq \ell$. By the construction, α and T are recursive.

Each cable T_e^w is in I by the definition of trap equipment, and so by the properties of the merger M_0 (see Definition 1 (ii)) T^w is in I. Now we verify the conditions for a trap.

(T1) Follows immediately from the construction.

(T2) Let $e \leq \ell h(w) \wedge T^{w\restriction e} \subseteq H \subseteq T^w \wedge Q(H,r)$ be true. By construction of T^w we have that ${}^{f_0^e}S = T_e^{w\restriction e} \subseteq T^{w\restriction e} \wedge \forall j \leq \ell\colon f_0^e(c_j) = \alpha(w\restriction e) + j$. It follows from condition (iii) for the start chip that $\forall j \leq \ell\colon f_0^e(c_j) = \alpha(w\restriction e) + j < \ell h(r) \wedge r_{\alpha(w\restriction e)+j} \neq 0$, and $\tau({}^{f_0^e}S,{}_0{}^{\alpha(w\restriction e)}*r[\alpha(w\restriction e),\alpha(w\restriction e)+\ell],\ {}^{f_0^e}\vec{s}) \neq \emptyset$. This proves (T2).

(T3) Let $e \leq n < \ell h(w) \wedge w_n = v*\langle e\rangle \wedge T^{w\restriction n+1} \subseteq H \subseteq T^w \wedge Q(H,r)$ be satisfied. This implies the hypotheses of (T2), so the conclusions drawn above hold. In particular let i be the least element of $\tau({}^{f_0^e}S,{}_0{}^{\alpha(w\restriction e)}*r[\alpha(w\restriction e),\alpha(w\restriction e)+\ell],\ {}^{f_0^e}\vec{s})$.

Now we assume that $r[\alpha(w\restriction e),\alpha(w\restriction e)+\ell] = v$. By definition $T_e^{w\restriction n+1} = T_e^{w\restriction n} \cup {}^{f_{n+1-e}^e}T_i$, and so $T_e^{w\restriction n+1}$ is an i-trapped cable and $T_e^w = T_e^{w\restriction n+1}$. We see that the hypotheses of the Trap Lemma are satisfied, and so $i \notin \tau({}^{f_0^e}S,{}_0{}^{\alpha(w\restriction e)}*v,\ {}^{f_0^e}\vec{s})$. This contradicts the definition of i and proves (T3). □

§ 5 Applications

In this section we apply the preceding theorem to several finitely based problems to obtain highly recursive graphs for which these problems are recursively unsolvable. In all cases we define a trap equipment, but leave the verification, that these data form indeed a trap equipment, to the reader.

Theorem 1 (Bean [B1]): For every $k \geq 3$ there is a highly recursive, k-colorable, and connected graph for which no k-coloring algorithm exists.

Proof: Let P^* be k-COLCON, and $P = (I,Q) = \text{fin } P^*$.
We define the recursive total function
$\tau: \{G \mid G \text{ finite graph}\} \times \{r \mid r \text{ finite sequence of natural numbers}\} \times \mathbb{N}^{k+1} \rightarrow$
$\rightarrow \{M \mid M \subseteq \mathbb{N} \wedge M \text{ finite}\}$ by $\tau(G,r,s_0,\ldots,s_k) =$
$= \{<i,j> \mid r_{s_i} = r_{s_j} \neq 0 \wedge 0 \leq i < j \leq k\}$.

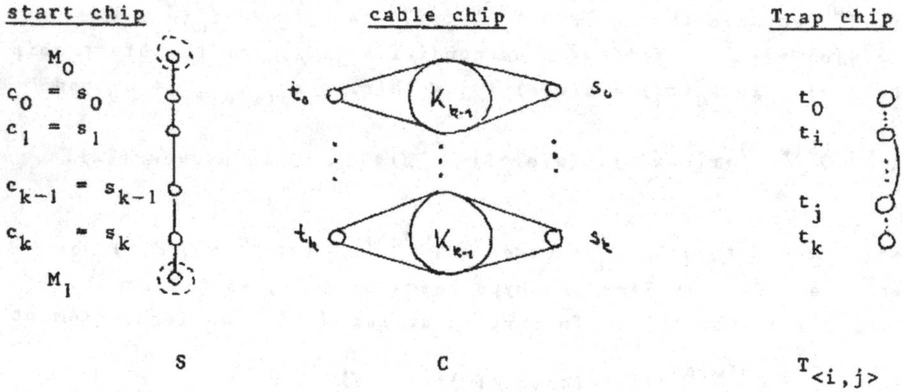

□

Let us remark that the analogous result is true for $k = 2$, if we forget about connectedness, whereas every connected 2-colorable highly recursive graph is obviously recursively 2-colorable.

Theorem 2: There is a highly recursive, connected, and planar graph, for which no planar embedding algorithm exists.

<u>Proof:</u> Let P^* be O-EMBEDCON, and $P = (I,Q) = \text{fin } P^*$.
We define a recursive total function $\tau:\{G|G \text{ finite graph}\} \times \{r|r \text{ finite sequence of natural numbers}\} \times \mathbb{N}^4 \to \{M|M \subseteq \mathbb{N} \wedge M \text{ finite}\}$.
Let $c_0,\ldots,c_3$ be elements of G s.t. $\#\{c_0,\ldots,c_3\} = 4$, and let r be a finite sequence. A path is called a <u>thread</u> from a to c, if its interior points have in G degree 2, and c has in G degree 1.

We search for the unique $a \in G\setminus\{c_0,\ldots,c_3\}$ s.t. there are 4 mutually disjoint threads from a to c_0, a to c_1, a to c_2, and a to c_3 in G. If such an a exists, we determine its neighbours $b_0,\ldots,b_3$ lying on the thread from a to $c_0,\ldots,$ and a to c_3 resp.

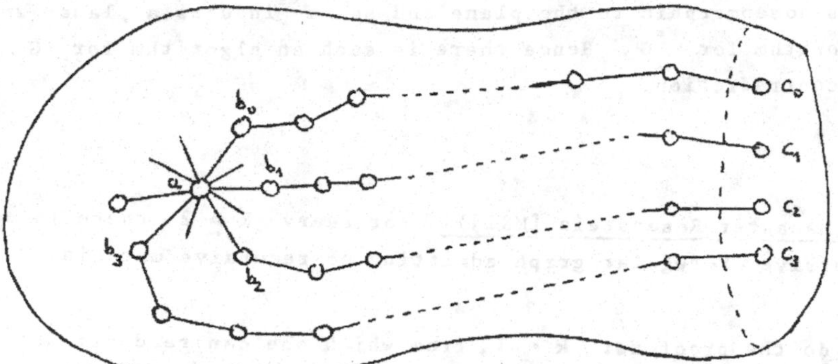

If $r_a \neq 0$ and $r_a - 1$ codes a sequence, in which every neighbour of a in G occurs exactly once, then we let φ be the permutation of $\{0,\ldots,3\}$ s.t. $b_{\varphi 0},\ldots,b_{\varphi 3}$ is the order, in which $b_0,\ldots,b_3$ occur in $r_a - 1$. Finally we let $\tau(G,r,c_0,c_1,c_2,c_3) := \{<\varphi 0,\varphi 1,\varphi 2,\varphi 3>\}$ if $\varphi \in \mathfrak{S}_4$ with these properties exists, and $= \emptyset$ else.

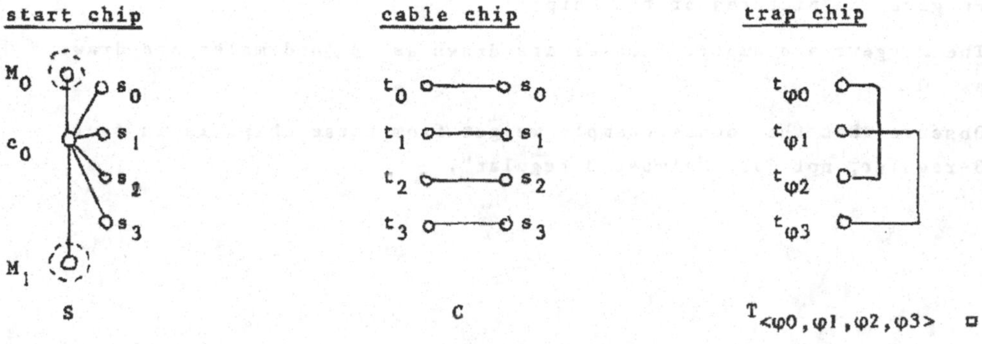

□

<u>Corollary:</u> For every g there is a highly recursive, connected graph of genus g, which has no embedding algorithm for the orientable surface of genus g.

<u>Proof:</u> Let E_g be a finite connected graph of genus g s.t. $D(E_g)$ is an initial segment, $a = \max D(E_g)$. Let G be a highly recursive graph constructed using the trap equipment from the preceding proof s.t. M_0 has number 0. $f(n) := n + a$, $H := {}^fG \cup E_g$. Suppose H has an embedding algorithm α for the orientable surface of genus g. Then α embeds E_g. Since ${}^fG$ is connected, ${}^fG$ has to be embedded by α in a face of E_g. Due to the fact that E_g has genus g, any such face is homeomorphic to the plane and so α induces a planar embedding algorithm for ${}^fG$. Hence there is such an algorithm for G, which is a contradiction. □

<u>Theorem 3 (Manaster-Rosenstein [MR2]):</u> For every $k \geq 2$ there is a highly recursive k-regular graph admitting no recursive matching.

<u>Proof:</u> We do the proof for $k = 3$, from which one can read off the general case.
Let P^* be MATCHING(k), and $P = (I,Q) = \text{fin } P^*$. We define a recursive total function $\tau: \{G \mid G \text{ finite graph}\} \times \{r \mid r \text{ finite sequence of natural numbers}\} \times \mathbb{N}^{18} \to \{M \mid M \subseteq \mathbb{N} \wedge M \text{ finite}\}$

$$\tau(G,r,s_{000},s_{001},\ldots,s_{220},s_{221}) := \{\langle i,j,k\rangle \mid i,j,k \leq 2 \wedge$$

$$\wedge \forall s \in \{s_{0i0},s_{0i1},s_{1j0},s_{1j1},s_{2k0},s_{2k1}\}: \exists a < \ell h(r): r_a \neq 0 \wedge s = r_a - 1\}$$

We give the pictures of the chips:

The mergers are empty, females are drawn as ●, and males are drawn as O.

Observe that the counterexample we get from these chips is in fact 3-regular, not just "almost 3-regular".

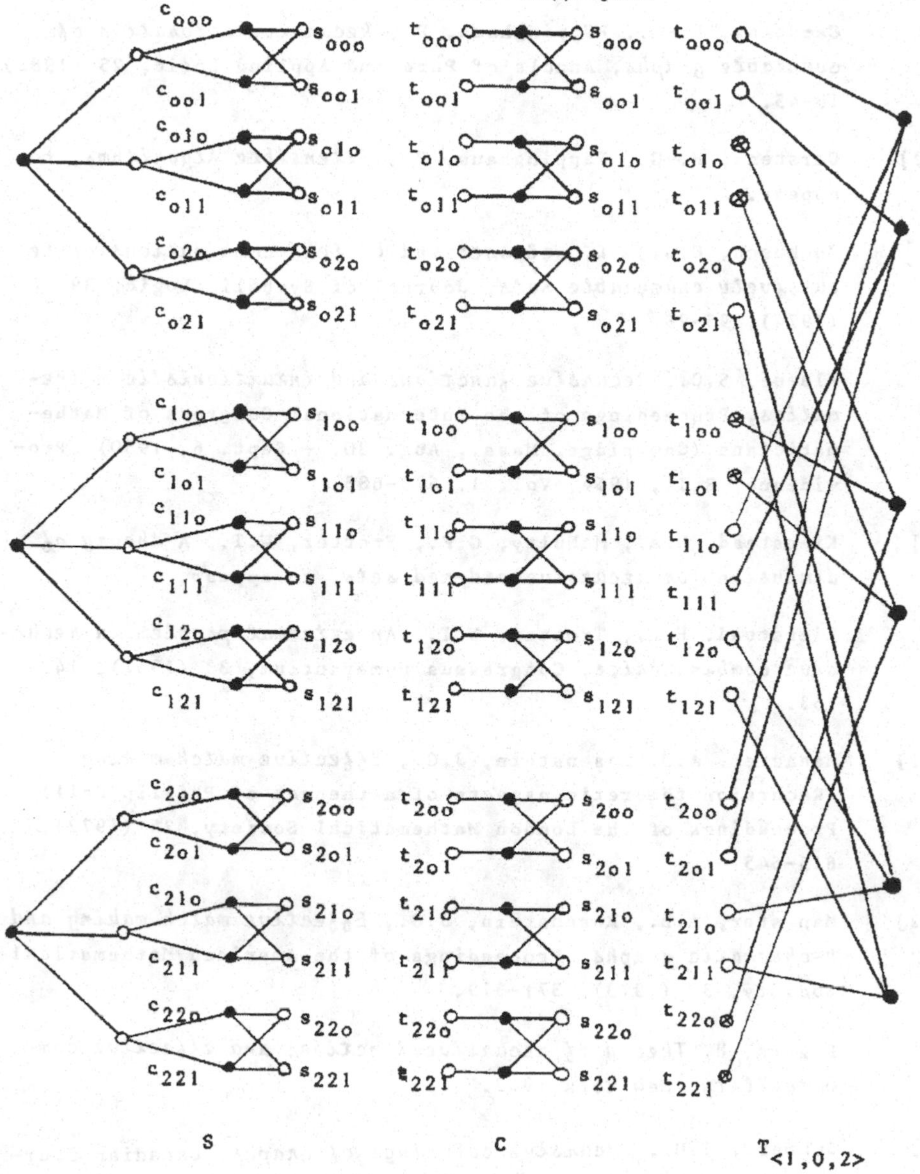

References

[B 1] Bean, D.R., *Effective coloration*, Journal of Symbolic Logic, 41 (1976), 469-480.

[B 2] Bean, D.R., *Recursive Euler and Hamilton paths*, Proceedings of the American Mathematical Society, 55 (1976), 385-394.

[CP 1] Carstens, H.-G., Päppinghaus, P., *Recursive coloration of countable graphs*, Annals of Pure and Applied Logic, 25 (1983), 19-45.

[CP 2] Carstens, H.-G., Päppinghaus, P., *Extensible Algorithms*, to appear.

[J] Jockusch, C.G., Π_1^0 *classes and boolean combinations of recursively enumerable sets*, Journal of Symbolic Logic, 39 (1974), 95-96.

[Kl] Kleene, S.C., *Recursive functions and intuitionistic mathematics*, Proceedings of the International Congress of Mathematicians (Cambridge, Mass., Aug. 30. - Sept. 6, 1950), Providence, R.I., 1954, Vol. 1, 679-685.

[KMT] Kierstead, H.A., McNulty, G.F., Trotter, W.T., *A theory of dimension for recursive ordered sets*, to appear.

[KT] Kierstead, H.A., Trotter, W.T., *An extremal problem in recursive combinatorics*, Congressus numerantium, 33 (1981), 143-153.

[MR 1] Manaster, A.B. Rosenstein, J.G., *Effective match making* (Recursion theoretic aspects of a theorem of Phillip Hall), Proceedings of the London Mathematical Society, 25 (1972), 615-645.

[MR 2] Manaster, A.B., Rosenstein, J.G., *Effective match making and k-chromatic graphs*, Proceedings of the American Mathematical Society, 39 (1973), 371-379.

[R] Rogers, H. *Theory of recursive functions and effective computability*, New York 1967.

[Sch] Schmerl, J.H., *Recursive colorings of graphs*, Canadian Journal of Mathematics, 32 (1980), 821-830.

[Sp] Specker, E., *Eine Verschärfung des Unvollständigkeitssatzes der Zahlentheorie*, Bulletin de l'Academie Polonaise des Sciences, Sér., Sci. Math. Astronom. Phys., 5 (1957), 1041-1045.

Minimal degrees and 1-generic sets below $\underset{\sim}{0}'$

C.T. Chong

C.G. Jockusch[1]

Sets which are Cohen-generic for 1-quantifier arithmetic (known as 1-generic sets) are easily constructed below $\underset{\sim}{0}'$ by the Kleene-Post method, and are known to have a number of interesting recursion-theoretic properties (see [J2,C1]). We show that no degree (of unsolvability) below $\underset{\sim}{0}'$ which contains a 1-generic set can bound a minimal degree. This gives a Kleene-Post type proof of the existence of a nonzero degree below $\underset{\sim}{0}'$ with no minimal predecessor which is far simpler than proofs based on embedding given atomless upper semilattices as initial segments of the degrees below $\underset{\sim}{0}'$. (See Lerman [L]. A direct construction using a difficult priority argument of a nonzero degree below $\underset{\sim}{0}'$ with no minimal predecessor was announced earlier by Yates [Y1].) Our result may be combined with known existence theorems for 1-generic sets to yield a number of corollaries demonstrating the ubiquity of degrees below $\underset{\sim}{0}'$ without minimal predecessors. Most of these corollaries do not seem accessible to the initial segment approach.

Our work is related to earlier work on n-generic sets for $n \geq 2$, and we pause to define this notion. For $n \geq 1$ and $A \subseteq \omega$, call A <u>n-generic</u> if for every Σ_n^0 set S of strings there is a string σ_0 extended by the characteristic function of A such that either σ_0 is in S or no extension of σ_0 is in S. See [J2] for information on the properties of n-generic sets and see Lemma 2.7 of [J2] for a proof that

[1]This research was supported by the National Science Foundation of the USA and the Science and Engineering Research Council of the UK.

this definition of n-genericity is equivalent to the original one in terms of Cohen genericity for n-quantifier sentences of first-order arithmetic. For $n = 1$ this definition is due to D. Posner [P] and is of course equivalent to the "jump forcing" definition of 1-genericity used in [J1] and [JP1].) A degree $\underset{\sim}{a}$ is called n-generic if some n-generic set has degree $\underset{\sim}{a}$.

The existence of a nonzero degree without minimal predecessors was first proved by D.A. Martin [M] as a corollary of a general result about Baire category. This general result was refined by Yates [Y2] in terms of "effective Baire category", and a closely related result in terms of n-genericity was proved in [J2, Theorem 4.1]. The latter result asserts that, for $n \geq 2$, if $\underset{\sim}{a}$ is n-generic, then every nonzero degree $\underset{\sim}{b} \leq \underset{\sim}{a}$ bounds an n-generic degree. All of these formulations give at once the existence of a nonzero degree below $\underset{\sim}{0}''$ without minimal predecessors, but none of the proofs seem amenable at first to modification to produce such a degree below $\underset{\sim}{0}'$. The main result of the present paper is that if $\underset{\sim}{a}$ is 1-generic and $\underset{\sim}{a} \leq \underset{\sim}{0}'$, then every nonzero degree $\underset{\sim}{b} \leq \underset{\sim}{a}$ bounds a 1-generic degree. This implies immediately that no 1-generic degree below $\underset{\sim}{0}'$ bounds a minimal degree. The overall idea and motivation of our proof is similar to that of the various versions of Martin's proof. However an important preliminary reduction (stated as Lemma 4.3 in [J2]) is no longer available since it made essential use of the 2-genericity of the given degree $\underset{\sim}{a}$. To overcome this obstacle we make an extra application of the 1-genericity of $\underset{\sim}{a}$ which has no counterpart in the proof of Martin's theorem and also utilize a recursive approximation to a 1-generic set A of degree $\underset{\sim}{a}$. Familiarity with Martin's theorem, especially the version in terms of n-genericity [J2], is helpful but not essential for understanding the present paper.

We view our main result mainly as a tool for obtaining strong existence theorems for degrees without minimal predecessors, as illustrated by the corollaries. It is possible to obtain such applications because of the availability of strong existence theorems for 1-generic degrees. Most of these existence theorems are of greatest interest below $\underset{\sim}{0}'$ so our restriction to degrees below $\underset{\sim}{0}'$ is not overly detrimental. Nonetheless it would be good to know whether our results hold without this restriction.

Our terminology is quite standard. Let $\omega = \{0,1,2,\ldots\}$. Other lower case Greek letters always denote <u>strings</u>, i.e. functions from finite initial segments of ω into $\{0,1\}$. For strings σ,τ, $\sigma \subseteq \tau$ denotes that τ extends σ, and σ,τ are said to be <u>compatible</u> if either extends the other. We write $\sigma\frown\tau$ for the usual concatenation of σ and τ, and $|\sigma|$ for the length of σ. We identify 0,1 with the corresponding strings of length 1 and write $\emptyset$ for the empty string. Let $Str(<k)$ be the set of strings of length less than k. An element σ of a set R of strings is called <u>$\subseteq$-maximal</u> in R if no proper extension of σ is in R. A set S of strings is called <u>dense</u> if every string has an extension in S. We identify sets $A \subseteq \omega$ with their characteristic functions. Thus $\sigma \subseteq A$ means that the characteristic function of A extends the string σ, and in this case we also say that σ is a <u>beginning</u> of A. Let $A\upharpoonright s$ denote the unique beginning of A of length s.

We use upper case Greek letters to denote partial maps from strings to strings. Such a map Φ is called <u>consistent</u> if $\Phi(\sigma) \subseteq \Phi(\tau)$ whenever $\sigma \subseteq \tau$ and σ,τ are both in the domain of Φ (denoted $dom(\Phi)$). If Φ is consistent and $B \subseteq \omega$, we write $\Phi(B)$ for $\cup\{\Phi(\sigma) : \sigma \subseteq B\}$. Thus $\Phi(B)$ is either a string or (the characteristic function of) some set C. A <u>tree</u> is a partial recursive map T from strings to strings such that $T(\sigma\frown 0)$, $T(\sigma\frown 1)$ are incompatible extensions of $T(\sigma)$ whenever $T(\sigma\frown 0)$ or $T(\sigma\frown 1)$

is defined. A tree T is called <u>total</u> if every string is in $\mathrm{dom}(T)$, and T is said to have <u>height</u> s if $\mathrm{dom}(T) = \mathrm{Str}(<s)$.

Let $\{e\}^{\sigma}(x) = y$ mean that the eth reduction procedure with oracle information σ and input $x < |\sigma|$ yields output y in at most $|\sigma|$ steps and further (by induction) that $\{e\}^{\sigma}(u)$ is defined for each $u < x$. Thus $\{e\}^{\sigma}$ is a string and can be effectively computed given e and σ. Strings σ, τ are called <u>e-split</u> if $\{e\}^{\sigma}$ is incompatible with $\{e\}^{\tau}$. A tree T is called <u>e-splitting</u> if $T(\sigma\frown 0)$, $T(\sigma\frown 1)$ are e-split whenever they are defined. Of course $\{e\}^{A}(x) = y$ means that $\{e\}^{\sigma}(x) = y$ for some $\sigma \subseteq A$, and B is recursive in A if and only if $B = \{e\}^{A}$ for some e. We use $\underset{\sim}{a}, \underset{\sim}{b}, \ldots$ to denote degrees of unsolvability and $\underset{\sim}{\cup}$ for the least upper bound operation on degrees.

We now state our main result.

<u>Theorem</u>. If $\underset{\sim}{a}$ is a 1-generic degree and $\underset{\sim}{0} < \underset{\sim}{b} \leq \underset{\sim}{a} \leq \underset{\sim}{0}'$, then there is a 1-generic degree $\underset{\sim}{c} \leq \underset{\sim}{b}$.

As indicated in the introduction, it is easy to obtain a large number of corollaries from this and known results on 1-generic degrees. Before proving the theorem, we give a sample of these.

<u>Corollary 1</u>. No 1-generic degree below $\underset{\sim}{0}'$ bounds a minimal degree.

<u>Proof</u>. Immediate since every 1-generic degree $\underset{\sim}{c}$ bounds an incomparable pair of degrees [JP1,Lemma 2] and hence is neither $\underset{\sim}{0}$ nor minimal. Of course the construction of a 1-generic set below $\underset{\sim}{0}'$ is an extremely easy Kleene-Post construction (implicit in Friedberg [F]), and this corollary shows that this construction automatically yields a degree below $\underset{\sim}{0}'$ with no minimal predecessor.

Corollary 2. Every nonzero r.e. degree bounds a nonzero degree with no minimal predecessor.

Proof. Immediate from Corollary 1 and the fact that every nonzero r.e. degree bounds a 1-generic degree. The latter is reasonably well-known as folklore, but we sketch here an elegant proof of it due to R. Shore. Let a nonrecursive r.e. set E be given in order to construct a 1-generic set $A \leq_T E$. We obtain A as $\bigcup_s \sigma_s$ via a construction recursive in E. Fix a recursive enumeration E^s of E. Let $f(s)$ be the least number t such that $E^t \upharpoonright s = E \upharpoonright s$. Clearly f has the same degree as E. Let W_e be the eth r.e. set (of strings) and let W_e^s be the finite subset of W_e enumerated in s steps. Let $\sigma_0 = \emptyset$. If σ_s has been defined, let e_s be the least number $e \leq s$ such that σ_s extends no element of $W_e^{f(s)}$ but $W_e^{f(s)}$ contains a string $\sigma \supseteq \sigma_s$. If e_s does not exist, let $\sigma_{s+1} = \sigma_s\frown 0$. If e_s exists, let $\sigma_{s+1} = \sigma\frown 0$ where σ is the least extension of σ_s in $W_{e_s}^{f(s)}$. Clearly $\bigcup_s \sigma_s$ is the characteristic function of some set A. Suppose for a contradiction that A fails to be 1-generic, and let W_e be an r.e. set of strings which witnesses this failure. Since e_s is 1-1 as a partial function of s, we may choose $s_0 \geq e$ so that $e_s \leq e$ holds for no $s \geq s_0$. We claim now that σ_{s+1} and $f(s)$ may be effectively computed from σ_s for $s \geq s_0$. This claim implies that f and hence E are recursive and thus yields the desired contradiction. To prove the claim, let σ_s be given. Then σ_s has some extension $\sigma \in W_e$. Let $g(s)$ be the least number t such that W_e^t contains an extension of σ_s. Then $g(s) > f(s)$ since otherwise we would have $e_s \leq e$. We then compute $f(s)$ as the least number t such that $W_e^t \upharpoonright s = W_e^{g(s)} \upharpoonright s$. Once $f(s)$ is known, the construction determines σ_{s+1} effectively.

Since initial segment constructions can be done below arbitrary nonzero r.e. degrees [L], Corollary 2 may also be obtained by the initial segment approach.

Corollary 3. Every degree $\underset{\sim}{d} \le \underset{\sim}{0}'$ with $\underset{\sim}{d}'' > \underset{\sim}{0}''$ bounds a nonzero degree with no minimal predecessors.

Proof. This follows from Corollary 1 since by [JP1,Lemmas 1 & 3] every such degree $\underset{\sim}{d}$ bounds a 1-generic degree $\underset{\sim}{a}$.

Corollary 4. If $\underset{\sim}{0} < \underset{\sim}{d} \le \underset{\sim}{0}'$, there is a degree $\underset{\sim}{a}$ with no minimal predecessors such that $\underset{\sim}{d} \underset{\sim}{\cup} \underset{\sim}{a} = \underset{\sim}{0}'$.

Proof. By the proof of [PR,Theorem 1], there is a 1-generic degree $\underset{\sim}{a}$ such that $\underset{\sim}{d} \underset{\sim}{\cup} \underset{\sim}{a} = \underset{\sim}{0}'$, and $\underset{\sim}{a}$ has no minimal predecessors by Corollary 1.

Corollary 5. The degrees below $\underset{\sim}{0}'$ without minimal predecessors generate the degrees below $\underset{\sim}{0}'$ under $\underset{\sim}{\cup}$ and $\underset{\sim}{\cap}$ (the partial inf operation).

Proof. This is immediate from Corollary 1 and [JP2,Corollary 4.6], which asserts that the 1-generic degrees below $\underset{\sim}{0}'$ generate the degrees below $\underset{\sim}{0}'$.

Corollary 1, and hence also the other corollaries, are proved using the basic fact that 1-generic degrees are neither $\underset{\sim}{0}$ nor minimal. However it is known that all 1-generic degrees are "far from minimal" in various senses. For instance, the degrees below a 1-generic degree never form a lattice [J2,Theorem 3.1], and every 1-generic degree $\underset{\sim}{c}$ is r.e. in some degree $\underset{\sim}{d} < \underset{\sim}{c}$, [J2,Theorem 5.1]. Using these stronger results in place of the basic fact above, one could in an obvious way considerably strengthen the corollaries at the cost of somewhat complicating the form of their statements. For instance, one could show along the lines of Corollary 3 that for every degree $\underset{\sim}{d} \leq \underset{\sim}{0}'$ with $\underset{\sim}{d}'' > \underset{\sim}{0}''$ there is a nonzero degree $\underset{\sim}{a} \leq \underset{\sim}{d}$ such that for every nonzero degree $\underset{\sim}{b} \leq \underset{\sim}{a}$ the degrees $\leq \underset{\sim}{b}$ do not form a lattice. We omit further examples and now proceed to the proof of our main result.

<u>Proof of the theorem.</u> Let A be a 1-generic set of degree $\underset{\sim}{a}$, and let B be any set of degree $\underset{\sim}{b}$. Fix a number e so that $B = \{e\}^A$. We shall construct a partial recursive consistent map Φ from strings to strings such that $\Phi(B)$ is a 1-generic set C. To explain the basic motivation of the construction of Φ we first give a rather drastically oversimplified description of the goal of the construction. Let $\Theta(\sigma) = \Phi(\{e\}^\sigma)$, so that $\Theta(A) = C$. Suppose that Φ could be constructed so that Θ is "onto" in the following strong sense:

$$(1)\ \mu \supseteq \Theta(\sigma) \Rightarrow (\exists\tau)[\tau \supseteq \sigma \ \&\ \Theta(\tau) = \mu]$$

The 1-genericity of C could then easily be deduced from that of A as follows. Let S be an r.e. set of strings. It must be shown that there is a beginning γ_0 of C such that either γ_0 is in S or no extension of γ_0 is in S. Let $T = \{\sigma : \Theta(\sigma) \in S\}$. Since T is r.e. and A is 1-generic, there is a beginning σ_0 of A such that σ_0 is in T or no extension of σ_0 is in T. By (1), we may take $\gamma_0 = \Theta(\sigma_0)$.

The nonrecursiveness of B implies that every beginning of A has e-split extensions. Suppose that in fact there exists a total recursive e-splitting tree $\hat{T}$ with A a branch of $\hat{T}$. We could then satisfy (1) by setting $\Phi(\{e\}^{\hat{T}(\nu)}) = \nu$ for all ν, so that $\Theta(\hat{T}(\nu)) = \nu$.

The above approach has two serious defects. First, the assumption that every beginning of A has e-split extensions is too weak to allow us to construct a total recursive e-splitting tree. Second, even if we had such a tree $\hat{T}$ the 1-generic set A could not be a branch of it (unless there were a string σ such that every set $D \supseteq \sigma$ were a branch of $\hat{T}$). Thus $\Phi(\{e\}^A)$ as defined above would not be total. If A were 2-generic the firs defect could be overcome by appropriate choice of e (see [J2,Lemma 4.3]) and then the second objection could be overcome by "fattening" the splitting tree $\hat{T}$ (see [J2,Lemma 4.4]) while only requiring (1) to hold for σ in a certain dense recursive set of strings. In the current proof we use the 1-genericity of A to obtain arbitrarily large finite splitting trees. The function Φ is constructed by stages using a finite splitting tree of height s at stage s and is defined roughly along the lines previously indicated. However, there is a slight change needed to ensure the consistency of Φ because the splitting trees used at different stages are unrelated to each other. Then (1) holds for sufficiently many pairs (σ,μ) that C can be shown 1-generic along the general lines already indicated but with an additional application of 1-genericity. The assumption that A is recursive in $0'$ enables a recursive approximation to A to be used in the construction, by the limit lemma. This recursive approximation guides the construction so that for each beginning σ of A, all but finitely many strings in the domain of Θ are extensions of σ. This property of the construction is useful both in showing that $\Theta(A)$ is total and that $C = \Theta(A)$ is 1-generic. In fact the clause $\tau \supseteq \sigma$ in (1) is taken care of by means of this property.

This concludes our admittedly vague overall description of the argument. Before constructing Φ we show that the required finite splitting trees exis

Lemma 1. For any $\sigma \subseteq A$ and any $s > 0$ there exists an e-splitting tree T of height s with $T(\emptyset) \supseteq \sigma$.

<u>Proof</u> As remarked in the preceding intuitive discussion, every string $\sigma \subseteq A$ has an e-split pair of extensions because $B = \{e\}^A$ is nonrecursive. This is equivalent to the case $s = 2$ of the lemma. The general proof uses the assumption that A is 1-generic but not that $A \leq_T 0'$.

For this proof only, call an e-splitting tree T <u>special</u> if there is a $k \in \omega$ such that $Str(<k) \subseteq dom(T) \subseteq Str(<k+1)$. Note that if T is special and $|dom(T)| = 2^s-1$, then T has height s. Thus it suffices to prove the following:

<u>Claim</u> For all odd numbers t and all $\sigma \subseteq A$, there exists a special e-splitting tree T such that $|dom(T)| = t$ and $T(\emptyset) \supseteq \sigma$.

The claim is proved by induction on odd t. It is trivially true for $t = 1$. Assume that the claim is true for t in order to prove it for $t + 2$. Let $\sigma_0 \subseteq A$ be given, so that we must find a special e-splitting tree T_0 with $T_0(\emptyset) \supseteq \sigma_0$ and $|dom(T_0)| = t + 2$.

Let k be such that $2^k - 1 \leq t < 2^{k+1} - 1$. Let

$$S = \{T(\nu) : T \text{ is a special e-splitting tree, } T(\emptyset) \supseteq \sigma_0, \; |dom(T)| = t, \; |\nu| = k, \text{ and } \nu\frown 0 \notin dom(T)\}.$$

Clearly S is an r.e. set of strings. In addition every string $\sigma \subseteq A$ has an extension $T(\nu) \in S$. To see this, apply the inductive hypothesis to obtain a special e-splitting tree T such that $T(\emptyset)$ extends both σ and σ_0 and $|dom(T)| = t$. Since $2^k - 1 \leq dom(T) < 2^{k+1} - 1$ and T is special, there is a string ν of length k with $\nu \in dom(T)$ and $\nu\frown 0 \notin dom(T)$. Clearly $T(\nu) \in S$ and $T(\nu) \supseteq T(\emptyset) \supseteq \sigma$. Since S is r.e. and A is 1-generic, it now follows that there is a string $\gamma \in S$ such that $\gamma \subseteq A$. Let $\gamma = T(\nu)$, where T, ν are as in the definition of S. Let γ_0, γ_1 be any two e-split extensions of γ. Let T_0 coincide with T on $dom(T)$ and further define $T_0(\nu\frown i) = \gamma_i$ for $i = 0,1$. Clearly T_0

is a special e-splitting tree, $T_0(\emptyset) = T(\emptyset) \supseteq \sigma_0$, and $|\mathrm{dom}(T_0)| = t + 2$. This completes the proof of the claim and the lemma.

We now give the construction by stages of the consistent partial recursive map Φ from strings to strings so that $\Phi(B)$ will be the desired 1-generic set $C \leq_T B$. We will write Φ^s for the finite part of Φ determined by the beginning of stage s. Fix a recursive approximation $\{A^s\}$ to the Δ_2^0 set A.

<u>Stage 0.</u> Let $\Phi^1(\emptyset) = \emptyset$.

<u>Stage s.</u> (s>0). Search for $t \geq s$ and an e-splitting tree T of height s such that $T(\emptyset) \supseteq A^t \upharpoonright s$ and $|\{e\}^{T(\emptyset)}| > |\mu|$, all $\mu \in \mathrm{dom}(\Phi^s)$. Let T_s be the first such tree found (if any), and let μ_s be the longest string μ in $\mathrm{dom}(\Phi^s)$ with $\{e\}^{T_s(\emptyset)} \supseteq \mu$. (The string μ_s exists by the action at stage 0 and is unique because all such μ are pairwise compatible.) Let $\Phi^{s+1} = \Phi^s$ on $\mathrm{dom}(\Phi^s)$, and set

$$\Phi^{s+1}(\{e\}^{T_s(\nu)}) = \Phi^s(\mu_s)^\frown\nu$$

for all $\nu \in \mathrm{Str}(<s)$. (Φ^{s+1} is single-valued because T_s is e-splitting and $|\{e\}^{T(\nu)}| \geq |\{e\}^{T(\emptyset)}| > |\mu|$ for $\mu \in \mathrm{dom}(\Phi^s)$, $\nu \in \mathrm{Str}(< s)$.)

Observe that T_s exists by Lemma 1. (In detail, apply Lemma 1 to $\sigma \subseteq A$ with $|\{e\}^\sigma| > |\mu|$, all $\mu \in \mathrm{dom}(\Phi^s)$, and $|\sigma| > s$. For the resulting T, $T(\emptyset) \supseteq \sigma \supseteq A^t \upharpoonright s$ holds for all sufficiently large t.) Thus the search at stage s is completed.

To show that Φ is consistent, it suffices to show by induction on s that Φ^s is consistent. This is obvious for s = 0,1. We now assume that Φ^s is consistent (s>0) and prove that Φ^{s+1} is consistent. Let $\sigma,\tau \in \mathrm{dom}(\Phi^{s+1})$ satisfy $\sigma \subsetneq \tau$. To show that $\Phi^{s+1}(\sigma) \subseteq \Phi^{s+1}(\tau)$ we consider various cases determined by which of σ,τ are in $\mathrm{dom}(\Phi^s)$. If $\sigma,\tau \in \mathrm{dom}(\Phi^s)$, the conclusion is immediate from our inductive hypothesis. If $\sigma,\tau \notin \mathrm{dom}(\Phi^s)$, let $\sigma = \{e\}^{T_s(\nu_1)}$, $\tau = \{e\}^{T_s(\nu_2)}$. Then $\nu_1 \subseteq \nu_2$ since

T_s is an e-splitting tree and $\sigma \subseteq \tau$. It follows that $\mu_s^\frown\nu_1 \subseteq \mu_s^\frown\nu_2$, i.e. $\Phi^{s+1}(\sigma) \subseteq \Phi^{s+1}(\tau)$. Now suppose that $\sigma \in \mathrm{dom}(\Phi_s)$, $\tau \notin \mathrm{dom}(\Phi_s)$. By construction, $|\sigma| < |\{e\}^{T_s(\emptyset)}|$ and $\tau = \{e\}^{T_s(\nu)}$ for some ν. It follows that $\sigma \subseteq \{e\}^{T_s(\emptyset)}$ since otherwise σ and $\{e\}^{T_s(\emptyset)}$ would be incompatible, in contradiction to the fact that τ extends both. By choice of μ_s, we conclude that $\sigma \subseteq \mu_s$. Hence $\Phi^s(\sigma) \subseteq \Phi^s(\mu_s) \subseteq \Phi^{s+1}(\nu)$, by inductive hypothesis and construction. Finally suppose that $\tau \in \mathrm{dom}(\Phi_s)$, $\sigma \notin \mathrm{dom}(\Phi_s)$. As above, we obtain $|\tau| < |\{e\}^{T_s(\emptyset)}| \leq |\sigma|$ which is incompatible with the assumption $\sigma \subseteq \tau$. Thus this fourth case cannot arise, and the proof that Φ is consistent is complete.

Let $\Theta^s(\sigma) = \Phi^s(\{e\}^\sigma)$ and $\Theta(\sigma) = \Phi(\{e\}^\sigma)$. The next lemma summarizes some useful properties of Θ which follow directly from the construction of Φ.

<u>Lemma 2.</u> (i) Rng Θ contains arbitrarily long strings.

(ii) If $\sigma_0 \subseteq A$, then $T_s(\emptyset) \supseteq \sigma_0$ for all sufficiently large s. Hence if $\sigma_0 \subseteq A$, then all but finitely many elements of $\mathrm{dom}(\Theta)$ are extensions of σ_0.

(iii) If $\mu \supseteq \Theta_s(\sigma)$, then either $\mu = \Theta_s(\tau)$ for some τ or μ extends an $\subseteq$-maximal element $\Theta_s(\sigma')$ of $\mathrm{rng}(\Theta_s)$.

(iv) If $\Theta_s(\sigma')$ is $\subseteq$-maximal in $\mathrm{rng}(\Theta_s)$ but has a proper extension in $\mathrm{rng}(\Theta)$, then every string $\mu \supseteq \Theta_s(\sigma')$ with $|\mu| < s$ is in $\mathrm{rng}(\Theta)$.

(Note that (iii) and (iv) are remnants of (1) from the intuitive discussion of the motivation of the proof.)

<u>Proof.</u> Although we specify details below, all parts should become clear by visualizing the construction.

(i) If $|\nu| = s-1$, then $|\Theta(T_s(\nu))| = |\Phi(\{e\}^{T_s(\nu)})| = |\mu_s\widehat{\ }\nu| \geq s-1$

(ii) Choose $s_0 \geq |\sigma|$ so that $A^t \supseteq \sigma$ whenever $t \geq s_0$. For $s \geq s_0$ there exists $t \geq s$ such that $T_s(\emptyset) \supseteq A^t \upharpoonright s \supseteq \sigma$. The second sentence of (ii) follows from the first since every string in $\mathrm{dom}(\Theta_{s+1}) - \mathrm{dom}(\Theta_s)$ extends $T_s(\emptyset)$.

Part (iii) is proved by induction on s. It is obvious for $s = 0,1$. We now assume it for s $(s>0)$ and prove it for $s+1$. Suppose $\mu \supseteq \Theta_{s+1}(\sigma)$.

Consider first the case where $\sigma \in \mathrm{dom}(\Theta_s)$. Then by inductive hypothesis, either $\mu = \Theta_s(\tau)$ for some $\tau \supseteq \sigma$ or $\mu \supseteq \Theta_s(\sigma')$ for some σ' such that $\Theta_s(\sigma')$ is $\subseteq$-maximal in $\mathrm{rng}(\Theta_s)$. The desired conclusion is immediate unless $\mu \supseteq \Theta_s(\sigma')$ where $\Theta_s(\sigma')$ is maximal in $\mathrm{rng}(\Theta_s)$ but not in $\mathrm{rng}(\Theta_{s+1})$. In this case we must have $\Theta_s(\sigma') = \Phi_s(\mu_s)$ since $\Phi_s(\mu_s)$ is the unique string which is $\subseteq$-maximal in $\mathrm{rng}(\Phi_s)$ but $\subseteq$-maximal in $\mathrm{rng}(\Phi_{s+1})$. Therefore $\mu = \mu_s\widehat{\ }\nu$ for some string ν. If $|\nu| < s$, then $\mu \in \mathrm{rng}(\Theta_s)$ by construction. If $|\nu| \geq s$, then $\mu \supseteq \mu_s\widehat{\ }(\nu\upharpoonright s-1)$, and $\mu_s\widehat{\ }(\nu\upharpoonright s-1)$ is $\subseteq$-maximal in $\mathrm{rng}(\Phi_{s+1}) = \mathrm{rng}(\Theta_{s+1})$ by construction.

To complete the proof of (iii), suppose now that $\mu \supseteq \Theta_{s+1}(\sigma)$, where $\sigma \in \mathrm{dom}(\Theta_{s+1}) - \mathrm{dom}(\Theta_s)$. Hence $\Theta_{s+1}(\sigma) \supseteq \mu_s$ by construction. Therefore μ is of the form $\mu_s\widehat{\ }\nu$ for some ν and the argument can be completed as in the previous case above.

Now assume the hypothesis of (iv). Let t be the first stage that $\mathrm{rng}(\Theta_{t+1})$ contains a proper extension of $\Theta_s(\sigma')$. Clearly $t > s$. As in part (iii), $\Theta_s(\sigma') \supseteq \mu_t$, and every string $\mu_t\widehat{\ }\nu$ with $|\nu| < t$ is in $\mathrm{rng}(\Theta_{t+1})$. This completes the proof of Lemma 2.

We now show that $\Theta(A)$ is total. We must show for each k there exists $\sigma \subseteq A$ such that $|\Theta(\sigma)| \geq k$. Given k, let $S = \{\sigma : |\Theta(\sigma)| \geq k\}$. We claim that every string $\sigma_0 \subseteq A$ has an extension $\sigma \in S$. This is immediate from Lemma 2, parts (i) and (ii). Since S is r.e. and A is 1-generic, there exists $\sigma \subseteq A$ with $\sigma \in S$. This completes the proof that $\Theta(A) = \Phi(B)$ is total.

Since $\Theta(A)$ is total and 0-1 valued, it is the characteristic function of a set C. Clearly $C \leq_T B$, since $C = \Phi(B)$ and Φ is partial recursive. It remains to be shown that C is 1-generic. Let S be an r.e. set of strings. Assume for a contradiction that every beginning of C has an extension in S yet no beginning of C is in S. As in the intuitive sketch before the construction, let $T = \{\sigma : \Theta(\sigma) \in S\}$. Then no beginning of A is in T because no beginning of C is in S. Since T is r.e. and A is 1-generic, there is a string $\sigma_0 \subseteq A$ which has no extension in T. By Lemma 2(ii) there is a stage s_0 so that $T_s(\emptyset) \supseteq \sigma_0$ for all $s \geq s_0$. Let

$$U = \{\sigma' \supseteq \sigma_0 : (\exists s \geq s_0)[\Theta_s(\sigma') \text{ is } \subseteq\text{-maximal in } \operatorname{rng}(\Theta_s) \ \& \ (\exists \mu \in S)[|\mu| < s \ \& \ \mu \supseteq \Theta_s(\sigma')]]\}.$$

We claim first that no string $\sigma' \subseteq A$ is in U. Suppose otherwise, and let s, μ witness that $\sigma' \in U$, where $\sigma' \subseteq A$. Then $\Theta_s(\sigma')$ is not $\subseteq$-maximal in $\operatorname{rng}(\Theta)$ because $\Theta(A)$ is total. By Lemma 2(iv), $\mu \in \operatorname{rng}(\Theta)$. Let $\mu = \Theta(\sigma'')$. Then σ'' enters $\operatorname{dom}(\Theta)$ after stage s and hence after s_0. Hence $\sigma'' \supseteq \sigma_0$. Therefore σ_0 has an extension $\sigma'' \in T$, in contradiction to the choice of σ_0.

Since U is r.e. and A is 1-generic, it follows that there is a string $\sigma_1 \subseteq A$ which has no extension in U. We may assume that $\sigma_1 \supseteq \sigma_0$ without loss of generality. By Lemma 2(ii) there is a stage s_1 so large that $T_s(\emptyset) \supseteq \sigma_1$, all $s \geq s_1$. Choose $\sigma_2 \subseteq A$ so that $|\sigma_2| \geq s_1$, $\Theta(\sigma_2)$ is defined, and $\Theta(\sigma_2)$ is longer than all strings in $\operatorname{rng}(\Theta^{s_1})$. Then $\Theta(\sigma_2) \subseteq C$ so there exists $\mu \in S$ with $\mu \supseteq \Theta(\sigma_2)$. For this μ, choose s_3 sufficiently large that $\Theta_{s_3}(\sigma_2)$ is defined, and $s_3 \geq s_2, |\mu|$. By Lemma 2(iii), applied with $s = s_3$, $\sigma = \sigma_2$, either $\mu = \Theta_{s_3}(\tau)$ for some τ, or μ extends an $\subseteq$-maximal element $\Theta_{s_3}(\sigma')$ of rng (Θ_{s_3}). Suppose first that $\mu = \Theta_{s_3}(\tau)$. Then $\tau \notin \mathrm{dom}(\Theta_{s_1})$ because $\Theta(\sigma_2)$, and hence μ, is longer than all strings in rng (Θ_{s_1}). Thus τ enters dom (Θ) after s_1, so $\tau \supseteq \sigma_1$ by choice of s_1. But $\sigma_1 \supseteq \sigma_0$, so τ is an extension of σ_0 and $\tau \in T$ since $\Theta(\tau) = \mu \in S$. This contradiction rules out the case that $\mu \in$ rng (Θ). The only remaining alternative is that μ extends some $\subseteq$-maximal element $\Theta_{s_3}(\sigma')$ of rng Θ_{s_3}. Since $\mu \supseteq \Theta_{s_3}(\sigma_2)$ we have that $\Theta_{s_3}(\sigma') \supseteq \Theta_{s_3}(\sigma_2)$. But $\Theta_{s_3}(\sigma_2)$ is longer than all strings in rng Θ_{s_1}. Thus $\Theta_{s_3}(\sigma')$ is not in rng (Θ_{s_1}) so $\sigma' \notin \mathrm{dom}(\Theta_{s_1})$. By choice of s_1 it follows that $\sigma' \supseteq \sigma_1$. Thus, by definition of U, σ' is an extension of σ_1 which is in U. This contradicts the choice of σ_1 and completes the proof of the theorem.

We conclude by conjecturing that the following two generalizations of our main result are false:

(F1) If $\underset{\sim}{a}$ is 1-generic and $\underset{\sim}{0} < \underset{\sim}{b} < \underset{\sim}{a}$, then there is a 1-generic degree $\underset{\sim}{c} \leq \underset{\sim}{b}$.

(F2) If $\underset{\sim}{a}$ is 1-generic and $\underset{\sim}{0} < \underset{\sim}{b} \leq \underset{\sim}{a} \leq \underset{\sim}{0}'$, then $\underset{\sim}{b}$ is also 1-generic.

Related results for 2-generic degrees may be found in [J2,Theorem 4.1 and Corollary 5.6].

References

C1 C.T. CHONG, "Generic sets and minimal covers", preprint.

F R.M. FRIEDBERG, "A criterion for completeness of degrees of unsolvability" J. Symbolic Logic 22 (1957), 159-160.

J1 C.G. JOCKUSCH, Jr., "Simple proofs of some theorems on high degrees of unsolvability", Canad. J. Math. 29 (1977), 1072-1080.

J2 C.G. JOCKUSCH, Jr., "Degrees of generic sets", in Recursion Theory: Its Generalisations and Applications, (F.R. Drake and S.S. Wainer, eds.), Cambridge University Press, Cambridge, 1980.

JP1 C.G. JOCKUSCH, Jr. and D.B. POSNER, "Automorphism bases for degrees of unsolvability", Israel J. Math. 40 (1981), 150-164.

JP2 C.G. JOCKUSCH, Jr. and D.B. POSNER, "Double jumps of minimal degrees", J. Symbolic Logic 43 (1978), 715-724.

L M. LERMAN, Degrees of Unsolvability, Springer Verlag, 1983.

M D.A. MARTIN, "Measure, category and degrees of unsolvability", unpublished preprint, 1967.

P D.B. POSNER, High Degrees, Doctoral Dissertation, University of California Berkeley, 1977.

PR D.B. POSNER and R.W. ROBINSON, "Degrees joining to 0'," J. Symbolic Logic 46 (1981), 714-722.

Y1 C.E.M. YATES, "Density and incomparability in the degrees less than $0^{(1)}$" (abstract), J. Symbolic Logic 31 (1966), 301.

Y2 C.E.M. YATES, "Banach-Mazur games, comeager sets, and degrees of unsolvability", Math. Proc. Camb. Phil. Soc. 79 (1976), 195-220.

National University of Singapore,
Singapore 0511, Singapore.

University of Illinois,
Urbana, IL 61801,
USA

UNDECIDABILITY AND RECURSIVE EQUIVALENCE II

J.N. CROSSLEY
Monash University, Clayton,
Victoria, 3168/Australia
and
J.B. REMMEL
University of California,
La Jolla, CA 92093/USA

1. *Introduction.* Various theories of recursive equivalence types (RETs) have been proved undecidable by Manaster and Nerode (1970, 1971), Nerode and Shore (1980) and Crossley and Remmel (1983). In the papers noted above it has been shown that these theories are all recursively isomorphic to second order arithmetic.

We continue this work here and the present paper consists of four parts. In the first few sections (§§2-4) we extend the results of Crossley and Remmel (1983) to theories of constructive order types with $\leq$ and $\leq^*$ as the only non-logical constants. The second part (§§5-7) treats the theory of RETs of matroids and acquaintance with Crossley and Nerode (1981) would be useful. In the third §§8-10 we deal with the theory of RETs of vector spaces over a recursive field. (For basic details of this theory see Dekker (1969).)

It had been thought by us that we could produce uniform proofs for matroids, vector spaces and algebraically closed fields but the notions of dependence involved are so different that this has turned out to be impossible. However we plan to treat vector spaces and algebraically closed fields in more detail in joint work with C.J. Ash.

Finally, in §11 we turn to the theory of *co-simple* COTs. Here we show that the theories with $+$ or $\leq$ and $\leq^*$ as the only non-logical constants are undecidable. In the co-simple case, however, we have been unable to establish the strength of the undecidability for we cannot use the techniques of Manaster, Nerode and Shore. On the other hand this is the first time a theory of *co-simple* recursive equivalence types (of any class of structure) with just + or just $\leq$ had been shown undecidable: even the question for co-simple isols (with $\leq$ or $+$ only (cf. Hay 1966), Remmel (1976)) has not been proved undecidable though it would be very surprising if it were not. However the theories of co-simple RETs, isols and regressive isols with both $+$ and $\cdot$ have been shown to be recursively isomorphic to first order arithmetic by Ellentuck (1973).

2. *Constructive Order Types (COTs).* In their (1970) Manaster and Nerode established the undecidability of the theories of RETs (of sets) with $+$ as the only non-logical constant and then quickly deduced similar results for $\leqslant$. At the time of writing Crossley and Remmel (1983) we were unable to push through the necessary reduction and this we remedy now. However, in the literature of COTs there are two definitions of $\leqslant$ (and of $\leqslant^*$). The first, based on $+$, is simply

(s) $A \leqslant B$ if, and only if, $\exists C(A + C = B)$

with $\leqslant^*$ defined dually by $A \leqslant B$ if, and only if, $\exists C(C + A = B)$. Since $+$ is not commutative for COTs two ordering relations appear to be essential.

The first definition may be found in Crossley (1965) amongst other places. The second definition which is to be found in the monograph Crossley (1969) does not require the separability properties of the definition above. Recall that if $\underline{\underline{A}}$ is a linear ordering contained in a standard representation of the rationals then we write $A = \text{COT}(\underline{\underline{A}})$ and A is the equivalence of $\underline{\underline{A}}$ under one-one, partial recursive functions which preserve order (strictly) from their domain $\delta p \subseteq \underline{\underline{Q}}$ to their range $\rho p \subseteq \underline{\underline{Q}}$. In order to distinguish these other ordering relations on COTs we attach a subscript "w" for "weak". We define

(w) $A \leqslant_w B$ if, and only if, there exist $\underline{\underline{A}} \in A$ and $\underline{\underline{B}} \in B$ with $\underline{\underline{A}}$ an initial segment of $\underline{\underline{B}}$

and dually define $A \leqslant_w^* \underline{\underline{B}}$ simply replacing "initial" by "final" in the definition above.

Our results for $\leqslant_w$ and $\leqslant_w^*$ will follow very easily from those for $\leqslant$ and $\leqslant^*$ since for most of the COTs with which we are concerned comparability under one order will imply comparability under the corresponding weak ordering. In the next section we treat only the strong ordering.

3. *Undecidability for* $\leqslant$ and $\leqslant^*$. First we observe that "X is a part of Y" is definable in the theory of COTs with $\leqslant$ and $\leqslant^*$ since

X is a part of Y if, and only if, $\exists U\, \exists V(U + X + V = Y)$

if, and only if, $\exists Z(X \leqslant^* Z \;\&\; Z \leqslant Y)$. To see this just take $Z = X + V$.

An examination of sections 4 and 5 of Crossley and Remmel (1983) shows that we now need only prove the following theorem and need not consider other contexts for "finitely different from".

<u>Theorem 1</u>. "X is an ω-indepecomposable" and "X is finitely different from the indecomposable Y and both are only finitely different from ω-indecomposables" are definable in the theory of COTs with $\leq$ and $\leq^*$ as its only non-logical constants.

The proof will be accomplished by a series of lemmata. First observe that "1-<u>pseudo-finite</u>" and "2-<u>pseudo-finite</u>" are definable in terms of $\leq$ and $\leq^*$ without using $+$. For convenience we now repeat the definition of 1-pseudo-finite and 2-pseudo-finite which clearly shows that these are definable.

<u>Definition 1</u>. X is said to be 1-pseudo-finite if, for all $Y \neq 0$, $Y \leq^* X$ implies $Y \leq X$. X is said to be 2-pseudo-finite if, for all $Y \neq 0$, $Y \leq X$ implies $Y \leq^* X$.

Lemma 2. X is finite if, and only if,

(i) X is both 1-pseudo-finite and 2 pseudo-finite

(ii) X is comparable under $\leq$ with every 1-pseudo-finite COT and

(iii) X is comparable under $\leq^*$ with every 2-pseudo-finite COT.

Proof. Suppose X is finite. Then $Y \leq^* X$ implies Y is finite and the classical order type of Y $\leq$ that of X. Hence $Y \leq X$ too. Therefore X is 1 pseudo-finite. By symmetry X is also 2-pseudo-finite.

By lemma 4.3 of Crossley and Remmel (1983) if Y is 1-pseudo-finite then Y is either finite or has an initial segment of type ω. In either case X is comparable with Y under $\leq$. This proves (ii), and (iii) follows by symmetry.

Now, for the other direction, suppose X satisfies conditions (i) - (iii). It follows from (i) and lemma 4.3 of Crossley and Remmel (1983) that either X is finite or has both an initial segment of type ω and a final segment of type ω^*.

Suppose X is not finite. By lemma 4.2 of Crossley and Remmel (1983) W is 1-pseudo-finite. Hence by (ii), since X is not finite, for some $Y, X = W + Y$. By the symmetrical argument using (iii) $X = Z + W^*$. Hence by the directed refinement theorem 2.3.2 of Crossley (1969) it follows that $X = W + A + W^*$ for some COT A. From this we have $W \leq X$ and by (i) X is 1-pseudo-finite, hence $W \leq^* X$. But this implies W and W^* are comparable under $\leq^*$ which is a contradiction. □

Lemma 3. "X has order type ω" is definable (in the theory of COTs with $\leq$ and $\leq^*$ as its non-logical constants).

Proof. We claim "X has order type ω" is equivalent to the conjunction of (i) "if Y is finite then $Y \leqslant X$ and $Y \nleqslant^{*} X$" and (ii) "if $Y \leqslant X$ and $Y \neq X$ then Y is finite". For classical order types the assertion is clearly true. The difficulty arises since there exist COTs Y of order type $> \omega$ for which there is no COT Z of order type ω such that $Z < Y$. So suppose X satisfies (i) and (ii).

By condition (i) X has an initial segment of order type ω and so X itself has order type $\omega + \alpha$ where $\alpha \geqslant 0$. If $\alpha > 0$ then we can take $\underline{\underline{X}} \in X$ and $x \in \underline{\underline{X}}$ such that the initial segment $\underline{\underline{X}}\,(\leqslant x)$ has order type of the form $\omega + \beta$ for some β, possibly zero. But $\underline{\underline{X}}$ is a sub-ordering of $\underline{\underline{Q}}$ hence $\mathrm{COT}(\underline{\underline{X}}(\leqslant x)) = Y$, say, such that $Y \leqslant X$ because $\underline{\underline{X}}\,(\leqslant x)$ and $\underline{\underline{X}}\,(>x)$ are separable by recursive segments of $\underline{\underline{Q}}$. This contradicts condition (ii).

The converse direction is clear. □

Lemma 4. (i) "X is dense" is definable in the theory of COTs with $\leqslant$ and $\leqslant^{*}$ as the only non-logical constants.

(ii) Similarly for "X is discrete."

Proof. First recall that by our convention in the third paragraph of Section 4 of Crossley and Remmel (1983), "X is dense" means "every $\underline{\underline{X}} \in X$ is dense." Now X is dense if, and only if, $X \neq 0, 1$ and for every part Y of X with $Y \neq 0, 1$ there is a part of Y, Z say, such that $Z \neq 0$ and

(i) if $1 \leqslant Y$ then $1 \nleqslant Z$,

(ii) if $1 \leqslant^{*} Y$ then $1 \nleqslant^{*} Z$ and

(iii) if $1 \nleqslant Y$ and $1 \nleqslant^{*} Y$ then $1 \leqslant Z$, $1 \leqslant^{*} Z$ and $Z \neq 1$.

To see this simply recall that (a) $\underline{\underline{X}}$ is dense if, and only if, $\underline{\underline{X}}$ has at least two elements and between any two distinct points of $\underline{\underline{X}}$ there is a third point of X and (b) $1 \leqslant Y$ means that if $\underline{\underline{Y}} \in Y$ then $\underline{\underline{Y}}$ has a first element (and dually if $1 \leqslant^{*} Y$ there is a last).

Finally the right hand side of the equivalence above is clearly expressible in the theory of COTs with $\leqslant$ and $\leqslant^{*}$ as its only non-logical constants.

(ii) Simply note that X is discrete if, and only if, every part of X is not dense. □

Lemma 5. X has order type $\omega + n$ for some finite n is equivalent to the conjunction of

(i) X is discrete,

(ii) X has an initial segment $Y \leqslant X$ of order type ω,

(iii) if $\underline{\underline{X}} \in X$ then $\underline{\underline{X}}$ contains no interval of type ω^* and

(iv) if Z is a part of X with initial segments of all finite types and Z has no last element then $Z \lessdot^* Y$ (where Y is as in (ii) above).

Proof. First suppose X has order type $\omega + n$, then there exists $Y \lessdot X$ with $|Y| = \omega$ since every finite set is separable.

(iii) is trivial since X is well-ordered.

For (iv) note that the unique initial segment of X with no last element is Y which has order type ω and moreover if $\underline{\underline{Y}} \in Y$ and $\underline{\underline{Z}} \subseteq \underline{\underline{Y}}$ then $\underline{\underline{Z}}$ has no last element only if $\underline{\underline{Z}}$ is cofinal in $\underline{\underline{Y}}$. Since Z is a part of X we must therefore have $\underline{\underline{Z}}$ is a final segment of $\underline{\underline{Y}}$. But then $Y = m + Z$ for some finite m and hence $Z \lessdot^* Y$. Thus (i) - (iv) hold.

Next suppose X satisfies (i) - (iv) and let Y be as in (ii). Then X has order type $\omega + \alpha$ for some order type α. Since X is discrete by (i), α is a discrete order type. Let $\underline{\underline{X}} \in X$ then each $x \in \underline{\underline{X}}$ lies in an interval I of order type k (for some finite k), ω, ω^* or $\omega + \omega^*$ which is maximal with respect to each of its points being only finitely distant from x. By (iii) the last two cases cannot occur. Hence α is a linearly ordered sum of the form $k_1 + \Sigma + k_2$ where k_1 and k_2 are finite and Σ is a linearly ordered sum of copies of ω.

If Z is any part of X without a last element then it is therefore of the form $(\omega +)k_1 + \Sigma'$ where the initial ω may be absent and Σ' is an initial segment of Σ and is itself a linearly ordered sum of copies of ω.

If, in addition, Z has initial segments of all finite order types then Z has an initial segment of order type $\omega + k_1 + \Sigma'$ or Σ' where in the latter case the sum Σ' must have a first copy of ω (which absorbs k_1). In the former case condition (iv) implies Σ' is empty so $\alpha = k_1 + k_2$ which is finite. In the latter case $\Sigma' = \omega$. But then Z has order type $\omega + \omega$ which contradicts $Z \lessdot^* Y$ since $|Y| = \omega$. □

Lemma 6. "X is indecomposable" is definable (in the theory of COTs with $\lessdot$ and $\lessdot^*$).

Proof. X is indecomposable if, and only if, X has order type $\omega + n$ for some finite n or is finite by lemma 2.1 of Crossley and Remmel (1983). It therefore suffices, by lemma 2, to show that the conditions (i) - (iv) of lemma 5 are definable.

(i) is lemma 4. (ii) is definable by $\exists Y(Y \leqslant X \ \& \ |Y| = \omega)$ and $|Y| = \omega$ is definable by lemma 3. (iv) is equivalent to (Z is part of X & $\forall F$(F is finite $\to F \leqslant X$) & $1 \nleqslant^* Z) \to Z \leqslant^* Y$. That "F is finite" is definable follows readily from lemma 2 and, by the remarks at the beginning of this section, "Z is part of X" is also definable.

So it only remains to consider (iii).

Let $\underline{\underline{X}} \in X$ and suppose $b \in \underline{\underline{X}}$ is such that there is an interval of order type ω^* whose last element is b. Let $\underline{\underline{V}} = \underline{\underline{X}}\ (<b)$ and V = COT ($\underline{\underline{V}}$). Then for all finite F we have $F \leqslant^* V$ and also $V \leqslant X$. Hence saying X has no interval of type ω^* is equivalent to $\forall V(V \leqslant X \to \neg(V \leqslant X \ \& \ \forall F(F \text{ is finite} \to F \leqslant^* V)))$ since the converse direction is clear.

Note that we have here used the element b to separate $\underline{\underline{V}}$ and $\underline{\underline{X}} - \underline{\underline{V}}$ by recursive intervals (of $\underline{\underline{Q}}$).

Finally, since as noted above, "F is finite" is definable, the proof is complete. □

Lemma 7. "X is finitely different from an ω-indecomposable" is definable (in the theory of COTs with $\leqslant$ and $\leqslant^*$).

Proof. The expression in quotation marks is equivalent to "X has order type $\omega + n$" or to "X is indecomposable and X is not finite" so this lemma follows from lemma 5 or lemma 6 and lemma 2.

Now note that "being finitely different from" is an equivalence relation. So from lemma 7 we get the second part of theorem 1 and from lemma 3 we get the first part. So we have established theorem 1. □

An examination of our proofs shows that in fact all the results go through when $\leqslant_w$ and $\leqslant_w^*$ are used in place of $\leqslant$ and $\leqslant^*$. Moreover all COTs of order type $\omega + n$ have exactly the same $\leqslant$ and $\leqslant_w$ predecessors.

4. Undecidability results for COTs.

So we can now immediately deduce all the undecidability results below from our work above and the work of Crossley and Remmel (1983) using, as before, the results of Nerode and Shore (1980).

Theorem 1. (i) The theory of constructive order types with $\leqslant$ and $\leqslant^*$ as its only non-logical constants is undecidable and recursively isomorphic to second order arithmetic.

(ii) As for (i) with $\leqslant_w$ and $\leqslant_w^*$ replacing $\leqslant$ and $\leqslant^*$. □

Theorem 2. (i) Each of the following theories with $\leq$ and $\leq^*$ as the only non-logical constants is undecidable and recursively isomorphic to second order arithmetic: (a) co-ordinals, (b) quords, (c) quasi-finite COTs, (d) losols.

(ii) As for (i) with $\leq_w$ and $\leq_w^*$ replacing $\leq$ and $\leq^*$. □

For definitions of these classes see either Crossley and Remmel (1983) or Crossley (1969).

5. RETs of matroids

The study of RETs of matroids was begun in Crossley and Nerode (1981). In this section we review the basic definitions and then turn our attention to definitions of addition for matroids. After this we reduce the various cases of the general theory to two other theories in order to establish strong undecidability results.

We define matroids in terms of independent sets. For other definitions which are classically (but not effectively) equivalent see Welsh (1976). The prime example of a matroid is the set of elements in a finite dimensional vector space over a finite field together with all its independent subsets. However we shall be almost entirely concerned with infinite matroids. For an example delete both occurrences of "finite" in the penultimate sentence. We treat these infinite matroids as independence spaces of finite character (Welsh (1976) p. 385-7)).

Card (X) denotes the cardinality of X.

Definition 1. A matroid M is a pair (U,I) such that U is a set and I is a set of subsets of U satisfying (I1 - I3) and (FC) below.

(I1) $\phi \in I$

(I2) If $X \in I$ and $Y \subseteq X$ then $Y \in I$

(I3) If X,Y are finite sets in and card $X = 1 + \text{card } Y$, then there exists $x \in X - Y$ such that $Y \cup \{x\} \in I$.

(FC) (Finite Character). If $X \subseteq U$ and every finite subset of X is in I, then $X \in I$.

A subset of U in I is said to be *independent*, otherwise *dependent*. x *depends on* A if $A \cup \{x\}$ is dependent. The closure of A, written $c\ell$ (A), is $\{x : x \text{ depends on } A\}$. A is *closed* if $A = c\ell(A)$.

It follows from (I1) - (I3) (see Welsh (1976) p.8-9) that $c\ell$ satisfies the usual conditions for a closure operator, namely (Cℓ1) - (Cℓ4) and (FC Cℓ) below.

(Cℓ1) $X \subseteq c\ell X$.

(Cℓ2) $X \subseteq Y$ implies $c\ell X \subseteq c\ell Y$.

(Cℓ3) $c\ell c\ell X = c\ell X$.

(Cℓ4) (Global exchange property). If $y \in c\ell X$ and $Y \in c\ell(X \cup \{x\})$, then $(X \cup \{y\})$.

(FC Cℓ) if $x \in c\ell X$, then there exists a finite subset Y of X such that $x \in c\ell Y$.

The terminology from vector spaces carries over in the obvious way. Thus a *base* (or *basis*) for X is a maximal independent subset of X or equivalently an independent spanning set.

Definition 2. A matroid $M = (U,I)$ is said to be *recursively presented* if U is a recursive set (of natural numbers) and M has a *dependence algorithm*, i.e. an algorithm which given a finite set $X \subseteq U$ will decide whether X is dependent or not.

Corollary 1. If $M = (U,I)$ is a recursively presented matroid, then, if X is recursive (r.e.), $c\ell X$ is recursively enumerable.

Proof. $x \in c\,Y$ if, and only if, $Y \cup \{x\}$ is not in I. So enumerate the finite subsets Y of U and enumerate U and put x into $c\ell X$ whenever $x \in c\ell Y$ for some such finite set $Y \subseteq X$. Then (FC) establishes the result.

As in Crossley and Nerode (1981) p.75 we shall adopt

Assumption 1. $M = (U,I)$ is a matroid such that for each natural number n there is a set in I containing at least n elements.

This assumption implies there are sets of all finite cardinalities and also infinite sets in I. The next lemma is well-known (see for example Metakides and Nerode (1977)).

Lemma 1. If M is a recursively presented matroid (satisfying assumption 1), then M has a recursive infinite basis.

Proof. Enumerate U and let b_{n+1} be the first element independent of $b_0,\ldots,b_n$. Then $b_0,b_1,\ldots$ is an r.e. basis. To see that $b = b_0,b_1,\ldots$ is in fact recursive note that $U - B = \{u \in U \mid \exists n(\{b_0,\ldots b_n\} \cup \{u\}$ is dependent$)\}$ and hence by the dependence algorithm $U - B$ is also r.e. □

We shall use such a basis to define addition in M below. We shall also require a notion derived from the classical notion of independence between sets, namely:

Definition 3. (cf. Jacobson (1953) Vol. II, p. 28).
Closed subsets $\{A_j : j \in J\}$ of a matroid M are said to be *independent* if $A_k \cap c\ell \cup(\{A_j : \neq k\}) = c\ell\emptyset$.

This clearly extends the definition of independence for elements (take $A_j = c\ell \cup\{a_j\}$).

Next we turn to recursive equivalence for independent sets in matroids.

Definition 4. If $M = (U,I)$ is a recursively presented matroid and $A,B \in I$ then A is said to be *recursively equivalent* to B by the one-one partial recursive function p if domain $p \supseteq A$, $p(A) = B$ and p preserves independence.

As usual we write $p : A \cong B$ or $A \cong B$ in this case.

This relation is an equivalence relation (Crossley and Nerode (1981) p.81) and we call the equivalence classes M-RETS.

We shall also use $c\ell$ M-RETS which are defined as in definition 4 above except that we require domain p and range p to be (recursively enumerable) *closed* subsets of U. In fact it often happens that maps of this kind are induced by maps of the former kind. Thus for vector spaces (over a recursive fiels) an independence preserving map $p : x_i \to y_i$ where the x_i, y_i are independent sets induces a (vector space) morphism from $c\ell\{x_i : i \in I\}$ onto $c\ell\{y_i : i \in I\}$ by linearity where the induced map sends $\Sigma\alpha_i x_i$ to $\Sigma\alpha_i y_i$ (where the α_i are scalars). However this does not happen for all matroids and it can certainly happen that taking closures does not commute with an independence preserving map.

Now when it comes to defining addition we have two possibilities. We shall treat these separately in the succeeding sections.

6. Addition for matroids

We now give the first definition of addition which seems appropriate to the matroid context. As in Dekker and Myhill (1960) we require a notion of (recursive) separability.

Definition 1. If $A, B \in I$, then A is said to be *separable* from B if there exist r.e. sets $A', B' \in I$ such that $A \subseteq A'$, $B \subseteq B'$, $A' \cup B' \in I$ and $A' \cap B' = \emptyset$.

Now by lemma 5.1 a recursively presented matroid $M = (U,I)$ which satisfies assumption 1 has a recursive basis $E = \{e_0, e_1, \ldots\}$.
Let $u_0 < u_1 < \ldots$ be an effective list of U.
So now let $f(u_i) = e_{2i}$ and $g(u_i) = e_{2i+1}$ then $f(E)$ and $g(E)$ are

r.e. independent sets whose intersection is empty. We therefore make the following

Definition 2. If $A, B \in I$, then $A + B = f(A) \cup g(B)$. If $\underline{A} = RET(A)$, $\underline{B} = RET(B)$ then $\underline{A} + \underline{B} = RET(A + B)$.

The next lemma shows this latter is a good definition.

Lemma 1. If M is a recursively presented matroid (satisfying assumption 1) then

(i) if A is separable from B then $A \cup B \simeq A + B$,
(ii) if $A_i \simeq B_i$ for $i = 1,2$, then $A_1 + A_2 \simeq B_1 + B_2$.

Proof. (i) If A, B are separable by r.e. sets A', B' then define p by

$$p(x) = f(x) \text{ if } x \in A', \quad = g(x) \text{ if } x \in B',$$

then it is readily verified that $p : A \cup B \simeq A + B$. (Cf. Dekker and Myhill (1960) p.73, theorem 9(b))).

(ii) If $p : A_1 \cong B_1$ and $q : A_2 \cong B_2$ then $r : A_1 + A_2 \quad B_1 + B_2$ where $r(y) = f(px)$ if $y = f(x)$ $= g(qx)$ if $y = g(x)$

since range f is recursively separable from range g. □

Now it is clear that the set of independent sets in a recursively presented matroid together with the independence preserving map forms an appropriate category in the sense of Crossley and Nerode (1974). However we are primarily concerned with properties of addition and for our purposes the following theorem, due to Crossley's student B. Redgen will suffice.

Theorem 1. (B. Redgen) Let X be any sentence of the form

$$\forall \underline{X}_1 \dots \underline{X}_p \; \underline{Y}_1 \dots \underline{Y}_r \left[\mathop{\&}_{j=1}^{n} \left(\sum_{i=1}^{p} a'_{j,i} \underline{X}_i = \sum_{i=1}^{p} a_j'' \underline{X}_i \right) \right.$$

$$\left. + \mathop{\&}_{j=1}^{m} \left(\sum_{i=1}^{p} b'_{j,i} \underline{X}_i + \sum_{i=1}^{r} c'_{j,i} \underline{Y}_i = \sum_{i=1}^{p} b''_{j,i} \underline{X}_i + \sum_{i=1}^{r} c_{j,i} \underline{Y}_i \right) \right.$$

then if X is true in the natural numbers it is true for Dedekind M-RETs (and conversely).

(Note that a Dedekind M-RET is defined, as in Crossley and Nerode (1974), as an M-RET with an isolated representative.)

The proof is virtually identical with that of Manaster (1966) using the fact that if $A,B \in I$ and A is separable from B then $A \cup B \in I$ by definitions 1 and 5.3. We leave the formal details to the reader. □

7. Undecidability for matroids with addition

In his (1969) Dekker first drew attention to α-spaces. In Crossley and Nerode (1974) we generalized this idea from vector spaces to more general structures and we called the corresponding objects and their appropriate RETs "sound". Their definition reduces to the following in the context of matroids.

Definition 1. An independent set S in a matroid $M = (U,I)$ is said to be *sound* if there is a r.e. $R \in I$ such that $S \subseteq R$. In this case the M-RET of S is also said to be *sound*.

Lemma 1. An M-RET $\underline{S}$, where M is a recursively presented matroid, is sound if, and only if, there exist $\underline{T}$, $\underline{V}$ such that $\underline{S} = \underline{T} + \underline{V}$.

Proof. If $\underline{S} = \underline{T} + \underline{V}$ let $T \in \underline{T}$ and $V \in \underline{V}$. Then $S \simeq T + V = f(T) + g(V) \subseteq E = c\ell\{e_0, e_1, \ldots\}$. Suppose $p : S \simeq T + V \subseteq E$ then $S \subseteq \underline{p}^{-1}(E)$ which is a r.e. independent set.

Conversely, if $\underline{S}$ is sound there exists $S \in \underline{S}$ with $S \subseteq E$ so $\underline{S} = \underline{S} + 0$ where 0 is the M-RET of the empty set. □

Note that (provided M satisfies assumption 1) $+$ is not defined for all M-RETs since by Crossley and Nerode (1981) theorem 5.5 there exist un-sound M-RETs.

Theorem 1. The theory of (Dedekind) M-RETs with addition as the only non-logical constant is undecidable and recursively isomorphic to second order arithmetic (provided M satisfies assumption 1).

Proof. In Crossley and Nerode (1981) we showed that ordinary RETs (called S-RETs in Crossley and Nerode (1974) and (1981)) are embeddable in the M-RETs and that their image is precisely the sound M-RETs. Moreover, from the definition of X given above it is clear that this embedding preserves X.

Now the previous lemma shows that "$\underline{S}$ is sound" is definable in the theory of M-RETs with X by the formula $\exists \underline{T} \, \exists \underline{V} \, (\underline{S} = \underline{T} + \underline{V})$. Further, by Manaster and Nerode (1970, 1971) the theory of S-RETs is recursively undecidable and recursively isomorphic to second order arithmetic. Hence so to is the theory of M-RETs with X.

For the case of Dedekind M-RETs then the result again follows from Manaster and Nerode (1970, 1971) since if the S-RET $\underline{X}$ is Dedekind then the embedding given in Crossley and Nerode (1981) takes $\underline{X}$ to a Dedekind M-RET $\underline{V}(\underline{X})$ and again it is easy to see that this map $\underline{V}$ is onto the Dedekind M-RETs. □

8. The other definition of addition for matroids.

In section 6 we gave what seems an obvious definition for + in terms of matroids qua matroids. However, when thinking in terms of matroids arising from vector spaces it seems natural to define addition in the following manner.

Definition 1. If $M = (U,I)$ is a recursively presented matroid and $A,B \in I$ then A is said to be *strongly recursively separable* from B if there exist independent r.e. closed sets A',B' such that $A \subseteq A'$, $B \subseteq B'$ (where A' and B' are independent if whenever $E_1, E_2 \in I, c\ell(E_1) = A'$, and $c\ell(E_2) = B'$, then $E_1 \cup E_2 \in I$.

Definition 2. If A,B are strongly recursively separable then $A + B = A \cup B$. If also $\underline{A} = \text{M-RET}(A)$ and $\underline{B} = \text{M-RET}(B)$ then $\underline{A} \oplus \underline{B} = \text{M-RET}(A \cup B)$.

The next lemma shows this is a good definition.

Lemma 1. If, for $i = 1,2$, $A_i, B_i \in I$, $p_i : A_i \simeq B_i$, A_1, A_2 are strongly recursively separated by the r.e. closed sets A_1', A_2' and similarly for the B's, then $A_1 \cup A_2 \simeq B_1 \cup B_2$.

Proof. First note that $c\ell\emptyset$ is recursive for $x \in c\ell\emptyset$ if, and only if, $\{x\}$ is dependent. Then the map q defined by

$$\begin{aligned} q(x) &= p_1(x) \text{ if } x \in (A_1' \cap \text{domain } p_1) - c\ell\emptyset \\ &= p_2(x) \text{ if } x \in (A_2' \cap \text{domain } p_2) - c\ell\emptyset \\ &\text{undefined if } x \in c\ell\emptyset \end{aligned}$$

is clearly partial recursive. From the properties of p_1 and p_2 it is one-one and independence preserving separately on A_1' and A_2'. Moreover, since A_1', A_2' are independent, and likewise B_1', B_2' are independent, q preserves independence from $A_1' \cup A_2'$ to $B_1' \cup B_2'$. □

In the same way as for the previous definition of addition we now have

Theorem 1. (B. Redgen) Let ϕ be any sentence of the form

$$\forall \underline{X}_1 \ldots \forall \underline{X}_p \ \exists \underline{Y}_1 \ldots \exists \underline{Y}_r$$

$$\left[\mathop{\&}_{j=1}^{n}\left(\sum_{i=1}^{p} a'_{j,i}\,\underline{X}_i = \sum_{i=1}^{p} a''_{j,i}\,\underline{X}_i\right) \rightarrow \mathop{\&}_{j=1}^{n}\left(\sum_{i=1}^{p} b'_{j,i}\,\underline{X}_i \oplus \sum_{i=1}^{r} c'_{j,i}\underline{Y}_i\right.\right.$$
$$\left.\left. \sum_{i=1}^{p} b''_{j,i}\,\underline{X}_i \oplus \sum_{i=1}^{r} c''_{j,i}\,\underline{Y}_i\right)\right]$$

then if ϕ is true in the natural numbers it is true for Dedekind M-RETs (and conversely) where Σ now denotes repeated $\oplus$.

We leave the details of the proof to the reader as above. □

9. Undecidability for M-RETs with $\oplus$.

The proof follows exactly the same strategy as Manaster and Nerode (1970) but there are slight changes because of the different notion of strong recursive separability. We shall therefore merely point out the changes and the reader is advised to have a copy of Manaster and Nerode (1970) at hand.

First we define $X \leq Y$ by $\exists Z(\underline{X} \oplus \underline{Z} = \underline{Y})$. Next we shall work with the basis $E = \{e_0, e_1, e_2, \ldots\}$ instead of the natural numbers half the time.

The definitions of *cover* and *indecomposable cover* are the obvious notational variants of those in Manaster and Nerode (1970) p.52. Theorem 1 of the preceding section gives us the required AE-metatheorem.

For the theorem of Manaster and Nerode's (1970) Section 1 we first let $\{\langle\omega_n, \theta_n\rangle\}_{n=0}^{\infty}$ be a list of all pars of r.e. independent closed sets in M. Let $\xi^i = \{e_{j(x,i)} : e_x \in \xi\}$ where j is the standard recursive pairing function, $j : N \times N \to N$. In fact each ξ^i and α will be a subset of E.

The construction now proceeds as in Manaster & Nerode (1970) noting that for each i,k, $\alpha_{i,k}$ and $\alpha^i_{i,k}$ are r.e. infinite, *independent* sets. Further I is independent. What in fact is happening is that inequality is replaced by the stronger condition of independence.

In the latter part of the proof note that conditions such as (x_1, x_2) (recursive separability) are now replaced by x_1, x_2 are contained in r.e. independent closed sets.

This suffices for the theorem. The theorem however can be used to produce $2^{\aleph_0}$ mutually comparable, indecomposable M-RETs by finding indecomposable covers of $\langle 1,1,1,\ldots\rangle$ where $1 = $ M-RET $(\{e_0\})$ and for the

Dedekind property noting that if a set $S \subseteq E$ contains an infinite recursive subset R then R can be split further into to infinite recursive subsets R_1 and R_2 such that $c\ell R_1$, $c\ell R_2$ and $c\ell E - (R_1 \cup R_2)$ are strongly recursively separable (and M-RET(S) is not indecomposable). So we now have the required $\underline{P}_0, \underline{P}_1, \underline{P}_2, \underline{X}_0, \underline{X}_1, \underline{X}_2, \ldots$ for p.54.

Now let $n = \text{M-RET}\{e_i : i < n\}$ and define $\underline{X} =_1 \underline{Y}$ in the obvious way.

The rest of the proof now follows verbatim except that $+$ is replaced by $\oplus$. Similarly adding the additional remarks of Manaster and Nerode (1971) with Nerode and Shore (1980) we have established:

Theorem 1. (i) The theory of M-RETs with $\oplus$ (for a recursively presented matroid M satisfying assumption 1) is recursively undecidable and recursively isomorphic to second order arithmetic.

(ii) Similarly for the theory of Dedekind M-RETs with $\oplus$.

(iii) Similarly for the theory of M-RETs with $\leqslant$ as the only non-logical constant.

(iv) Similarly for the theory of Dedekind M-RETs with $\leqslant$ as the only non-logical connective.

10. Undecidability of the theory of sound vector space RETs with $\oplus$.

Since sound vector space RETs are isomorphic to the RETs of their sound bases (see Crossley and Nerode (1975)) we only need to check that this isomorphism preserves direct sum, $\oplus$. But this is clear since for soundly based vector spaces A, B there are bases contained in r.e. independent sets α, β. Simply map α onto $\{e_{2i} : i = 0, 1, 2, \ldots\}$ and β onto $\{e_{2i+1} : i = 0, 1, 2, \ldots\}$ by partial recursive one-one functions p and q, say, (whose domains are α and β respectively) then $A \subseteq c\ell p(\alpha)$, $B \subseteq c\ell q(\beta)$ and $c\ell p(\alpha)$ and $c\ell p(\beta)$ are recursively separable independent spaces. So $c\ell(p(A) \cup q(B)) = \underline{A} \oplus \underline{B}$ where $\underline{A} = \text{RET}(A)$ and $\underline{B} = \text{RET}(B)$. Finally, we note that for soundly based RETs any two sound bases are recursively equivalent by Hamilton (1970). Hence $\oplus$ commutes with $c\ell$ up to recursive equivalence in the sense that if A, B are sound bases then $c\ell(A \oplus B) \simeq c\ell A \oplus c\ell B$ where the $\oplus$ on the left is in the matroid of bases and the $\oplus$ on the right is for vector subspaces of the universal space.

We can therefore deduce the required undecidability results from theorem 9.1.

Theorem 1. (i) The theory of soundly based vector space RETs (V-RETs) over a recursively presented field with $\oplus$ as the only non-logical constant is recursively undecidable and recursively isomorphic to second order arithmetic.

(ii) Similarly for the subclass of Dedekind (i.e. Dekker's (1969) isolic α-dimensional) vector space RETs.

(ii) and (iv) As for theorem 9.1 (iii) and (iv) but this time for soundly based vector space RETs. □

11. Co-simple COTs

In this section we establish undecidability results for various theories of co-simple COTs. Ellentuck (1973) established that the theory of co-simple isols with + and $\cdot$ is recursively isomorphic to first order arithmetic by defining $(N, +\cdot)$ in these theories (where N is the natural numbers). It therefore follows that we should not be able to use the techniques of Manaster and Nerode (since their method gives second order arithmetic undecidability). Instead we use the fact that the theory of a binary relation is finitely inseparable (for definition see below).

Definition 1. A set $B \subseteq \underline{\underline{Q}}$ is said to be simple if B is r.e. and $\underline{\underline{Q}} - B$ contains no infinite r.e. subset.

Definition 2. A COT $\underline{\underline{A}}$ is said to be co-simple if for some $A \in \underline{\underline{A}}$ we have $\underline{\underline{Q}} - A$ is a simple set.

Definition 3. A sequence of co-simple COT's $\underline{\underline{A}}_0, \underline{\underline{A}}_1, \ldots$ is said to be r.e. if there exists an r.e. sequence of r.e. sets $E_0, E_1, \ldots$ such that for all i, $\underline{\underline{Q}} - E_i \in \underline{\underline{A}}_i$.

Theorem 4. There exists an infinite r.e. sequence of pairwise incomparable co-simple ω-indecomposable COT's.

Proof. For each $i \geq 0$, let R_i be a recursive set of order type ω in $(i, i+1)$. Throughout this proof < will denote the order of $\underline{\underline{Q}}$. We shall construct an r.e. sequence of r.e. sets $E_i \subseteq R_i$ such that $\{R_i - E_i\}$ is an r.e. sequence of incomparable sets of order type ω. At each stage s of our construction and for each i, we shall specify an effective sequence $d^s_{0,i} < d^s_{1,i} < \ldots$ of elements of R_i such that $\{R_i - d^s_{0,i}, d^s_{1,i}, \ldots\} = E^s_i$ = the finite set of elements enumerated in E_i by the end of stage s. Our construction will ensure that $\lim_s d^s_{e,i} = d_{e,i}$ exists for all e and i and that $R_i - E_i = \{d_{0,i} < d_{1,i} < \ldots\}$ for all i.

Let $W_0, W_1, \ldots$ be an effective list of all r.e. sets of $\underline{\underline{Q}}$ and let W^s_e denote the finite set of elements enumerated into W_e by the end of stage s. Let $\phi_0, \phi_1, \ldots$ be an effective list of all 1:1 partial recursive functions. We write $\phi^s_e(x)\downarrow$ if the effective procedure to compute $\phi_e(x)$ gives an output in s or fewer steps and write $\phi^s_e(x)\uparrow$ otherwise. Also, we write $\phi_e(x)\downarrow$ if there exists an s such that $\phi^s_e(x)\downarrow$ and write $\phi_e(x)\uparrow$ otherwise. We let $\langle\ ,\ ,\ \rangle : \omega \times \omega \times \omega \to \omega$ and $\langle\ ,\ \rangle$ be some fixed recursive pairing functions.

To ensure that each $R_i - E_i$ is co-simple, we shall meet the following set of requirements.

$S_{\langle e,i\rangle}$: If W_e is infinite, then $[\underline{Q} - (R_i - E_i)] \cap W_i \neq \emptyset$.

We shall use the even stages of our construction to meet the requirements $S_{\langle e,i\rangle}$.

To ensure that $\{R_i - E_i\}_{i \in \omega}$ are pairwise incomparable, it is enough to meet the following set of requirements.

$P_{\langle e,i,j\rangle}$: There exists a $d \in R_i - E_i$ such that either $\phi_e(d)\uparrow$ or $\phi_e(d) \in R_j - E_j$.

To help us meet the requirements $P_{\langle e,i,j\rangle}$, we shall employ a set of movable markers $\Gamma_{\langle e,i,j\rangle}$. The idea is that we will want $\Gamma_{\langle e,i,j\rangle}$ to rest on some element x at stage s, denoted by $\gamma(e,i,j,s)$, such that $\phi^s_e(x)\downarrow$ and x $R_i - E^s_i$. Then if $\phi_e(x) \in R_j - E^s_j$ we shall enumerate $\phi_e(x)$ into E^{s+1}_j and attempt to ensure that x remains an element of $R_i - E_i$ so that x will be the required witness for requirement $P_{\langle e,i,j\rangle}$. We shall use the odd stages of our construction to meet the requirements $P_{\langle e,i,j\rangle}$.

We rank our requirements with those of highest priority first by $S_0, P_0, S_1, P_1, \ldots$. We ensure that $\lim_s d^s_{e,i}$ exists for all e and i by ensuring that $d^s_{e,i} \neq d_{e,i}$ only if forced by one of the requirements $S_0, P_0, \ldots, S_e, P_e$. There are conflicts between the requirements $\{S_e\}_{e \in \omega}$ and $\{P_k\}_{k \in \omega}$. That is, if $\gamma(e,i,j,s)$ is defined, then for the sake of requirement $P_{\langle e,i,j\rangle}$, we want to keep $\gamma(e,i,j,s)$ out of E_i. However at later stages, we may want to enumerate $\gamma(e,i,j,s)$ into E_i for the sake of meeting some requirement S_n or some requirement P_m where $m \neq k = \langle i,i,j\rangle$. By our priority ranking, we will allow $\gamma(e,i,j,s)$ to be put into E_i only for the sake of higher priority requirements, i.e. $S_0, P_0, \ldots S_{k-1}, P_{k-1}, S_k$. If $x = \gamma(e,i,j,s)$ is put into E_i, we shall remove the $\Gamma_{\langle e,i,j\rangle}$ marker from x. Thus, we say that a $\Gamma_{\langle e,i,j}$ marker is *active* at stage s if $\Gamma_{\langle e,i,j\rangle}$ rests on some element $x \in R_i - E^s_i$ (iff $\gamma(e,i,j,s)$ is defined) and is *inactive* otherwise. It will follow by a rather standard priority argument that we will have enough freedom to meet all the requirements.

Construction

Stage 0. For each $i \geq 0$, let $E_i = 0$ and for each e and i, let $d^0_{e,i} = r_{e,i}$ where $R_i = \{r_{0,i} < r_{1,i} < \ldots\}$.

Stage 2s + 1. Look for a $k \leq 2s + 1$ such that Γ_k is inactive at stage s and if $k = \langle e,i,j\rangle$, then there is an $x = d^{2s}_{e,i}$ with $e \leq 2s + 1$ such that $\phi^{2s+1}_e(x)\downarrow$ and either

(i) $\phi_e(x) \quad R_j \notin E^{2s}_j$ or

(ii) $\phi_e(x) = d^{2s}_{r,j}$ where $n > k$ and $d^{2s}_{r,j}$ has no Γ_m marker on it with $m < k$.

If there is no such k, let $E^{2s+1}_\ell = E^{2s}_\ell$ and $d^{2s+1}_{\ell,k} = d^{2s}_{\ell,k}$ for all k and ℓ. Otherwise, let $k(2s + 1)$ be the least such k and $x(2s + 1)$ be the least x corresponding to $k(2s + 1)$. Place the $\Gamma_{k(2s+1)}$ marker on $x(2s + 1)$ and let $y(2s + 1) = \phi_e(x(2s + 1))$ where $k(2s + 1) = \langle e,i,j\rangle$. Then if $y(2s + 1) \notin R_j - E^{2s}_j$, let $E^{2s+1}_\ell = E^{2s}_\ell$ and $d^{2s+1}_{\ell,k} = d^{2s}_{\ell,k}$ for all k and ℓ. Otherwise $y(2s + 1) = d^{2s}_{n,j}$ for some $n > k(2s + 1)$ and we set $E^{2s+1}_\ell = E^{2s}_\ell$ for $\ell \neq j$, $E^{2s+1}_j = E_j \cup \{y(2s + 1)\}$, $d^{2s+1}_{\ell,k} = d^{2s}_{\ell,k}$ for all $\ell \neq j$ and all k, and

$$d^{2s+1}_{m,j} = \begin{cases} d^{2s}_{m,j} & \text{if } m > n \\ d^{2s}_{m+1,j} & \text{if } m \geq n \end{cases}$$

Finally, we remove all Γ_m markers from $y(2s + 1)$ and make such markers inactive at stage 2s + 1.

Stage 2s > 0. Look for the least $p \leq 2s$ such that if $p = \langle e,i\rangle$, then $W^{2s}_e \cap [Q - (R_i - E^{2s-1}_i)] = \emptyset$ and there exists a $z = d^{2s-1}_{n,i} \in W^{2s}_e$ with $n > p$ such that z has no Γ_m markers on it with $m < p$. If there is no such p, let $E^{2s}_\ell = E^{2s-1}_\ell$ and $d^{2s}_{\ell,k} = d^{2s-1}_{\ell,k}$ for all ℓ and k. Otherwise, let $p(2s)$ denote the least such p and $z(2s)$ be the least such z corresponding to p. Set $E^{2s}_\ell = E^{2s-1}_\ell$ $\ell \neq i$ where $p(s) = \langle e,i\rangle$ and $E^{2s}_i = E^{2s-1}_i \cup \{z(2s)\}$. Also set $d^{2s}_{\ell,k} = d^{2s-1}_{\ell,k}$ for all $\ell \neq i$ and all k and if $z(2s) = d^{2s-1}_{n,i}$, then set

$$d^{2s}_{m,i} = \begin{cases} d^{2s}_{m,i} & \text{if } m < n \\ d^{2s}_{m+1,i} & \text{if } m \geq n \,. \end{cases}$$

Finally, we remove all Γ_m markers from $z(2s)$ and make such markers inactive at stage 2s.

This completes the construction. It is easy to see the construction is completely effective and that if $E_i = \bigcup_s E_i^s$, then $\underline{R_0 - E_0, R_1 - E_1, \ldots}$ is an r.e. sequence of COT's. To complete the proof, we need only show by a simultaneous induction that $\lim_s d_{e,i}^s = d_{e,i}$ exists for all i and that we meet all the requirements. So assume by induction that there is a stage t_0 large enough so that

(1) $d_{n,i}^s = d_{n,i}^{s+1}$ for all $s \geq t_0$, $n < e$, and i,

(2) we never take any action for any of the requirements $S_0, P_0, \ldots, S_{e-1}, P_{e-1}$ after stage t_0 (i.e. $k(s) \geq e$ and $p(s) \geq e$ for all $s \geq t_0$), and

(3) if $m = \langle n,i,j\rangle < e$ and Γ_m is active at stage t_0, then Γ_m remains active at all $s > t_0$ and $\gamma(m,i,j,s) = \gamma(m,i,j,t_0)$.

Then by construction $d_{e,i}^s = d_{e,i}^{s+1}$ for all $s \geq t_0$ since we have ensured that no requirement S_n or P_n with $n \geq e$ can force $d_{e,i}^s$ to change. We note that if $p(2s) = e$ for any $s \geq t_0$, then we act for requirement S_e at stage $2s$ and ensure that if $e = \langle n,i\rangle$, then $W_n^{2s} \cap E_i^{2s} \neq \emptyset$ so that requirement S_e will be satisfied. Moreover, if $W_n \subseteq R_i - E_i$ and W_e were infinite, then for some $2s > t_0$ and $m > n$, there would be $d_{m,i}^s \in W_e^{2s}$ with no Γ_m marker on it with $m < e$. Hence we would act for requirement S_e at stage $2s$ unless $W_n^{2s} \cap [Q - (R_i - E_i^{2s-1})] \neq \emptyset$ in which case S_e is automatically statisfied. Thus, we see that requirement S_e must be satisfied and there is a stage $t_1 > t_0$ such that $d_{e,i}^s = d_{e,i}^{s+1}$ for all i and $s \geq t_1$ and we never act for S_e after stage t_1. Finally, we consider requirement P_e. Let $e = \langle n,i,j\rangle$ and note that if Γ_e is ever active after stage t_1, say at some stage $s \geq t_1$, then at stage s, $\gamma(n,i,j,s) = x$ is in $R_i - E_i^s$ and is such that $\phi_n(x)\downarrow$ and $\phi_n(x) \notin R_j - E_j^s$. Moreover Γ_e remains on x and x remains in $R_i - E_i^t$ for all $t \geq s$ since only requirements $S_0, P_0, \ldots, S_{e-1}, P_e, S_e$ can force x into E_i once x has a Γ_e marker on it and by assumption such requirements never act after stage t_1. It follows that we can have $k(2s+1) = e$ at most once for $2s + 1 \geq t_1$ and hence there is a $t_2 > t_1$ such that (1) - (3) hold for t_0 replaced by t_2 and e replaced by $e + 1$. Moreover, it follows that requirement P_e must also be satisfied. That is, we have shown that $\lim_s d_{n,i}^s = d_{n,i}$ exists for all i. Hence if ϕ_n was really defined on all of $R_i - E_i$, there would be an m such that $\phi_n(d_{m,i}) \notin \{d_{0,j}, \ldots, d_{e,j}\}$ and $\phi_n(d_{m,i})$ had no Γ_q marker on it for $q < e$ since ϕ_n is one-one and there are only finitely many elements which ever have Γ_q marker on it with $q < e$. Thus there would be some stage $2s \geq t_1$ such that $\phi_n^{2s+1}(d_{m,i}^{2s})$ and hence $d_{m,i}^{2s}$ witnesses that it is possible to act for requirement P_e at stage $2s + 1$. Thus, we would act for requirement P_e at stage $2s + 1$ unless Γ_e were already active.

Thus, we have shown that all the requirements S_n and P_n are satisfied. Hence $\{\underline{\underline{R_i - E_i}}\}_{i \in \omega}$ is a sequence of pairwise incomparable co-simple COT's. Moreover $R_i - E_i$ has order type ω for all i since $\lim_s d^s_{e,i} = d_{e,i}$ exists for all e and $d^s_{0,i} < d^s_{1,i} < \ldots$ for all s. □

Theorem 5. Suppose that $\underline{\underline{A_0}}, \underline{\underline{A_1}}, \underline{\underline{A_2}}, \ldots$ is an r.e. sequence of co-simple COT's, then there is a co-simple indecomposable cover for $\underline{\underline{A_0}}, \underline{\underline{A_1}}, \underline{\underline{A_2}}, \ldots$.

Proof. Let f_i be a recursive isomorphism from $\underline{\underline{Q}}$ onto $(i,i+1)$ for $i \geq 0$. Let $E_0, E_1, \ldots$ be an r.e. sequence of r.e. sets such that $\underline{\underline{Q}} - E_i \in \underline{\underline{A_i}}$. Note that $f_i(E_i)$ is r.e. and that $(i,i+1) - f_i(E_i) \in \underline{\underline{A_i}}$. Moreover, $X = \underline{\underline{Q}} - (-\infty,0) - \bigcup_{i \geq 0} f_i(E_i)$ co-simple set and $\underline{\underline{X}}$ is an indecomposable cover of $\underline{\underline{A_0}}, \underline{\underline{A_1}}, \ldots$ just as in Crossley-Remmel (1983). □

We can now follow Crossley-Remmel (1983), Section 3 and let $\langle P, X_0, X_1, \ldots \rangle$ be an ω-sequence of pairwise incomparable co-simple ω-indecomposable COT's. Then

$$U = \langle X_i : i < \omega \rangle$$

$$R = \langle X_{i_0}, X_{j_0}, P, X_{i_1}, X_{j_1}, P, \ldots \rangle$$

are co-simple COT's for any recursive $\mathcal{R} = \{\langle i_k, j_k \rangle : k \in \omega\}$.

Thus using the translation machinery, we get the following analogue of lemma 5.3 of Crossley-Remmel.

Theorem 6.

(i) If CF is true in the co-simple COT's, then F is true in all finite models of a binary relation.

(ii) If F is false in some finite model of a binary relation, then CF is false in the co-simple COT's. □

Proof. We have shown that every recursive and hence every finite model of a binary relation is interpretable in the co-simple COT's.

Recall that a theory T is *finitely inseparable* if $\{\phi : T \models \phi\}$ and $\{\phi :$ there exists a finite model of M of T such that $M \models \neg\phi\}$ are recursively inseparable. Now the theory T of one binary relation is finitely inseparable. (See for example chapter 6 of Monk (1976).)

Theorem 7. The theory T' of the co-simple COT's (quords, quasi-finite COT's, losols) with + or with $\leqslant$ and $\leqslant^*$ as the only non-logical connectives is undecidable.

Proof. Let T be the theory of a single binary relation. Suppose T' were decidable. Then

(1) $\{\phi : T \models \phi\} \subseteq \{\phi : \phi$ is true in every finite model of $T\}$
$\subseteq \{\phi : T' \models C\phi\}$ and

(2) $\{\phi :$ there is some finite model of M of T such that $M \models \neg\phi\}$
$\subseteq \{\phi : T' = \neg C\phi\}$

by Theorem 6. But if T' were decidable, then $\{\phi : T' \models C\phi\}$ and $\{\phi : T' \models \neg C\phi\}$ would be recursive sets violating the fact that T is finitely inseparable. □

REFERENCES

CROSSLEY, J.N.

(1965) Constructive order types, I, pp. 189-264 in Formal Systems and Recursive Functions, ed. J.N. Crossley and M.A.E. Dummett, North-Holland Publishing Co., Amsterdam, 1965.

(1969) Constructive order types. North-Holland Publishing Co., Amsterdam (1969)

CROSSLEY, J.N. & A. NERODE

(1974) Combinatorial functors. Ergebnisse der Math. und ihrer Grenzgebiete, 81. Springer, Berlin (1974).

(1975) Combinatorial functors, pp. 1-21 in Logic Colloquium (Boston, Mass., 1972-1973), Springer Lecture Notes in Math. 453 Springer, Berlin (1975).

(1981) Recursive equivalence on matroids, pp. 69-86 in Aspects of effective algebra, ed. J.N. Crossley, U.D.A. Book Co., Steel's Creek, Australia (1981).

CROSSLEY, J.N. & J.B. REMMEL

(1983) Undecidability and recursive equivalence, I. To appear in Proc. Southeast Asian Logic Conference (Singapore, 1981), ed. Chong Chi-Tat.

DEKKER, J.C.E.

(1969) Countable vector spaces with recursive operations, Part 1. J. Symbolic Logic 34 (1969), 363-387.

DEKKER, J.C.E. & J. MYHILL

(1960) Recursive equivalence types. Univ. of Calif. publications in mathematics, n.s. 3 (1960), 67-214.

HAMILTON, A.G.

(1970) Bases and α-dimensions of countable vector spaces with recursive operations. J. Symbolic Logic, 35 (1970), 85-96.

HAY, L.

(1966) The co-simple isols. Ann. of Math. 83 (1966), 231-256.

JACOBSON, N.

(1953) Lectures in abstract algebra, Vol. II. D. Van Nostrand Co., New York (1953).

MANASTER, A.B.

(1966) Higher order indecomposable isols. Trans. Amer. Math. Soc. 125 (1966), 363-383.

MANASTER, A.B. & A. NERODE

(1970) A universal embedding property of the RETs. J. Symbolic Logic 35 (1970), 51-59.

(1971) The degree of the theory of addition of isols (abstract). Notices Amer. Math. Soc. 18 (1971), 68-OZ-3, 86.

METAKIDES, G. & A. NERODE

(1977) Recursively enumerable vector spaces. Ann. Math. Logic. 11 (1977), 147-171.

MONK, J.D.

(1976) Mathematical logic. Springer, Berlin (1976).

NERODE, A & R. SHORE

(1980) Second order logic and first order theories of reducibility orderings, pp. 181-200 in the Kleene Symposium ed. J. Barwise, H.J. Keisler and K. Kunen, North-Holland Publishing Co., Amsterdam (1980).

REMMEL, J.B.

(1976) Combinatorial functors on co-r.e. structures. Ann. Math. Logic 10 (1976), 261-287.

WELSH, D.J.A.

(1976) Matroid theory. Academic Press, London (1976).

LOGICAL SYNTAX AND COMPUTATIONAL COMPLEXITY

Larry Denenberg*
Harry R. Lewis*
Harvard University
Aiken Computation Laboratory
Cambridge, MA 02138 USA

*Research supported by NSF Grant MCS80-05386-A01.

Introduction

Special cases of the predicate calculus provide a rich source of problems of particular degrees of computational complexity. In this survey paper we sketch some results about formulas classified by quantificational and truth-functional structure. No attempt is made to give complete technical details, which in most cases can be found in the cited references. Instead we try to build the reader's intuition about how certain features contribute to the complexity of the decision problem.

Predicate logic is a descriptive system. To the extent that a subset S of its language can be used to describe the answers to some computationally difficult set of questions Q, that subset is a strong system, since resolving the truth of statements in S must be at least as hard as answering the questions in Q. Conversely, to the extent that the sentences in S are subject to analysis and resolution by some effective computational procedure, its descriptive strength is limited. Of course, it is a fundamental result of mathematical logic that the system of predicate logic as a whole admits no decision procedure. Propositional logic is a subsystem that is susceptible to an exhaustive decision procedure, and was therefore once viewed as computationally trivial in the sense of recursion theory, if somewhat puzzling from the viewpoint of practical computation. In the light of modern computational complexity theory the satisfiability problem for this system is now seen to be a 'hard core' problem for the subrecursive class NP. That is, propositional satisfiability is a complete problem for the class of all problems that can be solved by nondeterministic Turing machines that operate in a time bound polynomial in their input size, just as satisfiability in predicate logic is a complete problem for class of recursively enumerable (r.e.)

sets. The possibility of quantification in the logical language makes the difference in the difficulty of the decision procedure; this is consistent with our intuition about the semantic significance of quantification as a logical vehicle.

Throughout the subsequent discussion, unless otherwise specified, we mean by a 'formula' a closed first-order formula without identity or function signs, and by the 'decision problem' for a class of formulas the problem of determining whether an arbitrary given member of that class is satisfiable.

Monadic Predicate Calculus

Clearly any extension of propositional logic that falls short of full predicate logic must have a decision problem somewhere intermediate between (though perhaps not properly between) NP and r.e. In fact, natural intermediate examples have been identified. The quantified logic of monadic predicates only is complete for a class roughly one exponential up from NP, namely the problems solvable by nondeterministic Turing machines operating in time $c^{n/\log n}$ for some c [L80]. (The variable n always represents the length of the input in bits. The exact form of the exponent in this case arises because a formula of length n bits can contain only O(n/log n) distinct symbols, since no matter how the symbols are encoded as bit strings, to represent even a single occurrence of each of s symbols requires at least s log s bits in all. So the bound $c^{n/\log n}$ is actually exponential in the number of distinct symbols the formula contains.) An idea of the lower bound proof can be gotten by observing that with m = O(n/log n) monadic predicates B_0, ..., B_{m-1} and a monadic function sign f, in a formula of size O(m) one can stipulate that the universe has at least 2^m distinct elements. This is done by interpreting the terms 0, f(0), f(f(0)), ... as the integers 0, 1, ... 2^m-1, and using B_ix to mean that the i'th bit of the binary representation of x is a 1. The formula need simply describe the way the m bits of x+1 = f(x) depend on the n bits of x, i.e. the way the bits of an m-bit number change when counting in binary. Here is one set of conditions. (Bit 0 of an m-bit number is the most significant bit; bit m-1 the least significant bit.)

if bit m-1 of x is 0, then bit m-1 of f(x) is 1 and there is no carry into bit m-2.

if bit m-1 of x is 1, then bit m-1 of f(x) is 0 and there is a carry into bit m-2.

if bit m-2 of x is 0 and there is no carry into bit m-2, then bit m-2 of f(x) is 0 and there is no carry into bit m-3.

if bit m-2 of x is 0 and there is a carry into bit m-2, then bit m-2 of f(x) is 1 and there is no carry into bit m-3.

if bit m-2 of x is 1 and there is no carry into bit m-2, then bit m-2 of f(x) is 1 and there is no carry into bit m-3.

if bit m-2 of x is 1 and there is a carry into bit m-2, then bit m-2 of f(x) is 0 and there is a carry into bit m-3.

... (four conditions for each bit position m-3, m-4, ... 0).

Additional predicates (m-1 of them) are needed for the conditions 'there is a carry into bit i.' Given these O(m) conditions, for various elements of the universe all 2^m different combinations of the values of the B_i predicates must be realized, so the universe must have at least 2^m elements. Getting a formula that describes arbitrary computations of length 2^m, not just counting, is a technicality. In fact, rather than encoding Turing machine computations directly, it is easier to describe domino or tiling problems [LP81].

Explicit use of the monadic function sign f in this formula is incidental and can be eliminated if the function sign is viewed as an indicial (Skolem) function. The corresponding upper bound, i.e. the proof that any monadic formula without function signs can be decided nondeterministically in time $O(c^{n/\log n})$ for a suitable c, is a quantifier elimination argument of a basically classical flavor. (Here and below, when we claim to have 'corresponding' lower and upper exponential bounds, we do not claim that the bases of the exponentials are the same. And a 'complete' problem for a class such as nondeterministic $c^{n/\log n}$ time is, to be precise, complete for union over all c of nondeterministic $c^{n/\log n}$ time, with respect to logarithmic space reducibility.)

Extending the syntax a bit by allowing dyadic predicates as well as monadic ones increases the power of the language a great deal. It has long been known that the formulas with even just one dyadic predicate letter form an undecidable class. So classification along this one parameter only, degree of the relations, is rather coarse.

The $\exists^*\forall^*$ Class

Let us return to the propositional case and add quantification piecemeal. That is, we are now unconcerned about the predicates used in the logical language but insist on only restricted kinds of quantification. No quantification at all is essentially propositional logic. Purely existential and purely universal sentences are not much different from propositional sentences, and so fall into NP. The first interesting case is the class of sentences with prefixes of the form $(\exists x_1)\ldots(\exists x_e)(\forall y_1)\ldots(\forall y_a)$ for some e and a. (If e and a are allowed to range over all integers, we denote this class by $\exists^*\forall^*$. If e is unlimited but a is a fixed integer, we write $\exists^*\forall^a$ for this class, and so on.) An $\exists^e\forall^a$ sentence can be reduced to a propositional conjunction of e^a instances of the matrix, so such a formula can be tested nondeterministically in time c^n for some c, where n is its length in bits, by expanding it and then submitting the result to a satisfiability tester for the propositional calculus. (Both e and a are $O(n/\log n)$, so $e^a = O(c^n)$.) This turns out to be the lower bound as well: With a predicate letter of 2a or more arguments, one can talk about the successor relation between different a-tuples of the e digits named by the existentially quantified variables and thus 'count' in a way similar to that sketched above for the monadic class.

Yet sharper results can be gotten by considering the $\exists^*\forall^*$ prefix with specific numbers of $\exists$'s and $\forall$'s. The $\exists\forall^*$ prefix, with but one $\exists$, reduces again to the propositional case; with two $\exists$'s but an unlimited number of $\forall$'s, almost the full complexity of the $\exists^*\forall^*$ prefix is obtained (the exact complexity is $O(c^{n/\log n})$). A more interesting set of cases arises if the number of universal quantifiers is fixed. For example, consider the sentences of the form $\exists x_1\ldots\exists x_e\forall y_1\ldots\forall y_5 M$. Such a formula reduces to a propositional formula which is a conjunction of e^5 substitution-instances of M and can therefore be decided nondeterministically in time which is a fixed (roughly quadratic) polynomial in the length of this expansion. To be precise, a formula with prefix $\exists^*\forall^a$ can be decided by a nondeterministic Turing machine in time $O(n^{2a+2}/(\log n)^{2a-1})$. A corresponding lower bound can be obtained as well, by describing the computations of linear-time bounded nondeterministic a-dimensional iterative arrays, which were shown by Seiferas to be equivalent to nondeterministic n^a-time-bounded Turing machines. The best result known is that $\exists^*\forall^a$-satisfiability is hard for nondeterministic Turing machine time $(n/\log n)^{a-1}$ [LD81],

[D84].

This set of results is interesting for several reasons. First, it leaves the question of whether the actual complexity lies closer to n^a or to n^{2a}. We do not seem to have the tools to settle this, though the difficult constructions of Lynch [L82] may be relevant. Second, and more fundamentally significant, is the fact that these classes provide the only concrete examples we know of problems in NP which are hard for <u>specific</u> degrees of nondeterministic polynomial time. The existence of such problems is guaranteed by the nondeterministic time hierarchy theorems, but we know of no other natural examples where the degree corresponds to some natural numerical parameter of the problem.

<u>The $\exists^*\forall\exists^*$ and $\exists^*\forall\forall\exists^*$ Classes</u>

Classical investigations showed that the only other prefix classes for which the satisfiability problem is decidable are the $\exists^*\forall\exists^*$ and $\exists^*\forall\forall\exists^*$ classes; if the prefix cannot be fit into one of these forms then the decision problem for all sentences with that prefix is undecidable. For these two decidable prefixes as well a sharp classification is possible; the first class is hard for deterministic time $O(c^{n/\log n})$ and the second is complete for the corresponding nondeterministic time class [L80]. (We do not have a matching upper bound on the $\exists^*\forall\exists^*$ class; the best bound we know is deterministic time $O(c^{(n/\log n)^2})$ for a suitable c. See [F81], which presents a number of related results.) To get the lower bounds we use reductions out of automata theory. The reduction for the $\exists^*\forall\forall\exists^*$ class uses the same argument as was previously cited for the monadic class (so that even the <u>monadic</u> formulas with prefix $\exists^*\forall\forall\exists^*$ form a complete class for nondeterministic time $c^{n/\log n}$.)

The reduction for the $\exists^*\forall\exists^*$ class uses alternating pushdown automata, which have an acceptance problem complete for deterministic exponential time. This reduction uses the monadic indicial function signs associated with the trailing existential variables to represent the symbols of the pushdown alphabet of the automaton. A complete configuration of the machine can be described by its state, the position of the input head within the input string, and the contents of the pushdown store; such a configuration is represented by an atomic formula of the form Pt, where P is a monadic predicate representing the first two parameters and t is a term f(g(...(c)...)) representing the stack. The atomic formulas represent configurations from which the machine accepts (not, as is usual in such constructions, configurations reachable from the initial state). An existential branch from C to A or B is then represented by a formula of the form (A v B => C), which states that if either A or B represents an accepting configuration then so does C; a universal branch by a clause (A & B => C). A general description of the transition rules of the machine is obtained by quantifying universally over a variable x representing the portion of the stack below the top symbol, which does not affect the validity of any single transition.

Even these results on the decidable prefix classes can be sharpened a bit. For example, the $\exists^*\forall\exists$ class (only one trailing $\exists$) can be decided in polynomial space [F81] and hence is presumably strictly easier than the full $\exists^*\forall\exists^*$ class. This is because there is a c such that a formula with prefix $\exists^*\forall\exists^*$ is satisfiable only if the Herbrand expansion up to height c^n is satisfiable. (This is the set of formulas that results from substituting terms with functional nesting of c^n levels at most for the universal variable in the functional form.) If there is only a single monadic indicial function sign then a satisfying truth-assignment, if one exists, can be guessed in stages. First guess the truth-values of all atomic formulas of height 0; once guessed, these values can be forgotten, except for the constraints they put on the truth-values of atomic formulas of height 1; then guess the truth-values of atomic formulas of height 1, and forget them except for the constraints they put on atomic formulas of height 2; and so on. Only a polynomial amount of information need be retained at each stage since there are only a linear number of terms of each height, and if this process can be carried out for c^n stages then the formula is satisfiable. By Savitch's theorem [S70], this nondeterministic polynomial-space decision procedure can be turned into a deterministic polynomial-space procedure. The same argument does not work if there are two or more indicial function signs, since there are then an exponential number of terms of a given height.

A lower bound construction matching the upper bound of the last paragraph is also possible. We need merely construct a formula describing the computation of arbitrary deterministic linear-space bounded Turing machine. To do this, introduce a monadic predicate P_{ia} for every tape position i=1, ..., n and every symbol a of the tape alphabet, plus a monadic predicate S_k for each state k of the Turing machine and a monadic predicate H_i for each i=1, ..., n. An atomic formula $P_{ia}x$ signifies that at time x the symbol on tape square i is a, an atomic formula $S_k x$ that at time x the machine is in state k, and an atomic formula $H_i x$ that at time x the head is over tape square i. A polynomial-size formula can then be written describing the truth-values of the $P_{ia}x+1$, $S_k x+1$, and $H_i x+1$ in terms of the $P_{ia}x$, $S_k x$, and the $H_i x$. Thus satisfiability for $\exists^*\forall\exists$ formulas is complete for polynomial space.

Herbrand, Krom, and Horn Formulas

These constructions and algorithms, plus a few specialized arguments for extreme cases, complete the program of determining the complexity of sentences classified by the form of their quantification only. A fair amount of interest has been focussed, however, on sentences specialized in another way, namely by their truth-functional structure. Specifically, we now report in summary fashion on what is known about Herbrand, Horn, and Krom formulas of the predicate calculus. These are special classes of formulas in conjunctive normal form. Of course, every formula is equivalent to a CNF formula with the same prefix, so simply restricting to CNF does not yield any simplification of the decision problem. An Herbrand formula is one whose matrix is a conjunction of literals (a literal is an atomic formula or the negation of an atomic formula). A Horn formula is one whose matrix is a conjunction of Horn clauses, where a Horn clause is a disjunction of literals, of which all but possibly one are negative literals. Put another way, a Horn clause is an implication of the form $(A_1 \& \ldots \& A_n \Rightarrow B)$, where $A_1, \ldots, A_n$, and B are atomic and either the antecedent or the consequent may be missing. A Krom formula is a conjunction of Krom clauses, where a Krom clause is a disjunction of at most two literals.

It is easy to see that an Herbrand formula is satisfiable if and only if no pair of complementary literals is has unifiable atomic formulas. Dwork et al. [DKM84] have recently shown that unification of first-order terms is complete for P, and it follows almost immediately that satisfiability of Herbrand formulas is also complete for P.

The special characteristics of Krom and Horn formulas with respect to computational complexity were noticed some time ago in the propositional case. In his original NP-completeness paper [C71], Cook observed that Krom formulas can be decided in polynomial time; one way to see this is to notice that the resolvent of two Krom clauses is a Krom clause and that fewer than $4n^2$ Krom clauses can be constructed from n propositional variables and their complements. This observation was sharpened by Jones et al. [JLL76], who showed that the unsatisfiability problem for propositional Krom formulas is complete for nondeterministic logarithmic space. The lower bound can be seen by noticing that a log-space bounded Turing machine has but a polynomial number of configurations, and that if each configuration is represented by a propositional variable and each possible transition by an implication between such variables then the variable representing a final configuration is derivable from the variable representing the initial configuration just in case a final configuration is reachable from the initial configuration. On the other hand the decision problem for Horn formulas of the propositional calculus is complete for deterministic polynomial time; this was shown by Jones and Laaser [JL76], again by a fairly straightforward reduction. Given a polynomial time-bounded Turing machine and an input string, create a propositional variable for each combination of time step, tape square, and tape symbol (to represent the contents of the tape); each combination of time step and tape square (to represent the position of the head); and each combination of time step and state (to represent the machine state). It is then easy to write a formula of polynomial size describing how the variables for time t+1 depend on those for time t. The determinism of the machine makes the result a Horn formula; the state, head position, and tape contents at time time t+1 are *uniquely* determined by those at time t.

An interesting exercise is to find a construction giving Cook's NP-hardness result for the propositional calculus which automatically yields the P-hardness of propositional Horn satisfiability when the machines encoded are deterministic.

The upper bounds for Krom and Horn formulas are also easy. For example, a propositional Horn formula is satisfiable if and only if the set of propositional variables it implies is a model for it. This set can be enumerated in polynomial time by a procedure known as the chase: start with the propositional variables which are conjuncts of the formula, and add to this set a variable B whenever $A_1, \ldots, A_n$

have been obtained and (A_1 & ... & A_n => B) is a conjunct of the formula. To check satisfaction under this truth-assignment it suffices to verify that no conjunct consists wholly of the negations of variables from this set.

In quantificational logic both the Krom and the Horn formulas form undecidable classes with respect to satisfiability, and yet there is a sense in which these truth-functional restrictions simplify the decision problem. For Horn formulas the $\forall\exists\forall$ class is decidable; otherwise, each undecidable prefix class of unrestricted formulas is also undecidable for Horn formulas. For Krom formulas, both the $\forall\exists\forall$ and the $\exists^*\forall^*\exists^*$ classes are decidable; the $\exists\forall\exists\forall$, $\forall\exists\exists\forall$, $\forall\exists\forall\forall$, and $\forall\forall\exists\forall$ classes are undecidable; and the decidability of the $\forall\exists\forall\exists^*$ class, and the $\forall\exists\forall\exists^k$ classes for each $k \geq 1$, remain an open problem.

Thus new decidable classes emerge when truth-functional restrictions are imposed. Likewise, some problems that were already decidable become easier. Most of the results that follow are from [DL84] and [D84].

Monadic Krom and Horn Formulas

For example, let us restrict the monadic predicate calculus to Horn or Krom formulas. Whereas the monadic class is complete for nondeterministic time $c^{n/\log n}$, the Horn subclass is complete for deterministic time $c^{n/\log n}$. The lower bound is by exactly the construction used above for the $\exists^*\forall\exists^*$ prefix class; that construction happens to produce a Horn formula. The upper bound uses a refinement of the method given for the full monadic class.

On the other hand, the Krom subclass of the monadic predicate calculus is complete for deterministic polynomial time. The lower bound uses the same construction as given above for the $\exists^*\forall\exists^*$ class, but without universal states; thus instead of alternating pushdown automata, only nondeterministic pushdown automata can be represented, and their acceptance problem is complete for P. The upper bound uses a specialized version of the resolution principle, similar to that used for the $\exists^*\forall^*\exists^*$-Krom class described below. Application of this principle yields from a monadic Krom formula F an equivalent formula F', also Krom, with the same variables and predicate letters as F, such that F' is satisfiable if and only if its matrix is truth-functionally consistent. Since F' has size polynomial in the size of F and is Krom, F' can be tested for satisfiability in time polynomial in the size of F.

$\exists^*\forall^*$ Krom and Horn

Now let us consider the prefix restrictions of the Krom and Horn formulas. Let us start by looking at the $\exists^*\forall^*$ prefix class, which we observed was complete for nondeterministic exponential time. If we restrict the matrix to be a Horn formula, the expected happens: just as in the propositional case the decision problem drops back from NP to P, in this case it drops back from nondeterministic c^n time to deterministic c^n time. (This and several other of the results we mention were noticed by Plaisted [P].) If we look at Krom formulas with this prefix, then just as in the propositional case the decision problem simplifies from NP to nondeterministic logarithmic space, in this quantified case it simplifies from nondeterministic c^n time to polynomial space. Both the upper and the lower bounds for both of these arguments are very straightforward. For example, given a polynomial space bounded Turing machine and an input string, we can construct a corresponding $\exists^*\forall^*$ Krom formula as follows: Choose a predicate letter with a number of arguments equal to the space bound and write, for each tape position, a few Krom clauses describing the applicable transitions when the head is in the vicinity of that tape position. Constants play the role of the tape symbols themselves, and the preservation of the contents of the tape squares distant from the tape head can be succinctly described using a universally quantified variable in each such position. To get the lower bound for the Horn formulas with this prefix, use the same construction for an alternating linear-space bounded Turing machine. With the same kinds of clauses as were used above in the case of alternating pushdown automata, we obtain a linear-sized $\exists^*\forall^*$ Horn formula after distributing (A v B => C) as (A => C) & (B => C), and the result follows since alternating linear space is the same as deterministic exponential time.

If the number of universal quantifiers is held to a fixed constant a, the $\exists^*\forall^a$-Horn class is solvable in polynomial time (in time $O(n^{2a+2})$) and is hard for a specific degree of polynomial time (for deterministic time $O(n^{a/2-1})$). (Because we do not have a deterministic analogue to the result of Seiferas on nondeterministic iterative arrays mentioned earlier, the gap between the lower and upper bounds is quartic rather than quadratic.) Thus these classes have a role somewhat like that of the games of [AIK81]; they provide examples of problems hard for specific deterministic polynomial time classes. Unlike these results on the $\exists^*\forall^a$-Horn classes, however, the results of [AIK81] apply only to one-tape Turing machines. On the other hand, $\exists^*\forall^a$-Krom is for each fixed a complete for nondeterministic logarithmic space.

This leaves the decidable $\forall\exists\forall$-Krom and $\exists^*\forall^*\exists^*$-Krom classes. By an intricate argument involving much manipulation of the presentations of semilinear sets of integers, $\forall\exists\forall$-Krom can be shown to be solvable in nondeterministic logarithmic space, i.e. it is in a sense no harder than the class of propositional Krom formulas. We refer to [DL84] and [D84] for a presentation of this proof and the next one.

The class $\exists^*\forall^*\exists^*$-Krom is a complete problem for deterministic exponential time. (The original decidability proof is due to Maslov [M64].) Both the lower and the upper bounds are nontrivial and interesting. The algorithm used to obtain the upper bound is based on a carefully controlled variant of the resolution procedure, apparently first used in a similar context by Joyner [J76]. In essence, formation of resolvents is restricted to cases in which the resulting clause does not have nested function signs. There are only exponentially many of such clauses and so the procedure halts in exponential time.

To get a lower bound of deterministic exponential time for $\exists^*\forall^*\exists^*$-Krom we encode the computations of alternating linear space Turing machines; this will establish the result since alternating linear space is the same as deterministic exponential time. Recall from the discussion above that nondeterministic linear space is reducible to $\exists^*\forall^*$-Krom satisfiability: the universal variables play the role of tape positions, the existential variables the role of tape symbols, and there are a few Krom clauses for each transition the Turing machine can make. The question then remains of how to use the final existential variable (one is enough) of an $\exists^*\forall^*\exists$ formula to encode universal branching. If there are n+1 universal variables, the indicial function f corresponding to the final existential variable has arity n+1. A term with outermost function sign f can then be used to represent an entire configuration of length n (by a sequence of n constants) and also have a similar term with function sign f as the last argument. A deeply nested term of this type can be used to represent a <u>stack</u> of configurations. Let us assume that in our alternating Turing machine each universal state has but two successors. Then in the encoding formula the single n+1-place predicate letter R is to have the following interpretation: $Rt_1 \ldots t_n s$ holds if $t_1 \ldots t_n$ represents a configuration reachable from the initial configuration in such a way that s represents a stack of universal configurations whose right children have yet to be explored. In other words, our formula will describe a backtrack search of the configuration space which proceeds nondeterministically through existential configurations and which stacks a universal configuration when exploring its left child. Provided that such a search actually reaches an accepting configuration along some path, it backs up, retrieves the most recently stacked universal configuration, and explores its right child. When <u>that</u> search ends successfully, the next most recent universal configuration is restored off the stack, and so on. The formula is written to assert that such a search cannot end with an empty stack, and this condition will be unsatisfiable if and only if the alternating Turing machine accepts from the initial configuration. The whole trick is to see that if the universal variables are $y_1, \ldots, y_{n+1}$ and the existential variable is x and $y_1 \ldots y_n$ represents a configuration and y_{n+1} a stack, then a clause of the form $R_1 y_1 \ldots y_{n+1} \Rightarrow R_2 y_1 \ldots y_n x$ represents a pushing operation.

Many details have to be taken care of to obtain a tight result. In fact, however, the use of initial existential variables to represent tape symbols can be avoided by a suitable encoding, and only a single final existential variable is needed: $\forall^*\exists$-Krom is complete for deterministic exponential time.

Conclusion

Ideas similar to some of those presented here have been applied to areas of computer science where predicate logic provides a useful model. For example, [CLM] contains some complexity results for relational database dependencies of essentially the same ilk as some of the proofs presented here. Shapiro [S82] has analyzed the complexity of Prolog programs, viewing them as Horn formulas which can be used for describing Turing machine computations. The various positive results for quantificationally restricted Horn formulas may be of further interest in the analysis of Prolog programs.

The figure below summarizes the containment and relative complexity of some of the classes mentioned here.

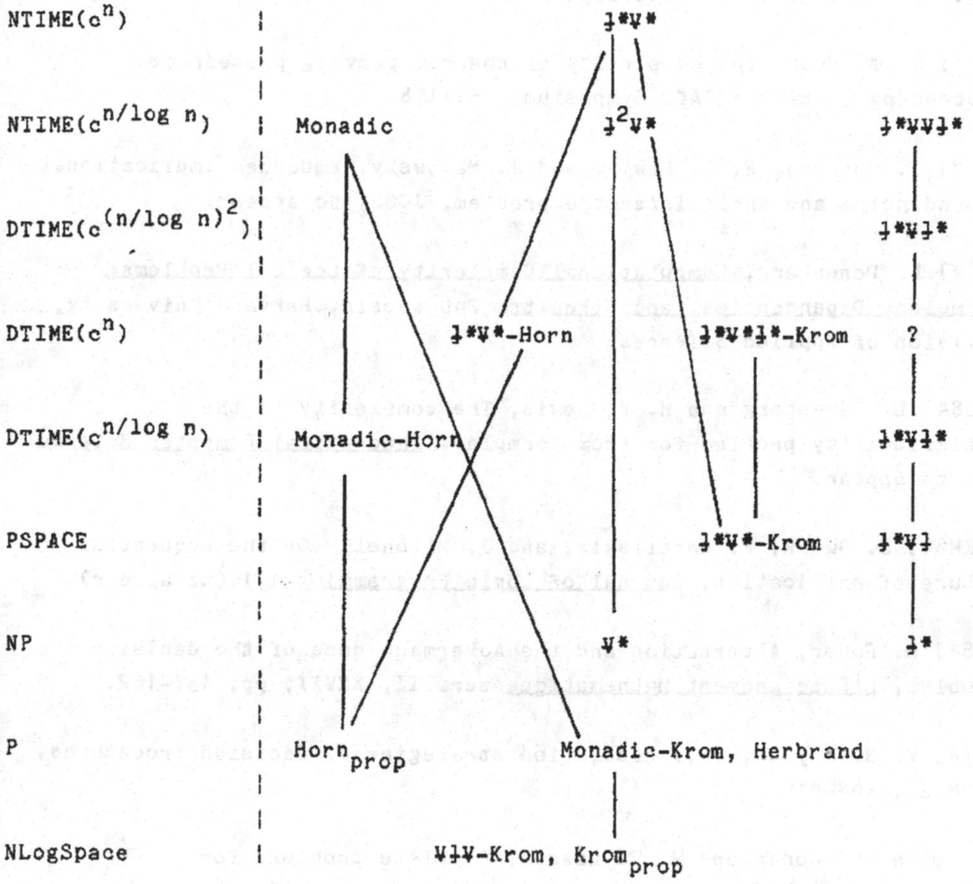

REFERENCES

(Notation: Paper [XYZ77] was published in 1977.)

[AIK81] A. Adachi, S. Iwata, and T. Kasai, Low level complexity for combinatorial games, Proceedings of 13th SIGACT Symposium, 228-237.

[C71] S. A. Cook, The complexity of theorem-proving procedures, Proceedings, Third SIGACT Symposium, 151-178.

[CLM] A. Chandra, H. R. Lewis, and J. Makowsky, Embedded implicational dependencies and their inference problem, JCSS, to appear.

[D84] L. Denenberg, Computational Complexity of Logical Problems: Formulas, Dependencies, and Circuits, PhD thesis, Harvard University, Division of Applied Sciences.

[DL84] L. Denenberg and H. R. Lewis, The complexity of the satisfiability problem for Krom formulas, Theoretical Computer Science 30, to appear.

[DKM84] C. Dwork, P. Kanellakis, and J. Mitchell, On the sequential nature of unification, Journal of Logic Programming 1,1 (to appear).

[F81] M. Furer, Alternation and the Ackermann case of the decision problem, L'Enseignement Mathematique ser. II, XXVII, pp. 137-162.

[J76] W. H. Joyner, Jr., Resolution strategies as decision procedures, JACM 23, 398-417.

[JL76] N. D. Jones and W. T. Laaser, Complete problems for deterministic polynomial time, Theoretical Computer Science 3, 105-117.

[JLL76] N. D. Jones, Y. E. Lien, and W. T. Laaser, New problems complete for nondeterministic log space, Mathematical Systems Theory 10, 1-17.

[L80] H. R. Lewis, Complexity results for classes of quantificational formulas, JCSS 21, 317-353.

[L82] J. F. Lynch, Complexity classes and theories of finite models, Mathematical Systems Theory, to appear.

[LD81] H. R. Lewis and L. Denenberg, A hard problem for NTIME(n^d), Proceedings, 19th Allerton Conference on Control, Communication, and Computing.

[LP81] H. R. Lewis and C. H. Papadimitriou, Elements of the Theory of Computation, Prentice-Hall Publishing Company.

[M64] S. Ju. Maslov, An inverse method of establishing deducibilities in the classical predicate calculus, Soviet Mathematics Doklady 5, 1420-1424.

[P] D. A. Plaisted, Complete problems in the first-order predicate calculus, manuscript.

[S70] W. J. Savitch, Relations between deterministic and nondeterministic tape complexities, JCSS 7, pp. 177-192.

[S82] E. Y. Shapiro, Alternation and the computational complexity of logic programs, Proceedings of First International Logic Programming Conference, 154-163.

SUBRECURSIVE HIERARCHIES VIA DIRECT LIMITS

E.C. Dennis-Jones and S.S. Wainer
Manchester University Leeds University

Introduction

The subrecursive classification problem is to find 'natural' ordinal assignments for (classes of) recursive functions, which reflect as accurately as possible their computational complexity.

The most common approach is to first view the problem in reverse, associating a hierarchy of increasing functions f_α with ordinals α so that f_α dominates f_β whenever $\beta < \alpha$. Then the complexity of a function h is measured by the least α such that h is computable within f_α-bounded time or space. The naturalness of such an assignment depends upon the acceptability of f_α as a 'functional representation' of ordinal α.

A typical example - and one which provides useful classifications of many subrecursive classes - is the transfinitely extended <u>Grzegorczyk Hierarchy</u>, first developed to level ω^ω by Robbin and then more generally to level ε_0 and beyond by Löb and Wainer [4, 7] and Schwichtenberg [6]. Details vary inessentially from author to author but the basic scheme is

$$F_0 = \text{initial function, e.g. exponential}$$
$$F_{\alpha+1} = \mathrm{It}(F_\alpha)$$
$$F_\lambda = \text{Diagonal } (F_{\lambda_x})_{x<\omega}$$

where It is some suitable iteration functional and $(\lambda_x)_{x<\omega}$ is a chosen fundamental sequence to limit λ. This is a standard kind of recursion-theoretic construction, motivated by the fact that the iteration functional plays an analogous role (w.r.t elementary recursiveness) to that of the jump operator in the generation of the hyperarithmetic hierarchy. However, the F_α-hierarchy is unsatisfactory on two counts:

(i) it is not sufficiently refined - $h_1(x) = F_\alpha(x^x)$ is clearly more complex than $h_2(x) = F_\alpha(x^3)$ and yet they will both appear at level $\alpha + 1$.

(ii) There seems to be no good mathematical reason for accepting F_α as a natural 'representation' of ordinal α.

If, on the other hand, one merely demands that a hierarchy should be as *refined* as possible - ignoring for the moment any consideration of its potential usefulness - then there is another, very obvious candidate: namely, the Slow Growing Hierarchy:

$$G_0 = \text{constant } 0$$

$$G_{\alpha+1} = G_\alpha + 1$$

$$G_\lambda = \text{Diagonal } (G_{\lambda_x})_{x<\omega}, \text{ i.e. } G_\lambda(x) = G_{\lambda_x}(x).$$

Furthermore, this seems to give a quite natural assignment of functions to ordinals, since diagonalisation is the functional counter part to ordinal supremum in the following sense: if the fundamental sequence $(\lambda_x)_{x<\omega}$ is chosen so that $x \le y \Rightarrow G_{\lambda_x}(y) \le G_{\lambda_y}(y)$, as will normally be the case (see later), then G_λ is the *least* function h (w.r.t. domination) such that $x \le y \Rightarrow G_{\lambda_x}(y) \le h(y)$.

Our aims here are in §1 to give a mathematically convincing reason why the G_α's can genuinely be regarded as natural representations of ordinals α, and in §2 as illustration, to show that the Slow Growing Hierarchy eventually produces an extremely refined form of the Grzegorczyk Hierarchy in such a way that the F_α's themselves turn out to represent certain 'large' Bachmann ordinals. In fact, G collapses the Bachmann Hierarchy of ordinal functions onto a version of the Grzegorczyk Hierarchy, thus providing a more delicate measure of the complexity of the F_α's.

The underlying ideas in §1 and indeed many of the results in §2, are already to be found - either explicit or implicit - in Girard [2] though in a different form and within a more elaborate framework. Our more down-to-earth approach, following on from [1], [8], is perhaps more closely related to Jervell [3].

§1. Direct Limit Representation of Ordinals

Girard's theory of dilators and their associated denotation systems is based on the simple observation that any ordinal can be represented as the direct limit of the partially ordered system of inclusions between all its finite subsets. For a countable ordinal one can go a step further and extract direct systems of finite embeddings which are ordered by the integers N and can therefore be coded in a canonical fashion by number-theoretic functions! It is this representation of *countable* ordinals which we shall develop.

The direct system one obtains for a given countable limit ordinal will depend upon the choice of a fundamental sequence for it, and will be built up inductively. So we need to consider ordinals not simply as set-theoretic objects, but rather as well-founded tree-structures which specify the particular choices of fundamental sequences used in their generation. This means that there will be many such structures corresponding to the same set-theoretic ordinal and that when thinking of an ordinal, one must have in mind a particular way in which it is built up.

<u>Definition 1</u>. The set Ω of countable *ordinal structures* or *tree-ordinals* consists of the infinitary terms generated inductively by:

(i) $0 \in \Omega$

(ii) $\alpha \in \Omega \Rightarrow \alpha +_0 1 \in \Omega$

(iii) $\forall x \in N(\alpha_x \in \Omega) \Rightarrow (\alpha_x)_{x\in N} \in \Omega$.

Lower-case Greek letters $\alpha,\beta,\gamma,\delta,\lambda$ henceforth denote tree-ordinals and λ will always denote a 'limit' i.e. $\lambda = (\lambda_x)_{x\in N}$. Each tree-ordinal α is a 'notation' for a set-theoretic ordinal $|\alpha|$ defined by $|0| = 0, |\alpha +_0 1| = |\alpha| + 1$ and $|(\alpha_x)_{x\in N}| = \sup_{x\in N} |\alpha_x|$. Thus, we will often write $\sup \alpha_x$ to denote a sequence $(\alpha_x)_{x\in N} \in \Omega$. Ω is partially-ordered by the obvious tree-ordering $\prec$ which is just the transitive closure of $0 \preccurlyeq \alpha$, $\alpha \prec \alpha +_0 1$ and $\alpha_n \prec \sup \alpha_x$ for each n.

<u>Definition 2</u>. Make Ω into a category by taking as morphisms the *embeddings* $g:\alpha \to \beta$, i.e. $\gamma \preccurlyeq \delta \prec \alpha \Leftrightarrow g(\gamma) \preccurlyeq g(\delta) \prec \beta$.

<u>Definition 3</u>. (a) A system $(\alpha_x, g_{xy})_{x<y\in N}$ of morphisms $g_{xy}:\alpha_x \to \alpha_y$ is *directed* if

$$x < y < z \Rightarrow g_{xz} = g_{yz} \circ g_{xy}.$$

(b) A direct system $(\alpha_x, g_{xy})_{x<y\in N}$ has $(\alpha, g_x)_{x\in N}$ as a *direct limit*, written

$$(\alpha, g_x)_N = \lim_{\to}(\alpha_x, g_{xy})_N$$

if (i) $g_x:\alpha_x \to \alpha$

(ii) $x < y \Rightarrow g_x = g_y \circ g_{xy}$

(iii) given any other $(\alpha', g'_x)_{x\in N}$ satisfying (i) and (ii) w.r.t.

$(\alpha_x, g_{xy})_N$ there is a unique $h:\alpha \to \alpha'$ such that for all $x \in N$, $g'_x = h \circ g_x$.

Note 1. Although a direct system $(\alpha_x, g_{xy})_N$ may in general have many different (though isomorphic) direct limits $(\alpha, g_x)_N$ in Ω, the ordinal $|\alpha|$ is *uniquely* determined (since the existence of a morphism $h:\alpha \to \alpha'$ implies that $|\alpha| \leq |\alpha'|$).

Definition 4. For each $\alpha \in \Omega$ and each $x < y \in N$, define the integer n_x^α and the maps $g_{xy}^\alpha, g_x^\alpha$ with domain n_x^α by the recursions

$$n_x^0 = 0 \qquad g_{xy}^0 = \emptyset \qquad g_x^0 = \emptyset$$

$$n_x^{\alpha+1} = n_x^\alpha + 1 \qquad g_{xy}^{\alpha+1} = g_{xy}^\alpha \cup \{<n_x^\alpha, n_y^\alpha>\} \qquad g_x^{\alpha+1} = g_x^\alpha \cup \{<n_x^\alpha, \alpha>\}$$

$$n_x^\lambda = n_x^{\lambda x} \qquad g_{xy}^\lambda = g_{xy}^{\lambda_x} \qquad g_x^\lambda = g_x^{\lambda_x} .$$

Definition 5.(a) For each fixed x the "x-ordering" $\prec_x$ is the transitive closure of $\alpha \prec_x \alpha +_0 1$ and $\lambda_x \prec_x \lambda$.

(b) $\alpha[x] = \{\beta \mid \beta +_0 1 \preccurlyeq_x \alpha\}$.

In Ω we shall identify the integers with the tree-ordinals $0 +_0 1 +_0 1 +_0 \ldots +_0 1$. It is easy to see that the cardinality of $\alpha[x]$ is n_x^α and that $\alpha[x]$ is the range of g_x^α. Thus g_x^α can be considered as a morphism: $n_x^\alpha \to \alpha$ and g_{xy}^α can be considered as a morphism from n_x^α to n_y^α provided $g_{xy}^{\lambda_x}(\ ) < n_y^\lambda$ at limits λ.

Lemma 1. The following are equivalent for each $\alpha \in \Omega$:

(a) $(n_x^\alpha, g_{xy}^\alpha)_{x<y\in N}$ is a direct system and

$$x < y \Rightarrow g_x^\alpha = g_y^\alpha \circ g_{xy}^\alpha ,$$

(b) $x < y \Rightarrow \alpha[x] \subseteq \alpha[y]$.

Proof. The second part of (a) implies that range $(g_x^\alpha) \subseteq$ range (g_y^α) whenever $x < y$. But this is just (b) because range $g_x^\alpha = \alpha[x]$ for each x. Now assume (b). From the above definitions, we note that

$$g_{xy}^\alpha(n) = m \Leftrightarrow \exists \beta \in \alpha[x](n = n_x^\beta \;\&\; m = n_y^\beta)$$

$$g_x^\alpha(n) = \beta \Leftrightarrow \beta \in \alpha[x] \;\&\; n = n_x^\beta .$$

Then from (b) we have (i) $n_y^\beta < n_y^\alpha$ when $\beta \in \alpha[x]$ and so $g_{xy}^\alpha : n_x^\alpha \to n_y^\alpha$; (ii) if $x < y < z$ then for each $\beta \in \alpha[x] \subseteq \alpha[y]$,

$$g_{yz}^\alpha(g_{xy}^\alpha(n_x^\beta)) = g_{yz}^\alpha(n_y^\beta) = n_z^\beta = g_{xz}^\alpha(n_x^\beta),$$

hence $(n_x^\alpha, g_{xy}^\alpha)_{x<y\in N}$ is a direct system; and (iii) if $x < y$ then for each $\beta \in \alpha[x]$,

$$g_y^\alpha(g_{xy}^\alpha(n_x^\beta)) = g_y^\alpha(n_y^\beta) = \beta = g_x^\alpha(n_x^\beta),$$

so $g_x^\alpha = g_y^\alpha \circ g_{xy}^\alpha$.

□

Lemma 2. For each $\alpha \in \Omega$,

if $x < y \Rightarrow \alpha[x] \subseteq \alpha[y]$

and $\beta \prec \alpha \Rightarrow \exists x(\beta \in \alpha[x])$

then $(\alpha, g_x^\alpha)_N = \underset{\rightarrow}{\lim}(n_x^\alpha, g_{xy}^\alpha)_N$.

Proof. It only remains, from Lemma 1, to check that given any system of morphisms $g_x' : n_x^\alpha \to \alpha'$ with $g_x' = g_y' \circ g_{xy}^\alpha$ there is a unique $h : \alpha \to \alpha'$ satisfying $g_x' = h \circ g_x^\alpha$ for each x. Given any β α we have $\beta \in \alpha[x]$ for all but finitely many x. Choose any such x, so $\beta = g_x^\alpha(n_x^\beta)$, and then define $h(\beta) = g_x'(n_x^\beta)$. If $x < y$ and $\beta = g_x^\alpha(n_x^\beta) = g_y^\alpha(n_y^\beta) = g_y^\alpha(g_{xy}^\alpha(n_x^\beta))$ then $g_x'(n_x^\beta) = g_y'(g_{xy}^\alpha(n_x^\beta)) = g_y'(n_y^\beta)$ so h is well-defined and it must be unique.

□

Towards the converse of Lemma 2, suppose $(\alpha, g_x^\alpha)_N$ is a direct limit for the system $(n_x^\alpha, g_{xy}^\alpha)_N$. Then by Lemma 1 we certainly have $x < y \Rightarrow \alpha[x] \subseteq \alpha[y]$. Let $\alpha' = \sup \alpha_x'$ where α_x' is the top element of $\alpha[x]$ and define $g_x' : n_x^\alpha \to \alpha'$ by $g_x'(n_x^\beta) = g_x^\alpha(n_x^\beta)$ for each $\beta \in \alpha[x]$. Then there must be a unique $h : \alpha \to \alpha'$ with $g_x' = h \circ g_x^\alpha$ for each x. Clearly, h must be the identity on $\bigcup_x \alpha[x]$. So for each $\beta \prec \alpha$, $h(\beta) \prec \alpha'$ and therefore $h(\beta) \preccurlyeq \alpha_x' = h(\alpha_x')$ for some x. Since h is an embedding, we then have $\beta \preccurlyeq \alpha_x' \prec \alpha$. Now if we were to assume, inductively, the direct limit representation of α_x' and the converse of Lemma 2 for α_x', then in the case of $\beta \prec \alpha_x'$ there would be a $y > x$ such that $\beta \in \alpha_x'[y]$. But since $\alpha_x' \in \alpha[x] \subseteq \alpha[y], \beta +_0 1 \preccurlyeq_y \alpha_x' \prec_y \alpha$ and so $\beta \in \alpha[y]$. We have now proved:

Theorem 1. The following are equivalent for each $\alpha \in \Omega$:

(a) For every $\gamma \preccurlyeq \alpha$, $(\gamma, g_x^\gamma)_N = \underset{\rightarrow}{\lim}(n_x^\gamma, g_{xy}^\gamma)_N$.

(b) For every $\gamma \preccurlyeq \alpha$,

$$x < y \Rightarrow \gamma[x] \subseteq \gamma[y]$$

and

$$\beta \prec \gamma \Rightarrow \exists x(\beta \in \gamma[x]).$$

Note 2. Looking back at Definition 4, one sees that $(g^{\alpha}_{xy})_{x<y\in N}$ and $(g^{\alpha}_{x})_{x\in N}$ are completely determined by α and the sequence $(n^{\alpha}_{x})_{x\in N}$. Furthermore, it is obvious that $(n^{\alpha}_{x})_{x\in N}$ *is just the Slow-Growing function* G_{α}. In order to emphasise the 'pointwise-at-x' definition of G *we shall henceforth denote* G_{α} *instead by* $G(\alpha)$ *and its value at* x *by* $G_x(\alpha)$. Thus $G_x(\alpha) = n^{\alpha}_{x}$ and instead of writing

$$(\alpha, g^{\alpha}_{x})_N = \underset{\rightarrow}{\lim}(n^{\alpha}_{x}, g^{\alpha}_{xy})_N$$

we may simply write

$$\boxed{\alpha = \underset{\rightarrow}{\lim}\, G(\alpha)}$$

Note 3. The inductive generation of Ω imposes little structure on tree-ordinals other than that $\prec$ is a well-founded partial ordering. Condition (b) of Theorem 1 gives α a more ordinal-like structure - if $\beta_1 \prec \alpha$ and $\beta_2 \prec \alpha$ then $\beta_1, \beta_2 \in \alpha[x]$ for all but finitely many x, so $\beta_1 \preccurlyeq_x \beta_2$ or $\beta_2 \preccurlyeq_x \beta_1$. Therefore, α is well-ordered by $\prec$ and $\beta \prec \alpha \Rightarrow \beta +_0 1 \preccurlyeq \alpha$. However, if $\alpha = \sup \alpha_x$ it still need not be the case that $(\alpha_x)_{x\in N}$ is a *fundamental* sequence, i.e. $\alpha_0 \prec \alpha_1 \prec \alpha_2 \prec \ldots$. Related to this is a further problem - it still may not be the case that $G(\alpha)$ dominates each $G(\alpha_x)$, since we may have $n^{\alpha_x}_{y} \geq n^{\alpha}_{y}$ for infinitely many $y > x$. Clearly, we need to ensure that the sequences $(\lambda_x)_{x\in N} \preccurlyeq \alpha$ mesh together in some suitable way.

The direct-limit picture suggests a natural additional requirement on the way in which $\lambda = \sup \lambda_x$ is built up by the enumeration $\lambda[0] \subseteq \lambda[1] \subseteq \lambda[2] \subseteq \ldots$.

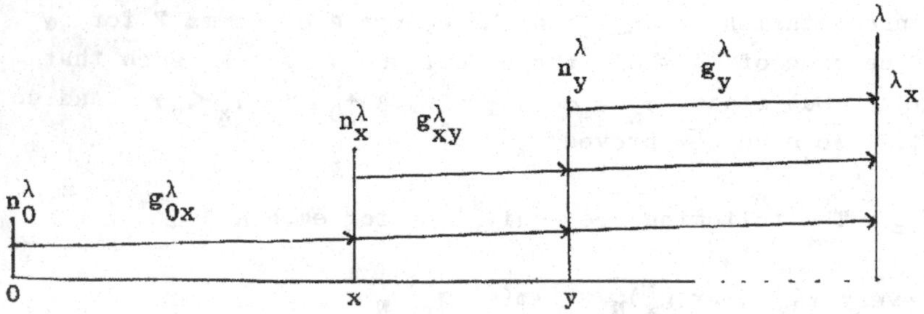

We shall demand that λ_x appears at the earliest possible stage, namely $y = x + 1$.

Definition 6. Write $\alpha = \underset{\rightarrow}{\mathrm{Lim}}\, G(\alpha)$ with a capital 'L' to mean that

(i) for every $\gamma \preccurlyeq \alpha$, $(\gamma, g_x^\gamma)_N = \underset{\rightarrow}{\lim}(n_x^\gamma, g_{xy}^\gamma)_N$ and

(ii) for every $\lambda = \sup \lambda_x \preccurlyeq \alpha$, $x < y \Rightarrow \lambda_x \in \lambda[y]$.

Definition 7. Call α a *nice* tree-ordinal if it satisfies condition (ii) above, i.e. if $\lambda = \sup \lambda_x \preccurlyeq \alpha$ then $x < y \Rightarrow \lambda_x +_0 1 \preccurlyeq_y \lambda_y$. (See Schmidt [5] for other conditions on fundamental sequences which are similar to, though not the same as, niceness.)

Lemma 3. If α is nice, then for every $\gamma \preccurlyeq \alpha$

$$x < y \Rightarrow \gamma[x] \subseteq \gamma[y]$$

and

$$\beta \prec \gamma \Rightarrow \exists\, x(\beta \in \gamma[x]).$$

Proof. By induction on $\gamma \preccurlyeq \alpha$ noting that if $\gamma = \sup \gamma_x \preccurlyeq \alpha$ then $x < y \Rightarrow \gamma_x \in \gamma[y] \Rightarrow \gamma_x[y] \subset \gamma[y]$.

□

Theorem 2. For each $\alpha \in \Omega$,

$\alpha = \underset{\rightarrow}{\mathrm{Lim}}\, G(\alpha)$ if and only if α is nice.

Proof. Immediate from Theorem 1, Lemma 3, and the Definitions.

□

Theorem 3. Each nice α is well-ordered by $\prec$ in such a way that

$$\beta \prec \gamma \preccurlyeq \alpha \Rightarrow G(\beta) \text{ is dominated by } G(\gamma).$$

Proof. We already know from Note 3 and Lemma 3, that $\prec$ well-orders α if it's nice. The rest follows by induction on γ for if $\gamma = \sup \gamma_x \preccurlyeq \alpha$ then since α is nice

$$x < y \Rightarrow \gamma_x[y] \subset \gamma[y] \Rightarrow n_y^{\gamma_x} < n_y^\gamma$$

and so $G(\gamma)$ dominates each $G(\gamma_x)$.

□

Thus, not only does a nice $\alpha \in \Omega$ provide a system of unique notations $\beta \preccurlyeq \alpha$ for the ordinals $\leq |\alpha|$, but in addition each such β naturally induces a direct-limit representation $G(\beta)$ of the ordinal $|\beta|$.

An obvious question at this point is whether *every* proper initial segment of the countable ordinals can be so represented? The answer is easily seen to be 'yes' .... .

First define addition $+_0$ on Ω by

$$\alpha +_0 0 = \alpha$$
$$\alpha +_0 (\beta +_0 1) = (\alpha +_0 \beta) +_0 1$$
$$\alpha +_0 \sup \lambda_x = \sup(\alpha +_0 \lambda_x)$$

Lemma 4. (i) $+_0$ is associative

(ii) $\gamma \preccurlyeq \alpha +_0 \beta \Rightarrow \gamma \preccurlyeq \alpha$ or $\exists \delta \preccurlyeq \beta (\gamma = \alpha +_0 \delta)$

(iii) $\delta \prec_y \beta \Rightarrow \alpha +_0 \delta \prec_y \alpha +_0 \beta$.

(iv) If α and β are nice, then $\alpha +_0 \beta$ is nice.

Now given any sequence $\alpha_0, \alpha_1, \alpha_2, \ldots$ of tree-ordinals, define

$$\Sigma\alpha_x = \sup_x(\alpha_0 +_0 \alpha_1 +_0 \ldots +_0 \alpha_x)$$

Lemma 5. If $\alpha_0, \alpha_1, \alpha_2, \ldots$ are all nice and $\neq 0$, then $\Sigma\alpha_x$ is nice.

Proof. By Lemma 4, since if $\alpha \neq 0$ is nice and $y > 0$, then $1 \preccurlyeq_y \alpha$.

□

Theorem 4. For every countable ordinal τ there is a nice $\alpha \in \Omega$ such that $|\alpha| = \tau$.

Proof. Clearly, 0 is nice and $\alpha +_0 1$ is nice if α is. If τ is a limit we can choose, inductively, a sequence $\alpha_0, \alpha_1, \alpha_2, \ldots$ of non-zero nice tree-ordinals such that $\tau = \sup|\alpha_x|$. But then, $\tau \leq |\Sigma\alpha_x|$ and Σ_x is nice by Lemma 5, so there is a nice $\beta \preccurlyeq \Sigma\alpha_x$ such that $\tau = |\beta|$.

□

Remark. Since tree ordinals are just countable well-founded trees with a certain structure, they can be coded in a standard way as reals. Thus it makes perfectly good sense to talk about 'recursive tree ordinals', 'Δ^1_2 tree ordinals', etc. It should be clear that if $\alpha_0, \alpha_1, \alpha_2, \ldots$ is a recursively-given sequence of recursive tree ordinals, then $\Sigma\alpha_x$ is also recursive. So a special case of Theorem 4 is that τ is a recursive ordinal if and only if there is a nice recursive $\alpha \in \Omega$ such that $|\alpha| = \tau$.

§2. The Bachmann and Grzegorczyk Hierarchies

Here we compute out some examples of

$$\alpha = \underset{\rightarrow}{\mathrm{Lim}}\, G(\alpha)$$

for certain well-known recursive ordinals α.

First some very simple examples:

Define multiplication and exponentiation on Ω by

$$\begin{array}{lll} \alpha.0 = 0 & & \alpha^0 = 1 \\ \alpha.(\beta +_0 1) = \alpha.\beta +_0 \alpha & & \alpha^{\beta+1} = \alpha^\beta.\alpha \\ \alpha.\lambda = \sup(\alpha.\lambda_x) & & \alpha^\lambda = \sup(\alpha^{\lambda_x}) \end{array}$$

Lemma 6. (i) If $\alpha \neq 0$ is nice then for each $y > 0$,

$$\delta \prec_y \gamma \Rightarrow \alpha.\delta \prec_y \alpha.\gamma .$$

Hence, if $\alpha \neq 0$ is nice and β is nice, then $\alpha.\beta$ is nice.

(ii) If α is nice and $\forall y > 0(2 \preccurlyeq_y \alpha)$ then for each $y > 0$

$$\delta \prec_y \gamma \Rightarrow \alpha^\delta \prec_y \alpha^\gamma .$$

Hence if α and β are nice and $\forall y > 0(2 \preccurlyeq_y \alpha)$ then α^β is nice.

Proof. By quite straightforward inductions over Ω. The additional conditions on α are needed to ensure that (i) $\alpha.\gamma +_0 1 \preccurlyeq_y \alpha(\gamma +_0 1)$, (ii) $\alpha^\gamma +_0 1 \preccurlyeq_y \alpha^{\gamma+1}$.

□

The following is also easily proved by inductions on $\beta \in \Omega$.

Lemma 7. (i) $G_x(\alpha +_0 \beta) = G_x(\alpha) + G_x(\beta)$

(ii) $G_x(\alpha.\beta) = G_x(\alpha).G_x(\beta)$

(iii) $G_x(\alpha^\beta) = G_x(\alpha)^{G_x(\beta)}$.

If we now take, as our standard representation of the first limit ordinal,

$$\omega = \sup(x+1) \in \Omega$$

then ω is nice, $\forall y > 0(2 \preccurlyeq_y \omega)$ and for each x, $G_x(\omega) = x + 1$. Thus, from Lemmas 6,7 and Theorem 2, we see immediately that *if* α *is in Cantor Normal Form to base* ω *then* $\alpha = \underset{\rightarrow}{\mathrm{Lim}}\, G(\alpha)$ *where* $G_x(\alpha)$ *is the result of replacing* ω *by* $x + 1$ *throughout the normal form.*

To see how the direct limit representation extends to higher ordinals, define the finite levels $\phi_n : \Omega \to \Omega$ of the Bachmann-Veblen Hierarchy on Ω by:

$$\begin{aligned} \phi_0(\alpha) &= \omega^\alpha \\ \phi_{n+1}(0) &= \sup_x \phi_n^x(1) \\ \phi_{n+1}(\beta +_0 1) &= \sum_x \phi_n^x(\phi_{n+1}(\beta) +_0 1) \\ \phi_{n+1}(\lambda) &= \sup_x \phi_{n+1}(\lambda_x) \end{aligned}$$

where $\phi^0(\alpha) = \alpha$ and $\phi^{x+1}(\alpha) = \phi(\phi^x(\alpha))$.

Define also $\phi_\omega(0) = \sup_x \phi_{x+1}(0)$.

Then, $|\phi_n(\alpha)|$ = Bachmann-Veblen $\varnothing_n(|\alpha|)$
= $|\alpha|$-th fixed point of $\varnothing_{n-1}$.

So, for example, $|\phi_1(\alpha)| = \varepsilon_{|\alpha|}$, $|\phi_2(\alpha)|$ is the $|\alpha|$-th critical epsilon-number, and $|\phi_\omega(0)|$ is the first 'prim-closed' ordinal.

Lemma 8. (i) $\phi_n(\beta) +_0 1 \preccurlyeq_y \phi_n(\beta +_0 1)$.

(ii) $\gamma \prec_y \beta \Rightarrow \phi_n(\gamma) \prec_y \phi_n(\beta)$.

(iii) $x < y \le z \Rightarrow \phi_n^x(1) +_0 1 \preccurlyeq_z \phi_n^y(1)$.

Lemma 9. (i) For each n, if α is nice, then $\phi_n(\alpha)$ is nice.

(ii) $\phi_\omega(0)$ is nice.

Lemma 10. For each n, x and each $\alpha \in \Omega$,

$$G_x(\phi_n(\alpha)) = f_{n,x}(G_x(\alpha))$$

where the functions $f_{n,x}$ are given by:

$$\begin{aligned} f_{0,x}(a) &= (x+1)^a \\ f_{n+1,x}(0) &= f_{n,x}^x(1) \\ f_{n+1,x}(b+1) &= \sum_{i \le x} f_{n,x}^i(f_{n+1,x}(b)+1). \end{aligned}$$

(again, f^x denotes the x-th iterate of f).

Notice that $f_{n+1,x}$ is defined from $f_{n,x}$ by what is essentially an *iteration*, i.e. $f_{n+1,x} = It_x(f_{n,x})$ where

$$It_x(f)(0) = f^x(1), \; It_x(f)(b+1) = \sum_{i \le x} f^i(It_x(f)(b)+1).$$

For each $x > 0$, $\{f_{n,x}\}_{n<\omega}$ is a version of the Grzegorczyk Hierarchy - every $f_{n,x}$ is primitive recursive and each primitive recursive function is dominated by some $f_{n,x}$.

From this family of hierarchies we can extract a 1-variable version $\{F_n\}_{n<\omega}$ by setting

$$F_n(x) = f_{n,x}(0).$$

Because of the uniformity in the definition of $f_{n,x}$ it is clear that each F_n is primitive recursive. Furthermore, it is not difficult to check that for each n and x, and all $a > x$

$$f_{n,x}(a) \le f_{n,a}(a) \le f_{n,a}(f^{a-1}_{n,a}(1) = f_{n+1,a}(0) = F_{n+1}(a).$$

Thus, if we place 'on top' of the F_n-hierarchy the function

$$F_\omega(x) = F_{x+1}(x) = f_{x+1,x}(0)$$

then F_ω is our version of the Ackermann function, and

$$G_x(\phi_\omega(0)) = G_x(\phi_{x+1}(0)) = F_\omega(x).$$

By Lemmas 9,10 and Theorem 2, we then get

<u>Theorem 5</u>. (i) For each n, $\phi_n(0) = \underset{\rightarrow}{\mathrm{Lim}}\, F_n$.

(ii) $\phi_\omega(0) = \underset{\rightarrow}{\mathrm{Lim}}\, F_\omega$.

Therefore, in terms of the Slow-Growing hierarchy G, we see that the ordinal complexity of the 'first' non-elementary function

$$F_1(x) = (x+1)^{(x+1)^{\cdot^{\cdot^{(x+1)}}}} \Big\} \; x \text{ times}$$

is its direct limit ε_0, and the complexity of the 'first non-primitive recursive function F_ω is the first primitive recursively closed ordinal $\phi_\omega(0)$.

The results of [1], [8] suggest how Theorem 5 extends to higher transfinite levels of the Bachmann and Grzegorczyk hierarchies, so that, for example, $\underset{\rightarrow}{\mathrm{Lim}}\, F_{\varepsilon_0}$ = Howard Ordinal. However, the hierarchies originally developed paid no attention to 'niceness' and need to be modified slightly in order to incorporate this property. The second author plans a more detailed treatment of these higher-level results.

References

[1] E.A. Cichon and S.S. Wainer 'The Slow-Growing and the Grzegorczyk Hierarchies', J. Symb. Logic 48 (1983) 399-408.

[2] J-Y. Girard, ' Π_2^1-Logic Part I', Annals. Math. Logic 21 (1981) 75-219.

[3] H.R. Jervell, 'Introducing Homogeneous Trees', Proc. Herbrand Symposium, Ed. J. Stern, North-Holland (1982).

[4] M.H. Löb and S.S. Wainer, 'Hierarchies of Number-Theoretic Functions I, II', Archiv. für math. Logik 13 (1970) 39-51, and 97-113.

[5] D. Schmidt, 'Built-Up Systems of Fundamental Sequences and Hierarchies of Number-Theoretic Functions', Archiv. für math. Logic 18 (1976) 47-53. Postcript 145-146.

[6] H. Schwichtenberg, 'Eine Klassifikation der ε_0-rekursiven Funktionen', Zeitschrift für math. Logik 17 (1971) 61-74.

[7] S.S. Wainer, 'A Classification of the Ordinal Recursive Functions Archiv. für math. Logik 13 (1970) 136-153.

[8] S.S. Wainer, 'The "Slow-Growing" Π_2^1 Approach to Hierarchies', to appear in Proc. AMS Summer Research Inst. on Recursion Theory, 1982, Eds. A. Nerode and R.A. Shore.

A STAR-FINITE RELATIONAL SEMANTICS FOR PARALLEL PROGRAMS

E. J. Farkas and M. E. Szabo *
Department of Mathematics
Concordia University
Montreal, Canada

0. Introduction

The essential difficulty in developing an intuitively plausible mathematical semantics for parallel programs derives from the fact that for a single input a parallel program may produce an infinite set of both finite and infinite computations. By translating such programs into equivalent non-deterministic sequential programs it is possible to structure these sets of computations as rooted trees and to study the properties of parallel programs in terms of the properties of such trees. The basic idea of <u>nonstandard</u> <u>computation theory</u> as developed in RICHTER and SZABO [1983, 1984] is to study <u>infinite computations</u> in terms of <u>star-finite</u> <u>computations</u> by extending the domain of computation from the standard model of Peano arithmetic to a nonstandard model and embedding infinite computations in computations that have the same formal properties as finite ones. The theory of star-finite sets and relations guarantees the existence of such computations. The purpose of this paper is to apply this technique to parallel programs by embedding infinite computation trees in star-finite ones and to show that we can use our intuition about finite trees, together with such nonstandard tools as the star-map and the Łos-Robinson transfer principle to develop a plausible

* The research of both authors is supported in part by the Natural Sciences and Engineering Research Council of Canada and by the Fonds F.C.A.C. of the Province of Quebec.

relational semantics for parallel programs. For an analysis of the intuitive nature of these techniques we refer to the paper by FARKAS and SZABO [1983].

The class of parallel programs considered in this paper is essentially the class of programs studied in OWICKI [1975] and OWICKI and GRIES [1978], with assignment statements restricted to successor and predecessor arithmetic as in RICHTER and SZABO [1983]. We show that every program in this class can be translated into an equivalent <u>bounded</u> star-finitely terminating program and on the basis of this result we give a nonstandard characterization of several of the properties of parallel programs considered in the definitive paper by ASHCROFT and MANNA [1971]. We then introduce the concept of the <u>graph</u> of a bounded parallel program and develop an <u>extensional</u> semantics for parallel programs in terms of their graphs. We conclude the paper with the presentation of a nonstandard generalization of finite automata and with the construction, for each bounded parallel program, of a star-finite non-deterministic automaton. We prove that every such automaton is canonically equivalent to a star-finite deterministic automaton recognizing the same language and take the next-state function of this deterministic automaton to be the <u>intensional</u> meaning of the given program. This semantics specializes to yield the interpretive models for parallel programs constructed in OWICKI [1975].

1. Parallel programs

For the purposes of this paper, we assume that $*\underline{N}$ is a fixed nonstandard extension of the standard model $\underline{N}$ of Peano arithmetic, and we assume that a program consists of a structured set of instructions designed to process vectors $(x_1,\ldots,x_n)$ of elements of $*\underline{N}$, for an arbitary but fixed index n. The precise definition of a program requires a set _T_ of _terms_ ranging over *_N_ and a set _B_ of Boolean test formulas defined over _T_.

1.1. Definition. The set _T_ of _arithmetic_ _terms_ is the smallest set containing the variables $x_1,\ldots,x_n$, the elements of the standard model _N_, and is closed under the operations (t+1) and (t-1).

1.2. Definition. The set _B_ of _Boolean_ _test_ _formulas_ is the smallest set containing the formulas (t=t') and (t<t'), for arbitary arithmetic terms t and t', and is closed under the operations (¬A) and (A&B).

1.3. Definition. A _bounded_ _test_ _formula_ is an expression (B/b), where B is a Boolean test formula and b is an element of *_N_.

1.4. Definition. An _assignment_ _statement_ is an expression of the form [x:=t], where x is one of the variables $x_1,\ldots,x_n$, and where t is either x, (x+1), or (x-1).

1.5. Definition. The set of _(unbounded_ _parallel)_ _programs_ is the smallest set satisfying the following conditions:

(1) The statement _null_ is a program.

(2) An assignment statement [x:=t] is a program.

(3) The statement (_begin_ P1;...;Pm _end_) is a program.

(4) The statement (_if_ B _then_ P1 _else_ P2) is a program.

(5) The statement (_while_ B _do_ P1) is a program.

(6) The statement (_cobegin_ P1//...//Pm _coend_) is a program.

(7) The statement (_await_ B _then_ P1) is a program,

provided that B is a Boolean test formula and P1, P2, ..., Pm are programs and are not _null_, that (7) occurs only inside a _cobegin_ statement, and that the program P1 in (7) is constructed only by means of (1)-(5). It is understood that the variables in the test formulas B in (4), (5), and (7) are among the variables in P1 and P2.

1.6. Definition. The set of bounded (parallel) programs is obtained from the set of unbounded parallel programs by replacing the Boolean test formulas B in (5) and (7) above by bounded test formulas (B/m).

The operational meaning of the unbounded programs is the usual one, whereas the programs (**while** (B/m) **do** P1) and (**await** (B/m) **then** P1) are interpreted as follows: Run program P1 as long (soon) as B is (becomes) true, but no more (later) than m times (steps after the beginning of the computation of the program in which the given statements occur). It is important for the work that follows that m ranges over $*\underline{N}$ and therefore may be also an infinite natural number. As mentioned in the Introduction, **cobegin** statements are executed as non-deterministic sequential statements.

2. Concurrent computations

Our semantics for parallel programs is based on the notion of a rooted finitely branching tree whose nodes consist of pairs (P,c), where P is a program and c is a vector $(c_1,\ldots,c_n)$ of elements of $*\underline{N}$. We refer to such trees as concurrent computations. Their definition requires the concept of a derived program.

2.1. Definition. The derived program P′ of a program P is defined inductively as follows:

(1) **null**′ is undefined.

(2) [x:=t]′ is **null**.

(3) (**if** B **then** P1 **else** P2)′ is P1′ if B is true and P2′ if B is false.

(4) (**while** B **do** P1)′ is (**while** B **do** P1) if B is true and **null**

if B is false.

(5) (__begin__ P1;...;Pm __end__)' is (__begin__ P1;...;Pm' __end__)';

(__begin__ P1;...;Pi;__null__ __end__)' is (__begin__ P1;...;Pi __end__)'.

(6) (__await__ B __then__ P1)' is (__await__ B __then__ P1) if B is false and P1' if B is true.

(7) (__cobegin__ P1//...//Pm __coend__)' consists of m programs obtained by priming each Pi in turn and keeping the remaining programs unchanged.

If the derivation process in (5) and (7) leads to the expressions (__begin__ P1 __end__), (__cobegin__ P1 __coend__), (__cobegin__ P1//__null__ __coend__), or (__cobegin__ __null__ //P1 __coend__), we replace them by P1 and thus obtain another well-formed program. Similarly, we replace (__cobegin__ __null__//__null__ __coend__) by __null__ to obtain a well-formed program.

__2.2.__ __Definition.__ The __derived__ __program__ P' of a bounded program P is defined as in 2.1, with the following modifications:

(5) (__while__ (B/m) __do__ P1)' is (__while__ (B/m-1) __do__ P1) if B is true and m > 0, and is __null__ if B is false or m = 0.

(7) (__await__ (B/m) __then__ P1)' is (__await__ (B/m-1) __then__ P1) if B is false and m > 0, and is P1' if B is true or m = 0.

Using the notion of a derived program, we can define the __next__ __nodes__ (P',c(P)) of a node (P,c) in a computation tree. If P is not a __cobegin__ statement and is not __null__, then the next node (P',c(P)) is unique. If P is __null__, there is no next node, and if P is (__cobegin__ P1//...//Pm __coend__) and none of the programs P1 to Pm has become __null__ during the process of computation, then (P,c) has exactly m next nodes. Otherwise (P,c) has fewer than m next nodes.

__2.3.__ __Definition.__ The __next__ __nodes__ of a node (P,c) in a computation tree are defined as follows:

(1) (null,c) has no next node.

(2) ([x:=x]',c([x:=x])) is (null,c).

(3) ([x:=x+1]',(..,c,..)(P)) is (null,(..,c+1,..)).

(4) ([x:=x-1]',(..,c,..)(P)) is (null,(..,c-1,..)).

(5) ((begin P1;...;Pm end)',c(P)) is (begin P1;...;Pm' end, c(Pm)) if Pm' is not null, and (begin P1;...;Pm-1 end,c) otherwise.

If m = 1, we put ((begin P1 end)',c(P)) = (P1',c(P)) if P1 is not null and (null,c) otherwise.

(6) ((while B do P1)',c(P)) is (while B do P1,c(P1)) if B is true and (null,c) otherwise.

(7) ((cobegin P1//P2 coend)',c(P)) denotes the pair of nodes ((cobegin P1'//P2 coend),c(P1)) and ((cobegin P1//P2' coend),c(P2)).

(8) ((while (B/m) do P1)',c(P)) is ((while (B/m-1) do P1), c(P1)),

etc.

For a program P containing the variables $x_1,\ldots,x_n$ and an input vector $c = (c_1,\ldots,c_n)$, we construct the computation tree T(P,c) for P at c as follows: At level 0 we put (P,c) as the root, and if (Q,d) is a node of level s and Q is not null, we connect it to all nodes of the form (Q',d(Q)) at level s+1. In this way we obtain the obvious rooted and finitely branching computation tree for P at the input c. In a terminating tree, all leaves are of the form (null,c) for some vector c. The concepts of a branch and of a maximal branch of a tree have their usual meaning.

Using the notion of a computation tree, we can now define an equivalence of programs. This relation compares the performance of programs only at standard inputs, i.e., for vectors $c = (c_1,\ldots,c_n)$

with c_i ranging over N, and considers only the standard parts of computation trees, i.e., disregards all nodes of the trees at infinite levels.

2.4. Definition. Two programs P and Q are equivalent at c if the standard parts of the computation trees T(P,c) and T(Q,c) coincide.

2.5. Definition. Two programs P and Q are equivalent if P and Q are equivalent at c for all standard c.

The following result is basic for our work:

2.6. Theorem (Upper bound theorem). Every parallel program is equivalent to a star-finitely bounded program.

The proof is a straighforward induction on programs. ▯

Since all bounded programs are, by definition, loop-free and contain only bounded await statements, they are star-finitely terminating and Theorem 2.6 therefore implies the following result:

2.7. Corollary (Termination theorem). Every parallel program is equivalent to a terminating program. ▯.

3. Properties of bounded parallel programs

We interpret each bounded parallel program $P = P(x_1,...,x_n)$ as a star-finite relation on $*\underline{N}^n$. The definition makes sense because P is bounded and because every branch of P is therefore star-finite. We call this relation the graph of P and denote it by graph(P). For any c in $\underline{N}^n$, we first use the computation tree T(P,c) to define the auxiliary notion of the local graph of P, denoted by graph(P,c):

<u>3.1. Definition.</u> graph(P,c) = { (c,y) | y is a leaf of T(P,c) }.

The <u>(global) graph</u> of P is then defined as follows:

<u>3.2. Definition.</u> graph(P) = the union of all local graphs of P.

We now define a number of properties of <u>bounded</u> parallel programs in terms of their computation trees, local graphs, and graphs, and we say that an unbounded parallel program has a certain property if one of its equivalent bounded programs has that property. The fact that this definition leads to well-defined properties of unbounded parallel programs will be self-evident.

<u>3.3. Definition</u>. A bounded program P <u>really terminates</u> if every T(P,c) has a finite branch.

<u>3.4. Definition</u>. A bounded program P is <u>partially correct</u> if some local graph of P contains only a single point.

<u>3.5. Definition</u>. A bounded program P is <u>really partially correct</u> if for some c, T(P,c) has a finite branch and graph(P,c) = { (c,y) } and y is standard.

<u>3.6. Definition</u>. A bounded program P is <u>totally correct</u> if graph(P) is functional.

<u>3.7. Definition</u>. A bounded program P is <u>really totally correct</u> if P is totally correct and is really partially correct for all c.

Next we compare these properties of parallel programs with the corresponding properties defined in ASHCROFT and MANNA [1971]. Their definitions translate into the present context as follows:

3.8. Definition (ASHCROFT and MANNA [1971]). A parallel program P is universally defined at c if all branches of T(P,c) are finite.

3.9. Definition (ASHCROFT and MANNA [1971]). A parallel program P is existentially defined at c if some branch of T(P,c) is finite.

3.10. Definition (ASHCROFT and MANNA [1971]). A parallel program P is partially universally correct (with respect to a predicate Q(x,y) on ($\underline{N}^n$ X $\underline{N}^n$)) if { (c,y) | (null,y) is a leaf on a finite branch of T(P,c) } is a subset of Q.

3.11. Definition (ASHCROFT and MANNA [1971]). A parallel program P is partially existentially correct (with respect to a predicate Q(x,y) on ($\underline{N}^n$ X $\underline{N}^n$)) if (c,y) is in Q for some leaf (null,y) of T(P,c).

3.12. Definition (ASHCROFT and MANNA [1971]). A parallel program P is partially determined at c if all finite branches of T(P,c) have the same leaves.

We note that 3.3 and 3.9 are equivalent in the sense that unbounded parallel programs really terminate if and only if they are existentially defined for all c. It is also immediate that 3.5 and 3.11 are equivalent with respect to the restriction of the graph of P to standard points. For terminating programs, Definition 3.4 corresponds to Definition 3.12 if we replace finite by star-finite since all branches of a computation tree of a bounded parallel program are star-finite. Moreover, 3.7 and 3.10 are equivalent in the sense that if we specialize the predicate Q(x,y) to the restriction of the graph of P to standard points and assume that graph(P) is functional, then an unbounded parallel program P is really totally correct if and only if it is partially universally correct with respect to Q(x,y). Definition 3.6 is significant for parallel programs precisely because of Corollary 2.7 and has no analogue in ASHCROFT and MANNA [1971] since infinite computations cannot be adequately dealt with in a standard setting. Definition 3.8, finally, is a special case of 3.3, yet is inappropriate for bounded parallel programs with <u>await</u> statements: The following program is clearly a really totally correct bounded program which is universally defined in the obvious sense and yet has a computation tree with an infinite branch at some standard point:

Let b and b' be infinite elements of *<u>N</u> and let P = (<u>cobegin</u> P1//P2 <u>coend</u>), with P1 = (<u>while</u> (0<x/b) <u>do</u> [x:=x-1]) and P2 = (<u>await</u> (x=0/b') <u>then</u> [y:=y+1]). If c = (x,y) = (1,2), for example, then T(P,c) contains the infinite branch

(P,c) -> (P,c) -> ... -> (P,c) -> (<u>null</u>,(0,1))

of length b'. This program should clearly satisfy the definition of a standard <u>universally</u> <u>defined</u> program.

If we replace P2 in P above by P3 = (<u>while</u> (0<x/b') <u>do</u> [y:=y+1]), we get a program P' which really terminates and should satisfy the definition of a nonstandard <u>universally</u> <u>defined</u> program.

<u>4. Star-finite automata</u>

In conclusion we show that the computational process itself can be incorporated into our semantics of parallel programs by thinking of star-finite computations of bounded parallel programs as words in a language accepted by a star-finite deterministic automaton. In the applications we have in mind, our words are star-finite analogues of the kinds of computation found in OWICKI [1975]. Our construction relies on a variety of properties of star-finite sets and relations. We assume, in particular, familiarity with the property of being an <u>internal</u> set or relation, and require the fact that for any infinite element k of $*\underline{N}$, the cube $[0,k]^n$ (which includes all of $\underline{N}^n$) is star-finite, and that the set $ß(K)$ of internal subsets of a star-finite set K is itself star-finite. In addition, we use the fact that star-finite unions of star-finite sets are star-finite, and that for any infinite element b of $*\underline{N}^n$, the set of internal sequences over $[0,k]^n$ of length no greater than b is star-finite. For a detailed discussion of these properties, we refer to STROYAN and LUXEMBURG [1976].

For any two disjoint star-finite sets K and S and any infinite element b of $*\underline{N}$, we define a star-finite deterministic and a star-finite non-deterministic automaton as follows:

<u>4.1. Definition.</u> A <u>star-finite</u> <u>deterministic</u> <u>automaton</u> over K and S is a system

$$\underline{A} = (K,\ S,\ \mu : (K \times S) \to K,\ q_0,\ F)$$

consisting of a nonempty star-finite set of states K, a star-finite input-output alphabet S, an internal next-state function μ, an element q_0 of K designated as the initial state of $\underline{A}$, and a star-finite subset F of K designated as the set of final states of $\underline{A}$.

For any fixed infinite element b of $*\underline{N}$, we let $Words_{<b}(S)$ be the set of star-finite sequences (<u>words</u>) over S of length no greater than b, and extend μ to a function $(K \times Words_{<b}(S)) \to K$ by putting $\mu(q,\emptyset) = q$ and $\mu(q,xa) = \mu(\mu(q,x),a)$ for each $x \in Words_{<b}(S)$ of length $< b$ and each $a \in S$. This definition makes sense by virtue of the Łos-Robinson transfer principle applied to the recursive definition of the next-state functions of finite automata. The symbol $\emptyset$ denotes the empty word. The verification that this extension is internal is routine.

4.2. Definition. A star-finite non-deterministic automaton over K and S is a system

$$\underline{A} = (K,\ S,\ \mu : (K \times S) \to \beta(K),\ q_0,\ F)$$

satisfying the conditions of 4.1.

We extend the given function μ to an internal function on the set $(\beta(K) \times \mathrm{Words}_{<b}(S))$ by defining $\mu(q,\emptyset) = \{ q \}$, $\mu(q,xa)$ as the union over all $\mu(p,a)$, with $p \in \mu(q,x)$, and put $\mu(\{ p_1,\ldots,p_r \},x)$ equal to the union of all $\mu(p_i,x)$. The fact that this extension is internal is easily verified.

We call a subset of the set $\mathrm{Words}_{<b}(S)$ a b-bounded language over S, and by applying the star-map and the Łos-Robinson transfer principle to the usual proof of the equivalence theorem for finite deterministic and non-deterministic automata (cf. HOPCROFT and ULLMAN [1969]) we obtain the following star-finite analogue:

4.3. Theorem (Equivalence theorem). A b-bounded language is accepted by a star-finite non-deterministic automaton if and only if it is accepted by a star-finite deterministic automaton. ▯

For any bounded parallel program $P = P(x_1,\ldots,x_n)$ we now construct a star-finite non-deterministic automaton Aut(P) which recognizes precisely the language consisting of the "computations of P". We then take the next-state function of the associated deterministic automaton Aut'(P) to be the structural meaning of P.

For any $c \in [0,k]^n$ and the corresponding computation tree $T(P,c)$ of P, we put

$K_c = \{ Q \mid (Q,x) \in T(P,c) \}$, and

$S_c = \{ x \mid (Q,x) \in T(P,c) \}$.

We let K be the union of the K_c and call its elements the program components of P, and let S be the union of the S_c and call its elements the input-output components of P. The fact that c ranges over the star-finite set $[0,k]^n$ guarantees that K and S are star-finite.

We let $q_0 = \{ P \}$ and $F = \{ \underline{null} \}$, and define $\mu(Q,x)$ to be the set of program components of the next nodes of (Q,x). If (Q,x) is not a node of $T(P,c)$, we put $\mu(Q,x) = \{ \underline{null} \}$. By combining these objects, we obtain the desired star-finite non-deterministic automaton

$$Aut(P) = \langle K, S, \mu : (K \times S) \to \beta(K), q_0, F \rangle.$$

An induction on bounded programs shows that the upper bound b in the definition of a b-bounded language can be taken large enough so that the language accepted by Aut(P) consists precisely of the sequences of input-output vectors determined by the maximal branches of the computation trees of P.

By letting $K' = \beta(K)$ and applying the Łos-Robinson transfer principle to the usual construction of the deterministic automaton referred to in 4.3, we obtain canonically the star-finite deterministic automaton

$$Aut'(P) = \langle K', S, \mu : (K' \times S) \to K', q'_0, F' \rangle$$

and take the extension

$$\mu : (K' \times Words_{<b}(S)) \to K'$$

of the next-state function of Aut'(P) to be the structural (intensional) meaning of P.

References

ASHCROFT, E.A. and MANNA, Z.

[1971] Formalization of properties of parallel programs, Machine Intelligence, 6, pp. 17-41.

FARKAS, E.J. and SZABO, M.E.

[1983] On the plausibility of nonstandard proofs in analysis, Dialectica. To appear.

HOPCROFT, J.E. and ULLMAN, J.D.

[1969] Formal languages and their relation to automata, Addison-Wesley, Reading, Mass.

OWICKI, S.

[1975] Axiomatic proof techniques for parallel programs, Ph.D. thesis, Cornell University.

OWICKI, S. and GRIES, D.

[1976] Verifying properties of parallel programs, Comm. ACM, 19, pp. 279-284.

RICHTER, M.M. and SZABO, M.E.

[1983] Towards a nonstandard analysis of programs, in: Nonstandard analysis - recent developments, A.E. Hurd (editor), Lecture Notes in Computer Science, 983, pp. 186-203.

RICHTER, M.M. and SZABO, M.E.

[1984] Nonstandard computation theory, Proceedings of the Colloquium on Algebra, Combinatorics, and Logic in Computer Science, Gyoer, Hungary. To appear.

STROYAN, K.D. and LUXEMBURG, W.A.J.

[1976] Introduction to the theory of infinitesimals, Academic Press, New York.

BETWEEN CONSTRUCTIVE AND CLASSICAL MATHEMATICS

Solomon Feferman 1/
Department of Mathematics
Stanford University
Stanford, CA 94305 USA

1. Introduction. In constructive mathematics restrictions are generally placed both on the objects studied and on the methods of proof which may be applied; these restrictions are dictated by a fundamentalist position as to the nature of mathematics. Quite opposed to this, classical mathematics also succeeds in arriving at constructively meaningful results with no such restrictions, though only on an *ad hoc* case-by-case basis. Here we propose a middle ground between the two in which restrictions are placed only on the objects studied, but in such a way that all results have direct constructive (computational) significance; this permits a systematic pursuit of constructivity without ideological constraints on methods of proof. A suitable formal framework for this is provided by the system T_0 for representing Bishop-style constructive mathematics (Feferman 1975 and 1979:hereinafter F 1979). The system T_0 has both recursive and classical models and each theorem of T_0 is thus a generalization of a classical theorem which has computational content. In this paper we only look at one example that cannot be explained adequately in ordinary constructive terms, and show how it can be handled in T_0. This is the work of Pour-El / Richards 1983, characterizing those linear operators on Banach spaces which preserve computability of elements. Many further (and more crucial) problems ought to be examined with respect to the program suggested here, so this is just a beginning 2/. The main facts we need to know about T_0 are reviewed in §2. The

1/ Research supported by a grant from the National Science Foundation.

2/ In the talk at Aachen on which this paper is based, I had mentioned two other problem areas - one having to do with categorical axiomatization of fundamental structures, and the second having to do with *prima-facie* impredicative class concepts such as the power class. I have not had a chance to treat these completely and intend to pursue them on another occasion. See also §4 below.

reformulation of the Pour-El/Richards work is given in §3. The paper concludes in §4 with some remarks on the potential computational value of work in the Bishop school from the present standpoint.

There are a number of reasons for pursuing this program and its particular form proposed here. Explanation of these would require extensive discussion and detailed comparison with the main systematic approaches which have been taken to constructivity 3/. The following is only intended to indicate some of the lines of thought involved.

Constructivist philosophy is persuasive in its denial of the platonistic view of mathematics as having for its subject matter independently existing abstract entities. What is less persuasive is the constructivist attempt to reduce mathematics to entirely subjective elements. I believe a coherent case can be made for the human source of mathematical conceptions resulting in intersubjective notions of objective character. Some of these concepts are well-determined to such an extent that questions of truth concerning them are recognized to be meaningful and definite (e.g. the natural numbers, finite graphs, etc.) This justifies laws of classical reasoning (e.g. the law of excluded middle (L.E.M.) and consequently proof by contradiction, least element principle etc.). Other concepts are only partly determined (e.g. <u>sets</u> of natural numbers, tree ordinals, etc.) but still have objective features. The application of classical logic to such may be considered problematic but not for <u>logically simple closure conditions</u> (e.g. that any two sets have an intersection, or that any sequence of tree ordinals has a supremum) 4/. Given a philosophical basis of the type indicated, one seeks a formal system which is recognized to be correct for concepts of constructed objects but does not necessarily restrict the means of reasoning employed.

3/ Such comparisons can be found in Troelstra 1977, Feferman 1979 and (at length) Beeson 1984a.

4/ Logically complicated closure conditions may involve essentially impredicative features, implicitly presuming a definite completed totality of all objects of the given kind, such as full 2nd order comprehension for sets of numbers. These are <u>prima-facie</u> problematic for the constructivist position, but not necessarily to be rejected out-of-hand.

The minimal requirement for constructivity is that all objects considered must be capable of being presented, e.g. functions are presented by rules, sets are presented by defining properties, etc. Moreover, this should be hereditary, i.e. the objects on (and to) which a function operates and the members of a set are all to be presented. The minimal requirement for a formal theory of constructive mathematics is that it should have an interpretation in which all objects considered are hereditarily presented [5]. This insures that existential results are in fact witnessed by explicit solutions.

While the platonistic assumption of independently existing abstract mathematical entities may be unwarranted, it must be admitted that the resulting intuitions are (so far as everyday mathematics is concerned), persuasive, powerful and coherent [6]. In one way or another, constructive mathematics takes the classical developments as a point of departure and seeks to reformulate them in concretely meaningful terms. The use of formal theories which have both constructive and classical interpretations acknowledges and utilizes this source and, by refining the classical results, repays a debt.

The pursuit of recursive analogues of classical mathematics can have constructively meaningful conclusions. The defect is that these are not (necessarily) generalizations of both classical and constructive results. In addition, the preponderance of work on recursive analogues is of a negative character (telling what cannot be obtained recursively), while dedication to the constructive redevelopment of mathematics leads one to concentrate on positive results. The psychological difference in orientation is of prime importance for how subjects are to be developed.

---

[5] Brouwer's notion of choice sequence creates a problem for this requirement. There are two outs: One is to say that it is just a manner of speaking, to be eliminated in favor of strictly constructive principles (cf. Kreisel/Troelstra 1970); the other would be to try to reformulate just what is presented in terms of partial information about infinite sequences.

[6] That is taken by some as the basis for accepting a platonist view, but doing so is by no means a necessary consequence of the admission.

Work in metamathematics of constructivity has been dominated by results of "constructivity in principle", e.g. E-theorems (if $\exists x\,\phi(x)$ is provable then $\phi(t)$ is provable for some term t) which are extracted by proof theory or realizability arguments. Among all constructive approaches, the one inaugurated by Bishop 1967 offers the most direct reading in computational terms, and the theory T_0 embodies that; the recursive model of T_0 avoids having to apply proof theory or realizability. Even so, it still remains to carry out the passage to the practical implementation of theoretical constructive results. What is done here is regarded as a framework in which that can be represented and facilitated (cf. the remarks at the conclusion of this paper).

Finally, a more technical point, which will be brought out in the work of §3: it is crucial there to be able to distinguish between effectively given Cauchy sequences and those which (in addition) have an effective rate of convergence. Logically, the distinction lies in the difference between two properties, one of the form $\forall n\,\exists m\,\phi(n,m)$ and the other of the form $\exists M\,\forall n\,\phi(n,M(n))$ (where 'n', 'm' range over $\mathbb{N}$). This difference cannot be brought out in theories whose constructivity is implicit in the restriction to intuitionistic logic and which thus admit the axiom of choice $\forall n\,\exists m\,\phi(n,m) \rightarrow \exists M\,\forall n\,\phi(n,M(n))$. But this is a distinction which is readily maintained in T_0 and is also of computational significance.

2. <u>The system</u> T_0 . The following presents the main features of T_0 ; for full details cf. F 1979, pp. 179 ff. The language of T_0 is two-sorted, with individual variables $a,b,c,\ldots,x,y,z$ and class variables $A,B,C,\ldots,X,Y,Z$. There are also certain individual constants (to be described) and the class constant $\mathbb{N}$. The atomic relations are $=$, App and $\in$, where App is 3-placed, Atomic formulas are obtained from these, using either sort of variable or constant in any position. Formulas $\phi, \psi, \ldots$ are built up by the propositional operations together with quantifiers $\forall$, $\exists$ applied to either sort of variable. The basic logic was taken to be intuitionistic in F 1979 but here, in accordance with §1, we allow use of the law of excluded middle (L.E.M.) as well - in other words, full classical logic. There is, in addition, a <u>basic ontological axiom</u> relating the two sorts of variables

as follows: $\forall X \exists x (X=x)$.

The intended constructive interpretation is that individual variables range over a universe V of finite symbolic expressions, and the class variables over the subuniverse of V which define properties of individuals. Then $x \in A$ means that x has the property defined (or given) by (the expression) A. This explanation justifies the basic ontological axiom. The relation App (x, y, z) is supposed to hold when x is an expression in V which is a defining rule for computing a partial function, whose value at y is defined and equals z; we write $xy \simeq z$ for App (x, y, z). Formal "application" terms t built up by repeated formation of $t_1 t_2$ need not represent an object. The relation $t \simeq z$ is defined inductively, with

$$t_1 t_2 \simeq z \leftrightarrow \exists x, y [t_1 \simeq x \wedge t_2 \simeq y \wedge xy \simeq z].$$

We write $t\downarrow$ for $\exists z(t \simeq z)$. Partial functions of several variables are treated by iterated application, $x(y_1,\ldots,y_n) := xy_1 \ldots y_n$. For $n=1$ we thus also write $x(y)$ for xy and for $n=2$, $x(y_1,y_2)$ for $xy_1 y_2$.

The axioms of T_o fall into five groups I - V which we explain informally as follows.

I. APP(Applicative axioms). In addition to the unicity statement, $xy \simeq z \wedge xy \simeq w \rightarrow z=w$, these axioms specify the action of two ("combinatory") constants k and s which serve to generate all partial functions explicitly definable by application. In addition, there are total pairing and projection functions given by constants p, p_1, p_2. One writes (x,y) for pxy so $p_1(x,y) = x \wedge p_2(x,y) = y$.

II. $\mathbb{N}$ (Axioms for $\mathbb{N}$). Here there are additionally four constants 0, $s_{\mathbb{N}}$ (for successor on $\mathbb{N}$), $p_{\mathbb{N}}$ (for predecessor on $\mathbb{N}$) and $d_{\mathbb{N}}$ (for definition by cases on $\mathbb{N}$) with usual axioms. Write x' for $s_{\mathbb{N}}(x)$. Closure is expressed by the axiom $0 \in \mathbb{N} \wedge \forall x(x \in \mathbb{N} \rightarrow x' \in \mathbb{N})$. Induction is taken as the usual scheme, except that for restricted systems it is taken only as the axiom

(Ind) $\quad 0 \in X \wedge \forall x (x \in X \rightarrow x' \in X) \rightarrow \forall x(x \in \mathbb{N} \rightarrow x \in X).$

When Ind is taken in place of the scheme we write $(\mathbb{N}\!\!\restriction)$ for this axiom group.

III. ECA(Elementary comprehension). For each of a certain class of elementary formulas $\phi(x, y_1, \ldots, y_n, Z_1, \ldots, Z_m)$ there is a constant c_ϕ denoting a partial $(n+m)$-ary function defined at every tuple $(y_1, \ldots, y_n, Z_1, \ldots, Z_m)$. Writing $\{x \mid \phi(x, y_1, \ldots, y_n, Z_1, \ldots, Z_m)\}$ for $c_\phi(y_1, \ldots, y_n, Z_1, \ldots, Z_m)$ the axiom is

$$\exists X(\{x \mid \phi(x, \underline{y}, \underline{Z})\} \simeq X \wedge \forall x[x \in X \leftrightarrow \phi(x, \underline{y}, \underline{Z})]).$$

Elementary formulas are those which contain no bound class variables and in which all free class variables occur only to the right of '$\in$' (and no other terms occur to the right of '$\in$').

IV. J (Join) This axiom specifies a constant j satisfying:

$$\forall x \in A\, \exists Y[f(x) \simeq Y] \rightarrow \exists X\{j(A,f) = X \wedge \forall z(z \in X \leftrightarrow \exists x,y(z = (x,y) \wedge x \in A \wedge y \in f(x))\}.$$

We write $\sum_{x \in A} f(x)$ for $j(A,f)$; this represents the disjoint union of the classes $Y_x = f(x)$.

V. IG(Inductive generation). This axiom specifies a constant i which for each A, R gives a class $i(A,R)$, interpreted as the class of R-accessible elements of A. The details of IG are not needed in the following.

Axioms I-IV serve most purposes of constructive analysis of the Bishop school. In fact by F 1979, p.193 (15.5) the system $EM_0\!\restriction$ which consists of APP + $(\mathbb{N}\!\restriction)$ + ECA suffices for most of these purposes. By F 1979, p.218, $EM_0\!\restriction$ is conservative over PA (Peano arithmetic). As noted there, Beeson showed that if the logic is restricted to be intuitionistic, $EM_0\!\restriction$ is conservative over HA (Heyting arithmetic). By F 1979, p.220, T_0 itself is interpretable in the system $(\Delta^1_2 - CA) + (BI)$ of analysis. (Jäger and Pohlers have subsequently shown that these systems are of the same strength.)

Remarks. (i) Beeson 1984 has recently given a reformulation of T_0 which is attractive in certain respects. Application terms are permitted explicitly within the base formalism and for each such term t there is a formula $(t\downarrow)$ which expresses that t is defined. The logic is modified accordingly, e.g. one takes $\forall x\,\phi(x) \wedge t\downarrow \to \phi(t)$. In addition, classes are treated extensionally, as a separate kind of object. But there is a relation of "representation" for certain individuals to classes.

(ii) We write $V := \{x | x=x\}$, so that $\forall x(x \in V)$ by ECA. It is possible to modify T_0 mildly by formulating axioms only for "bounded" classes or "sets" so that there is no universal class; cf. Feferman 1978. (That makes it easier to construct set-theoretical models of T_0.)

Returning now to T_0 as indicated, we can form for any classes A, B the classes $A \cap B$, $A \cup B$, $-A$ and $A \times B$ in the obvious way. Further we can define $(B^A) := \{f | \forall x \in A\, \exists y \in B\,(f(x) \simeq y)\}$. When $\forall x \in A\, \exists Y(f(x) \simeq Y)$ we write B_x for $f(x)$ and $\sum_{x \in A} B_x$ for $j(A,f)$ (when the join axiom is assumed). Then one writes $\prod_{x \in A} B_x$ for $\{f : \forall x \in A(f(x) \in B_x)\}$, where $f(x) \in B_x$ can be replaced by $(x, f(x)) \in Z$, for $Z = \sum_{x \in A} B_x$; thus $\prod_{x \in A} B_x$ exists by ECA. We write $A \subseteq B$ for $\forall x(x \in A \to x \in B)$.

In Bishop-style practice, it is usual to consider classes A equipped with an "equality" relation $=_A$, i.e. simply an equivalence relation $E \subseteq A \times A$. Given $(A, =_A)$ and $(B, =_B)$, by a <u>function</u> from A to B is meant $f \in B^A$ such that $\forall x_1, x_2 \in A[x_1 =_A x_2 \to f(x_1) =_B f(x_2)]$; we write $f : A \to B$ in this case. By a <u>partial function</u> from A to B is meant any f such that

$$\forall x_1, x_2 \in A[x_1 =_A x_2 \wedge f(x_1)\downarrow \wedge f(x_2)\downarrow \to f(x_1) =_B f(x_2)];$$

we write $f : A \rightharpoondown B$ in this case. Then dom (f) is written for $\{x | x \in A \wedge f(x)\downarrow\}$ and range(f) for $\{y | y \in B \wedge \exists x \in A(f(x) \simeq y)\}$.

In the following 'i','j','k','n','m','p' range over $\mathbb{N}$. The construction of the integer and rational number systems $\mathbb{Z}$ and $\mathbb{Q}$ from $\mathbb{N}$ proceeds as usual (cf. F 1979, p.167). For any A, the class of $\mathbb{N}$-indexed sequences from A is just $A^{\mathbb{N}}$; typical such sequences are denoted $a = \langle a_n \rangle_{n \in \mathbb{N}}$ or even more simply $\langle a_n \rangle_n$ or even $\langle a_n \rangle$ (where it is understood that $a_n = a(n)$). Then range $(\langle a_n \rangle) = \{x | x \in A \wedge \exists n(x = a_n)\}$ is denoted $\{a_n\}_n$. For double sequences, i.e. elements of

$A^{\mathbb{N}\times\mathbb{N}}$, the notation taken is $a = \langle a_{nk} \rangle$, i.e. this is a function $a: \mathbb{N}\times\mathbb{N} \to A$ with $a_{nk} = a(n,k)$.

With each application term t having free variables among $x, \underline{y}$, is associated another such term t^*, with variables among $\underline{y}$, built up using just k, s and the variables in $\underline{y}$ and such that (for any $\underline{y}$), $t^*\downarrow \wedge \forall x[t^*(x) \simeq t]$. We denote t^* by $\lambda x.t$ or $\lambda x.t[x]$. It is then a direct matter to establish a general recursion theorem (F 1979 p. 185). In particular, this shows the partial functions on $\mathbb{N}$ to $\mathbb{N}$ to be closed under all partial recursive definitions. But also primitive recursion can be applied to define functions from $\mathbb{N}$ into any class A.

The axioms APP can be modeled as follows: the individual variables are taken to range over the set ω of natural numbers and $App(x,y,z)$ is interpreted as $\{x\}(y) \simeq z$. Then an interpretation of the class variables by codes in ω for defining properties is defined inductively to satisfy the axioms II-V (cf. F 1979, pp. 199-200). The model thus obtained is called here the <u>recursive model</u> for T_0. At the opposite extreme, any model of ZFC in the cumulative hierarchy gives rise to a model of T_0 in which each function from ω to ω is represented by an $f: \mathbb{N} \to \mathbb{N}$, and similarly for higher-type notions. This is called here the <u>classical model</u> for T_0. In that model we have

$$\forall A \forall X \subseteq A \exists f[f : A \to \{0,1\} \wedge \forall x \in A (fx = 0 \leftrightarrow x \in X)],$$

in other words every class has a characteristic function. This statement is far from true in the recursive model. Given $g : \mathbb{N} \to \mathbb{N}$ with range (g) recursively enumerable but not recursive, take $X = \mathrm{range}(g)$. Then

$$\neg (\exists f : \mathbb{N} \to \{0,1\}) \forall n[f(n) = 0 \leftrightarrow n \in \mathrm{range}\,(g)]$$

holds in the recursive model.

These are all the features of T_0 that are needed to make sense of the following.

3. Non-computability results for linear operators formulated in T_0. The purpose of this section is to reformulate the main theorem of Pour-El / Richards 1983 in T_0. This concerns the conditions for computability of linear operators $f: X \to Y$ between Banach spaces X, Y which carry "computation theories" in a suitable sense. The main result is that if f acts computably on a countable subset with dense span in X and f is a closed operator then f preserves computability - i.e. f maps computable elements of X to computable elements of Y - if and only if f is bounded. This theorem is proved in classical mathematics, using the notion of recursiveness to explain the notions of computation theory on X and computable element of X. For example, with $X = \mathbb{R}$, the computable elements of X are just the recursive reals having a recursive rate of convergence. When $X = C[a,b]$ with the uniform (sup)norm, the computable elements are those continuous functions which are limits of recursive sequences of polynomials, again with recursive rate of (uniform) convergence. Similar examples are provided by the various ℓ^p and L^p spaces.

The question was raised (by Pour-El) as to what, if any, constructive significance can be given to their work. An answer is provided here using the Bishop approach to constructivity as represented in T_0. Actually, we proceed informally, much as in the style of the Bishop school, referring as we go along to formal representation in T_0 only where necessary to explain certain distinctions. The conclusion will be that a certain reformulation of the Pour-El / Richards theorem is provable in T_0. Then, using the recursive and classical models of T_0 indicated in §2, we obtain various consistency and independence results.

Usual independence results from T_0 are obtained simply from counter-examples to statements $\phi^{(r)}$ which are recursive analogues to classical theorems ϕ. The relationship is that $\phi^{(r)}$ is equivalent to the interpretation of ϕ in the recursive model of T_0. Hence if $\neg\phi^{(r)}$ holds we cannot prove ϕ from T_0. For instance, when ϕ is the theorem that every continuous function on $[0,1]$ takes on its maximum at some point, Specker's example of a recursively continuous function on $[0,1]$ which does not take its maximum at any recursive point shows that ϕ (as expressed in the language of T_0) is independent of T_0.

On the other hand, the mere fact that a classical theorem ψ also has a true recursive analogue $\psi^{(r)}$ is not sufficient to establish its constructivity. Such is obtained in our setting by actually proving ψ within T_0, thus generalizing both the classical ψ and recursive $\psi^{(r)}$.

The problem raised by Pour-El/ Richards 1983 is that the very formulation of their theorem contains within it, side-by-side, notions in both their ordinary and recursive interpretation, so the preceding general considerations cannot apply directly. For example, one talks of both reals and computable reals, and about functions on the former which may or may not preserve computability when applied to the latter. In T_0, where all notions must be susceptible of interpretation in the recursive model, there is no way to make the distinction between these two kinds of real numbers. However, closer examination of their argument reveals that the only distinction which is actually needed is that between two kinds of recursiveness as applied to real numbers, regarded as Cauchy sequences of rational numbers $\langle r_n \rangle$. The first (or weaker notion) merely requires that the sequence $\langle r_n \rangle$ is recursive, while the second requires in addition that $\langle r_n \rangle$ has a recursive rate-of-convergence function. In T_0, this difference is expressed more abstractly as the difference between those $\langle r_n \rangle$ which happen to be Cauchy sequences, i.e. for which $\forall p \exists m \forall k_1, k_2 \geq m \{ |r_{k_1} - r_{k_2}| \leq \frac{1}{p+1} \}$, and those for which there exists a function $M : \mathbb{N} \to \mathbb{N}$ such that $\forall p \forall k_1, k_2 \geq M(n) \{ |r_{k_1} - r_{k_2}| \leq \frac{1}{p+1} \}$. It is the latter notion which is used throughout Bishop-style analysis, in order to get workable constructive generalizations of classical theorems en masse. But there is no reason which prevents us from considering the weaker notion of Cauchy sequence side-by-side with the stronger one. Note that this distinction cannot be exhibited in formal theories for constructivity which accept some form of the axiom of choice on $\mathbb{N}$:

$$(AC_{\mathbb{N},\mathbb{N}}) \quad \forall n \exists m \phi(n,m) \to \exists M \forall n \; \phi(n, M(n)).$$

But the distinction is computationally significant and should be expressible in the formal framework. This is the main reason why I stress taking what Bishop says about witnessing information literally (see F 1979, pp.176-179, and §4 below).

We now proceed to the details of the reformulation, which is just an extended exercise, given the above idea.

In the following, X is any class with an equality relation $=_X$ and $d:X^2 \to A$ is a given function, i.e. $x_1 =_X x_2 \wedge y_1 =_X y_2 \to d(x_1,y_1) =_A d(x_2,y_2)$. We drop the subscripts to equality relations in the following.

3.1. Definition. d is a metric on X if for all $x,y,z \in X$:

(i) $d(x,y) \geq 0$,

(ii) $d(x,y) = d(y,x)$, and

(iii) $d(x,z) \leq d(x,y) + d(y,z)$.

In the following d is assumed to be a metric on X.

3.2. Definition. For $x \in X$, $\langle x_n \rangle \in X^{\mathbb{N}}$:

(i) $x = \lim_n x_n \leftrightarrow \forall p > 0\ \exists m\ \forall n \geq m\{d(x,x_n) \leq \frac{1}{p}\}$

(ii) $\langle x_n \rangle$ is Cauchy $\leftrightarrow \forall p > 0\ \exists m\ \forall n_1, n_2 \geq m\{d(x_{n_1},x_{n_2}) \leq \frac{1}{p}\}$.

As usual, if $x = \lim_n x_n$ then $\langle x_n \rangle$ is Cauchy. The corresponding explicit ("effective") notions are as follows.

3.3. Definition. For $x \in X$, $\langle x_n \rangle \in X^{\mathbb{N}}$ and $\langle x_{k,n} \rangle \in X^{\mathbb{N} \times \mathbb{N}}$:

(i) $x \underset{E}{=} \lim_n x_n \leftrightarrow (\exists M:\mathbb{N} \to \mathbb{N})\forall p > 0\ \forall n \geq M(p)\{d(x,x_n) \leq \frac{1}{p}\}$.

(ii) $\langle x_n \rangle$ is E-Cauchy $\leftrightarrow (\exists M:\mathbb{N} \to \mathbb{N})\ \forall p > 0\ \forall n_1,n_2 \geq M(p)\{d(x_{n_1},x_{n_2}) \leq \frac{1}{p}\}$

(iii) $\langle x_k \rangle \underset{E}{=} \langle \lim_n x_{kn} \rangle \leftrightarrow (\exists M:\mathbb{N}^2 \to \mathbb{N})\forall k\ \forall p > 0\ \forall n \geq M(k,p)\{d(x_k,x_{kn}) \leq \frac{1}{p}\}$

(iv) $\langle x_{kn} \rangle$ is (uniformly) E-Cauchy $\leftrightarrow$

$$(\exists M:\mathbb{N}^2 \to \mathbb{N})\forall k\ \forall p > 0\ \forall n_1, n_2 \geq M(k,p)\{d(x_{k_1 n}, x_{k_2 n}) \leq \frac{1}{p}\}.$$

Remarks. (i) If $x \underset{E}{=} \lim_n x_n$ then $\langle x_n \rangle$ is E-Cauchy.

(ii) Evidently if $AC_{(\mathbb{N} \to \mathbb{N})}$ holds then $(x = \lim_n x_n \leftrightarrow x \underset{E}{=} \lim_n x_n)$ and (x is Cauchy $\leftrightarrow$ x is E-Cauchy).

(iii) Constructively and computationally, the E-Cauchy sequences are taken to be pairs $(\langle x_n \rangle, M)$ where M is a rate-of-convergence function as on the r.h.s. of 3.3 (ii). Similarly for all the other E-notions.

3.4. Definition. Suppose $D \subseteq X$.

(i) $x \in \bar{D}$ if $x = \lim_n x_n$ for some $\langle x_n \rangle \in D^{\mathbb{N}}$.

(ii) $x \in \bar{D}^{(E)}$ if $x =_E \lim_n x_n$ for some $\langle x_n \rangle \in D^{\mathbb{N}}$.

(iii) Relative to any fixed D for which $X = \bar{D}$ we write $E(X)$ for $\bar{D}^{(E)}$ and $E(X^{\mathbb{N}})$ for those $\langle x_k \rangle \underset{E}{=} \langle \lim_n x_{k,n} \rangle$ with $\langle x_{k,n} \rangle \in D^{\mathbb{N} \times \mathbb{N}}$.

Other notions are explained directly in terms of these, e.g. $E(X^{\mathbb{N} \times \mathbb{N}})$ is reduced to $E(X^{\mathbb{N}})$ using the standard isomorphism $\mathbb{N} \times \mathbb{N} \cong \mathbb{N}$. Informally, relative to D, $E(X)$ is the set of "effective" limits from D and $E(X^{\mathbb{N}})$ is the set of "uniformly effective" sequences of limits from D.

Remark. Again, $AC_{(\mathbb{N} \to \mathbb{N})}$ implies that $\bar{D} = \bar{D}^{(E)}$. But a stronger form of choice is needed to prove that if D is dense in X, i.e. $\forall x \in X \, \forall p > 0 \, \exists y \in D\{d(x,y) \leq \frac{1}{p}\}$, then $X = \bar{D}$.

We now specialize by taking $A = \mathbb{Q}$, the rational number field, and $d(r,s) = |r-s|$ for $r, s \in \mathbb{Q}$, which is a metric on $\mathbb{Q}^2$ to $\mathbb{Q}$.

3.5. Definition.

(i) $\mathbb{R}$ is the set of Cauchy sequences from $\mathbb{Q}$.

(ii) $E(\mathbb{R})$ is the set of E-Cauchy sequences from $\mathbb{Q}$.

For $x = \langle r_n \rangle$, $y = \langle s_n \rangle$ in $\mathbb{R}$ we define as usual: $x \pm y = \langle r_n \pm s_n \rangle$, $x \cdot y = \langle r_n \cdot s_n \rangle$, $|x| = \langle |r_n| \rangle$; $x =_{\mathbb{R}} y$ is defined as $\lim_n (r_n - s_n) = 0$, $x \geq 0$ as $x =_{\mathbb{R}} |x|$ and $x \geq y$ as $(x-y) \geq 0$. Then $\mathbb{R}$ is an Archimedean-ordered field.

Remark. In constructivity it is usual to write $\mathbb{R}$ for what is here written $E(\mathbb{R})$, in fact more explicitly as the set of pairs $(\langle r_n \rangle, M)$ where M is a rate-of-convergence function for $\langle r_n \rangle$. In Bishop 1967, the reals are defined even more particularly as the regular Cauchy sequences of rationals, i.e. those $\langle r_n \rangle$ for which

$$|r_n - r_m| \le \frac{1}{n} + \frac{1}{m} \qquad (n, m > 0).$$

This determines the same limits as E-Cauchy sequences.

3.6. Lemma. Suppose $\langle x_k \rangle \underset{E}{=} \langle \lim_n r_{kn} \rangle$ with $\langle r_{k,n} \rangle \in \mathbb{Q}^{\mathbb{N}\times\mathbb{N}}$ and that $\langle x_k \rangle$ is E-Cauchy. Then $\lim_k x_k$ exists in $E(\mathbb{R})$.

This is the "effective" form of Cauchy-completeness for $E(\mathbb{R})$. Thus $AC_{\mathbb{N}\to\mathbb{N}}$ implies completeness of $\mathbb{R}$.

Now, having $\mathbb{R}$, we can deal with Banach spaces. Let $F = \mathbb{R}$ or $\mathbb{C}$ (the complex numbers, defined as usual from $\mathbb{R}$). We write $\mathbb{Q}(F)$ for the set of rational points in F. The notion of X being a vector space over F is defined as usual. For $Y \subseteq X$, Span(Y) denotes the linear span of Y and $\mathbb{Q}$-Span(Y) the set of all finite linear combinations $\Sigma_{i=0}^n r_i y_i$ with $r_i \in \mathbb{Q}(F)$ and $y_i \in Y$.

Assume now that X has a vector space structure over F, that $D \subseteq X$, and that $\|\cdot\| : X \to \mathbb{R}$ is a given function.

3.7. Definition. X is said to be a Banach space which is separable relative to D if:

(i) the function $d(x,y) = \|x-y\|$ is a metric on X to $\mathbb{R}$,

(ii) $a \in F \wedge x \in X \to \|ax\| = |a| \cdot \|x\|$,

(iii) D is countable, and

(iv) $X = \bar{D}$.

For such a space, the notions $E(X)$ and $E(X^{\mathbb{N}})$ and thence $E(X^{\mathbb{N}\times\mathbb{N}})$ are supplied by the definition 3.4.

3.8. Lemma. Suppose X is separable Banach; then the following hold.

(i) (Composition). If $a : \mathbb{N} \to \mathbb{N}$ and $\langle x_k \rangle \in E(X^{\mathbb{N}})$ then $\langle x_{a(k)} \rangle \in E(X^{\mathbb{N}})$.

(ii) (Insertion). If $\langle x_k \rangle, \langle y_k \rangle \in E(X^{\mathbb{N}})$ then $\langle z_k \rangle \in E(X^{\mathbb{N}})$ where

$$z_{2k} = x_k, \quad z_{2k+1} = y_k.$$

(iii) (Summation). If $\langle x_k\rangle \in E(X^{\mathbb{N}})$ and $\langle |a_{nk}|\rangle \in E(\mathbb{R}^{\mathbb{N}\times\mathbb{N}})$ and $b:\mathbb{N}\to\mathbb{N}$ then $\langle s_k\rangle \in E(X^{\mathbb{N}})$ where $s_k = \Sigma_{n=0}^{b(k)} a_{nk}x_n$.

(iv) (Norms). If $\langle x_k\rangle \in E(X^{\mathbb{N}})$ then $\langle \|x_k\|\rangle \in E(\mathbb{R}^{\mathbb{N}})$.

(v) (Limits). If $\langle x_k\rangle \underset{E}{=} \langle \lim_n x_{kn}\rangle$ and $\langle x_{kn}\rangle \in E(X^{\mathbb{N}\times\mathbb{N}})$, then $\langle x_k\rangle \in E(X^{\mathbb{N}})$.

The proofs of these are straightforward from the E-definitions. The conditions (i)-(v) are just the reformulation in the present setting of the axiomatic notion of "computation theory" for X given in Pour-El/Richards 1983.

Note. It follows from 3.8(iv) that if $x\in E(X)$ then $\|x\|\in E(\mathbb{R})$.

For applications it turns out to be more convenient to deal with Banach spaces supplied with what Pour-El and Richards call an effective generating set. Let $\langle e_n\rangle \in X^{\mathbb{N}}$ and $D_o = \{e_n\}_n$. We put $D = \mathbb{Q}\text{-span}(D_o)$ and then obtain a sequence $\langle d_n\rangle \in X^{\mathbb{N}}$ whose range is D in a standard way.

3.9. Definition. X is said to be a Banach space with generating set $D_o = \{e_n\}_n$ if it is separable relative to $D = \mathbb{Q}\text{-Span}(D_o)$ and if $\langle \|d_n\|\rangle \in E(\mathbb{R}^{\mathbb{N}})$ for the enumeration $\langle d_n\rangle$ obtained for D.

As an example of this notion, let $a,b\in E(\mathbb{R})$, and take $X = C[a,b]$, the class of continuous functions on $[a,b]$ to $\mathbb{R}$, with the sup norm $\|\cdot\|$. Let $e_n = \lambda x.x^n$, so $D = \mathbb{Q}\text{-Span}(D_o)$ is the set of polynomials with rational coefficients. By the Weierstrass approximation theorem X is a Banach space with generating set D_o. The subspace E(X) consists of those continuous functions which are "effectively" approximated by sequences from D.

Now suppose X, Y are two separable Banach spaces (over F), relative to $D_X, \|\cdot\|_X, D_Y, \|\cdot\|_Y$; the subscripts are dropped in the following. The notion of $f: X \rightharpoonup Y$ being a linear operator from X to Y is defined as usual; so also is the notion of f being a bounded linear operator. The following is standard, too.

3.10. Definition. $f : X \rightrightarrows Y$ is said to be closed if whenever $x = \lim_n x_n$ in X and $y = \lim_n y_n$ in Y and $f(x_n) = y_n$ for each n then $x \in \mathrm{dom}(f)$ and $f(x) = y$.

We can now give the reformulation in T_0 of the Main Theorem of Pour-El/ Richards 1983.

3.11. Theorem. Suppose X, Y are separable Banach spaces where X has generating set $D_0 = \{e_n\}_n$ and $\mathbb{Q}$-Span$(D_0) = \{d_n\}_n$. Suppose that $f : X \rightrightarrows Y$ is a closed linear operator such that $\langle f(e_n)\rangle \in E(Y^{\mathbb{N}})$.

(i) If f is bounded then $f : E(X) \to E(Y)$

(ii) If $(\exists a : \mathbb{N} \to \mathbb{N}) \neg (\exists g : \mathbb{N} \to \mathbb{N}) \forall n[g(n) = 0 \leftrightarrow n \in \mathrm{range}(a)]$ and $f : E(X) \to E(Y)$ then f is bounded.

Proof. This follows exactly the lines of the Pour-El/Richards proof, to which we refer the reader for full details. It is easy to show (i). For the proof of (ii), proceed by contradiction. Suppose f is unbounded. Then for each n we can find some finite rational combination d_{k_n} in $\mathbb{Q}$-Span(D_0) with $\|f(d_{k_n})\| > 10^n \|d_{k_n}\|$. Moreover a sequence $\langle d_{k_n}\rangle$ can be constructed with this property; let $p_n = d_{k_n}$. Since $\langle f(e_n)\rangle \in E(Y^{\mathbb{N}})$ it follows that $\langle \|f(p_n)\|\rangle \in E(\mathbb{R}^{\mathbb{N}})$ by 3.8(iv). Take $z_n = f(p_n)/\|f(p_n)\|$ and $u_n = p_n/\|f(p_n)\|$. Then $\|z_n\| = 1$ while $\|u_n\| < 10^{-n}$. Take $x = \sum_{k=0}^{\infty} 10^{-a(k)} u_k$, $y = \sum_{k=0}^{\infty} 10^{-a(k)} z_k$; these limits are seen to exist in X and Y respectively. Indeed $x \in E(X)$. Moreover since $f(u_k) = z_k$ and f is a closed linear operator it follows that $f(x) = y$. However, by the argument of Pour-El and Richards we cannot have $y \in E(Y)$, otherwise one could use that information to decide membership in range (a).

3.12. Corollary. It is consistent with T_0 to assume that for any X,Y,f satisfying the hypotheses of 3.11,

$f : E(X) \to E(Y)$ implies that f is bounded.

Proof. The hypothesis of 3.11(ii) is satisfied in the recursive model of T_0 by taking $\underline{a}$ to be a recursive function with non-recursive range.

Remark. We can conclude the following independence result from T_0: for any particular X, Y, f for which T_0 proves the hypotheses of 3.11 and in addition that f is unbounded, the statement that $f : E(X) \to E(Y)$ is not provable in T_0. Each of the counter-examples to computability of linear operators provided in Pour-El/Richards 1983 can now be converted into an independence result of this type. Take, as a simple instance, the linear operator of differentiation $D : X \rightharpoondown X$ where $X = C[0,1]$, and $\mathrm{dom}(D) = C^1[0,1]$. This is unbounded by $D(e_{n+1}) = ne_n$ for $e_n = \lambda x.x^n$. Note that $Dg(x) = \lim_{h \to 0} \frac{g(x+h) - g(x)}{h}$ makes perfectly good sense constructively for $g \in C^1[0,1]$, but does not always give a real number with rate-of-convergence. Bishop 1967 p.40 further restricts the domain of D to those g for which $Dg(x)$ can be supplied with this additional information.

To conclude, something should be said about the reading of 3.11 in the classical model for T_0. This does not directly return the Pour-El/Richards theorem, since $AC_{\mathbb{N} \to \mathbb{N}}$ is true there and hence the E-notions make no difference (e.g. $\mathbb{R} = E(\mathbb{R})$, etc.). To obtain the classical theorem, one generalizes the work of this section slightly. Working in T_0 assume that U is any class of functions satisfying the applicative axioms of T_0. Then relativize all E-notions to U, starting with 3.3, where instead of considering arbitrary sequences $\langle x_n \rangle \in X^{\mathbb{N}}$, one considers only those lying in U, and requires the rate-of-convergence function M to lie in U. Then all the results go through replacing 'E' throughout by 'E_U'. In the classical model, U can be interpreted to be any class of partial functions closed under partial recursion.

As has been mentioned, the Pour-El/Richards theorem is formulated still more abstractly in terms of the notion of a "computation theory" on a Banach space X. This can be done equally well in T_0. The main point of the above exercise can then be considered to be the verification in 3.8 that the axioms of a computation theory are provable in T_0 for the notion of computable sequence in X provided by $E_U(X^{\mathbb{N}})$. We have preferred to stay less abstract because this is the computation theory used in all the examples and because one can then talk more readily of the computational significance of the work.

Remark. In this connection, after explaining the (original) Pour-El/Richards result to an applied mathematics group at Stanford, I was told that it is quite frequent that one has to deal computationally in applied work with real numbers for which rate-of-convergence information is simply unavailable. In other words, their result is not devastating for applied work (where in any case one is familiar with the problems caused by unbounded operators); on the other hand, the restriction to work with E-notions in the Bishop school insists on too much information for actual practice.

4. Further work; mathematics and computation. Constructivist ideology officially pretends that classical mathematics is endowed with meaning only insofar as one can find constructive substitutes for classical theorems. The working reality is of course otherwise. For the platonist, the reason for this is plain. But the argument here is that one can be constructivist about objects without insisting on a verificationist (or operational) theory of meaning. This gives two layers of constructivist mathematical understanding: a broader, theoretical one, and a narrower operational one. By rejecting the former, constructivism ties its hands and averts its view unnecessarily. On the other hand, the claims for the latter with respect to computational significance have yet to be realized.

If one agrees with this, there are two general directions for further work. The first has to do with those mathematical results which have resisted a strict constructivist reinterpretation, at least as presently practiced. Some of these were brought out in Bishop 1967 and Bishop 1970; to my knowledge progress on them has not been marked (despite much excellent work in the Bishop school; cf. Bridges 1979 and its bibliography). In analysis, one could mention two particular problems: the closed-graph theorem for Banach spaces (Bishop 1967, p.296) and Birkhoff's ergodic theorem (Bishop 1970, p.55). Spectral theory of linear operators also raises various problems; the work of Pour-El/Richards 1984 may offer a good initial testing place. In topology, a proper treatment of singular cohomology theory apparently remains to be given (Bishop 1970 p.56). A number of open problems in probability theory are described in Chan 1981. Of course one stumbles on problems at every turn. 7/ The question is, which to pursue as a natural and viable extension of the constructivist approach.

In the other direction, what remains to be done is to demonstrate the actual computational significance of theoretical constructive mathematics. Bishop thought of his work as providing a high-level programming language (cf. his 1967, pp.354-357 and 1970). He expressed the hope that "each constructive result T [can be realized] as a computer program requiring minimal preparation and supervision by the operator of the computer" (Bishop 1967, p.355). This has yet to be carried out. I had thought specifically to examine Bishop's goal in terms of results formulated in T_0. Various realizability theorems for T_0 with intuitionistic logic have been claimed to have computational significance; cf. Beeson 1984 and Hayashi 1983 (the latter even gives an interpretation in LISP programs). But these do not go far enough, first because they depend on restriction of the reasoning employed, and second because they still only give computability "in principle". The latter criticism applies equally well to T_0 in classical logic as promoted here; the recursive model for T_0 again just gives computability in principle.

To see how (and how well) Bishop-style constructive mathematics may be turned into actual computer programs one must return to the mathematics itself and carry out detailed case studies. The value of the logical work on systems like T_0 should be to give one confidence that the programs are there to be extracted, and to draw the various formal distinctions which are glossed over in the mathematical work but are necessary for specifying programs. What this comes down to is being clear about exactly <u>what data is carried by constructively presented objects</u>, because this is the information with which one has to compute. I have argued that for this one

7/ A systematic introductory outline of current constructive practice and explanation of places where this runs into difficulty will be found in Chapter 1 of Beeson 1984a.

should read Bishop quite literally about witnessing data (F 1979, pp.176-177). Doing so makes metatheory dispensable for the task of turning actual constructive proofs into actual computer programs.

That is only the beginning: there is no assurance that the resulting programs would be useful. Indeed, it is likely that without further examination, the programs extracted directly from proofs would be horribly inefficient if not altogether infeasible. For example, one must constantly apply (partial) sorting to choose the maximum or minimum of a finite sequence of rational values. Computer scientists have spent much time on the analysis of algorithms for such problems, and which are entirely ignored in the original proofs. I conjecture that there are a small finite number of "black boxes" for which reasonably efficient programs have been obtained as a result of work in computer science, and which must be incorporated in the programs extracted directly from Bishop-style proofs in order to make them efficient as a whole. What those black boxes are will emerge only after doing a number of case studies.

If the foregoing project is sensible - and I believe it is - then its value will be seen in ways not yet handled in current programming practice or theory. To begin with, the locus of attention on correctness of programs would be shifted from the programs themselves to the high-level proofs from which they are extracted. It is the latter which carry conviction and which tell us what the programs are <u>for</u>, as one specific part of a much more substantial network of meanings and understanding. Relatedly, if one is to supply users of programs with reasons for answers on call, a hierarchy of appeal will have to be established, beginning with "By theorem so-and-so ...". This envisions a step-wise transition from high-level understanding and proofs to operational implementation and use which will require much work to establish, but about which not much more can be said without trying.

Bibliography

M. Beeson

1984 Proving programs and programming proofs, Proc. VIIth Int'l. Cong. Logic, Methodology and Phil. of Sci. (Salzburg, 1983)(t.a.).

1984a Foundations of constructive mathematics. Metamathematical studies. (Springer, Berlin)(t.a.).

E. Bishop

1967 Foundations of constructive analysis. (McGraw-Hill, New York).

1970 Mathematics as a numerical language, in Intuitionism and proof theory (North-Holland, Amsterdam), 53-71.

D.S. Bridges

1979 Constructive functional analysis, Research Notes in Maths. 28 (Pitman, London).

Y.K. Chan

1981 On some open problems in constructive mathematics, in Constructive mathematics, Lecture Notes in Maths. 873 (Springer, Berlin) 44-53.

S. Feferman

1975 A language and axioms for explicit mathematics, in Algebra and logic, Lecture Notes in Maths. 450 (Springer, Berlin) 87-139.

1978 Recursion theory and set theory: a marriage of convenience, in Generalized recursion theory II (North-Holland, Amsterdam) 55-98.

1979 Constructive theories of functions and classes, in Logic Colloquium 78 (North-Holland, Amsterdam) 159-224.

S. Hayashi

1983 Extracting LISP programs from constructive proofs, Pub. Res. Inst. Math. Sciences, Kyoto University, No.19, 161-191.

G. Kreisel and A.S. Troelstra

1970 Formal systems for some branches of intuitionistic analysis, Annals of Math. Logic 1, 229-387.

M. Pour-El and I. Richards

1983 Noncomputability in analysis and physics: a complete determination of the class of noncomputable linear operators, Advances in Maths 48(1983) 44-74.

1984 The eigenvalues of an effectively determined self-adjoint operator are computable, but the sequence of eigenvalues is not, (t.a.).

A.S. Troelstra

1977 Aspects of constructive mathematics, in Handbook of mathematical logic (North-Holland, Amsterdam) 973-1052.

PARTIAL CLOSURES AND SEMANTICS OF WHILE: TOWARDS AN ITERATION-BASED THEORY OF DATA TYPES

G. Germano and S. Mazzanti

Dipart. Informatica Università di Pisa - Corso Italia, 40 - 56100 PISA (ITALIA)

ABSTRACT. The present paper proposes first a generalization of closure theory and revisits Moore's theory in this framework. Afterwards closures of non cyclic functions are introduced and a method is given to transform cyclic into non cyclic functions. Eventually semantics of the **while** construct is found to be the closure of a function. Computability on inductive and non inductive data types is then studied with iterative means.

INTRODUCTION.

Every experienced programmer is very well aware of the difference between **while** style programming and recursive procedure style programming as much from the point of view of writing programs as from the point of view of executing them.
Historically, recursive procedures go back to generalized recursive equations, and in particular to λ-calculus [12, 15, 16, 18], whereas the **while** construct, though lacking any so clearly identifiable ancestor, goes back in some way to «iteration». It is worth noticing that what is meant [20] to be the beginning of the theory of recursive functions, the famous recursion theorem of Dedekind [5], concerns, more precisely, iterative functions. Later on Gödel [11] introduces his limited ϵ-operator (which already suggests repeating the successor function up to a certain bound) and, on the other hand, provides the β-function to bypass recursion iteratively. Eventually Kleene [15] uses the ϵ-operator (later μ-operator [18]) without bound, and offers in this way the first example of a **while** construct. By means of β-function and μ-operator, the class of recursive functions could be recovered by completely avoiding recursion, i.e. purely iteratively [17, 25, 26].
In any case, the computation of a recursive function or the execution of a recursive procedure implies operating iteratively. In fact, for recursive functions, Kleene's normal form theorem holds; on the other hand the execution of recursive procedures is done by iterating substitutions: older machines do this by building the well-known stack, whereas the newer reduction machines do this directly, according to their built-in reduction system [32]. Also the most classical abstract machines, like Turing machines, limited and unlimited register machines, Post systems and Markov algorithms, control the sequencing of steps by some kind of **goto** or **while** and a very easy way of proving the normal form theorem is reducing it to such machines [14,6]. Recent works seem to show that iteration is not weaker than recursion from the point of view of formal expressiveness [10] and of complexity [33].
Strangely enough recursive procedures (as well as λ-calculus) have received a careful and very elegant semantical treatment, in the framework of denotational semantics [4, 21, 29] via the recursion theorem of Kleene [19] and the fixpoint theorem of Scott [27, 28] whereas until now there is no satisfactory semantical treatment of the **while** construct independently from that of recursive procedures. This appears to be particulary inconvenient if one considers that denotational semantics is supposed to supply the guideline of implementation: this should mean that, according to the state of the art in semantics, **whiles** should be implemented by first simulating them through recursive procedures!
The present paper offers an independent semantical treatment of the **while** construct which should perfectly correspond to computing practice. This succeeds via a generalization of inductive closure theory. Inductive closures on a power set have been explicitly introduced by Moore [24] and constitute a classical chapter of most treatises on lattice theory and universal algebra [1, 3, 13, 2], but appear, less explicitly, already in [5].
Tarski has made use of them for his algebraic treatment of deduction [30] on which his classical work on undecidability [31] is based.
It is worthwhile noticing that the (inductive) closure we use here is less than topological closure. So we get a semantics for the **while** construct (and therefore for programs in general) without any use of continuity as is the case in the recursion theorem and in the fixpoint theorem. Therefore one can think of an iterative computability (via inductive closures as semantics of **while** constructs) as in [7,

8] for any data type. In particular it appears that a data type must not have an inductive carrier set, as we will show by way of the example of the integers. In fact the only inductive set, to which in any case computability refers, seems to be the set of computation steps.

In section 1 we generalize the classical closure theory for partial orderings. A partial function is said to be a partial closure on an ordering if it satisfies conditions analogous to the well know ones on extensivity, isotonicity and idempotence. The notion of a closure algebra is introduced and it is shown how the partial closure relatively to a closure algebra is constructed.

In section 2 we show how this generalized notion of closure applies to the case of a power set investigated by Moore and we recover his theory.

In section 3 we try to develop an induction theory. We introduce first the notion of «coalescence», a binary operation on sequences which is intermediate between concatenation and composition. Afterwards we introduce the notion of chain linked by a given function. Eventually we show that chains can be obtained by coalescence and that inductive sets can be obtained by taking last elements of chains.

In section 4 we consider the ordering generated by a non cyclic function; in this case the construction of a partial closure is always possible.

In section 5 we show how to cut any function to obtain an equivalent non cyclic function and eventually we are able to reduce the accepted meaning of the **while** construct to a closure.

In section 6 we show how to apply the methods above to construct data types by first considering some inductive examples and then a typically non inductive one, the data type of integers.

1. A THEORY OF PARTIAL CLOSURE ON A PARTIAL ORDERING.

Consider a universe V, let $x, y, u, v \in V$ and let R be a partial ordering on V.

1.1. Partial closure.

Let $C \subseteq V \times V$

DEF. C is a partial closure on R iff CL1. $xCu \Rightarrow xRu$

CL2. $xRy \wedge xCu \wedge yCv \Rightarrow uRv$

CL3. $xCu \Rightarrow uCu$

CL4. $\exists v\ (xRv \wedge vCv) \Rightarrow \exists u\ xCu$

Conditions CL1, CL2, CL3 state the extensivity, the monotonicity and the idempotence of C respectively. Condition CL4 assures the totality of C on the ideal of any fixpoint.

1.2. Closure algebras.

Let $P \subseteq V$.

DEF. (P,R) is a closure algebra iff $\forall x\ (\exists v\ v \in (P \cap \mathrm{filt}(x,R)) \Rightarrow \exists u\ u\ \min_R\ (P \cap \mathrm{filt}(x,R)))$

where $\mathrm{filt}(x,R) = \{y \mid xRy\}$ is the filter generated by x and R and where $\min_R$ is the relation of being the mininum with respect to R.
The condition above is a generalization of the Moore condition for a family of sets belonging to a power set as we will see later. A less general notion of closure algebra has been introduced in [22, 23].

1.3. Partial closures and closure algebras.

Consider the partial function

$$\mathrm{clos}(P,R) = \{(x,u) \mid u\ \min_R\ (P \cap \mathrm{filt}(x,R))\}$$

TH. 1. If (P,R) is a closure algebra then clos(P,R) is a partial closure on R.

PROOF. Suppose (P,R) is a closure algebra. For C = clos(P,R) we have to show that conditions CL1-4 hold.
As concerns CL1, it suffice to consider the definition of filt(x,R).
As concerns CL2, from xRy it follows that $\mathrm{filt}(y,R) \subseteq \mathrm{filt}(x,R)$ and that $P \cap \mathrm{filt}(y,R) \subseteq P \cap \mathrm{filt}(x,R)$.

Therefore the minimum of $P \cap filt(x,R)$ minorizes $P \cap filt(y,R)$.
Condition CL3 can be shown to hold analogously.
As concerns CL4, remark that if xRv and vCv then $v \in filt(x,R)$ and $v \in P$. So there is some u such that xCu, because (P,R) is a closure algebra.
Consider the set of fixpoints of C:

$$fixp(C) = \{u \mid uCu\}$$

Lemma. For any partial closure C on R

$$xCu \Rightarrow u \; min_R \; (fixp(C) \cap filt(x,R))$$

Proof. Suppose xCu. Then $u \in (fixp(C) \cap filt(x,R))$ by CL3 and CL1.
On the other hand, u minorizes $fixp(C) \cap filt(x,R)$ by CL2.

Remark. From this lemma we obtain that every partial closure is a partial function.

Th. 2. If C is a partial closure on R then (fixp(C),R) is a closure algebra on R.

Proof. Suppose C is a partial closure on R. So, if $fixp(C) \cap filt(x,R)$ is not empty, then there exist some u such that xCu (by CL4) and so u min_R $fixp(C) \cap filt(x,R)$ by the lemma.

Th. 3. fixp(clos(P,R))=P
It follows immediately from the definition.

Th. 4. If C is a partial closure on R then clos(fixp(C),R)=C

Proof. If $\langle x,y \rangle \in clos(fixp(C),R)$ then y min_R $(fixp(C) \cap filt(x,R))$ so $xRy \wedge yCy$. This implies that there is some u such that xCu by CL4 and so u min_R $(fixp(C) \cap filt(x,R))$ by the lemma. So y=u and we obtain xCy. Therefore $C \subseteq clos(fixp(C),R)$ by the lemma.

2. Some important closures.

2.1. Set theoretic closures and Moore families.

Let V be a universe; let $X,Y \in PV$ and $M,N \subseteq PV$.

We want now to recover the classical closure theory [24, 1, 3, 13, 2] in the framework of the general theory given in section 1.

Def. M is a Moore family iff $\forall N \subseteq M. \; \cap N \in M$.

Th. 1. If M is a Moore family then $(M,\subseteq)$ is a closure algebra.

Proof. Suppose M is a Moore family. Then $\cap(M \cap filt(X,\subseteq)) \in M$.

On the other hand $\cap(M \cap filt(X,\subseteq)) \in filt(X,\subseteq)$.

Therefore $\cap(M \cap filt(X,\subseteq)) \in (M \cap filt(X,\subseteq))$ and so $\cap(M \cap filt(X,\subseteq))$ $min_\subseteq$ $(M \cap filt(X,\subseteq))$

Remark. If M is a Moore family then $clos(M,\subseteq)$ is a total function because $V = \cap \emptyset \in M$

Example. Set $T = \{R \subseteq V \times V \mid R \text{ is transitive}\}$.

T is a Moore family, so $(T,\subseteq)$ is a closure algebra and $clos(T,\subseteq)$ is the (total) function assigning to each R its transitive closure.

2.2. Set theoretic closure under a set of functions.

Consider a set F of functions $f: U^k \rightarrow U$, with k depending on f.

Define: F-closed X iff $im_F(X) \subseteq X$,

where $im_F(X)$ is the image of X by the functions in F.

Th. 2. The family of F-closed sets is a Moore family.

The proof is immediate.
Therefore the family of F-closed sets defines a closure algebra by th. 1 and we agree that C_F is its closure function. For $F=\{f\}$ we set $C_f=C_F$

Example. An inductive set can be seen, following [5], as the closure of the set of some basis elements under some step-functions.

We set

$$\text{ind}(x,f)=C_f\{x\}$$

So, in particular, the set of natural numbers can be obtained from the zero element and the successor function as $N=\text{ind}(0,S)$ and the transitive closure of a relation can be obtained from that relation and the composition operator as $\text{ind}(R,\bigcirc)$.

3. Induction theory.

3.1. Concatenation, composition and coalescence.

Consider a free monoid U over a set A, with respect to a binary operation + and a neutral element 0, i.e. a monoid $U\supseteq A$ such that U can be uniquely decomposed in terms of A (this means that any mapping of A into another monoid U′ can be uniquely extended to a homomorphism of U into U′). Elements of U will be called «sequences», + will be called «concatenation» and 0 will be called «empty sequence».
The use of the sign «+» for concatenation is motivated by the fact that for natural numbers, as words over {|}, concatenation coincides with addition: 2+2=||+||=||||=4.

From now on let $r, s, t \in U$ and $X, Y \subseteq U$.

The semigroup of non empty sequences on X can be defined as $X^+=\text{ind}(X,+)$.

We introduce first an extended (associative) «cartesian product»: $X\times Y=\{s+t \mid s\in X \wedge t\in Y\}$.

Since X is closed with respect to + iff $X\times X\subseteq X$, we obtain $X^+=X\cup X\times X\cup\ldots$

Note that $A^+=A^*\times A=A\times A^*$.

We introduce now an extended «composition»:

$$\begin{array}{rl}(s+x)\bigcirc(x+t)=s+t & \\ (s+x)\bigcirc(y+t) & \text{undef. if } x\neq y\\ s\bigcirc t & \text{undef. if } s=0 \text{ or } t=0\end{array}$$

As we have already remarked, the transitive closure of a set X can be obtained as $X^{\bigcirc}=\text{ind}(X,\bigcirc)$

As concatenation induces the operation of cartesian product, composition induces a binary operation on sets: $X\circ Y=\{s\bigcirc t \mid s\in X \wedge t\in Y\}$

In analogy to what happened for concatenation and cartesian product, we obtain $X^{\bigcirc}=X\cup X\circ X\cup\ldots$

Eventually we introduce «coalescence», a binary operation which is somehow intermediate between concatenation and composition:

$$\begin{array}{rl}(s+x)\oplus(x+t)=s+x+t & \\ (s+x)\oplus(y+t) & \text{undef. if } x\neq y\\ s\oplus t & \text{undef. if } s=0 \text{ or } t=0\end{array}$$

Closure under coalescence is obtained as $X^{\oplus}=\text{ind}(X,\oplus)$

Also coalescence induces a binary operation on sets: $X\otimes Y=\{s\oplus t \mid s\in X \wedge t\in Y\}$

Also for coalescence: $X^{\oplus}=X\cup X\otimes X\cup\ldots$

For concatenation, composition and coalescence we introduce the respective monoids by adjoining the correspondent sets of neutral elements:

$$A^{+*}=A^{+}\cup\{0\}$$

$$A^{\bigcirc *}=A^{\bigcirc}\cup id_A$$

$$A^{\oplus *}=A^{\oplus}\cup A$$

We define analogonsly X^{+*}, $X^{\bigcirc *}$ and $X^{\oplus *}$ and according to tradition we set $X^{*}=X^{+*}$

3.2. Chains.

We introduce now functions K and L on sequences (generalizing the known functions on pairs) and the corresponding predecessor functions on sequences P_K and P_L:

$$K(x+s)=x,\ K0 \quad \text{undef.}$$

$$P_K(x+s)=s,\ P_K0 \quad \text{undef.}$$

$$L(s+x)=x,\ L0 \quad \text{undef.}$$

$$P_L(s+x)=s,\ P_L0 \quad \text{undef.}$$

Now the unique decomposition of U in terms of A can be described by means of

$$s_i=K(P_K^{i-1}s)$$

The set of subsequences of a given sequence can be obtained as

$$subw(s)=C_F\{s\},\ \text{where } F=\{P_K,\ P_L\}$$

and the set of its subsequences having length i can be obtained as

$$subw_i(s)=subw(s)\cap A^i,$$

where A^i is the cartesian product of A with itself i-times.

We can now characterize the set of chains linked by f as follows:

$$ch(f)=\{s \mid subw_2(s)\neq\emptyset \wedge subw_2(s)\subseteq f\}$$

On the other hand we can characterize the set of chains beginning whith x and linked by f as follows:

$$ch(x,\ f)=\{x\}\cup((\{x\}\times U)\cap ch(f))$$

The «length» of a sequence can be characterized as

$$lg(s)=\mu i\ subw_{i+1}(s)=\emptyset$$

3.3. Composition, coalescence and chains.

Consider a relation $R\subseteq A\times A$ and a partial function f: $A\rightarrow A$.

TH. 1. (Basic theorem) $R^{\bigcirc}=im_{K\wedge L}(R^{\oplus})$,

where $(K\wedge L)s=(Ks,Ls)$. See [7].

PROOF. We show first that $R^{\bigcirc}\subseteq im_{K\wedge L}(R^{\oplus})$ by induction on $R^{\bigcirc}$.

Induction basis: $R=im_{K\wedge L}(R)$.

Induction step: suppose $(x,y)=(K\wedge L)s\in R^{\bigcirc}$ and $(y,z)=(K\wedge L)t\in R^{\bigcirc}$ for s and $t\in R^{\oplus}$. Then

$$(x,y)\bigcirc(y,z)=(x,z)=(K\wedge L)(s\oplus t)\in im_{K\wedge L}(R^{\oplus})$$

Conversely we show that $R^{\bigcirc}\supseteq im_{K\wedge L}(R^{\oplus})$ by induction on $R^{\oplus}$.
The induction basis is the same as above.
Induction step: suppose $(x,y)=(K\wedge L)s\in R^{\bigcirc}$ and $(y,z)=(K\wedge L)t\in R^{\bigcirc}$ for s and $t\in R^{\oplus}$. Then

$$(K\wedge L)(s\oplus t)=(x,z)=(x,y)\bigcirc(y,z)\in R^{\bigcirc}$$

COROLL. 1. $\{x\}\otimes R^{\circ}=im_{K\wedge L}(\{x\}\otimes R^{\oplus})$

COROLL. 2. $\{x\}\otimes R^{\circ *}=im_{K\wedge L}(\{x\}\otimes R^{\oplus *})$

COROLL. 3. $im_L(\{x\}\otimes R^{\circ *})=im_L(\{x\}\otimes R^{\oplus *})$

We want now to put together chains and sequences generated by coalescence:

TH. 2. $ch(f)=f^{\oplus}$

PROOF. We show first that $ch(f)\subseteq f^{\oplus}$ by induction on $A\times A^{+}$.

Induction basis: if $(x,y)\in ch(f)$ then $(x,y)\in f\subseteq f^{\oplus}$.

Induction step: suppose $(s+t)\in ch(f)$. Then s and $t\in ch(f)$ whereas $(Ls,Kt)\in f$.
Therefore $s+t=s\oplus(Ls,Kt)\oplus t\in f^{\oplus}$.
Conversely one can immediatly see that $f^{\oplus}\subseteq ch(f)$ by induction on $f^{\oplus}$.

COROLL. 4. $ch(x,f)=\{x\}\cup(\{x\}\otimes f^{\oplus})=\{x\}\otimes f^{\oplus *}$

3.4. Inductive sets and chains.

As $f^{\circ *}$ is a preorder, we can consider its filters and we see that they coincide with inductive sets

TH. 1. $filt(x,f^{\circ *})=ind(x,f)$

PROOF. We show first that $filt(x,f^{\circ *})\subseteq ind(x,f)$. Obviously $x\in ind(x,f)$. On the other hand it is easy to see that, if $(x,y)\in f^{\circ}$, then $y\in ind(x,f)$ by induction on f°. Indeed, for $(x,y)\in f^{\circ}$ and $(y,z)\in f^{\circ}$, by hypothesis, $y\in ind(x,f)$ and $z\in ind(y,f)$. So $z\in ind(y,f)\subseteq ind(x,f)$.
Conversely one can show that $filt(x,f^{\circ *})\supseteq ind(x,f)$ by induction on $ind(x,f)$ immediately.
To conclude our induction theory we retrieve now Dedekind's relationship between inductive sets and chains:

TH. 2. $ind(x,f)=im_L(ch(x,f))$

PROOF. $ind(x,f)=filt(x,f^{\circ *})=\{x\}\otimes f^{\circ *}=im_L(\{x\}\otimes f^{\oplus *})=im_L(ch\ (x,f))$

by th. 1 and cor. 3 and 4 of 3.3.

4. FUNCTIONS, CHAINS AND ORDERINGS.

Consider a universe V; let x, y, u, v $\in$ V and let $P\subseteq V$ and $f: V\rightarrow V$.

4.1. Orderings on chains.

Chains can be ordered by length according to the following relation:

$$s\leqslant t \quad \text{iff} \quad lg(s)\leqslant lg(t)$$

and can be ordered also by the relation of «being a prefix»:

$$s \text{ pref } t \quad \text{iff} \quad \exists r\ s+r=t$$

The relation of being a prefix is obviously an ordering.
The relation $\leqslant$ is obviously reflexive and transitive and results to be also antisymmetric on ch(x,f):

$$s\in ch(x,f) \wedge t\in ch(x,f) \wedge s\leqslant t \wedge t\leqslant s \Rightarrow s=t.$$

So the relation $\leqslant$ is an ordering on ch(x,f) and from the well-oraering of N we obtain that $(ch(x,f),\leqslant)$ is a well-order.

We want now to show that, on ch(x,f), ordering by lenght and ordering by prefix coincide.

TH. 1. $(ch(x,f) \upharpoonright \leqslant)=(ch(x,f) \upharpoonright pref)$.

PROOF. We show first that $(ch(x,f) \upharpoonright \leqslant) \subseteq (ch(x,f) \upharpoonright pref)$ by induction on $d=(lg(t)-lg(s))\in N$.

The induction basis can be obtained from the antisymmetry above.
For the induction step note that, if s and $t\in ch(x,f)$ and $lg(t)-lg(s)=d+1$, then $lg(P_Lt)-lg(s)=d$ and so s pref P_Lt pref t.
Conversely one can immediately see that $(ch(x,f) \upharpoonright \leqslant) \supseteq (ch(x,f) \upharpoonright pref)$ by definition.

COROLL. 1. (ch(x,f) ↾ pref) is a well-order

4.2. Well-filtered orderings.

Let $g \subseteq V \times V$ and let S be a transitive relation on V.
For studying orderings generated by functions, we define:

$$g \text{ is cyclic iff } \exists x,y,s,t\ (x \neq y \wedge x \oplus s \oplus y \oplus t \oplus x \in g^{\oplus})$$

LEMMA 1. g is non cyclic iff g° is antisymmetric.

REMARK. Being reflexive does not imply being cyclic.

COROLL. 1. g is non cyclic iff $g^{\circ *}$ is a partial ordering.

DEF. S is well-filtered iff, for every x, filt(x,S) is well-ordered by S.

LEMMA 2. If S is a well-filtered partial ordering then, for every P, (P,S) is a closure algebra.

LEMMA 3. If g is a function then L: ch(x,g)→filt(x,$g^{\circ *}$) is a homomorphism relatively to pref and $g^{\circ *}$.

PROOF. If s pref t then there is an r such that s+r=t∈ch(g). So Ls+r∈ch(Ls,g).

By th. 2 and 1 of 3.4 L(Ls+r)∈filt(Ls,$g^{\circ *}$) and it results that (Ls,Lt)∈$g^{\circ *}$.

TH. 1. If g is a function then $g^{\circ *}$ is well-filtered.

The proof is immediate by Corollary 1 of 4.1 and the lemma above.

COROLL. 2. If g is a non cyclic function then $g^{\circ *}$ is a well-filtered partial ordering.

COROLL. 3. If g is a non cyclic function, then for every P, (P,$g^{\circ *}$) is a closure algebra.

REMARK. If g is a non cyclic function without fixpoints then L: ch(x,g)→ind(x,g) is an isomorphism.

In fact, in this case, L is injective because, for s and t∈ch(x,g) with Ls=Lt, there is an r such that Ls+r is a cycle.

TH. 2. If g is a non cyclic function then clos(P,$g^{\circ *}$)={(x,Ls) | s∈(ch(x,g)∩((−P)*×P)}

PROOF. Suppose g is a non cyclic function. Then, (P, $g^{\circ *}$) is a closure algebra by cor. 3, and by th. 1 and 2 of 3.4

$$\begin{aligned}(x,u)\in \text{clos}(P,g^{\circ *}) &\text{ iff } u \min (P \cap \text{filt}(x,g^{\circ *}))\\ &\text{ iff } u \in (P \cap \text{filt}(x,(-P \otimes g)^{\circ *}))\\ &\text{ iff } u \in (P \cap \text{ind}(x,-P \otimes g))\\ &\text{ iff } u \in (P \cap \text{im}_L(\text{ch}(x,-P \otimes g)))\\ &\text{ iff } u \in \text{im}_L(\text{ch}(x,g) \cap ((-P)^* \times P))\end{aligned}$$

TH. 3. If g is a non cyclic function then clos(P,$g^{\circ *}$)=clos(P,$(-P \otimes g)^{\circ *}$)

PROOF. By the theorem above

$$\begin{aligned}(x,u)\in \text{clos}(P,g^{\circ})^* &\text{ iff } u \in \text{im}_L(\text{ch}(x,g) \cap ((-P)^* \times P))\\ &\text{ iff } u \in \text{im}_L(\text{ch}\ (x,-P \otimes g) \cap ((-P)^* \times P))\\ &\text{ iff } (x,u) \in \text{clos}(P,(-P \otimes g)^{\circ *})\end{aligned}$$

because ch(x,g)∩((−P)*×P)=ch(x,−P⊗g)∩((−P)*×P).

5. PARTIAL CLOSURE OF A PARTIAL FUNCTION.

5.1. Closure of a (cyclic and non cyclic) function.

A function f generates the relation $f^{\circ *}$.

This relation is obviously transitive and reflexive and so it is an ordering iff it is antisymmetric. By

lemma 1 this is the case iff f is non cyclic. To avoid cycles we have in general to make some «cuts» on f. We consider first $-P\otimes f$ and the ideals with respect to $(-P\otimes f)^{\circ *}$.
On the ideal generated by P the function f turns out to be non cyclic.

We set $P{\downarrow}f=\mathrm{ideal}(P,(-P\otimes f)^{\circ *})$, where $\mathrm{ideal}(P,R)=\{x \mid \exists u\in P\ <x,u>\in R\}$ is the ideal generated by P with respect to R.

DEF. $P|f=-P\otimes f\otimes(P{\downarrow}f)$

TH. 1. P|f is non cyclic.

PROOF. Suppose that P|f is cyclic. Then there exist x, s, t such that $x\oplus s\oplus y\oplus t\oplus x\in(P|f)^{\oplus}$ for $y\neq x$ and $y\in\mathrm{filt}(x,(P|f)^{\circ *})$. For every such y, therefore, P|f is defined insofar as $(y,t_2)\in(P|f)$.
On the other hand there is a $u\in(P\cap\mathrm{filt}(x,(-P\otimes f)^{\circ *}))=P\cap\mathrm{filt}(x,(P|f)^{\circ *})$.

For such u, therefore, P|f should be defined, in contradiction to the fact that $u\in P$.
By corollary 2 of 4.2 we obtain

COROLL. 1. For every P, $(P|f)^{\circ *}$ is a well-filtered partial ordering.

By corollary 3 of 4.2 we obtain:

COROLL. 2. For every P, $(P,(P|f)^{\circ *})$ is a closure algebra.

REMARK. Other non cyclic functions which could be mentioned in this context are

$$\text{push } f:\ s \mapsto s+f(Ls)$$
$$S\times f\ :(n,x)\mapsto(n+1,fx)$$

Via these functions one could construct closure algebras connected with the closure algebra above, because $\mathrm{ind}(x,f)=\mathrm{im}_L(\mathrm{ind}(x,\text{push } f))=\mathrm{im}_L(\mathrm{ind}((0,x),S\times f))$.

We set

$$C(P,f)=\mathrm{clos}(P,(P|f)^{\circ *})$$

and we call it the «closure of the function f with respect to the predicate P».
By th. 1 of 1.3 C(P,f) is a partial closure on the partial ordering $(P|f)^{\circ *}$.

TH. 2. If f is non cyclic then $\mathrm{clos}(P,f^{\circ *})=C(P,f)$

PROOF. Suppose f is non cyclic. By th. 3 of 4.2 we know that $\mathrm{clos}(P,f^{\circ *})=\mathrm{clos}(P,(-P\otimes f)^{\circ *})$ $=\mathrm{clos}(P,(P|f)^{\circ *})$ because, if $(x,u)\in\mathrm{clos}(P,(-P\otimes f)^{\circ *})$, then $u\in P\subseteq(P{\downarrow}f)$.

TH. 3. $C(P,f)=\{(x,Ls) \mid s\in(\mathrm{ch}(x,f)\cap((-P)^{*}\times P))\}$

Proof. By th. 2 of 4.2 and th. 1, because $\mathrm{ch}(x,f)\cap((-P)^{*}\times P)=\mathrm{ch}(x,P|f)\cap((-P)^{*}\times P)$

TH. 4. $C(P,f)=(-P\otimes f)^{\circ *}\otimes P$

PROOF. By cor. 3 and 4 of 3.3 and th. 2 of 4.2

$$\begin{aligned}(x,Ls)\in\mathrm{clos}(P,(P|f)^{\circ *})\quad &\text{iff } s\in\mathrm{ch}(x,P|f)\cap((-P)^{*}\times P)\\ &\text{iff } s\in\mathrm{ch}(x,-P\otimes f)\cap((-P)^{*}\times P)\\ &\text{iff } s\in\{x\}\otimes(-P\otimes f)^{\oplus *}\cap((-P)^{*}\times P)\\ &\text{iff } s\in\{x\}\otimes(-P\otimes f)^{\oplus *}\otimes P\\ &\text{iff } (x,Ls)\in(-P\otimes f)^{\circ *}\otimes P\end{aligned}$$

5.2. Closure under functions and closure of a function.

In 2.2. we have considered a set F of functions and the closure function C_F defined by the closure algebra consisting of the family of F-closed sets and of the inclusion relation. We want now to obtain the same closure function as closure of the function

$$ext_F(X)=X \cup im_F(X),$$

where $im_F(X)$ is the image of X by the functions in F.
Obviously ext_F is a non cyclic function. Therefore by corollary 3 of 4.2 (F-closed, $ext_F^{\circ^*}$) is a closure algebra and by th. 2 we obtain

$$clos(\text{F-closed}, ext_F^{\circ^*})=C(\text{F-closed}, ext_F)$$

Set $C_F=C(\text{F-closed}, ext_F)$.

TH. 1. $(X,\mathcal{C}_F X)\in ext_F^{\circ^*} \Rightarrow C_F X=\mathcal{C}_F X.$

PROOF. Since $X\subseteq \mathcal{C}_F X$ and $\mathcal{C}_F X$ is F-closed, $C_F X\subseteq \mathcal{C}_F X$.

On the other hand also $C_F X$ is F-closed. So, if $(X,\mathcal{C}_F X)\in ext_F^{\circ^*}$, then $C_F X\in(\text{F-closed}\cap filt(X,ext_F^{\circ^*}))$, then $(C_F X,\mathcal{C}_F X)\in ext_F^{\circ^*}$ and therefore $C_F X\subseteq \mathcal{C}_F X$.

To see that $C_F X=\mathcal{C}_F X$ in general is not true, consider the successor function S and ext_S; now $C_S\{0\}=N$, whereas $\mathcal{C}_S\{0\}$ doesn't exist. In fact, if $\mathcal{C}_F X$ exists, then there is an $s\in ch(X,ext_F)$ such that $Ls=\mathcal{C}_F X$. Such s cannot exist for N. For it the situation looks like this:

$$ext_S: \{0\} \to \{0,1\} \to \dots N = C_S\{0\}$$

The existence of a FINITE chain from X to $\mathcal{C}_F X$ is a characteristic property of a closure of a function. In the next paragraph we will see, in fact, that closures of functions correspond to **whiles** and so (just because of their intrinsic FINITENESS) do not lead outside FINITE computability.

5.3. Semantics of **whiles** as closures of functions.

Partial closures of partial functions are defined on ordering relations generated by functions. Such ordering relations imply, as we just remarked, finiteness of the number of steps conducting from the argument to the value. If B is a predicate (the meaning of <boolean expression>) and f is a function (the meaning of <program>), then the meaning of

while <boolean expr.> **do** <program>

is the function which assigns, to every argument x, the last term of that sequence s which begins by x, which is linked by f and whose terms belong all to B except the last one; i.e. for

M(<boolean expr.>)=B
M(<program>)=f

we have

M(**while** <boolean expr> **do** <program>)

$$=\{(x,Ls) \mid Ks=x \wedge subw_2(s)\subseteq f \wedge subw_1(P_L s)\subseteq B \wedge Ls\in -B\}$$

$$=\{(x,Ls) \mid s\in(ch(x,f)\cap(B^*\times -B))\}$$

$$=\mathcal{C}(-B,f) \text{ by th. 3 of 5.1}$$

So we obtain the meaning of a **while** as the closure of the meaning of its body whith respect to the meaning of its boolean condition.
By th. 4 of 5.1 we are also able to express such a meaning in terms of coalescence and composition:

M(**while** <boolean expr.> **do** <program>)$=(B\otimes f)^{\circ^*}\otimes -B$

Remark that for a non cyclic f, (by th. 2 of 5.1) the meaning, we are discussing, can be obtained as $clos(-B,f^{\circ^*})$ instead of $clos(-B,(-B|f)^{\circ^*})$.
As an example we want to express the action of the μ-operator in terms of a closure. Consider the function $\mu x\ R(x,y)$. It is computable by the program

x:=0; **while** ¬R(x,y) **do** x:=x+1 **od**

This program has the meaning

$$(O\times I)\circ C(R,S\times I)\circ K$$

where O is the zero function and $I=id_N$.

6. COMPUTABILITY VIA CLOSURES ON SOME IMPORTANT (INDUCTIVE AND NON-INDUCTIVE) DATA TYPES.

6.1. Computabilty on sequences of natural numbers.

Computable functions on sequences of natural numbers i.e. computable functions $f: N^h\to N^k$ have been studied in [7, 8] and provide the functional meaning of limited register machines [9].

We have already mentioned composition of functions $f\circ g$. Cartesian product of functions $f: A\to C$ and $g: B\to D$ is defined as

$$f\times g : A\times B \to C\times D$$

$$s+t \mapsto (fs)+(gt)$$

A repetition operator of **while** type can be defined according to 5.3 for functions $f: A\to N$ and $g: A\to A$ as $f/g=C(coim_f\{0\},g)$

Consider now the following class of functions on sequences of natural numbers:

$$SN=C_F\{O,\ S,\ P,\ \Box\},$$

where $O: N^0\to N$ is the zero function, $S: N\to N$ is the successor function, $P: N\to N$ is the predecessor function, $\Box: N\to N$ is the cancellator function and $F=\{f\circ g,\ f\times g,\ f/g\}$.

REMARK. The class SN can be also obtained by using the weaker repetition operator $\bar{g}=C(\{0\}\times N^*,g)$

Let $\mathfrak{R}$ be the class of traditional recursive functions $f: N^h\to N$ and, for $f: N^h\to N^i$ and $g: N^h\to N^j$, let

$$f\wedge g : N^h \to N^i\times N^j$$

$$s \mapsto (fs)+(gs)$$

It is easy to prove the following

TH. 1. $SN=C_\wedge\mathfrak{R} \cup U_i hom(N^i, N^0)$

Assuming Church Thesis i.e. that any computable function $f: N^h\to N$ belongs to $\mathfrak{R}$, it can be easily shown that every computable function $f: N^h\to N^k$ belongs to SN [7].

6.2. Computability on natural numbers.

Computable functions on natural numbers have been studied in [26] and are the bulk of the main construction of [31].

Let $\mathfrak{R}_1=\mathfrak{R}\cap hom(N,N)$. According to [26] this class can be characterized as

$$\mathfrak{R}_1=C_F\{S,E\},$$

where E is the function giving the excess over a square and F is the set containing composition, the addition operator on functions $(f+g)x=fx+gx$ and the inversion operator on functions, we can now define as $f^{-1}=(O\times I)\circ C(f,S\times I)\circ K$.

This class can now be characterized using a while operator as $\mathfrak{R}_1=C_G(X_1)=C_G(X_2)$,

where $X_1=\{S,\ x^2,\ [\sqrt{x}]\}$, $X_2=\{S,\ E,\ x^2,\ [x:2]\}$ and $G=\{f\circ g,\ f+g,\ f\dot{-}g,\ f/g\}$.

REMARK. The class $\mathfrak{R}_1$ can be also obtained by using the weaker repetition operators

C(square numbers, g)

C(triangular numbers, g)

where square numbers$=\{n^2 \mid n\in N\}$ and triangular numbers$=\{(n^2+n):2 \mid n\in N\}$.

6.3. Computability on sequences of integers.

So far we have considered computable functions on well-ordered data.

On such data we have defined computable functions via **while**, but traditionally they are defined via recursion (which is based just on being the carrier set totally or partially well-ordered). We want now to show, by way of an example, that computability can be developed via **while** also on non well-ordered data: in fact it suffices that «computing steps» be well-ordered, i.e. that the «functional order», to which any **while** refers, be well fittered.

Let I be the set of integers and x, y, r $\in$ I. For f, g: $I^h \to I^k$ set

$$(f,g)_1 : I^h \to I^k$$

$$s \mapsto fs \quad \text{if } Ks>0$$

$$s \mapsto gs \quad \text{if } Ks\leq 0$$

$$(f,g)_2 : I^h \to I^k$$

$$s \mapsto fs \quad \text{if } Ks<0$$

$$s \mapsto gs \quad \text{if } Ks\geq 0$$

Consider now the following class of functions on sequences of integers:

$$\mathbf{SI}=C_F\{O, S, P, \Box\}=C_F\{O, S, C, \Box\},$$

where O: $I^0 \to I$ is the zero function, S: $I \to I$ is the successor function, P: $I \to I$ is the predecessor function, $\Box$: $I \to I^0$ is the cancellator function, C: $I \to I$ is the complement function and $F=\{f\circ g,\ f\times g,\ (f,g)_i\ ,\ f/g\}$ for $i=1$ or $i=2$.

Remark 1. The class **SI** can be also obtained by using other operators of the type C(B,g), where B can be the set of sequences whose first element is zero or whose first element is different from zero or whose first element is positive or non negative or non positive or negative.

Remark 2. As usually, the **if**-type operators $(f,g)_i$ can be eliminated if a test function like

$$\text{test}: x \mapsto 1,x \quad \text{if } x\geq 0$$

$$x \mapsto 0,x \quad \text{if } x<0$$

or combinatory functions like

$$\Delta : x \mapsto x,x$$

$$\Theta : x,y \mapsto y,x$$

are introduced.

Assuming Church Thesis, it can be easily shown that any computable function f: $I^h \to I^k$ belongs to **SI**. This can be done by showing that for every f$\in$**SN** there is a conjugate function g$\in$**SI** such that f$=$ c$\circ$g$\circ$d, where c and d are computable codifications c: $N^h \to I^h$ and d: $I^k \to N^k$.

7. Refenreces.

[1] G. Birkhoff, Lattice theory, AMS Colloq. Publ. 25, Providence 1967.

[2] S. Burris and H. P. Sankappanavar, A Course in Universal Algebra, New York 1981.

[3] P. M. Cohn, Universal Algebra, New York 1965.

[4] J. de Bakker, Mathematical Theory of Program Correctness, Englewood Cliffs 1980.

[5] R. Dedekind, Was sind und was sollen die Zahlen, Braunschweig 1888.

[6] G. Germano and A. Maggiolo-Schettini, Equivalence of partial recursivity and computability by algorithms without concluding formulas, Calcolo 8 (1971), 273-292.

[7] G. Germano and. A. Maggiolo-Schettini, Sequence-to-sequence recursiveness, Inform. Processing Lett. 4 (1975), 1-6.

[8] G. Germano and A. Maggiolo-Schettini, Computable stack functions for semantics of stack programs, J. Comput. System Sci. 19 (1979), 133-144.

[9] G. Germano and A. Maggiolo-Schettini, Sequence recursiveness without cylindrification and limited register machines, Theor. Comput. Sci. 15 (1981), 213-221.

[10] M. D. Gladstone, Simplifications of the recursion scheme, J. Symbolic Logic 36 (1971), 653-665.

[11] K. Gödel, Über formal unentscheidbare Sätze der Principia Mathematica und verwandter Systeme. I, Monatsh. Math. Phys. 38 (1931), 173-198.

[12] K. Gödel, On undecidable propositions of formal mathematical systems, Mimeography, Princeton 1934.

[13] G. Grätzer, Universal Algebra, New York 1968.

[14] H. Hermes, Aufzählbarkeit, Entscheidbarkeit, Berechenbarkeit, Berlin 1961.

[15] S. C. Kleene, General recursive functions of natural numbers, Math. Ann. 112 (1936), 727-742.

[16] S. C. Kleene, λ-definability and recursiveness, Duke Math. J. 2 (1936), 340-353.

[17] S. C. Kleene, A note on recursive functions, Bull. Amer. Math. Soc. 42 (1936), 544-546.

[18] S. C. Kleene, On notation for ordinal numbers, J. Simbolic logic 3 (1938), 150-155.

[19] S. C. Kleene, Introduction to Metamathematics, Amsterdam 1952.

[20] S. C. Kleene, The theory of recursive functions approaching its centennial, Bull. Amer. Math. Soc. 5 (1981), 43-60.

[21] Z. Manna, Mathematical Theory of Computation, New York 1974.

[22] J. C. C. Mc Kinsey and A. Tarski, The algebra of topology, Ann. of Math. 45 (1944), 141-191.

[23] J. C. C. Mc Kinsey and A. Tarski, On closed elements in closure algebras, Ann. of Math. 47 (1946), 122-162.

[24] E . H. Moore, Introduction to a form of a general analysis, AMS Colloq. Publ. 2, New Haven 1910.

[25] R. M. Robinson, Primitive recursive functions, Bull. Amer. Math. Soc. 53 (1947), 925-942.

[26] J. Robinson, General recursive functions, Bull. Amer. Math. Soc. 56 (1950), 703-717.

[27] D. S. Scott. The Lattice of Flow Diagrams, Technical Monograph PRG 3, Oxford Univ. Computing Laboratory, Oxford 1970.

[28] D. S. Scott, Lectures on a Mathematical Theory of Computation, Oxford Univ. PRG Tech. Monograph 1981.

[29] J. E. Stoy, Denotational Semantics: The Scott-Strachey Approach to Programming Language Theory, Cambridge, MA 1977.

[30] A. Tarski, Fundamentale Begriffe der Methodologie der deduktiven Wissenschaften I, Monatsh. Math. Phys. 37 (1930), 360-404.

[31] A. Tarski, A. Mostowski and R. M. Robinson, Undecidable Theories, Amsterdam 1953.

[32] P. C. Treleaven, D. R. Brownbridge and R. P. Hopkins, Data-driven and demand-driven computer architecture, ACM Computing Surveys 14 (1982), 93-143.

[33] T. R. Walsh, Iteration strikes back - at the cyclic towers of Hanoi, Inform. Processing Lett. 16 (1983), 91-93.

Toward logic tailored for computational complexity

by Yuri Gurevich[1]

Computer Science
The University of Michigan
Ann Arbor, Michigan 48109

Abstract. Whereas first-order logic was developed to confront the infinite it is often used in computer science in such a way that infinite models are meaningless. We discuss the first-order theory of finite structures and alternatives to first-order logic, especially polynomial time logic.

Introduction

Turning to theoretical computer science a logician discovers with pleasure an important role of first-order logic. One of the fashionable programming languages - PROLOG - is based on first-order logic; variants of first-order logic - Tuple Calculus, Relational Algebra, Domain Calculus - are used as query languages to retrieve information from relational databases; et cetera.

The database applications of first-order logic are of special interest to us here. In this connection let us mention that relational databases are not a side issue in the data field. The relational data model together with the network and the hierarchical data models are "the three most important 'data models', the models that have been used in the great bulk of commercial database systems" [Uℓ, Section 1.4]. The relational data model brought a Turing Award to its inventor E.F. Codd. The three query languages, mentioned above, were also introduced by Codd and are important: "A language that can (at least) simulate tuple calculus, or equivalently, relational algebra or domain calculus, is said to be complete" [Uℓ, Section 6.1].

Some of the new applications of first-order logic are unusual in that only finite structures are of interest. In particular, relational databases can be seen as finite first-order structures (for the purpose of this paper), and the query languages, mentioned above, express exactly the first-order properties of relational databases. The question arises how good is first-order logic in handling finite

[1]Supported in part by NSF grant MCS83-01022

structures. It was not designed to deal exclusively with finite structures. In a sense the contrary is true. It was developed as a tool in Foundations of Mathematics, especially when mathematicians and philosophers confronted paradoxes of the Infinite.

We do not question here the greatness of first-order logic of not necessarily finite structures. Taking into account how elegant, natural and expressive first-order logic is, it is actually amazing that formulas true in all structures (of an appropriate vocabulary) are exactly the ones for which there exist proofs in a specific formal system. (Let us also recall the unique character of first-order logic [Lin].) But what happens to recursive axiomatizability, compactness and other famous theorems about first-order logic in the case of finite structures? We address this question in §1. Our feelings about the answer are expressed in the title of §1: Failure of first-order logic in the case of finite structures.

In §2 we address a certain ineffectiveness of famous theorems about first-order logic. Consider for example Craig's Interpolation Theorem: for each valid implication $\phi \to \psi$ there is an interpolant θ such that

$$\text{vocabulary}(\theta) \subseteq \text{vocabulary}(\phi) \cap \text{vocabulary}(\psi)$$

and the implications $\phi \to \theta$ and $\theta \to \phi$ are valid. No total recursive function constructs an interpolant from the given implication [Kr]. There is no recursive bound on the size of the desired interpolant in terms of the size of the given implication [Fr]. Moreover, weaken the interpolation theorem by replacing "the implications $\phi \to \theta$ and $\theta \to \psi$ are valid" by "the implications $\phi \to \theta$ and $\theta \to \psi$ are valid in all finite structures of appropriate vocabularies". Still there is no recursive bound on the size of the desired interpolant in terms of the size of the given implication.

What is the use of criticizing first-order logic if we cannot come up with a reasonable alternative? We think here about applications where one needs at least the expressive power of first-order logic, like PROLOG or relational query languages. "It is the case that almost all modern query languages embed within them on of the three notations" [Uℓ, Section 6.1]. (The three notations are the tuple calculus, the relational algebra, and the domain calculus.)

One would like to enrich first-order logic so that the enriched logic fits better the case of finite structures. The first temptation of a logician would be to regain recursive axiomatizability. But no extension of the first-order theory of finite structures is recursively axiomatizable. (Satisfiability of first-order formulas on finite structures is recursively axiomatizable. But this axiomatizability provides only a criterion of existence of a formal proof for existence of a finite model. It is not interesting. The whole point of axiomatizability was to provide an existential criterion for a universal statement.)

Another temptation is to consider second-order logic (without third-order predicates or functions) or its fragments (like existential second-order logic) as an alternative to first-order logic. Confining ourselves to finite structures, we consider this alternative in §3.

Second-order logic is certainly elegant, natural and much more expressive than first-order logic. Second-order logic itself becomes more attractive in the case of finite structures: no nonstandard models, no distinction between the weak and the strong versions of second-order logic, etc. There is however one important - from the point of view of computer science - property of first-order logic that is lost in the transition to second-order logic. For every first-order sentence ϕ there is an algorithm that, given a presentation of a structure S of the vocabulary of ϕ, computes the truth-value of ϕ on S within time bounded by a polynomial in the cardinality $|S|$ of S (and within working space bounded by $\log|S|$). In other words, first-order properties are PTIME (and LOGSPACE) computable. Second-order properties and even existential second-order properties are not PTIME computable unless $P=NP$. If one takes the popular point of view that feasible computations are PTIME bounded and that P is probably different from NP then second-order logic is not a good alternative to first-order logic.

Let us mention that computer scientists do feel that first-order logic is unreasonably restrictive. PROLOG does have non-first-order features, and it was suggested to augment the essentially first-order query languages by different operators preserving feasible computability of queries. Of course the notion of feasibility varies with applications. From the point of view of PTIME computability, the least fixed point operator LFP [AU] appeared to be especially important. It preserves PTIME computability and has great expressive power.

A natural idea arises to extend first-order logic in such a way that exactly PTIME (LOGSPACE, etc.) computable properties of structures are expressible in the extended logic. Chandra and Harel [CH2] considered the extension FO + LFP of first-order logic by LFP from that point of view and discovered that FO + LFP does not capture PTIME. It turned out, however, that FO + LFP does capture PTIME in the presence of linear order [IM1, Var]. In §4 we discuss fixed points and logics with order (as a logical constant) tailored for PTIME.

In §5 we return to some of the famous theorems about first-order logic and consider whether their analogues hold in the case of logic specially designed for PTIME. More specifically, we consider the analogues of Craig's Interpolation Theorem, Beth's Definability Theorem and the Weak Beth Definability Theorem for polynomial time logic. These analogues happen to be equivalent to natural complexity principles whose status is unknown.

A lot of interesting problems arise. Design a logic that captures PTIME even in absence of linear order, or prove that there is no reasonable such logic if $P \neq NP$. What is a logic? What is a complexity class? Can every reasonable complexity class be captured by a logic in the presence of linear order? Capture LOGSPACE, NLOGSPACE, LOG^2SPACE, $LOG^2SPACE \cap PTIME$, etc. in the presence of linear order. What are complexity tailored logics good for? Are complexity bounded programming languages useful? Some answers can be found in [Im2] and [Gu3].

Our terminology is more or less standard. We use the term "vocabulary" rather than "signature" or "similarity type", and we use the term "structure" rather than "model" or "algebraic system". Our vocabularies are always finite.

Acknowledgements. I am very grateful to Andreas Blass and Neil Immerman for very useful discussions related to this paper.

§1. Failure of first-order logic in the case of finite structures

We examine famous theorems about first-order logic in the case when only finite structures are allowed. The terms formula and sentence will refer in this section to first-order formulas and first order sentences. As usual, a sentence is a formula without free individual variables.

Recall that a formula ϕ is called valid (or logically true) if it is true in every structure of the vocabulary of ϕ, a formula ϕ (resp. a set Φ of formulas) is said to imply a formula ψ logically if ψ is true in every model of ϕ (resp. of Φ) whose vocabulary includes that of ψ, and formulas ϕ,ψ are called logically equivalent if each of them logically implies the other. We will say that a formula ϕ is <u>valid in the finite case</u> if it is true in every finite structure of the vocabulary of ϕ, a formula ϕ (resp. a set Φ of formulas) <u>implies</u> a formula ψ <u>in the finite case</u> if ψ is true in every finite model of ϕ (respectively of Φ) whose vocabulary includes that of ψ, and formulas ϕ,ψ are <u>equivalent in the finite case</u> if each of them implies the other in the finite case.

The Soundness and Completeness Theorem is formulated usually for a specific logical calculus. It states that a formula is valid iff it is provable in the calculus. The calculus-independent meaning of this theorem is that first-order logic is recursively axiomatizable, which boils down to the fact that valid formulas are recursively enumerable. Trakhtenbrot [Tr] proved that the formulas valid in the finite case are not recursively enumerable. Therefore first-order logic is not recursively axiomatizable in the finite case, and the Soundness and Completeness Theorem fails for any logical calculus in the finite case.

<u>Remark</u>. Tiny fragments of first-order logic are not axiomatizable recursively in the case of finite structures. For example, let σ be a vocabulary that consists of one binary predicate symbol. The $\exists^3\forall^*$ σ-sentences (i.e. the prenex σ-sentences with prefixes $\exists^3\forall^n$), that are valid in the finite case, are not enumerable recursively [Gu1, Ko]. Summaries of results of that sort can be found in [Gu2]. Goldfarb claims that even $\exists^2\forall^*$ σ-sentences with equality, valid in the finite case, are not enumerable recursively [Go].

The Compactness Theorem for first-order logic states that if a set Φ of formulas logically implies another formula ψ then some finite subset of Φ logically implies ψ. The theorem fails in the finite case. Let for example $\Phi=\{\phi_n : n\geq 1\}$ where every sentence ϕ_n states existence of at least n different elements, and let ψ be any logically false formula. Then Φ implies ψ in the finite case; however no finite subset of Φ implies ψ in the finite case.

The Craig Interpolation Theorem states that if a formula ϕ logically implies a formula ψ then there is a formula θ (an interpolant) such that

$$\text{vocabulary}(\theta)\subseteq \text{vocabulary}(\phi)\cap \text{vocabulary}(\psi),$$

ϕ logically implies θ, and θ logically implies ψ.

The interpolation theorem implies the Beth Definability Theorem that states the following. Suppose that a sentence $\phi(P)$ defines an ℓ-ary relation P implicitly i.e. if P' is a new ℓ-ary predicate symbol then $\phi(P)$ and $\phi(P')$ imply

$$\forall x_1 \ldots \forall x_\ell (P(x_1,\ldots,x_\ell) \longleftrightarrow P'(x_1,\ldots,x_\ell)).$$

Then there is an explicit first-order definition of the same relation i.e. there is a formula $\theta(x_1,\ldots,x_\ell)$ such that

$$\text{vocabulary}(\theta)\subseteq \text{vocabulary}(\phi(P)) - \{P\}$$

and $\phi(P)$ logically implies

$$\forall x_1 \ldots \forall x_\ell (P(x_1,\ldots,x_\ell) \longleftrightarrow \theta(x_1,\ldots,x_\ell)).$$

If $\phi(P)$ and P' are as in the antecedent of the Beth Definability Theorem then $\phi(P)\&P(x_1,\ldots,x_\ell)$ logically implies $\phi(P')\longrightarrow P'(x_1,\ldots,x_\ell))$, and the corresponding interpolant is the desired explicit definition. The same proof shows that the finite case version of the interpolation theorem implies the finite case version of the definability theorem.

The Weak Definability Theorem is the result of strengthening the antecedent of the Beth Definability Theorem. The antecedent of the Beth Definability Theorem states that for every structure of the vocabulary

$$\sigma=\text{vocabulary}(\phi(P))-\{P\}$$

there is at most one relation P that satisfies $\phi(P)$. The antecedent of the Weak

Definability Theorem states that for every σ-structure there is a unique relation-P that satisfies $\phi(P)$.

Theorem 1. The Craig Interpolation Theorem, the Beth Definability Theorem and the Weak Definability Theorem fail in the finite case.

Proof. Let us recall the definition of the quantifier depth of a formula:

q.d.(a quantifier-free formula)=0

q.d.(a Boolean combination of formulas $\alpha_1,\dots,\alpha_m$)=$\max\{$q.d.$(\alpha_1),\dots,$q.d.$(\alpha_m)\}$

q.d.$(\forall x\alpha)$=q.d.$(\exists x\alpha)$=1 + q.d.(α).

Lemma.

(i) Suppose that α is a sentence in the vocabulary $\{<\}$ of order, n is the quantifier depth of α, and A,B are finite linear orders of cardinalities $|A|,|B|\geq 2^n$. Then α does not distinguish between A and B i.e. A satisfy α iff B satisfies α.

(ii) There is no sentence α in the vocabulary $\{<\}$ such that an arbitrary finite linear order S satisfies α iff the cardinality $|S|$ is even.

(iii) There is no formula $\theta(x)$ in the vocabulary $\{<\}$ such that if S is a finite order $a_1<a_2<\dots<a_n$ then $\{x:S\models\theta(x)\}=\{a_k: k \text{ is even}\}$.

Proof of Lemma.

(i) We use the Ehrenfeucht games [Eh]. It suffices to exhibit a winning strategy for player II in the Ehrenfeucht game $G_n(A,B)$. Without loss of generality no element is picked twice during the game. The proposed strategy is to ensure the following. Let $a_1<a_2<\dots<a_k$ and $b_1<b_2<\dots<b_k$ be the elements chosen in A and B respectively during the first k steps of the game. Let $A_0,A_1,\dots,A_k$ be the segments $[\min(A),a_1]$, $[a_1,a_2],\dots,[a_k,\max(A)]$ of A, and let $B_0,B_1,\dots,B_k$ be the respective segments of B. Then for every $i=1,\dots,k$ the elements a_i,b_i were chosen at the same step of the game, and either $|A_i|=|B_i|$ or $|A_i|,|B_i|>2^{n-k}$ for $0<i<k$, and either $|A_i|=|B_i|$ or $|A_i|,|B_i|\geq 2^{n-k}$ for $i\in\{0,k\}$.

(ii) The statement follows from (i).

(iii) If $\theta(x)$ defines the set of even elements in every finite linear order then the sentence

$$\exists x(x \text{ is maximal and } \theta(x))$$

holds in an arbitrary finite linear order S iff $|S|$ is even.□

Since the interpolation theorem implies the definability theorem and the definability theorem implies the weak definability theorem, it suffices to refute the weak definability theorem. It is easy, however, to construct separate counter-examples to all three theorems.

Write a sentence α stating that $<$ is a linear order. let P,Q be distinct unary predicates. Write a sentence $\beta(P)$ in the vocabulary $\{<,P\}$ such that if S is a finite linear order $a_1<a_2<\ldots<a_n$ and S satisfies $\beta(P)$ then $\{x:S \models P(x)\}=\{a_k:k \text{ is even}\}$. (Write that P does not contain the first element, and the successor of an element x belongs to P iff x does not belong to P.) Obviously $\alpha\&\beta(P)\&P(x)$ implies $\beta(Q)\longrightarrow Q(x)$ in the finite case. If the interpolation theorem were true in the finite case then the interpolant would violate the statement (iii) of the Lemma.

Obviously, $\alpha\&\beta(P)$ defines P implicitly in finite structures. If the definability theorem were true in the finite case then the explicit definition of P would violate the statement (iii) of the Lemma. Finally, the sentence

$$(\alpha \longrightarrow \beta(P))\&(\neg\alpha \longrightarrow \neg\exists x P(x))$$

defines P uniquely in the finite case. If the weak definability theorem were true in the finite case then the explicit definition would violate the statement (iii) of the Lemma. Theorem 1 is proved.

<u>Remark</u>. The formula $\beta(P)$ in the proof of Theorem 1 can be simplified if we use an individual constant for the first element in the order and an additional binary predicate symbol for the successor relation. The Lemma remains true for the richer vocabulary if 2^n is changed to 2^{n+1} (with an obvious change in the proof).

A sentence ϕ is said to be preserved under substructures if every substructure of a model of ϕ is a model of ϕ. According to the Substructure Preservation Theorem [CK, §3.2], a sentence ϕ is preserved under substructures iff it is logically equivalent to a universal sentence.

<u>Theorem</u> 2 (Tait). <u>The Substructure Preservation Theorem fails in the case of finite structures. In other words, there is a sentence ϕ such that any substructure of a finite model of ϕ is a model of ϕ, yet ϕ is not equivalent to any universal sentence in the finite case.</u>

Proof. Let ϕ_1 be the universal closure of the conjunction of the following formulas (where $x\leq y$ abbreviates $x=y \vee x<y$):

$$(x<y \mathbin{\&} y<z) \longrightarrow x<z,$$

$$\neg(x<x),$$

$$x\leq y \vee y\leq x,$$

$$0\leq x,$$

$$[S(x,y) \mathbin{\&} y\neq 0] \longrightarrow [x<y \mathbin{\&} (z\leq x \vee y\leq z)],$$

$$S(x,0) \longrightarrow y\leq x.$$

ϕ_1 states that $<$ is a linear order, 0 is the minimal element, and Sxy implies that either y is the successor of x or else x is the maximal element and $y=0$. Let ϕ_2 be $\forall x \exists y\, S(x,y)$, and let ϕ be $\phi_1 \mathbin{\&} (\phi_2 \longrightarrow \exists x P(x))$ where P is a unary predicate symbol.

First we check that ϕ is preserved under substructures of finite models. Suppose that A is a finite model of ϕ and B is a substructure of A. Then B contains 0 and satisfies ϕ_1. If B does not satisfy ϕ_2 then it satisfies the second conjunct of ϕ by default. If B satisfies ϕ_2 then $B=A$ and B satisfies ϕ.

Next, let α be a sentence $\forall x_1 \ldots \forall x_n \beta(x_1,\ldots,x_n)$ where β is quantifier-free. Let A be a model of $\phi_1 \mathbin{\&} \phi_2$ such that the vocabulary of A includes that of α, A has at least $n+2$ elements and P is empty in A, so that ϕ fails in A. If α is true in A then it is not equivalent to ϕ in the finite case. Suppose that α is false in A. Then $\beta(c_1,\ldots,c_n)$ is false in A for some $c_1,\ldots,c_n$. Choose $d\in A$ different from $0,c_1,\ldots,c_n$, and put d into P. The resulting structure B satisfies ϕ. However $\beta(c_1,\ldots,c_n)$ remains false in B. Hence α is false in B, and α is not equivalent to ϕ in the finite case.□

I did not perform an exhaustive study of important theorems about first-order logic in the finite case. Some theorems become meaningless in the finite case.

Some theorems do survive: the game criterion for two structures to be indistinguishable by sentences of a given quantifier depth [Eh], composition theorems of the sort found in [FV], etc. Moreover, some theorems were specifically proved for the finite case: the 0-1 Law Theorem for example [GKLT, Fa]. Too often however we see the familiar pattern: the proof uses a kind of compactness argument and the theorem fails in the finite case. Sometimes a weaker version of the theorem in question survives. Here is an example. Recall that an $\forall^*\exists^*$ sentence is a prenex sentence with a prefix $\forall^m\exists^n$.

Theorem 3 (Compton). Let ϕ be an $\forall^*\exists^*$ sentence without function symbols. If ϕ is preserved by substructures of its finite models then it is equivalent to some universal sentence in the finite case.

Proof. First let us recall a relativized version of the Substructure Preservation Theorem:

Let T_0 be a first-order theory, and α be a sentence in the language of T_0. Suppose that for every model A of T_0 and for every substructure B of A that is a model of T_0, if A is a model of α then B is a model of α. Then α is equivalent in T_0 to some universal sentence.

The usual proof of the Substructure Preservation Theorem is easily relativizable: just take Δ to be the set of all sentences, that are equivalent in T_0 to universal sentences, in the proof of Theorem 3.2.2 in [CK].

In our application T_0 is the first-order theory of finite structures of the vocabulary of ϕ. Let A be a (possibly infinite) model of T_0 that satisfies ϕ. Let B be a substructure of A that is also a model of T_0. It suffices to prove that B satisfies ϕ.

First, we show that an arbitrary finite substructure A_0 of A satisfies ϕ. Write an existential sentence α stating existence of elements that form a structure isomorphic to A_0. The sentence $\alpha\&\phi$ has a finite model A_1: otherwise $T_0 \vdash \neg(\alpha\&\phi)$ which contradicts the fact that A is a model of T_0, α and ϕ. Since A_1 satisfies α it has a substructure isomorphic to A_0. Now use the fact that ϕ holds in A_1 and is preserved by substructures of finite structures.

Recall that ϕ is

$$\forall x_1 \ldots \forall x_m \exists y_1 \ldots \exists y_n \psi(x_1,\ldots,x_m,y_1,\ldots,y_n)$$

for some quantifier-free formula ψ. We argue by reductio ad absurdum. Suppose that B fails to satisfy ϕ. Then there are elements $a_1,\ldots,a_m$ such that the universal formula

$$\forall y_1 \ldots \forall y_n \neg\psi(a_1,\ldots,a_m,y_1,\ldots,y_n)$$

holds in B. This universal formula logically implies $\neg\phi$ and holds in the substructure $A_0=\{a_1,\ldots,a_m\}$ of B (because universal formulas are preserved by substructure). Thus a finite substructure of A fails to satisfy ϕ which is impossible.□

Note that the counterexample to the Substructure Preservation Theorem, constructed in the proof of Theorem 2 is logically equivalent to an $\exists^*\forall^*$ sentence. Thus Theorems 2 and 3 delimit each other.

Historical Remarks. I am not the first to discover that Craig's Interpolation Theorem and Beth's Definability Theorem fail in the finite case. (A question of Steve Simpson led me from Craig's Theorem to Beth's Theorem.) Ron Fagin knew about the failure. It was probably discovered long ago though I do not have any reference.

Theorem 2 was proved in [Ta]. The proof above is due to Gurevich and Shelah (that were not aware [Ta]). Theorem 3 was formulated and proved by Kevin Compton in a letter [Co] to me.

§2. An ineffective side of first-order logic

We saw in §1 that Craig's Interpolation Theorem, Beth's Definability Theorem, the Weak Definability Theorem and the Substructure Preservation Theorem fail in the case of finite structures. One may be tempted to allow infinite structures (to allow infinite relational databases in database theory) in order to regain these wonderful theorems; see [Va] for example. There is however a catch there. Let us speak, for example, about the weak definability theorem. Even if you happen to know that $\phi(P)$ implicitly defines a relation P in every - finite or infinite - structure and even if you are interested in an explicit definition of the same relation P in finite structures only, still constructing the desired explicit definition from the given implicity definition may be most problematic. This is the point of the present section. Again, the terms formula and sentence mean first-order formulas and first-order sentences. The length of a formula ϕ is denoted $|\phi|$.

So then, how constructive are the wonderful theorems mentioned above? In a certain sense the interpolation theorem is very constructive. The desired interpolant for a valid implication $(\phi\to\psi)$ is easily constructible from a proof of $(\phi\to\psi)$ in an appropriate predicate calculus [Cr]. In the same sense the definability theorem is very constructive because the desired explicit definition can be found as an interpolant for an implication that is easily built from the given implicit definition, see §1.

There are also partial recursive functions f and g such that if $(\phi\to\psi)$ is a valid implication then $f(\phi\to\psi)$ is an interpolant for $(\phi\to\psi)$, and if $\phi(P)$ is an implicit definition of a relation P then $g(\phi(P))$ is an explicit definition of the same relation. However, there are no total recursive functions f and g with the same properties [Kr]. Moreover, there are no total recursive functions that bound the length of the desired interpolant or explicit definition in terms of the length of a given formula [Fr]. Even the weak definability theorem is ineffective in that sense: the length of the desired explicit definition is not bounded by any recursive function of the length of a given implicit definition. The next theorem gives a straightforward proof of this result of Friedman and strengthens it in a way related to finite structures.

<u>Theorem 1. For every total recursive function</u> f <u>there is a sentence</u> $\phi(P)$ <u>such that</u>

(i) $\phi(P)$ <u>implicitly defines a relation</u> P <u>in every structure of the vocabulary</u> σ=<u>vocabulary</u>$(\phi(P))-\{P\}$, <u>and</u>

(ii) <u>if</u> ψ <u>is an explicit definition of the same relation</u> P <u>in every finite</u> σ-<u>structure, then</u> $|\psi|\geq f(|\phi(P)|)$.

<u>Proof.</u> Given a total recursive function f we construct an auxiliary total recursive function g. The exact definition of g will be given later. Let M be a Turing machine that computes g. We suppose the following about M. Its internal states are $q_0,\ldots,q_m$ here q_0 is the initial state and q_m is the halting state. The only tape of M is one-way infinite, the tape alphabet is $\{0,1\}$ where 0 is also the blank. An instruction of M is a 5-tuple $q_i a q_j b d$ where $d\varepsilon\{-1,0,1\}$ indicates whether the head of M will move to the left, stay still or move to the right. If at moment 0 the state of M is q_0, the head is in cell 0 and the tape word is 1^n then M will eventually halt in the halting state q_m with the tape word $1^{g(n)}$.

In order to describe computations of M by formulas we introduce unary predicates $q_0(t),\ldots,q_m(t)$ to indicate the state at moment t, a binary predicate $H(x,t)$ to indicate that the head is in cell x at moment t, a binary predicate $C(x,t)$ to indicate that the content of cell x at moment t is 1, and unary predicates $D_{-1}(t)$, $D_0(t)$, $D_{+1}(t)$ to indicate the move of the head that the machine is instructed at moment t to perform.

In order to use all these predicates properly, we need binary predicates $<,S$ and an individual constant 0. Let a sentence ϕ_0 state that $<$ is a linear order, 0 is the minimal element, S is the corresponding successor relation, and every nonmaximal element has a successor.

A sentence ϕ_1^n describes the initial configuration of M with the input 1^n. It is the conjunction of sentences

$q_0(0)$, $H(0,0)$,

$\exists x_0\ldots\exists x_n[x_0=0 \text{ and } \bigwedge_{i<n} S(x_i,x_{i+1}) \text{ and } \bigwedge_{i<n} C(x_i,0) \text{ and } \neg C(x_n,0)]$,

$\forall x\forall y[\neg C(x,0) \text{ and } x<y \text{ imply } \neg C(y,0)]$.

A sentence ϕ_2 describes one computational step. It is the universal closure of a quantifier-free conjunction. Every instruction $q_i a q_j bd$ contributes the conjunct

$$[q_i(t) \& H(x,t) \& C_a(x,t) \& S(t,t') \text{ implies } q_j(t') \& C_b(x,t') \& D_d(t)]$$

where C_1, C_0 are $C, \neg C$ respectively. In addition the quantifier-free part of ϕ_2 has the following conjuncts:

$[\neg q_i(t) \text{ or } \neg q_j(t)]$ for $0 \leq i < j \leq m$,

$[H(x,t) \text{ and } H(y,t) \text{ imply } x=y]$,

$[\neg D_d(t) \text{ or } \neg D_e(t)]$ for $-1 \leq d < e \leq 1$,

$[D_0(t) \& H(x,t) \& S(t,t') \text{ implies } H(x,t')]$,

$[D_1(t) \& H(x,t) \& S(t,t') \& S(x,x') \text{ implies } H(x',t')]$,

$[D_{-1}(t) \& S(t,t') \& S(x,x') \& H(x',t) \text{ implies } H(x,t')]$,

$[\neg H(x,t) \text{ and } S(t,t') \text{ imply } (C(x,t') \longleftrightarrow C(x,t))]$.

A sentence ϕ_3 describes what happens after halting. It is the universal closure of the formula

$$[q_m(t) \text{ and } t<u \text{ imply } (\bigwedge_{i<m} \neg q_i(u) \text{ and } \neg H(x,u) \text{ and } \neg C(x,u) \text{ and } \bigwedge_{-1 \leq d \leq 1} \neg D_d(u))].$$

<u>Lemma. For every model A of ϕ_0 and for every natural number n there are unique predicates</u>

$$q_0, \ldots, q_m, H, C, D_{-1}, D_0, D_1$$

on A that satisfy $\phi_1^n \& \phi_2 \& \phi_3$.

<u>Proof</u> is straightforward. In particular, the sentences ϕ_0, ϕ_1^n, ϕ_2 and ϕ_3 imply that for every t there is a unique x with $H(x,t)$: the head does not slip from the tape because M computes a total function, and if $D_1(t)$, $H(x,t)$, $S(t,t')$ hold then $x \leq t < t'$ and there is x' such that $S(x,x')$, $H(x',t')$ hold.

Let P be a ternary predicate symbol. Write a sentence $\phi^n(P)$ that states the following. If ϕ_0 fails or there are at most m+3 elements then P is empty. If ϕ holds and there are more than m+3 elements then

(a) ϕ_1^n, ϕ_2, ϕ_3 hold where $q_i(t), D_d(t), H(x,t), C(x,t)$ abbreviate $P(0,i,t)$, $P(0,m+2+d,t), P(1,x,t), P(2,x,t)$ respectively, and

(b) $P(0,x,t)$ fails for $x>m+3$, and $P(x,y,t)$ fails for $x>2$.

When numbers 1,2,etc. appear as arguments of P they mean of course the successor of 0, the successor of the successor of 0, etc. It is easy to see that ϕ^n implicitly defines a relation P in every structure of the vocabulary $\sigma=\{<,0,S\}$. Let ψ^n be an explicit definition of the same relation in every finite σ-structure.

Note that ϕ_1^n and the quantifier depth of ϕ^n do not depend on the choice of g. Define g(n) to be the power of 2 such that

$$\log_2 g(n)=f(|\phi_1^n|+n)+\text{q.d.}(\phi^n)+1.$$

ϕ_1^n is the only part of ϕ^n that depends on n. It occurs in ϕ^n only once. Thus the number $k=|\phi^n|-|\phi_1^n|$ does not depend on n. Let $\phi=\phi^k$ and $\psi=\psi^k$. Then

$\log_2 g(k)=f(|\phi|)+\text{q.d.}(\phi)+1.$

Let α be the sentence

$[\phi(\psi) \text{ and } \exists t q_m(t)]$.

Every model of α reflects the whole computation M and has at least g(k) elements. By the Remark following the Lemma in §1, $g(k)\leq 2^{1+\text{q.d.}(\alpha)}$. Hence

$$\text{q.d.}(\alpha)+1\geq\log_2 g(k)=f(|\phi|)+\text{q.d.}(\phi)+1$$

But $\text{q.d.}(\alpha)\leq\text{q.d.}(\phi)+\text{q.d.}(\psi)$. Hence $|\psi|\geq\text{q.d.}(\psi)\geq f(|\psi|)$.□

<u>Remark</u> 1. It is easy to make the relation P of Theorem 1 unary. The idea is to use auxiliary elements to code triples of real elements.

<u>Remark</u> 2. Mundici exhibits in [Mu] short valid implications $(\phi\to\psi)$ whose interpolants are enormously long. The proof of Theorem 1 can be used for analogous purposes.

<u>Theorem 2. For every total recursive function f there is a sentence ϕ such that</u>

(i) <u>ϕ is preserved by substructures, and</u>

(ii) <u>if ψ is a universal sentence that is equivalent to ϕ in every finite structure of the vocabulary of ϕ then $|\psi|\geq f(|\phi|)$.</u>

<u>Proof.</u> Let f be a total recursive function. As in the proof of Theorem 1, let g be an auxiliary total recursive function (specified later) and let M be a Turing machine that computes g. Once again we describe computations of M by

first-order sentences. However, we take some additional care to make the desired description preserved under substructures.

Instead of the sentence ϕ_0 in the proof of Theorem 1 we use sentences ϕ_1,ϕ_2 from the proof of Theorem 2 in §1. We call them ϕ_{01} and ϕ_{02} here. We split the sentence ϕ_1^n from the proof of Theorem 1 into a conjunction $\phi_{11}\&\phi_{12}^n$ where ϕ_{12}^n is the existential conjunct of ϕ_1^n and ϕ_{11} is the conjunction of the two other conjuncts of ϕ_1^n. Let ϕ^n be the sentence

$$\phi_{01} \,\&\, \phi_{11} \,\&\, \phi_2 \,\&\, \phi_3 \,\&\, [\phi_{02} \,\&\, \phi_{12}^n \rightarrow \exists t \exists x(q_m(t) \,\&\, x \leq t \,\&\, Q(x))]$$

where ϕ_2 and ϕ_3 are the sentences from the proof of theorem 1 and ϕ is a new unary predicate.

First we check that every ϕ^n is preserved by substructures. Let A be a model of ϕ^n. Every substructure B of A contains 0 and satisfies the sentences $\phi_{01},\phi_{11},\phi_2,\phi_3$ because universal sentences are preserved by substructures. If B does not satisfy ϕ_{02} or ϕ_{12}^n then it satisfies the last conjunct of ϕ^n by default. Suppose that ϕ^n satisfies ϕ_{02} and ϕ_{12}^n. Since B satisfies ϕ_{02} it is closed in A under successors. If A is finite then B is equal to A and satisfies ϕ^n. Suppose A is infinite. Then B includes the least substructure of A closed under successors whose elements can be identified with natural numbers in the obvious way. Since B satisfies the existential sentence ϕ_{12}^n the structure A satisfies ϕ_{12}^n too. It is easy to see that A reflects the whole computation of the machine M on input 1^n. If M halts at moment T(n) then $q_m(T(n))$ holds in A. In virtue of ϕ_3 there is no element $u>T(n)$ in A that satisfies q_m. Since A satisfies ϕ^n there is some $x \leq T(n)$ in A that satisfies Q. Both T(n) and x belong to B; hence B satisfies ϕ^n.

Note that ϕ_{12}^n does not depend on the choice of g. Define $g(n)=f(|\phi_{12}^n|+n)$. Since ϕ_{12}^n is the only part of ϕ^n that depends on n, the number $k=|\phi^n|-|\phi_{12}^n|$ does not depend on n. Let $\phi=\phi^k$. Then $g(k)=f(|\phi|)$. The computation of M on input 1^k halts at certain moment that will be denoted T(k).

Finally let ψ be a universal sentence $\forall x_1 \ldots \forall x_\ell \psi'(x_1,\ldots,x_\ell)$ that is equivalent to ϕ in the finite case. Here ψ' is quantifier-free. Let A be the model of $\phi_{01} \,\&\, \phi_{02} \,\&\, \phi_{11} \,\&\, \phi_{12}^k \,\&\, \phi_2 \,\&\, \phi_3$ with the universe $\{0,1,\ldots,T(k)\}$ and the intended interpretation of the predicates. First we define Q to be empty in A. The resulting structure A_0 does not satisfy ϕ; hence it does not satisfy ψ and $\neg\psi'(c_1,\ldots,c_\ell)$ holds in A_0 for some $c_1,\ldots,c_\ell$. If $\ell<T(k)$ choose $c\epsilon A-\{0,c_1,\ldots,c_\ell\}$ and put c into Q. The resulting structure A_1 satisfies ϕ yet $c_1,\ldots,c_\ell$ still witness failure of ψ in A_1 which is impossible. Thus $|\psi|\geq\ell\geq T(k)\geq g(k)\geq f(|\phi|)$. ◻

§3. First-order logic versus second-order logic

In spite of the criticism in Sections 1 and 2, first-order logic is still a very good logic even in the case of finite structures. It is not without reason that first-order logic is used in computer science. It is elegant, natural and fairly expressive. However, if elegance, naturality and expressiveness are that important why wouldn't we turn to second-order logic? Second-order logic is elegant and natural as well, and it is much more expressive.

Second-order logic is not very popular among logicians. The objection against second-order logic is that it is not well manageable. However some fragments of second-order logic are much better manageable. One of them is weak second-order logic, which allows quantification over finite predicates only. In the finite case, of course, there is no difference between the two versions of second-order logic.

As we saw in §1 the theorems that made first-order logic so much preferable to second-order logic often fail or become meaningless in the finite case. Is there any important advantage of first-order logic versus second-order logic in the finite case? We take a computational point of view and answer this question positively.

Proviso 1. <u>The term "structure" refers to finite structures if the contrary has not been stated explicitly.</u>

A structure will be viewed as certain data, as an input to algorithms. A seeming difficulty is that elements of a structure are not necessarily constructive objects. We are interested however in the isomorphism type of a given structure rather than in the nature of its elements. Recall that $|S|$ is the cardinality of a structure S.

Proviso 2. <u>The universe of a structure S consists of numbers</u> $0,1,\ldots,|S|-1$.

Proviso 2 by itself does not turn structures into inputs. We still have to choose a way to represent basic relations and functions. For example, a graph (V,E) may be represented as the lexicographically ordered list of edges or as an array $A(i,j)$ where $A(i,j)=1$ if $(i,j)\in E$ and $A(i,j)=0$ otherwise.

Proviso 3. <u>A reasonable standard way to represent structures is chosen.</u>

We introduce global predicates. Let σ be a vocabulary. An ℓ-ary <u>σ-predicate</u> is a function π that assigns to each σ-structure S an ℓ-ary predicate π^S on S. (The superscript S will be usually omitted.) A zero-ary σ-predicate π assigns a truth value to each σ-structure and therefore can be viewed as the set $\{S: S \text{ is a } \sigma\text{-structure and } \pi^S \text{ is true}\}$. Every first-order formula in the vocabulary σ with ℓ free variables gives an ℓ-ary σ-predicate. A <u>global predicate</u> is a σ-predicate for some σ.

<u>Examples</u>. Let $\sigma=\{E\}$ where E is a binary predicate symbol. Note that σ-structures are graphs. The first example is a binary σ-predicate π_1 such that for any graph G and any elements x,y of G, $\pi_1(x,y)$ holds in G iff there is an E-path from x to y. A more usual way to describe π_1 is just to say that π_1 is the binary σ-predicate "There is an E-path from x to y". The second example is the set π_2 of symmetric graphs. In other words, π_2 is a zero-ary σ-predicate such that π_2 holds in a graph G iff G is symmetric. Note that every relational query is a global predicate.

With each global predicate π we associate the problem of computing (or recognizing) π. It is a decision problem. An instance of this decision problem is a pair $(S,\bar{x})$ where S is a structure of the vocabulary of π and $\bar{x}$ is a tuple of elements of S whose length is the arity of π. The corresponding question is whether $\pi(\bar{x})$ holds in S. In order to avoid trivialities we suppose that the length of the presentation of S is at least $|S|$.

<u>Theorem 1.</u> <u>A Boolean combination of PTIME recognizable global predicates is a PTIME recognizable global predicate. If $\pi(x_1,\ldots,x_\ell,y)$ is an $(\ell+1)$-ary PTIME recognizable global predicate then $\exists y\pi(x_1,\ldots,x_\ell,y)$ is an ℓ-ary PTIME recognizable global predicate (with an obvious meaning). Every first-order global predicate is PTIME recognizable.</u>

<u>Proof</u>. The first statement is obvious. The compute the truth value of $\exists y\pi(\bar{x},y)$ in S compute successively the truth values for $\pi(\bar{x},0),\pi(\bar{x},1),\ldots,$ $\pi(\bar{x},|S|-1)$ in S. Since atomic first-order predicates are PTIME computable (here

we need reasonable standard representationa of structures), the third statement follows from the first two. □

Some second-order global predicates are NP-complete. For example, the set of 3-colorable graphs - a well-known NP-complete set - is definable by a second-order sentence $\exists X \exists Y \exists Z \psi(X,Y,Z)$ where X,Y,Z are unary predicate variables and ψ is first-order. Attaching little gadgets to vertices it is easy to construct an NP-complete set of graphs definable by a second-order sentence $\exists X \psi(X)$ where X is a unary predicate variable and ψ is first-order. Thus there are second-order global predicates that are not PTIME recognizable unless $P=NP$.

It is almost a consensus in Theoretical Computer Science that PTIME computations are feasible wherease superpolynomial time computations are intractable, see [GJ], [HU]. In particular, Hopcoft and Ullman write the following. "Although one might quibble that an n^{57} step algorithm is not very efficient, in practice we find that problems in P usually have low-degree polynomial time solutions".

Thus first-order global predicates appear to be feasibly recognizable, whereas recognizing a second-order global predicate may be intractable. From our point of view, explicit PTIME recognizability is a decisive advantage of first-order logic versus second order logic.

<u>Remark</u>. Theorem 1 remains true if "PTIME" is replaced by "LOGSPACE". The same proof proves the new (and stronger) version of the theorem: just represent numbers in binary. Theorem 1 and the stronger version of it are well-known.

§4. Fixed points and polynomial time logic

Provisos 1-3 of §3 are in force.

As we saw above in §3, first-order global predicates are PTIME computable and even LOGSPACE computable. Unfortunately neither of these two statements can be reversed. For example, the property of graphs to be of even cardinality is recognizable by an obvious algorithm in linear time and logarithmic space. In virtue of the Lemma in §1 this property is not first-order.

A natural idea arises: to augment first-order logic by additional operators in order to express exactly the PTIME (LOGSPACE, etc.) computable global predicates. This is the idea reflected in the title: given a complexity level to tailor a logic expressing exactly the global predicates computable within the complexity level. Neil Immerman uses the word "capture" [Im2]. The problem is to capture a given complexity level by logical means. This section is devoted mainly to logic tailored for PTIME.

Remark. Actually it makes sense to generalize the notion of global predicate to the notion of global function and try to capture exactly the global functions computable within a given complexity level. Restricting attention to global predicates is even ridiculous if we see our logic as a notation system for algorithms or a potential programming language. Just imagine a programming language such that each program outputs only a boolean value. Global functions and functional (rather than predicate) logics are explored in [Gu3].

Let us start with a note that first-order expressible global predicates apparently do not form a natural complexity class. They certainly do not form a complexity class defined by Turing machines with bounds on time and/or space (see again the even cardinality example). A computational model which is much closer to first-order logic is that of uniform sequences of boolean circuits of constant depth, unbounded fan-in, and polynomial size. Modest extensions of first-order logic do capture natural circuit complexity classes, see [Im2] and especially [GL] in this connection.

If we consider *NP*, co-*NP* and higher levels of the polynomial hierarchy [St] as genuine complexity classes then second-order logic and some of its natural sublogics do capture complexity classes. (When we speak about second-order logic we suppose that there are no third-order predicates or functions.) Recall that an existential second-order formula is a second-order formula $\phi=\exists X_1 \ldots \exists X_k \psi$ where ψ is first-order and $X_1,\ldots,X_k$ are predicate (or function) variables. The formula ψ may have free predicate and function variables as well as free individual variables.

Theorem 1. A global predicate is computable in polynomial time by a nondeterministic Turing machine if and only if it is expressible by an existential second-order formula.

Theorem 1 is due to Fagin [Fa1] and is readily generalizable to capture co-*NP* and higher levels of the polynomial hierarchy. Actually Fagin did not seek to characterize *NP*. It was just the other way around. He sought to characterize existential second-order sentences (generalized spectra in his terminology). Theorem 1 grew from investigations on spectra of first-order sentences [Be, JS, Fa1, Bö]. It looks pretty obvious today, and nondeterministic polynomial time computable global predicates are not necessarily feasible. However existential second-order logic does capture exactly the nondeterministic polynomial time computable global predicates and this fact inspired attempts to capture in a similar way deterministic PTIME computable global predicates. (About extending Fagin's result to richer logics and higher complexity classes see [St] and [CKS].)

Meantime Codd proposed the relational database model and used variations of first-order logic (relational algebra, relational calculus) as query languages [Uℓ]. The relational model was a big success. However, the first-order query languages were proven to be too restrictive in many applications. Attempts were made to enrich those languages by additional operators, most notably by the transitive closure operator [Zℓ] and the least fixed point operator [AU].

The transitive closure of a binary global predicate $\alpha(x,y)$ of some vocabulary σ is a global σ-predicate $\beta(x,y)$ such that for every σ-structure S the relation β^S is the transitive closure of the relation α^S. More generally one can speak about the transitive closure of a global predicate $\alpha(\bar{x},\bar{y})$ where $\bar{x},\bar{y}$ are tuple of individual variables of the same length [Im2]. In addition to $\bar{x}$ and $\bar{y}$, α may have individual parameters. First-order expressible global predicates are not closed under the transitive closure operator; see Appendix 2.

In a conversation with Andreas Blass the question of notation for the transitive closure of $\alpha(\bar{p},\bar{x},\bar{y})$ was raised. The naive notation $TC\alpha(\bar{p},\bar{x},\bar{y})$ is ambiguous. A possible unambiguous notation is

$$TC(\bar{x},\bar{y};\ \alpha(\bar{p},\bar{x},\bar{y}),\bar{u},\bar{v}) \quad \text{or} \quad TC_{\bar{x},\bar{y}}(\alpha(\bar{p},\bar{x},\bar{y}),\bar{u},\bar{v}).$$

Here $\bar{x}$ and $\bar{y}$ are tuples of bound variables, $\bar{p}$ is a tuple of parameters, and $\bar{u},\bar{v}$ are tuples of new free variables.

Let us define the least fixed point operator for global predicates. It will be convenient to view global predicates as global sets: a global ℓ-ary predicate α of a vocabulary σ assigns a set $\alpha^S \subseteq S^\ell$ to each σ-structure S. We order global ℓ-ary σ-predicates by inclusion: $\alpha \le \beta$ if $\alpha^S \subseteq \beta^S$ for every σ-structure S. We say that a global σ-predicate α is <u>empty</u> if α^S is empty for every σ-structure S.

<u>Definition</u>. Let σ be a vocabulary, P be an additional predicate variable of some arity ℓ, and $\pi(P)$ be a global ℓ-ary predicate of the vocabulary $\sigma \cup \{P\}$. View $\pi(P)$ as an operator that, given a global ℓ-ary σ-predicate α, produces a global ℓ-ary σ-predicate $\pi(\alpha)$. A global ℓ-ary σ-predicate α is a <u>fixed point</u> for $\pi(P)$ if $\alpha=\pi(\alpha)$, and α is the <u>least fixed point</u> for $\pi(P)$ if it is a fixed point and $\alpha \le \beta$ for every fixed point β for $\pi(P)$.

Recall the notion of monotonicity of a first-order formula in a predicate variable defined in Appendix 1. This notion obviously generalizes to monotonicity of a global predicate in a predicate variable.

<u>Claim</u> 1. <u>Let</u> σ,P,ℓ <u>and</u> $\pi(P)$ <u>be as in the definition above</u>. <u>Suppose that</u> $\pi(P)$ <u>is monotone in</u> P. <u>Then there is a unique least fixed point for</u> $\pi(P)$. <u>Moreover, let</u> $\alpha_0,\alpha_1,\alpha_2,\ldots$ <u>be global</u> ℓ-<u>ary</u> σ-<u>predicates such that</u> α_0 <u>is empty and every</u> α_{m+1} <u>equals</u> $\pi(\alpha_m)$. <u>If</u> β <u>is the least fixed point for</u> $\pi(P)$ <u>and</u> S <u>is</u>

<u>a σ-structure then</u> $\beta^S=\alpha_m^S$ <u>where</u> $m=|S|^\ell$. <u>Thus the least fixed point for</u> $\pi(P)$ <u>is</u> PTIME <u>computable if</u> $\pi(P)$ <u>is</u>.

The proof is clear. The claim appears in [AU] in terms of relational algebra. A transfinite induction generalizes the claim to infinite structures. In either form the claim is a special case of the classical theorem of Tarski [Tar].

<u>Example</u> 1 [AU]. The transitive closure of a global predicate $E(x,y)$ is the least fixed point with respect to P for

$$E(x,y) \vee \exists z[P(x,z) \,\&\, P(z,y)].$$

<u>Example</u> 2. The semigroup generated by a set A is the least fixed point with respect to P for

$$A(x) \vee \exists y \exists z[P(y) \,\&\, P(z) \,\&\, x=y*z]$$

A possible notation for the least fixed point for a global ℓ-ary predicate $\pi(P)$ with free individual variables $x_1,\ldots,x_\ell$ is

$$LFP(P,x_1,\ldots,x_\ell;\ \pi,\ y_1,\ldots,y_\ell).$$

It reflects the fact that LFP binds P and $x_1,\ldots,x_\ell$. The new individual variables $y_1,\ldots,y_\ell$ are free.

By the definition, LFP applies only to global predicates that are monotone in a given predicate variable. By Claim 1 in Appendix 1 the decision problem whether a given first-order formula is monotone in a given predicate variable, is unsolvable. This poses a difficulty in defining the extension of first-order logic by LFP. To overcome this difficulty Chandra and Harel [CH2] use positivity instead of monotonicity.

Let FO + LFP be the extension of first-order logic by the following formation rule. (For the sake of definiteness we assume that substitution of terms for free occurrences of individual variables is one of the first-order formation rules.)

LFP <u>formation rule</u>. Let P be a predicate variable of some arity ℓ and let $\phi(P,x_1,\ldots,x_\ell)$ be a well-formed formula. If all free occurrences of P in ϕ are positive and $y_1,\ldots,y_\ell$ are new individual variables then

$$LFP(P,x_1,\ldots,x_\ell;\ \phi(P,x_1,\ldots,x_\ell),\ y_1,\ldots,y_\ell)$$

is a well-formed formula. All occurrences of P and $x_1,\ldots,x_\ell$ in the new formula are bound. If Q is a predicate variable different from P then every free (resp. bound) occurrence of Q in ϕ remains free (resp. bound), and every positive (resp. negative) occurrence of Q in ϕ remains positive (resp. negative). The only occurrences of individual variables $y_1,\ldots,y_\ell$ in the new formula are bound. (ϕ may have individual parameters. They remain free.) The meaning of the new formula is that the tuple $(y_1,\ldots,y_\ell)$ belongs to the least fixed point for $\phi(P,x_1,\ldots,x_\ell)$.

Remark. Allowing individual parameters does not increase the expressive power of FO + LFP. For example, the formula

$$\mathrm{LFP}(P,y;\ E(u,y) \vee \exists z(P(z)\ \&\ E(z,y)),\ x)$$

is equivalent to the formula

$$\mathrm{LFP}(Q,w,y;\ E(w,y) \vee \exists z(Q(w,z)\ \&\ E(z,y)),\ u,x).$$

More generally, $\mathrm{LFP}(P,y;\ \phi(P,u,y),\ x)$

is equivalent to

$$\mathrm{LFP}(Q,w,y;\ \phi(Q_w,w,y),u,x)$$

where $Q_w(z)=Q(w,z)$. However, parameters may be useful from the computational point of view.

Sometimes logicians speak about logic with equality. In those cases the equality relation is a logical constant. The equality sign is interpreted as the identity relation on the elements of a given structure and it is not listed as a member of a given vocabulary. By Proviso 2 our structures are built from natural numbers. This allows us to introduce the natural order of elements as a logical constant and to speak about logic with order.

Theorem 2 [Iml, Var]. A global predicate is PTIME computable if and only if it is expressible in FO + LFP with order.

The "if" implication of Theorem 2 follows from Theorem 1 in §3 and from Claim 1. A sketch of a proof of the "only if" implication can be found in [Iml]. An alternative proof of the "only if" implication will be indicated later in this section.

Aho and Ullman [AU] define a generalization of LFP whose application is not restricted by monotonicity. A similar idea was independently explored by Livchak [Li1]. Unaware of developments related to the least fixed point operator Livchak (who happened to be a former Ph.D. student of mine) proposes to augment the definition of first-order formulas by the following additional formation rule:

If $F(\bar{x})$, $G(\bar{x})$ and $H(\bar{x})$ are well-formed formulas with the same free individual variables $\bar{x}=(x_1,\ldots,x_\ell)$ then $L(F(\bar{x}), G(\bar{x}), H(\bar{x}))$ is a new well-formed formula whose meaning is the infinite disjunction

$$F_0(\bar{x}) \vee F_1(\bar{x}) \vee F_2(\bar{x}) \vee \cdots$$

where $F_0(\bar{x})$ is $H(\bar{x})$ and each $F_{i+1}(\bar{x})$ is the disjunction of $F_i(\bar{x})$ and the result of replacing each subformula $G(y_1,\ldots,y_\ell)$ of $F(\bar{x})$ by $F_i(y_1,\ldots,y_\ell)$.

The extension of first-order logic with order by Livchak's rule captures PTIME [Li2]. We incorporate this fact into Theorem 3. But first let us reformulate Livchak's rule.

Definition. Let σ be a vocabulary, P be an additional predicate variable of some arity ℓ, and $\pi(P)$ be a global ℓ-ary predicate of the vocabulary $\sigma \cup \{P\}$. View $\pi(P)$ as an operator that, given a global ℓ-ary σ-predicate α, produces a global ℓ-ary σ-predicate $\pi(\alpha)$. This operator $\pi(P)$ is inflationary if $\alpha \leq \pi(\alpha)$ for every global σ-predicate α. Let α_0, α_1, etc. be a sequence of global ℓ-ary σ-predicates where α_0 is empty and each α_{i+1} equals to $\pi(\alpha_i)$. A fixed point β for $\pi(P)$ is an iterative fixed point if for every σ-structure S there is an i with $\beta^S=\alpha_i^S$.

Claim 2. Let σ, P, ℓ, $\pi(P)$ and α_0, α_1, etc. be as in the definition above. Suppose that $\pi(P)$ is inflationary in P. Then there is a unique iterative fixed point β for $\pi(P)$. Moreover, for every σ-structure S, $\beta^S=\alpha_m^S$ where $m=|S|^\ell$. Thus the iterative fixed point for $\pi(P)$ is PTIME computable if $\pi(P)$ is.

The proof is clear. Note that if P and $\pi(P)$ are as in the definition above then $P \vee \pi(P)$ is inflationary. Let FO + IFP be the extension of first-order logic by the following formation rule.

IFP formation rule. Let P be a predicate variable of some arity ℓ, and let $\phi(P,\bar{x})$ be a well-formed formula whose free individual variables are all or some members of $\bar{x}=(x_1,\ldots,x_\ell)$. If $\bar{y}=(y_1,\ldots,y_\ell)$ is a tuple of new individual variables then

$$\mathrm{IFP}(P,\ \bar{x};\ P(\bar{x}) \vee \phi(P,\bar{x}),\bar{y})$$

is a well-formed formula. The meaning of the new formula is that $\bar{y}$ is in the iterative fixed point for $P(\bar{x}) \vee \phi(P,\bar{x})$.

Claim 3. FO + IFP expresses exactly the global predicates expressible in first-order logic augmented by Livchak's rule.

Proof. We consider the extension of first-order logic by both formation rules and show that either rule can be eliminated. The formula

$$\mathrm{IFP}(P,\bar{x};P(\bar{x}) \vee \phi(P,\bar{x}),\bar{y})$$

is equivalent to

$$L(\phi(P,\bar{y}),\ P(\bar{y}),\ \mathrm{FALSE}(\bar{y})).$$

Given a formula $L(F,G,H)$ with free individual variables $\bar{x}=(x_1,\ldots,x_\ell)$ and an additional ℓ-ary predicate variable P write down a formula $F'(P,\bar{x})$ such that $F(\bar{x})=F'(G,\bar{x})$. Using $P,F'(P,\bar{x})$, $H(\bar{x})$ and first-order means write down a formula $\phi(P,\bar{x})$ saying the following:

If $\neg\exists\bar{x}H(\bar{x})$ then $F'(P,\bar{x})$,

else if $\neg\exists\bar{x}P(\bar{x})$ then $H(\bar{x})$,

else $F'(P,\bar{x})$.

It is easy to check that $L(F(\bar{x}),G(\bar{x}),H(\bar{x}))$ is equivalent to

$$\mathrm{IFP}(P,\bar{y};P(\bar{y}) \vee \phi(P,\bar{y}),\ \bar{x})$$

where $\bar{y}$ is a tuple of new individual variables. □

Theorem 3. Let π be a global predicate. The following statements are equivalent:

(1) π is PTIME computable,

(2) π is expressible in FO + LFP with order, and

(3) π is expressible in FO + IFP with order.

Proof. The implication (1) → (2) follows from Theorem 2. The implication (3) → (1) follows from Claim 2. To prove the implication (2) → (3) note that if a global ℓ-ary predicate $\pi(P,\overline{x})$ is monotone in an ℓ-ary predicate variable P then

$$LFP(P,\overline{x};\pi(P,\overline{x}), \overline{y})$$

is equivalent to

$$IFP(P,\overline{x};P(\overline{x}) \vee \pi(P,\overline{x}), \overline{y}).\square$$

Chandra and Harel show that FO + LFP without order is not able to express the global zero-ary predicate "The cardinality of a given structure is even" [CH2]. Their argument can be extended to show that FO + IFP without order is not able to express the same global predicate. Our Appendix 3 gives an alternative proof of the fact that FO + IFP without order is not able to express some PTIME computable order-independent global predicates π.

We turn our attention to global functions.

Definition. A global partial function f of vocabulary σ, arity ℓ and co-arity r assigns to each σ-structure S a partial function f^S from S^ℓ to S^r.

Example 3. Let σ consist of one binary predicate variable E (for "edge"), so that σ-structures are (directed) graphs. Let f(x,y) be the length of a shortest path from x to y. If S is a graph and f^S is defined at (x,y) then $f^S(x,y)<|S|$ and therefore $f^S(x,y)$ is an element of S.

Example 4. Let again σ be the vocabulary of graphs. For every graph S and every $x \in S$ let f(x) be the pair $(y,z) \in S^2$ such that there are exactly $y \cdot |S|+z$ elements u with an edge from x to u.

As was mentioned above, we are interested in logics (or algebras) that capture PTIME (LOGSPACE, etc.) computable global functions. In a sense FO + LFP with order does capture PTIME computable functions: it allows one to speak about the graph of a PTIME computable function f and about digits in the binary notation for $f(\overline{x})$. We prefer to speak about global functions directly. See [Gu3] in this connection. Here we mention only the results related to PTIME.

Let us ingore singleton structures (altneratively we may allow boolean variables). See the definition of recursive global partial functions in [Gu3].

Theorem 4 [Gu3, Sa]. A global partial function is PTIME computable if and only if it is recursive.

Two algebras of recursive global partial functions were given in [Gu3] by some initial members and certain operations. Let ARF (for "Algebra of Recursive Functions") be either of them.

Theorem 5 [Gu3]. A global partial function is PTIME computable if and only if it belongs to ARF.

An important advantage of (the proof of) Theorem 5 versus (the proof of) Theorem 2 is preserving essential time bounds.

Remark. It is easy to prove directly that the graph of every function in ARF is expressible in FO + LFP with order. This together with Theorem 5 gives an alternative proof of the "only if" implication of Theorem 2. Let $\pi(x)$ be a PTIME computable global predicate and let $f(\bar{x})$ be the characteristic function for $\pi(x)$ i.e. $f(\bar{x})=1$ if $\pi(\bar{x})$ holds and $f(\bar{x})=0$ otherwise. By Theorem 5, f is in ARF. Hence the predicate $f(\bar{x})=y$ is expressible in FO + LFP with order. Hence the predicate $f(\bar{x})=1$ is expressible in FO + LFP with order.

§5. Interpolation and definability for polynomial time logic

According to §1, many famous and important theorems about first-order logic fail in the case of finite structures. What happens to those theorems in the case of logic tailored for polynomial time? We concentrate here on the interpolation and definability principles for polynomial time logic and show that these principles are equivalent to natural complexity principles whose status is unknown.

Let PTL (for Polynomial Time Logic) be the logic FO + LFP with order, or the logic FO + IFP with order, or an algebra of PTIME computable functions from [Gu3]. It will be important that PTL expresses precisely PTIME computable global predicates. The exact syntax of PTL will not be important.

Definition. A partial function f from $\{0,1\}^*$ to $\{0,1\}^*$ is polynomially bounded if there is a natural number k such that $|f(x)| \leq |x|^k$ for all $x \in \text{Domain}(f)$ with $|x|>1$. Here $|x|$ is the length of x. More generally, a binary relation B over $\{0,1\}^*$ is polynomially bounded if there is k such that B(x,y) and $|x|>1$ imply $|y| \leq |x|^k$.

We identify a nonempty word $x=a_0a_1\ldots a_{\ell-1}$ in $\{0,1\}^*$ with the structure with universe $\{0,1,\ldots,\ell-1\}$ and one unary predicate $X=\{i: a_i=1\}$. If $\ell \geq 2$, $m=\ell^k$ for some k and y is a word $b_0,b_1,\ldots,b_{m-1}$ in $\{0,1\}^*$, we identify the pair (x,y) with the extension of the structure x by a k-ary predicate

$Y=\{(i_1,\ldots,i_k)$: if j is the number whose notation in the positional system of base ℓ is $i_1\ldots i_k$ then $b_j=1.\}$

Lemma 1. For every *NP* set A of nonempty words over $\{0,1\}^*$ there is a PTL sentence ϕ such that

$A=\{x\in\{0,1\}^*$: x is the reduct of a model of $\phi\}$.

Proof. Without loss of generality, every $x \in A$ is of length at least 2. There are a natural number k and a PTIME computable polynomially bounded binary relation B over $\{0,1\}^*$ such that $A=\{x: (x,y)\in B$ for some $y\}$, and $(x,y)\in B$ implies $|y|=|x|^k$. The desired sentence ϕ expresses $(x,y)\in B$. □

The analogue of Craig's Interpolation Theorem for PTL will be called the Interpolation Principle for PTL. This principle states that for every valid (in all relevant finite structures) PTL sentence $\phi_1 \to \phi_2$ there is a PTL sentence θ such that

$$\text{vocabulary}(\theta) \subseteq \text{vocabulary}(\phi_1) \cap \text{vocabulary}(\phi_2)$$

and the implications $\phi_1 \to \theta$, $\theta \to \phi_2$ are valid.

Theorem 1. The following two statements are equivalent:

(1) The Interpolation Principle for PTL, and

(2) The following separation principle for NP: for every pair of disjoint NP subsets A_1, A_2 of $\{0,1\}^*$ there is a P subset B of $\{0,1\}^*$ such that B includes A_1 and avoids A_2.

Proof. First suppose (1) and let A_1, A_2 be disjoint NP subsets of $\{0,1\}^*$. Without loss of generality neither A_1 nor A_2 contains the empty word. By Lemma 1 there are PTL sentences ϕ_1, ϕ_2 such that $A_i = \{x \in \{0,1\}^*: x$ is the reduct of a model of $\phi_i\}$ for $i=1,2$. Without loss of generality, the only common non-logical constant of ϕ_1, ϕ_2 is the unary predicate variable X. Obviously, the implication $\phi_1 \to \neg\phi_2$ is valid. Let θ be an appropriate interpolant. The set of models of θ is the desired set B.

Next suppose (2) and let $\phi_1 \to \phi_2$ be a valid PTL sentence. Let σ be the common part of the vocabularies of ϕ_1, ϕ_2. For $i=1,2$ let

$$A_i = \{x: x \text{ is the binary code for the } \sigma\text{-reduct of a model of } \phi_i\}.$$

By (2) there is a P set B that includes A_1 and avoids A_2. The desired interpolant θ expresses $x \in B$. □

Note that the Interpolation Principle for PTL implies $NP \cap \text{co-}NP = P$.

The analogue of Beth Definability Theorem for PTL will be called the Definability Principle for PTL. This principle states the following. Let σ be a vocabulary, P be an additional predicate variable of some arity ℓ and $\phi(P)$ be a PTL sentence of the vocabulary $\sigma \cup \{P\}$. Suppose that for every σ-structure S and all $P_1, P_2 \subseteq S^\ell$, $\phi^S(P_1)$ and $\phi^S(P_2)$ imply $P_1 = P_2$. Then there is a PTL formula ψ of the vocabulary σ with ℓ free variables such that for every

σ-structures S and every $P_1 \subseteq S^{\ell}$,

$$\phi^S(P_1) \text{ implies } P_1=\psi^S.$$

The <u>Weak Definability Principle</u> for PTL is the result of strengthening the antecedent of the Definability Principle for PTL as follows: for every σ-structure S there is a unique $P_1 \subseteq S^{\ell}$ such that $\phi^S(P_1)$ holds.

<u>Definition</u> (cf. [Vaℓ]). A nondeterministic Turing machine M is <u>unambiguous</u> if for every input x there is at most one accepting computation of M on x. An NP subset A of $\{0,1\}^*$ is UNAMBIGUOUS if there is an unambiguous Turing machine that accepts A.

<u>Theorem 2. The following statements (1)-(4) are equivalent.</u>

(1) <u>The Definability Principle for</u> PTL.

(2) <u>For every polynomially bounded partial function</u> f <u>from</u> $\{0,1\}^*$ <u>to</u> $\{0,1\}^*$, <u>if the graph of</u> f <u>is in</u> P <u>then</u> f <u>is</u> PTIME <u>computable</u>.

(3) <u>For every polynomially bounded partial function</u> f <u>from</u> $\{0,1\}^*$ <u>to</u> $\{0,1\}^*$, <u>if the graph of</u> f <u>is in</u> P <u>then the domain of</u> f <u>is in</u> P.

(4) UNAMBIGUOUS = P.

<u>Proof.</u> (1) $\rightarrow$ (2). Suppose (1) and let f be a polynomially bounded partial function from $\{0,1\}^*$ to $\{0,1\}^*$ with PTIME computable graph. It suffices to construct a PTIME algorithm for calculating f(x) for x of length at least 2. Let x range over words of length at least 2. Without loss of generality there is k such that $|f(x)|=|x|^k$ for all x in Domain(f). There is a PTL sentence $\phi(X,Y)$ with a unary predicate variable X and a k-ary predicate variable Y that expresses $(x,y)\in\text{Graph}(f)$. By (1) there is a PTL formula ψ such that if $\phi(X,Y)$ holds in the structure (x,y) then

$$Y=\{(i_1,\ldots,i_k):\ \psi(i_1,\ldots,i_k) \text{ holds in } x\}.$$

Here is a PTIME algorithm for calculating f(x). View x as a structure in the vocabulary $\{X\}$. Compute

$$Y=\{(i_1,\ldots,i_k):\ \psi(i_1,\ldots,i_k) \text{ holds in } x\}.$$

The extension of the structure x by the predicate Y corresponds to a pair (x,y) for some word y of length $|x|^k$. Check whether $\phi(X,Y)$ holds in the extended structure. If yes then $y=f(x)$.

The implications (2) → (3) and (3) → (4) are trivial.

(4) → (1). Suppose (4) and let σ be a vocabulary variable of some arity ℓ and $\phi(P)$ be a PTL sentence such that for every σ-structure S there is at most one $P \subseteq S^\ell$ satisfying $\phi^S(P)$. Set

$$K=\{(S,\bar{c}): S \text{ is a } \sigma\text{-structure}, \bar{c}\in S^\ell \text{ and there is } P\subseteq S^\ell \text{ such that } \phi^S(P) \text{ holds and } \bar{c}\in P\}.$$

Obviously, K is UNAMBIGUOUS. By (4), K is $\mathcal{P}$. The desired PTL formula $\psi(v_1,\ldots,v_\ell)$ expresses $(S,v_1,\ldots,v_\ell)\in K$. □

The following theorem was established in a discussion with Neil Immerman. (It succeeded Theorem 1 but preceded Theorem 2.)

Theorem 3. The following statements (1) - (3) are equivalent.

(1) The Weak Definability Principle for PTL.

(2) For every polynomially bounded function $f:\{0,1\}^* \to \{0,1\}^*$, if the graph of f is in $\mathcal{P}$ then f is PTIME computable.

(3) UNAMBIGUOUS ∩ co-UNAMBIGUOUS = $\mathcal{P}$.

Proof. The case (1) → (2) is similar to the case (1) → (2) in the proof of Theorem (2).

(2) → (3). Suppose (2) and let A_0, A_1 be complementary UNAMBIGUOUS subsets of $\{0,1\}^*$. There are unambiguous nondeterministic Turing machines M_0, M_1 accepting A_0, A_1 respectively. For $i=0,1$ and $x\in A_i$ let $f(x)$ be the digit i followed be the binary code for the accepting computation of M_i on x. By (2), f is PTIME computable. Hence A_0 and A_1 are $\mathcal{P}$.

(3) → (1). Suppose (3) and let $\sigma, P, \phi(P)$ and K be as in the case (4) → (1) of the proof of Theorem 2 except now for every σ-structure S there is a unique $P\subseteq S^\ell$ satisfying $\phi^S(P)$. Obviously K is UNAMBIGUOUS and co-UNAMBIGUOUS. By (3), K is $\mathcal{P}$. The desired PTL formula $\psi(v_1,\ldots,v_\ell)$ expresses $(S,v_1,\ldots,v_\ell)\in K$. □

As we saw in §1, the Interpolation Principle for first-order logic implies the Definability Principle for first-order logic. The same proof shows that the Interpolation Principle for PTL implies the Definability Principle for PTL. If $P=NP$ then the Interpolation Principle for PTL is obviously true. It is easy however to construct an oracle for which even the Weak Decidability Principle for PTL fails.

Claim (Andreas Blass). There is an oracle for which the Weak Definability Principle for PTL fails.

Proof. By Theorem 3 it suffices to construct an oracle A and a function f from $\{0,1\}^*$ (or from a P subset of $\{0,1\}^*$) to $\{0,1\}^*$ such that f is not PTIME computable relative to A whereas the graph of f is. We construct $A \subsetneq \{0,1\}^*$, containing exactly one word w_n of each length n, such that the function $f(0^n)=w_n$ is not PTIME computable relative to A.

Enumerate all (deterministic) PTIME bounded query machines. Let p_k be the time bound for a machine M_k. We define A in stages, choosing finitely many w_n's at each stage. The kth stage will ensure that M_k^A does not compute f.

Stage k. Fix a natural number d that is larger than any n for which w_n has already been chosen and so large that $p_k(d) \leq 2^d-2$. Set $w_n=0^n$ for all $n<d$ for which w_n was not previously chosen. Run M_k with input 0^d and oracle $\{w_n: n<d\}$. Let B be the set of queries during the computation. By the time bound, $|B| \leq p_k(d) \leq 2^d-2$. Choose w_d in $\{0,1\}^d-B$ such that w_d differs from the output (if any) of M_k. If $d<\ell$ and $\ell<|x|$ for some $x \in B$ chose w_ℓ in $\{0,1\}^\ell-B$. It is easy to see that M_k^A will not compute w_d on input 0^d.□

The computational status of Craig's Interpolation Theorem for propositional logic was explored by Mundici [Mu].

Appendix 1. Monotone versus positive.

This appendix is devoted to an important theorem about first-order logic whose status in the finite case is unknown.

Definition. Let σ be a vocabulary, P be an additional predicate variable of arity ℓ and $\phi(P,x_1,\ldots,x_r)$ be a first-order formula in the vocabulary $\sigma\cup\{P\}$ with free individual variables as shown. The formula ϕ is monotonically increasing in P if every σ-structure S satisfies the following for every ℓ-ary predicates P_1,P_2 on S,

if $\forall x_1\ldots\forall x_\ell[P_1(x_1,\ldots,x_\ell) \to P_2(x_1,\ldots,x_\ell)]$

then $\forall x_1\ldots\forall x_r[\phi(P_1,x_1,\ldots,x_r) \to \phi(P_2,x_1,\ldots,x_r)]$.

Define in the obvious way the following: ϕ is monotonically decreasing in P, ϕ is monotonically increasing in P on finite structures (or, in the finite case), ϕ is monotonically decreasing in P on finite structures. We restrict our attention to monotonically increasing behavior; the generalization for monotonically decreasing behavior will be obvious. We say "monotone" for "monotonically increasing".

We say that a first-order formula ϕ is positive in a predicate symbol P if every appearance of P in ϕ is positive. A precise definition of positive appearances of a predicate symbol in a first-order formula can be found in [CK]. It is easy to see that ϕ is monotone in P if ϕ is positive in P.

Theorem 1. If a first-order formula ϕ is monotone is a predicate symbol P then there is a first-order formula ϕ' such that ϕ' is equivalent to ϕ and positive in P.

I do not know who was the first to formulate this theorem but it is an obvious consequence of the Lyndon Interpolation Theorem [CK].

Conjecture. Theorem 1 fails in the case of finite structures.

The rest of this appendix contains a few remarks related to the conjecture. First we exhibit a sentence which is monotone in a unary predicate symbol P on finite structures but which is not monotone in P in general. Let f be a unary function symbol. The desired sentence (with equality) says that f is one-to-one and that P is closed under f-predecessors if P is closed under f-successors.

Claim 1. *Let* P *be a predicate variable. The following problems are undecideable:*

(i) *Given a first-order sentence* ϕ *tell whether* ϕ *is monotone in* P, *and*

(ii) *Given a first-order sentence* ϕ *tell whether* ϕ *is monotone in* P *on finite structures.*

Proof. Without loss of generality P is just a propositional variable. Let ψ be a first-order sentence that does not mention P. It is valid (resp. valid in the finite case) iff the sentence $P \rightarrow \psi$ is monotone (resp. monotone on finite structures) in P. □

Corollary 1. *Let* P *be a predicate variable. There is no recursive function* f *from first-order sentences to first-order sentences such that an arbitrary first-order sentence* ϕ *is monotone in* P *if and only if the sentence* $f(\phi)$ *is positive in* P.

Corollary 2. *Let* P *be a predicate variable. There is no partial recursive function* f *from first-order sentences to first-order sentences such that an arbitrary first-order sentence* ϕ *is monotone in* P *on finite structures if and only if* $f(\phi)$ *is positive in* P.

In the case of a propositional variable P there is a simple function that, given a first-order sentence $\phi(P)$, produces a first-order sentence $\phi'(P)$ such that $\phi'(P)$ is positive in P and $\phi'(P)$ is logically equivalent to $\phi(P)$ if $\phi(P)$ is monotone in P. The desired $\phi'(P)$ is $\phi(False) \vee [P \,\&\, \phi(True)]$.

On the other hand, a routine coding, given an arbitrary first-order formula with a predicate variable P, produces a first-order formula ϕ' in the vocabulary $\{E,Q\}$ such that E is a binary predicate symbol, Q is a unary predicate symbol and ϕ is monotone (respectively, positive) in P iff ϕ' is monotone (respectively, positive in Q. Moreover, it can be ensured that ϕ' has a conjunct saying that E is symmetric and irreflexive.

I have also checked that Theorem 1 remains true in the case of finite structures if ϕ is an existential sentence, a universal sentence, a prenex sentence with prefix $\exists^n\forall$ or a prenex sentence with prefix $\forall^n\exists$. Andreas Blass observed that if $\phi(P)$ is positive (resp. monotone, monotone on finite structures) in P then so is $\neg\phi(\neg P)$. Thus, if Theorem 1 is true in the finite case for prenex sentences ϕ with certain prefixes then it is true in the finite case for prenex sentences with the dual prefixes.

Appendix 2. Transitive closure is not first-order expressible.

Theorem. Connectivity of a given graph is not first-order expressible in the case of finite structures (even in the presence of linear order). Hence the transitive closure of a given binary relation is not first-order expressible in the case of finite structures (even in the presence of linear order).

Proof. For every positive integer n let S_n be the set $\{0,1,\ldots,n-1\}$ with the natural order and $E_n(x,y)$ be the following binary relation on $\{0,1,\ldots,n-1\}$: either $y=x+2$ or $x=n-1$ and $y=0$. Note that a graph (S_n,E_n) is connected iff n is even, and a relation E_n is obviously and uniformly in n expressible in the first-order language of order. Now use Lemma(ii) in §1.□

The theorem (without mentioning linear order) is due to Fagin [Fa2] and was reproved several times. Gaifman and Vardi wrote even a special paper [GV] with a specially short proof and a brief review of other known proofs. Fagin's proof actually gives more: connectivity is not expressible by an existential second-order sentence where all quantified predicate variables are unary. Yiannis Moschovakis noticed that all those proofs do not work in the presence of linear order and expressed an interest in such results in the presence of linear order.

Appendix 3. The fixed point operators can be powerless.

Provisos 1-3 of §3 are in force here because we will consider structures as inputs for algorithms. On the other hand the natural order of elements of a given structure is not a logical constant here, so that isomorphisms can break the order of elements. We are interested more in isomorphism types of structures than in specific representations.

Definition. A global ℓ-ary predicate π of some vocabulary σ is invariant if for every isomorphism f from a σ-structure A onto a σ-structure B and every ℓ-tuple $\bar{a}\varepsilon A$, $\pi^A(\bar{a})$ is equivalent to $\pi^B(f\bar{a})$.

Any first-order expressible global predicate is invariant as well as any global predicate expressible in FO + LFP or FO + IFP. All these global predicates are PTIME computable. They do not exhaust, however, the PTIME computable global predicates. The following theorem provides plenty of counterexamples. It speaks about FO + IFP because FO + IFP subsumes FO + LFP. Recall that a first-order theory T is called ω-categorical if all countable models of T are isomorphic.

Theorem 1. Let T be an ω-categorical first-order theory of some vocabulary σ. Then for every formula $\phi(\bar{x})$ in FO + IFP there is a first-order formula $\phi'(\bar{x})$ such that the global predicates $\phi(\bar{x})$ and $\phi'(\bar{x})$ coincide on the finite models of T.

Remark. Actually, the global predicates $\phi(\bar{x})$ and $\phi'(\bar{x})$ will coincide on all models of T but we do not care here about infinite models.

Proof. Without loss of generality the given formula $\phi(\bar{x})$ is

$$IFP(P,\bar{y};\ P(\bar{y}) \vee \psi(P,\bar{y}),\ \bar{x})$$

where ψ is first-order. Let $\alpha_0(\bar{x})$=FALSE and let each $\alpha_{i+1}(\bar{x})=\alpha_i(\bar{x}) \vee \psi(\alpha_i,\bar{x})$. By Ryll-Nardzewski's Theorem, there is a finite p such that $\alpha_{p+1}(\bar{x})$ is equivalent to $\alpha_p(\bar{x})$ on the infinite models of T. By the compactness theorem, there is a finite m such that $\alpha_{p+1}(\bar{x})$ is equivalent to $\alpha_p(\bar{x})$ on all models of T of size at least m. Hence there is a finite q such that $\alpha_{q+1}(\bar{x})$ is equivalent to $\alpha_q(\bar{x})$ in T. The first-order formula α_q is the desired ψ. □

Now we are ready to show that some polynomial time computable invariant predicates are not expressible in FO + IFP.

Example 1. The first-order theory of equality is ω-categorical. The predicate

$$\pi(A) = \begin{cases} 0 & \text{if } |A| \text{ even} \\ 1 & \text{otherwise} \end{cases}$$

is not first-order (see Lemma in §1). By Theorem 1, π is not expressible in FO + IFP. (Cf. [CH2, Theorem 6.2].)

Example 2. Let F be a finite field. Let T be the first-order theory of vector spaces over F. Obviously, T is ω-categorical. Let π be the global predicate such that for every structure A of the vocabulary of T, $\pi^A=1$ if A is a model of T and dimension(A) is even, and $\pi^A=0$ otherwise. Using Ehrefeucht games [Eh] it is easy to check that π is not first-order. By Theorem 1 it is not expressible in FO + IFP.

It is not difficult to extend FO + LFP in such a way that the predicates of Examples 1 and 2 are expressible in the extended logic and only polynomial time computable invariant predicates are expressible in the extended logic. Still, the problem to design a logic that expresses exactly polynomial time computable invariant global predicates is open.

References

[Au] A.V. Aho and J.D. Ullman, "Universality of Data Retrieval Languages", Proc. of 6th ACM Symposium on Principles of Programming Languages, 1979, 110-117.

[Be] J. Bennet, "On Spectra", Doctoral Dissertation, Princeton University, N.J., 1962.

[Bö] E. Börger, "Spektralproblem and Completeness of Logical Decision Problems", in "Logic and Machines: Decision Problems and Complexity (ed. E. Börger, G. Hasenjaeger, D. Rödding), Springer Lecture Notes in Computer Science, to appear.

[CH1] A.K. Chandra and D. Harel, "Computable Queries for Relational Databases", Journal of Computer and System Sciences 21 (1980), 156-178.

[CH2] A.K. Chandra and D. Harel, "Structure and Complexity of Relational Queries", Journal of Computer and System Sciences 25 (1982), 99-128.

[CK] C.C. Chang and H.J. Keisler, "Model Theory", North-Holland, 1977.

[CKS] A.K. Chandra, D.C. Kozen and L.J. Stockmeyer, "Alternation", Journal of the ACM 28 (1981), 114-133.

[Co] K. Compton, A private communication.

[Cr] W. Craig, "Three Uses of the Herbrand-Gentzen Theorem in Relating Model Theory and Proof Theory", J. Symbolic Logic 22 (1957), 250-268.

[Eh] A. Ehrenfeucht, "An Application of Games to the Completeness Problem for Formalized Theories", Fund. Math. 49 (1961), 129-141.

[Fa1] R. Fagin, "Generalized First-Order Spectra and Polynomial-Time Recognizable Sets", in "Complexity of Computation" (ed. R. Karp), SIAM-AMS Proc. 7 (1974), 43-73.

[Fa2] R. Fagin, "Monadic generalized spectra", Zeitschr. f. math. Logik und Grundlagen d. Math. 21, 1975, 89-96.

[Fr] H. Friedman, "The Complexity of Explicit Definitions", Advances in Mathematics 20 (1976), 18-29.

[FV] S. Feferman and R.L. Vaught, "The First-Order Properties of Algebraic Systems", Fund. Math. 47 (1959), 57-103.

[GJ] M.R. Garey and D.S. Johnson, "Computers and Intractability: A Guide to the Theory of NP-completeness", Freeman, 1979.

[GKLT] Yu. V. Glebskii, D.I. Kogan, M.I. Liogon'kii and V.A. Talanov, "Range and Degree of Relizability of Formulas in the Restricted Predicate Calculus", Cybernetics 5:2 (1969), 17-28; translation (1972), 142-154.

[GL] Y. Gurevich and H.R. Lewis, "A Logic for Constant-Depth Circuits", Manuscript, 1983.

[GV] H. Gaifman and M.Y. Vardi, "A Simple Proof that Connectivity is not First-Order", Manuscript, 1983.

[Go] W. Goldfarb, "The Gödel Class with Identity is Unsolvable", Journal of Symbolic Logic, to appear.

[Gu1] Y. Gurevich, "Existential Interpretation", Algebra and Logic, 4:4 (1965), 71-85 (or "Existential Interpretation II", Archiv Math. Logik 22 (1982), 103-120).

[Gu2] Y. Gurevich, "Recognizing Satisfiability of Predicate Formulas", Algebra and Logic 5:2 (1966), 25-55; "The Decision Problem for Logic of Predicates and Operations", Algebra and Logic 8 (1969), pages 160-174 of English translation; "The Decision Problem for Standard Classes", Journal of Symbolic Logic 41 (1976), 460-464.

[Gu3] Y. Gurevich, "Algebras of Feasible Functions", Proc. of 24th IEEE Symposium on Foundations of Computer Science, 1983, 210-214.

[HU] J.E. Hopcroft and J.D. Ullman, "Introduction to Automata Theory, Languages and Computation", Addison-Wesley, 1979.

[Im1] N. Immerman, "Relational Queries Computable in Polynomial Time", Proc. of 14th ACM Symposium on Theory of Computing, 1982, 147-152.

[Im2] N. Immerman, "Languages which Capture Complexity Classes", Proc. of 15th ACM Symposium on Theory of Computing, 1983, 347-354.

[JS] N.G. Jones and A.L. Selman, "Turing Machines and the Spectra of First-Order Formulas", Journal of Symbolic Logic 39 (1974), 139-150.

[Ko] V.F. Kostyrko, "To the Decision Problem for Predicate Logic", Ph.D. Thesis, Kiev, 1965 (Russian).

[Kr] G. Kreisel, Technical Report No. 3, Applied Mathematics and Statistics Labs., Stanford University, January 1961.

[Li1] A.B. Livchak, "The Relational Model for Systems of Automatic Testing", Automatic Documentation and Mathematical Linguistics 4 (1982), pages 17-19 of Russian original.

[Li2] A.B. Livchak, "The Relational Model for Process Control", Automatic Documentation and Mathematical Linguistics 4 (1983), pages 27-29 of Russian original.

[Lin] P. Lindström, "On Extensions of Elementary Logic", Theoria 35 (1969), 1-11.

[Mu] D. Mundici, "Complexity of Craig's Interpolation", Annales Societ. Math. Polonae, Series IV: Fundamenta Informaticae, vol. 3-4 (1982); "*NP* and Craig's Interpolation Theorem", Proc. of Florence Logic Colloquium 1982, North Holland, to appear.

[Sa] V. Sazonov, "Polynomial Computability and Recursivity in Finite Domains", Elektronische Informationsverarbeitung and Kybernetik 16 (1980), 319-323.

[Sℓ] L.J. Stockmeyer, "The Polynomial - Time Hierarchy", Theoretical Computer Science 3 (1977), 1-22.

[Tai] W.W. Tait, "A Counterexample to a Conjecture of Scott and Suppes", Journal of Symbolic Logic 24 (1959), 15-16.

[Tar] A. Tarski, "A Lattice-Theoretical Fixpoint Theorem and its Application", Pasific Journal of Mathematics 5 (1955), 285-309.

[Tr] B.A. Trakhtenbrot, "Impossibility of an Algorithm for the Decision Problem on Finite Classes", Doklady 70 (1950), 569-572.

[Uℓ] J.D. Ullman, "Principles of Database Systems", Computer Science Press, 1982

[Vaℓ] L.G. Valiant, "Relative Complexity of Checking and Evaluating", Information Processing 5 (1976), 20-23.

[Va] M.Y. Vardi, "Complexity of Relational Query Languages", Proc. of 14th ACM Symposium on Theory of Computing, 1982, 137-146.

[Zℓ] M.M. Zloof, "Query-by-Example: a Database Language", IBM Syst. Journal 16 (1977), 324-343.

ON A NEW NOTION OF PARTIAL CONSERVATIVITY

Petr Hájek, Math. Inst. ČSAV, 115 67 Prague

Abstract. Roughly, a bounded formula $\Phi(x)$ is $(2^c,c)$-conservative if assuming $\Phi(2^c)$ gives no new bounded information on c (c being a constant for a non-standard element in Peano arithmetic PA). Similarly for iterated powers of 2. This notion is analyzed, various existence theorems are proved and, as a corollary, we obtain a strengthening of Second Gödel's Incompleteness Theorem saying that for each non-standard model M of PA and each non-standard element $a \in M$ there is a model K of PA coinciding with M up to a and such that in K there is a very short proof of constradiction (bounded by 2 to the 2 to the 2 to the c).

§ 1. Introduction and statement of results.

PA denotes Peano arithmetic. Let c be a new constant and let PA_c be $PA + \{c \geq \bar{n},\ n \text{ natural}\}$ (PA with a constant for a non-standard element). A Δ-formula (i.e. bounded formula) $\Phi(x)$ with one free variable is called $(2^c,c)$-conservative if, for each Δ-formula $\Psi(x)$ with one free variable, $PA_c + \Phi(2^c) \vdash \Psi(c)$ implies $PA_c \vdash \Psi(c)$. (Thus $\Phi(x)$ is $(2^c,c)$-conservative if assuming Φ on 2^c gives no new bounded information on c.)

Let us present a model-theoretic characterization of $(2^c,c)$-conservativity. Let L denote the language of PA and let L' be the language consisting of two ternary predicates $\oplus$ and $\odot$ ($\oplus(x, y, z)$ being read " z is the sum of x and y " and similarly for $\odot$ and product). If $M \models PA$ and $a \in M$ then M_a denotes the set $\{b \in M,\ M \models b \leq a\}$ (the initial segment determined by a) endowed with the structure of M expressed using $\oplus$ and $\odot$ (addition and multiplication restricted to M_a as ternary relations).

Theorem 0. $\Phi(x)$ is $(2^c,c)$-conservative iff, for each countable non-standard $M \models PA$ and each $a \in M - N$, there is a $K \models PA$ and $b \in K$ such that

(1) M_a is isomorphic to K_b (as models of L'),

(2) $K \models \Phi(2^b)$.

(We can say that K coincides with M up to a ($= b$) and in K, 2^a satisfies Φ.)

Similarly we can define $(2^{2^c},c)$-conservativity and obtain the evident modification of the model-theoretic characterization and so on.

For typographical reasons, we shall write exp x instead of 2^x and similarly exp exp x etc.

Evidently if $\Phi(2^c)$ is provable in PA_c (i.e., for some $n \in N$, PA proves $(\forall x \geq \bar{n})$ (x is a power of $2 \to \Phi(x)$) then Φ is (exp c, c)-conservative; similarly for exp exp c . We offer the following results on <u>exponential conservativity</u> (as we could say) to the reader:

<u>Theorem 1</u>. There is a Φ such that Φ(exp exp c) is unprovable, moreover, $PA_c \vdash \Phi(\text{exp exp c}) \rightarrow \neg Con_{PA}$, and Φ is (exp exp c, c)-conservative.

<u>Theorem 2</u>. There is a Φ such that Φ(exp c) is unprovable and Φ is (exp c, c)-conservative.

<u>Theorem 3</u>. There is a Φ such that both Φ and $\neg\Phi$ are (exp exp c, c)-conservative.

<u>Theorem 4</u>. The set of all (exp exp c, c)-conservative sentences is Π_2^0-complete.

<u>Theorem 5</u>. There is a Φ such that $\Phi(x)$ implies "beneath x there is a proof of contradiction in PA " and Φ is (exp exp c, c)-conservative.

<u>Corollary 6</u>. For each $M \models PA$ and each $a \in M - N$ there is a $K \models PA$ identical with M up to a and such that in K , there is a PA-proof of contradiction beneath exp exp exp a .

Note that Corollary 6 may be viewed as a new strengthening of Second Gödel's Incompleteness Theorem, saying that Con_{PA} is unprovable in PA : Feferman [1] showed that $\neg Con_{PA}$ is interpretable in PA , thus each model of PA has an end-extension in which $\neg Con_{PA}$ holds. (Further nuances of Second Incompleteness Theorem are studied e.g. in [5].) If M′ is an end extension of M and $M' \models \neg Con_{PA}$ then the shortest inconsistency proof in M′ may be rather <u>long</u>, i.e. preceded by all elements of M . Our result can be understood as saying that for each M and a as above there is a K identical with M up to K such that in K there is a rather <u>short</u> proof of contradiction. (Obviously, K need not be an end extension of M .)

Our study of exponential conservativity was inspired by the paper

[6] of Paris and Dimitracopoulos. They show among other things that, for each non-standard $M \models PA$ and each $a \in M - N$ there is a K identical with M up to a and a Δ-formula $\Phi(x)$ which distinguishes exp exp a in M from exp exp a in K, i.e. in one of the models M, K, exp exp a satisfies Φ and in the other it satisfies $\neg\Phi$. We may ask whether there is a fixed Φ such that for each M and a, there is a K identical with M up to a Φ and such that Φ distinguishes exp exp a in M from exp exp a in K. This leads to our notion of (exp exp c, c)-conservativity: a Φ yielded by Theorem 3 works (by Theorem O).

Our definition of conservativity is similar to Σ_n-conservativity and Π_n-conservativity, which has been studied earlier (Guaspari [2], Hájek [3], Lindström [4]): a PA-sentence ϕ is Σ_n-conservative if, for each Σ_n-sentence ψ, $(PA + \phi) \vdash \psi$ implies $PA \vdash \psi$. A very important means for studying this notion of partial conservativity is partial truth definition for Σ_n-sentences in connection with self-reference. Now, for Δ-formulas we have also a certain partial truth definition and a possibility of self-reference (see below). This encourages us to study our present notion of conservativity in analogy to the older notion.

§ 2. Preliminaries.

(1) A truth definition for L'-formulas. Recall that L' is the language consisting of two ternary predicates $\oplus$, $\odot$. In PA, let $x \models y[z]$ mean that $x > 0$, y is a L'-formula and z codes a sequence of length, say, y, of numbers $\leq x$ such that z satisfies y in x. The condition of coding a sequence of length y of numbers $\leq x$ simply means $z < (x+1)^y$; each such z codes an y-tuple of numbers $u_i \leq x$, namely $z = \sum_{i=0}^{y-1} u_i(x+1)^i$. Concerning satisfaction, x is understood as a model of L' in the obvious way (its domain consists of all $u \leq x$ and $\oplus$, $\odot$ are interpreted using + and).

It is easy to see that $x \models y[z]$ is a Σ_1-formula: it says (cf. [6]) $(\exists w)$ (w is a satisfaction relation on x for formulas $\leq y$ and $(y,z) \in w$).

Now, a satisfaction relation on x for formulas $\leq y$ is a set of pairs (u,z) where $u \leq y$ and z is as above, thus, using the obvious coding, $w \leq \exp((x+1)^y \cdot y)$. A pedantical analysis shows that all necessary quantifications in the definition of a satisfaction relation can be bounded by $\exp((x+1)^y \cdot y)$ assuming $y \leq x$. (Formulas are coded as words in an alphabet with, say, 16 symbols - see also § 3; then " y is a formula" is expressed as $(\exists u \leq y^{\log_{16} y})\ \Psi(y,u)$ where Ψ is bounded. But $y \leq x$ implies certainly

$$y^{\log_{16} y} \leq \exp((x+1)^y \cdot y) .$$

Thus $y \leq x$ implies

$$x \models y[z] \equiv (\exists q \leq \exp((x+1)^y \cdot y))\,\omega(x,y,z,q)$$

where ω is an appropriate Δ-formula. If x is large enough and y is much smaller then x (say, $y \leq \log_2 x$) then this gives

$$x \models y[z] \equiv (\exists q \leq \exp\exp x)\,\omega(x,y,z,q) .$$

This gives the result of Lessan (cf. [6]) modified for our purposes:

<u>Fact 1</u>. There is a Δ-formula $\Gamma_0(u,y)$ such that, for some natural n, the following is PA-provable:

(1) $(\forall x \geq \overline{n})(\forall y \leq \log_2 x)(\, y$ is a L'-sentence $\rightarrow \{x \models y \equiv \Gamma_0(\exp\exp x,\ y)\})$

(2) $(\forall x \geq \overline{n})(\forall y)(\Gamma_0(\exp\exp x,\ y) \rightarrow y \leq \log_2 x$ & y is a L'-sentence$\}$.

We simply define $\Gamma_0(u,y) \equiv (\exists x \leq u)(u = \exp\exp x \ \&\ \omega(x,y,0,u))$; $u = \exp\exp x$ is Δ by Benett's result (used throughout the paper, see [6]).

<u>Fact 2</u>. (Corollary, Lessan). If $M \models PA$ and $a \in M - N$ then, for each standard ϕ, $M_a \models \phi \equiv M \models \Gamma_0(\exp\exp a,\ \overline{\phi})$.

<u>Remark</u>. Note that in Fact 1, $\Gamma_0(\exp\exp x,\ y)$ may be replaced by $\Gamma_0(\exp\exp(x/2),y)$ (or, more generally, by $\Gamma_0(\exp\exp(x/m),y)$ for a standard number m); the proof is the same as above. In Fact 2, $\Gamma_0(\exp\exp a,\ \overline{\phi})$ may be replaced by $\Gamma_0(\exp\exp(a/2),\overline{\phi})$. (Here $x/2$ is the biggest u such that $2u \leq x$.)

(2) Δ-<u>formulas versus</u> L'-<u>formulas</u>. Evidently, each L'-sentence ϕ can be transformed into a Δ-formula with one free variable $\phi^{\leq x}$ by (i) replacing each $+$ and $\cdot$ by formulas of the form $w = u + v$, $w = u \cdot v$ and (ii) restricting all quantifiers to x . Then

$$PA \vdash x \models \overline{\phi} \equiv \phi^{\leq x} .$$

Similarly, if $\phi(y_1,\ldots,y_n)$ is an L'-formula with $y_1,\ldots,y_n$ free and if x is a new variable then $\phi^{\leq x}$ is defined as above and we have

$$PA \vdash (\forall y_1,\ldots,y_n \leq x)(x \models \overline{\phi}[y_1,\ldots,y_n] \equiv (\phi(y_1,\ldots,y_n))^{\leq x}).$$

(Here $[y_1,\ldots,y_n]$ is shorthand for $[z]$ where z is a sequence having $y_1,\ldots,y_n$ on appropriate places and other members being 0 .)

<u>Fact 3</u>. Conversely, if $\phi(x,y_1,\ldots,y_n)$ is a Δ-formula then there is a L'formula $\check{\phi}(y_1,\ldots,y_n)$ (denoted by Transf(ϕ,x) if necessary) such that

$$PA \vdash (\forall x \geq 2)(\forall y_1,\ldots,y_n \leq x)(\phi(x,y_1,\ldots,y_n) \equiv x \models \overline{\check{\phi}}[y_1,\ldots,y_n]) .$$

Sketch of proof (cf. [6]): (i) ϕ is transformed into a formula without composed terms, i.e. having $+$ and $\cdot$ only in contexts $w = u + v$, $w = u\cdot v$, where u , v , w are variables; all new quantifiers are bounded by x^k where k is an appropriate natural number, computed from ϕ .

(ii) Subformulas of the form $(\forall u \leq x^n)\psi(u,\ldots)$ are successively replaced by $(\forall u_1,\ldots,u_n \leq x)\psi'(u_1,\ldots,u_n,\ldots)$ (n-tuples of numbers $\leq x$ are treated as $(x + 1)$-adic numbers).

(iii) Replace $w = u + v$ by $\oplus(u,v,w)$, analogously for $\odot$, delete the restrictions $\leq x$ in quantifiers. Remaining free occurences of x , if any, are replaced by speaking on "the maximal element of the universe".

Using Fact 3, we may reformulate Fact 1 as follows:

<u>Fact 4</u>. There is a L'-formula $\Gamma(y)$ such that, for some natural number n , the following is PA-provable:

(1) $(\forall x \geq \overline{n})(\forall y \leq \log_2 x)(y$ is a L'-sentence $\rightarrow$

$\rightarrow (x \models y \equiv \exp\exp x \models \Gamma[y]))$.

(2) $(\forall x \geq \overline{n})(\forall y)(\exp\exp x \models \Gamma[y] \rightarrow y \leq \log_2 x)$.

(Thus for x and y as said,

$x \models \neg y \equiv \exp\exp x \models \Gamma[\neg y] \equiv \exp\exp x \models \neg\Gamma[y]$.)

<u>Remark</u>. Also here we can replace $\exp\exp x$ by $\exp\exp(x/2)$.

(3) <u>Witnesses for Σ_1-formulas</u>. Many notions are defined as $(\exists y)\psi(x,y)$ where ψ is Δ ; e.g. Formula$(x) \equiv (\exists y)(y$ is a sequence of words, each member of y is either an atomic formula or ...) . If the notion in question is $P(x)$ then we call any pair (x,y) satisfying $\psi(x,y)$ a <u>witness</u> for P . For example we say that $z = (x,y)$ is a witness for a formula or that z is a witness for x being a for-

mula. Similarly, if F is a prim. rec. function then $x_1 = F(x_2)$ is equivalent to $(\exists y)\psi(x_1,x_2,y)$ where ψ is Δ ; then we can say that (x_1,x_2,y) is a witness for x_2 being $F(x_1)$ or is a witness for $F(x_2)$.

(4) Variants of Gödel's self-reference lemma.

(a) For Δ-formulas.

Fact 5. If $\phi(x,y)$ is Δ then there is a Δ-formula $\psi(x)$ and a natural n such that

$$PA \vdash (\forall x \geq \overline{n})\big(\psi(x) \equiv \phi(x,\overline{\psi})\big) .$$

Proof. The proof is a modification of the usual proof of the usual self-reference lemma: Put

$$\chi(x,y) \equiv (\exists z,u \leq x)\big(z \text{ is a witness for } u \text{ being } Sb^{vr_2}_{nm_y}(y) \ \& \ \phi(x,u)\big),$$

$$\psi(x) \equiv \chi(x,\overline{\chi}) .$$

Let n be the witness for $\psi(x)$ being the result of substitution of $\overline{\chi}$ for y into $\chi(x,y)$. Then we can proceed in PA as follows: Assume $x \geq \overline{n}$. Then $\psi(x) \equiv \chi(x,\overline{\chi}) \equiv (\exists z \leq x)\big(z$ is a witness for $\overline{\chi(x,\overline{\chi})}$ being $Sb^{vr_2}_{nm_{\overline{\chi}}}(\overline{\chi})$ & $\phi\big(x,\overline{\chi(x,\overline{\chi})}\big) \equiv \phi(x,\overline{\psi})$.

(b) For L'-formulas.

Fact 6. For each L'-formula $\phi(y)$ there is a L'-sentence ψ and an $n \in N$ such that

$$PA \vdash x \geq \overline{n} \rightarrow (x \models \overline{\psi} \equiv x \models \overline{\phi}[\overline{\psi}]) .$$

Proof. Let $\phi_0(x,y)$ be a Δ-formula such that $\phi(y) = \check{\phi}_0 = \mathrm{Transf}(\phi_0,x)$. Let $F(\alpha) = \mathrm{Transf}(\alpha,x)$ for each α ; F is prim. rec. Put

$$\phi_1(x,y) \equiv (\exists z,u \leq x)\big(z \text{ is a witness for } u \text{ being } F(y) \ \& \ \phi_0(x,u)\big).$$

By Fact 4, there is a Δ-formula $\psi_0(x)$ and an $n \in N$ such that

$$\begin{aligned} PA \vdash x \geq \overline{n} &\rightarrow \big(\psi_0(x) \equiv \phi_1(x,\overline{\psi}_0)\big) \\ &\rightarrow \big(\psi_0(x) \equiv \phi_0(x,\overline{\check{\psi}_0})\big) \\ &\rightarrow (x \models \overline{\psi} \equiv x \models \overline{\phi}[\overline{\psi}]) , \end{aligned}$$

where $\psi = F(\psi_0)$.

Fact 7. In Facts 5 and 6, PA can be replaced by PA_c .

§ 3. Proofs of Theorems 0, 1 and 5.

Remark. Here we prove Theorems 1, 0 and 5 (as well as Corollary 6); the proof of Theorem 5 uses Theorem 1. Next section contains a proof of Theorem 2 (which uses Theorem 1) and proofs of Theorems 3 and 4; the latter two proofs depend neither on any of previous theorems nor on each other.

Proof of Theorem 1. We shall construct an L'-sentence λ such that $\lambda^{\leq x}$ is (exp exp c, c)-conservative. Let $B(x,y)$ be a Δ-formula (without composed terms) equivalent to $x = \exp\exp y$ (existing by Benett's result). Let PA_c be the extension of PA by two constans $\hat{c}$, c and the axioms

$$B(\hat{c},c)\ ,\ \{\hat{c} \geq \overline{n},\ n \in N\}\ .$$

Evidently, PA_c is a conservative extension of PA_c, i.e. both theories prove the same sentences not mentioning c. (But we shall profit from the fact that our axioms express non-standardness of $\hat{c}$, not of c.) By Fact 5, let λ be a L'-sentence such that $PA_c \vdash \exp\exp c \models \overline{\lambda} \equiv \exp\exp c \models (\exists s,z)(\Gamma(\neg s)\ \&\ [z$ is a witness for a $PA_{\hat{c}}$-proof of $\ulcorner\overline{\lambda}^{\leq\hat{c}} \rightarrow s^{\leq c}\urcorner]^{\vee})$. Observe that we may understand λ as saying (in exp exp c) the following: There is a false formula s and a proof of s (converted to a Δ-formula) from me (converted to a Δ-formula) in PA_c; thus λ says "I imply a false formula". This is the usual trick for obtaining partially conservative formulas.

Now assume that ϕ is a L'sentence such that

$$PA_{\hat{c}} \vdash \lambda^{\leq\hat{c}} \longrightarrow \phi^{\leq c}\ .$$

Let q be a witness for a proof of the last formula in PA_c. Let us proceed in $PA_c + \exp\exp c \models \neg\overline{\lambda}$. Then we have

$$\exp\exp c \models (\forall s,z)(z \text{ a witness for } PA_c\text{-proof of } \ulcorner\lambda^{\leq\hat{c}} \longrightarrow s^{<c}\urcorner \longrightarrow \neg\Gamma(\neg s))\ ,$$

thus $\exp\exp c \models \neg\Gamma(\neg\phi)$, hence $\exp\exp c \models \Gamma(\phi)$ (since ϕ is standard and hence $\leq \log_2 c$). This implies $c \models \overline{\phi}$, thus in PA_c, $c \models \overline{\phi}$ follows both from $\exp\exp c \models \lambda$ and from its negation. We have proved $\lambda^{\leq x}$ to be (exp exp c, c)-conservative.

We now prove $PA_c \vdash \exp\exp c \models \overline{\lambda} \longrightarrow \neg Con_{PA}$. We have to distinguish carefully between c and its formalization; for pedagogical reasons, we shall work with an arbitrary model M of $PA_c + \exp\exp c \models \overline{\lambda}$, thus M is a model of PA, we have a $b \in M - N$ and $M \models \lambda^{\leq \exp\exp b}$. We have to prove $M \models \neg Con_{PA}$. From

$$M \models \lambda^{\leq \exp \exp b} \tag{1}$$

it follows that there are $s_0, z_0 \in M$ such that, in M, $\exp \exp b \models$ $\models \Gamma[\neg s_0]$, thus

$$M \models (\neg s_0) \leq \log b, \tag{2}$$

$$M \models b \models \neg s_0, \tag{3}$$

z_0 is a witness if a PA_c-proof $z_1 \leq z_0$ of $\ulcorner \overline{\lambda}^{\leq c} \rightarrow s_0^{\leq c} \urcorner$. (4)

By the property of Δ-formulas (1) and (2) imply

$$M \models Pr_{PA}(\overline{\lambda}^{\leq \exp \exp b}), \tag{5}$$

$$M \models Pr_{PA}(\neg s_0^{\leq b}). \tag{6}$$

Now, take the $PA_{\hat{c}}$-proof z_1 of $\ulcorner \overline{\lambda}^{\leq \hat{c}} \rightarrow s_0^{\leq c} \urcorner$ and substitute $\exp \exp b$ for $\hat{c}$ and b for c, obtaining z_2. We have

$$M \models Pr_{PA}\big(B(\exp \exp b, b)\big), \tag{7}$$

furthermore, $z_1 \leq \exp \exp b$, thus it uses only axioms $\ulcorner \hat{c} \geq \dot{\overline{a}} \urcorner$ for $a < \exp \exp b$, thus $\ulcorner \overline{\exp \exp b} \geq \overline{a} \urcorner$ is a true Δ-formula so that

$$M \models Pr_{PA}(\overline{\exp \exp b} \geq \overline{a}) \tag{8}$$

for each axiom $\ulcorner \hat{c} \geq \overline{a} \urcorner$ occuring in z_1. Consequently, z_2 can be prolonged in M to a PA-proof z_3 of $\ulcorner \lambda^{\leq \exp \exp b} \rightarrow s_0^{\leq b} \urcorner$, thus

$$M \models Pr_{PA}(\lambda^{\leq \exp \exp b} \rightarrow s_0^{\leq b}). \tag{9}$$

Evidently, (5), (6) and (9) give $M \models \neg Con_{PA}$, qed.

<u>Proof of Theorem 0</u>. We prove (i) $\equiv$ (ii) where

(i) $\lambda^{\leq x}$ is $(\exp c, c)$-conservative.

(ii) For each $M \models PA$ and each $a \in M - N$, there is a $K \models PA$ and $b \in K$ such that $K_b \simeq M_a$ and $K \models (\lambda^{\leq \exp b})$.

First assume (ii) and let $PA_c \models \lambda^{\leq \exp c} \rightarrow \phi^{\leq c}$. Let $M \models PA$ and $a \in$ $\in M \cdot - N$. By (ii), let $K_b = M_a$, $K \models \lambda^{\leq \exp b}$. Then $K \models \phi^{\leq b}$, thus $M \models \phi^{\leq a}$. This gives $PA_c \vdash \phi^{\leq c}$, qed.

Conversely, assume $\lambda^{\leq x}$ to be $(\exp c, c)$-conservative, let M and a be given. Let $T = (PA_c + Th_\Delta(a) + \lambda^{\leq \exp c})$ (where $Th_\Delta(a) =$ $= \{\phi^{\leq c}, M \models \phi^{\leq a}\}$). If T is contradictory then $PA_c + \lambda^{\leq \exp c} \vdash \neg\phi^{\leq c}$ for a ϕ such that $M \models \phi^{\leq a}$. Thus $PA_c \vdash \neg\phi^{\leq c}$ by conservativity of

$\lambda^{\leq x}$, which implies $M \models \neg\phi^{\leq a}$, a contradiction. Thus T is consistent.
Let (K,b) be a model of T such that SS(K) = SS(M) (standard sys-
tems; the model exists since T ϵ SS(M) and T is consistent). Take
K_b . It is recursively saturated (K_b is finite in K and therefore,
in K , K_b is saturated), and M_a is elementarily equivalent to K_b
(w.r.t. L'). Thus M_a and K_b are isomorphic, qed.

Proof of Theorem 5. Our formula $\phi(x)$ will say "beneath x , there is a witness for a proof of a contradiction". We leave to the reader
the detailed checking that if u is a proof (in PA or in a similar
theory) then there is a witness for this which is bounded by $u^{\log u}$.
We shall analyze our proof of Theorem 1 and get

Lemma 1. There is a standard n such that, if $M \models PA$, $b \epsilon M - N$
and $M \models \lambda^{\leq \exp \exp b}$ (where λ is as in Theorem 1) than, in M , there
is a PA-proof q of a contradiction such that

$$lh(q) \leq \exp(n \cdot \exp b) .$$

(lh denotes length).

To prove this we shall need the following

Lemma 2. There is a standard number n such that the following holds
in each model $M \models PA$:

If b > n , t is a L'-sentence in prenex normal form having u
quantifiers and if $b \models t$ then there is a PA-proof q of $\ulcorner t^{\leq b} \urcorner$ such
that

$$lh(q) \leq \left(b \cdot lh(t)\right)^{u+n} .$$

Before we start proving Lemma 2 let us make some comments on our
coding of formulas.

Formulas are built from 16 symbols
$$= + \cdot \ S \ \overline{0} \ v \ (\ ) \ 0 \ 1 \ \& \ v \longrightarrow \neg \ \forall \ \exists$$
coded by 0,...,15 ; 0 and 1 code themselves and may occur only in
the context v(D) where D is a dyadic number (the D-th variable).
Formulas are identified with certain hexadecimal numbers in the obvious
way. (I learned this coding from Solovay.) Thus the fifth variable is
v(101) and its length is 6 ; $\overline{m}$ has length m + 1 , the formula
$\overline{m} + \overline{n} = \overline{p}$ has m + n + p + 5 symbols. We shall not be explicit on our
choice of logical axioms and deduction rules; they can be extracted from
the following proofs and any reasonable choice would do.

Proof of lemma 2. We prove the following: There is an n such if $\phi(x_1,\ldots,x_h)$ is a L'-formula, $k_1,\ldots,k_h \leq m$ are natural numbers and if $\phi^{\leq m}(k_1,\ldots,k_h)$ is true then there is a PA-proof q of the last formula such that

$$\mathrm{lh}(q) \leq \{m\cdot \mathrm{lh}\phi(x_1,\ldots,x_h)\}^{n+i}$$

where i is the number of quantifiers in ϕ .

After the proof is finished the reader will see that it formalizes directly in PA . This gives our Lemma 2.

First consider atomic and negated atomic formulas. Let us inspect the proof of $\overline{m}+\overline{n}=\overline{m+n}$.

$$\begin{aligned}
x+\overline{0} &= x\\
x+Sy &= S(x+y)\\
\overline{m}+\overline{0} &= \overline{0}\\
\overline{m}+S\overline{0} &= S(\overline{m}+\overline{0})\\
\overline{m}+S\overline{0} &= \overline{m+1}\\
&\vdots\\
\overline{m}+\overline{n} &= S(\overline{m}+\overline{n-1})\\
\overline{m}+\overline{n} &= \overline{m+n}
\end{aligned}$$

This has $(2n+1)$ lines, each having $\leq 2(m+n+12)$ symbols, thus its length is bounded by a quadratic polynomial in $\max(m,n)$.

The reader may easily check that the usual proof of $\overline{m}\cdot\overline{n}=\overline{m\cdot n}$ has length bounded by a cubic polynomial and the usual proof of $\overline{m}\neq\overline{n}$ (for $m\neq n$) has length bounded by a quadratic polynomial. Similarly for $\overline{m}+\overline{n}\neq\overline{h}$, $\overline{m}\cdot\overline{n}\neq\overline{h}$, $\overline{m}\leq\overline{n}$, $\overline{m}\not\leq\overline{n}$. The argument of the polynomial is always the maximal number such that the corresponding numeral occurs in the formula. Thus we have a single cubic polynomial $p(x)$ such that if $\phi(x,y,z)$ is atomic or negated atomic, $k_1,k_2,k_3\leq m$ are natural numbers and $\phi(k_1,k_2,k_3)$ is true then there is a proof of the last formula of length $\leq p(m)\leq p\ m\cdot \mathrm{lh}\ \ \phi(x,y,z)$. Surely there is a $j\geq 3$ such that if $m\geq 2$ then $p(m)\leq m^j$. We have the following:

(*) There is a natural j such that if $\phi(x,y,z)$, k_1,k_2,k_3,m are as above and $\phi(k_1,k_2,k_3)$ is true then there is a proof of the last formula of length $\leq (m\cdot \mathrm{lh}\phi)^j$.

Now consider open formulas. If $\phi(x_1,\ldots,x_h)$ is open and $k_1,\ldots$ $\ldots,k_h\leq m$ are natural numbers then $\mathrm{lh}\{\phi(k_1,\ldots,k_h)\}\leq m\cdot \mathrm{lh}\{\phi(x_1,\ldots$ $\ldots,x_h)\}=m_*$. Suppose that $\phi(k_1,\ldots,k_h)$ is true and consider the sequence

$(')$ $\varepsilon_1\phi_1, \varepsilon_2\phi_2, \ldots, \varepsilon_i\phi_i$

where $\phi_1, \ldots, \phi_i$ are all subformulas of $\phi(k_1, \ldots, k_n)$ (in increasing order) and ε_p is negation if ϕ_p is false and is nothing otherwise. For each atomic ϕ_p produce a proof d_p of $\varepsilon_p\phi_p$; it has length $\leq m_*^j$. The sequence $(')$ has $\leq lh\{\phi(x_1, \ldots, x_h)\}$ members, each of length $\leq m_*$, and concatenation of all d_p together with $(')$ is a proof of $\phi(k_1, \ldots, k_h)$. Its length is estimated by

$$lh(\phi)\cdot m_*^j + \{lh(\phi)\}^2\cdot m \leq \{m\cdot lh(\phi)\}^{j+3} .$$

Let us consider quantifications. Let $\phi(y_1, \ldots, y_n)$ be an L'-formula of the form $(\forall x)\psi(x, y_1, \ldots, y_n)$ and assume that for each $h \leq m$ and each $k_1, \ldots, k_n \leq m$, if $\psi^{\leq m}(h, k_1, \ldots, k_n)$ is true then there is a proof of $\psi^{\leq m}(h, k_1, \ldots, k_n)$ of length $\leq (i+1)m_*^{i+j}$ where $m_* = m\cdot lh\{\psi(x, y_1, \ldots,)\}$ and ψ has i quantifiers. Assume $[(\forall x)\psi(x, k_1, \ldots, k_n)]^{\leq m}$ to be true. To prove the last formula (i.e. to prove $(\forall x \leq \overline{m})\psi^{\leq m}(x, k_1, \ldots, k_n))$ is suffices to prove each

$$\psi^{\leq \overline{m}}(h, k_1, \ldots, k_n) \ , \quad (h \leq m, \text{ proofs } d_h);$$

then the sequence

$(')$ $d_0, d_1, \ldots, d_m , \quad (\forall x \leq \overline{m})\psi^{\leq \overline{m}}(x, k_1, \ldots, k_n)$

is a proof of its last element. Now, the length of each d_p is estimated by $(i+1)\cdot m_*^{i+j} \leq (i+1)\cdot m_{**}^{i+j}$ where $m_{**} = m\cdot lh\{(\forall x)\psi(x, y_1, \ldots$ $\ldots, y_n)\}$ and evidently the formula being proved has length $\leq m_{**}$. Thus the whole proof $(')$ has length estimated by $(i+1)\cdot m_{**}^{i+1+j}$ + $+ m_{**} \leq (i+2)\cdot m_{**}^{i+1+j}$.

The existential quantifier is handled similarly. Thus for each L'-formula ϕ as assumed, there is a proof of $\phi^{\leq m}(k_1, \ldots, k_n)$ of length $\leq (i+1)\{m\cdot lh(\phi(x_1, \ldots))\}^{i+j}$. But certainly $i+1 \leq lh(\phi)$, thus putting $p = j+1$ we have the bound $\{m\cdot lh(\phi)\}^{i+p}$ as desired. This completes the proof.

<u>Proof of Lemma 1</u>. Let $M \models PA$, $b \in M - N$, $\hat{h} = \exp\exp b$, $M \models \lambda^{\leq \hat{b}}$ (λ from Theorem 1). Thus in M , there are $z, s \leq \hat{b}$ such that $\hat{b} \models_{\hat{}} \Gamma[\neg s]$ (hence $b \leq \neg s$) and z is a PA-proof of $\ulcorner \lambda^{\leq c} \to s^{\leq c}\urcorner$. Now, $\lambda^{\leq \hat{b}}$ implies that there is a PA-proof z_1 of $\ulcorner \lambda^{\leq \hat{b}} \urcorner$; since λ is standard (it has length h and i quantifiers) Lemma 2 gives $lh(z_1) \leq (\hat{b}\cdot h)^{n+i} \leq \hat{b}^{n_1} = \exp(n_1 \exp b)$ for some standard n_1 . Furthermore, $\hat{b} \models \Gamma[\neg s]$ implies $(\neg s) \leq \log_2 b$, thus $(\neg s) \leq b$; $b \models \neg s$

implies that there is a proof z_2 of $\ulcorner \neg s^{\leq b} \urcorner$ and, by Lemma 2, $lh(z_2) \leq (b \cdot b)^{b+n} \leq b^{3b} \leq \exp(3 \cdot \exp b)$.

Now take the proof z above and substitute the $\hat{b}$-th numeral for $\hat{c}$ and the b-th numeral for c. Call the resulting sequence z_{33}. Evidently, $lh(z_{33}) \leq lh(z) \cdot \exp \exp b \leq \exp(2 \cdot \exp b)$.
Let z_{32} be the sequence of proofs of $\ulcorner \bar{a} \leq \bar{\hat{b}} \urcorner$ for all $a \leq \hat{b}$; by Lemma 2, $lh(z_{32}) \leq \hat{b} \cdot (\hat{b} \cdot h)^n$ for some standard h, thus $lh(z_{32}) \leq$ $\leq \exp(n_2 \cdot \exp b)$ for some standard n_2. Finally, let z_{31} be a proof of $B(\hat{b}, b)$; by Lemma 2, its length is estimated by $(\hat{b} \cdot h)^{n+i}$ for some standard h, thus $lh(z_{31}) \leq \exp(n_3 \cdot \exp b)$ for some standard n_3.

The concatenation of z_1, z_2, z_{31}, z_{32}, z_{33} is a proof of contradiction (in M); its length is bounded by $\exp(n_1 \exp b) +$ $+ \exp(n_2 \exp b) + \exp(n_3 \exp b) + \exp(3 \cdot \exp b) + \exp(2 \cdot \exp b)$ which is bounded by $\exp(n_0 \cdot \exp b)$ for some standard n_0. (Observe that if $n_1 \leq n_2$ then $\exp(n_1 \exp b) + \exp(n_2 \exp b) \leq \exp((n_2 + 1) \exp b)$.)

This completes the proof of Lemma 1.

We shall now complete the proof of Theorem 5. Note that we could directly prove Corollary 6 from Lemma 1; the proof would be slightly simpler than that of Theorem 5.

Proof of Theorem 5 (continuation). Recall that $\phi(x)$ says "beneath x, there is a witness for a proof of contradiction". First we show that $\phi(x)$ is $(\exp \exp \exp (2c), c)$-conservative. Let $M \models PA$ and $b \in M - N$; by Theorem 1 there is a $K \models PA$ identical with M up to b and such that $K \models \lambda^{\leq \exp \exp b}$. We show that $K \models$ $\models \phi(\exp \exp \exp (2b))$. By Lemma 1, in K there is a proof q of a contradiction such that $lh(q) \leq \exp(n \cdot \exp b)$ for some standard n, thus $q \leq \exp \exp(n \cdot \exp b)$. There is a witness for q being a proof beneath $q^{\log q}$, but

$$q^{\log q} \leq (\exp \exp(n \cdot \exp b))^{\exp(n \cdot \exp b)} = \exp((\exp(n \cdot \exp b))^2) =$$
$$= \exp \exp(2n \cdot \exp b) \leq \exp \exp(\exp b \cdot \exp b) = \exp \exp \exp (2b) .$$

Now we show how to improve the preceding result to obtain $(\exp \exp \exp c, c)$-conservativity. First, using the remark following Fact 4, we may modify the proof of Theorem 1 to get a $\lambda^{\leq x}$ which is $(\exp \exp (c/2), c)$-conservative and implies $\neg Con_{PA}$. (Use $\hat{c}$ to denote $\exp \exp (c/2)$.) Then inspect the preceding part of the proof of Theorem 5: Lemma 2 is good as it stands and Lemma 1 now reads: If $M \models PA$, $b \in$ $\in M - N$ and $M \models \lambda^{\leq \exp \exp (b/2)}$ then, in M, there is a proof of contradiction of length $\leq \exp\{n_0 \cdot \exp(b/2)\}$. Then the proof q itself is estimated by $\exp\exp\{n \cdot \exp(b/2)\}$ and the corresponding withess is estimated by

$$q^{\log q} \leq \exp\exp\{n \cdot \exp(b/2)\}^{\exp(n \cdot \exp(b/2))} =$$

$$= \exp\{(\exp(n \cdot \exp(b/2)))^2\} = \exp\{\exp(2n \cdot \exp(b/2))\} \leq$$

$$\leq \exp\exp\{\exp(b/2) \cdot \exp(b/2)\} = \exp\exp\exp b \ .$$

This completes the proof.

§ 4. Proofs of Theorems 2, 3, 4.

First we shall prove Theorem 2 claiming that there is an unprovable (exp c, c)-conservative sentence. On the one hand, this is stronger than existence of an (exp exp c, c)-conservative unprovable sentence, guaranteed by Theorem 1; on the other hand, we shall not be able to exhibit our (exp, c)-conservative sentence explicitly, i.e. we prove a mere existential theorem.

Lemma 3. If Φ is (exp exp c, c)-conservative and $PA_c \nvdash \Phi(\exp\exp c)$ then either Φ is (exp exp c, exp c)-conservative or there is a Ψ which is (exp c, c)-conservative and $PA_c \nvdash \Psi(\exp c)$.

Proof. Assume that Φ is not (exp exp c, exp c)-conservative, i. e. there is a Δ-formula $\Psi(x)$ such that $PA_c \nvdash \Psi(\exp c)$ but $PA_c + \Phi(\exp\exp c) \vdash \Psi(\exp c)$. This Ψ must be (exp c, c)-conservative: Let $\Omega(x)$ be Δ and let $PA_c + \Psi(\exp c) \vdash \Omega(c)$. Then $PA_c + \Phi(\exp\exp c) \vdash \Omega(c)$ and hence $PA_c \vdash \Omega(c)$.

Hence in the proof of Theorem 2 we may assume that there is an (exp exp c, exp c)-conservative formula $\Phi(x)$ such that $PA_c \nvdash \Phi(\exp\exp c)$.

Proof of Theorem 2. Let PA_d be a copy of PA_c with c replaced by d . Interpret PA_d in PA_c by defining $d = \log c$. This extends PA_c to a theory PA_{cd} . Let Φ be (exp exp c, exp c)-conservative and unprovable. Note that $\exp d < 2c$, thus $\exp\exp d < (\exp c)^2$. Now, for each u , the L'-structure of u^2 is L'-definable in u (using pairs of elements of u to code elements of u^2); thus the L'-structure of exp exp d is L'-definable in exp c . Let λ be an L'-sentence such that $PA_d \vdash \Phi(\exp\exp d) \equiv \exp\exp d \models \overline{\lambda}$; there is an L'-sentence λ^* such that

$$PA_{cd} \vdash [(\exp c \models \overline{\lambda^*}) \equiv (\exp\exp d \models \overline{\lambda}) \ .$$

We shall show that the formula $x \models \overline{\lambda^*}$ is (exp c, c)-conservative and that $\exp c \models \overline{\lambda^*}$ is PA_c-unprovable. The latter is immediate, since $\exp \exp d \models \overline{\lambda}$ is PA_d-unprovable, hence PA_{cd}-unprovable (note that PA_{cd} extends both PA_c and PA_d conservatively); thus $\exp c \models \overline{\lambda^*}$ is PA_{cd}-unprovable and hence PA_c-unprovable.

Assume that μ is an L′-sentence such that $PA_c \vdash (\exp c \models \overline{\lambda^*}) \rightarrow (c \models \mu)$; then

$PA_{cd} \vdash (\exp \exp d \models \overline{\lambda^*}) \rightarrow (c \models \mu)$. But then

$PA_d \vdash (\exp \exp d \models \overline{\lambda^*}) \rightarrow (\forall x \leq \exp d)(\exp(d - 1) < x \rightarrow x \models \overline{\mu})$.

Clearly, this may be written as

$$PA_d \vdash (\exp \exp d \models \overline{\lambda^*}) \rightarrow (\exp d \models \overline{\mu^*})$$

for an appropriate L′-sentence μ^* . Since we assume the formula $x \models \overline{\lambda^*}$ to be (exp exp d, d)-conservative, we obtain $PA_d \vdash \exp d \models \overline{\mu^*}$, thus $PA_{cd} \vdash (c \models \overline{\mu})$, thus $PA_c \vdash (c \models \overline{\mu})$, which completes the proof of Theorem 2.

Proof of Theorem 3. We construct an appropriate diagonal formula (cf. the proof of Theorem 1, also proof of Lemma 7 in [6] and proof of Theorem 3 in [3]). Let λ be such that
$PA_{\hat{c}} \vdash \exp \exp c \models \overline{\lambda} \equiv \exp \exp c \models (\exists s,z)(\Gamma(\neg s)$ & [z is a witness for a $PA_{\hat{c}}$-proof of $\ulcorner \lambda^{\leq \hat{c}} \rightarrow s^{\leq c} \urcorner$ & $(\forall s',z' < s,z)$(if z' witnesses a $PA_{\hat{c}}$-proof of $\ulcorner \neg\lambda^{\leq \hat{c}} \rightarrow s^{\leq c} \urcorner$ then $\Gamma(s)$)]) . ($PA_{\hat{c}}$ is as in the proof of Theorem 1.)

Thus, roughly, λ says in exp exp c the following: There is a false s and a PA_c-proof of s from me such that whenever s' is a smaller formula and z' a smaller proof of s' from my negation, s' is true. We prove that both $\lambda^{\leq x}$ and $\neg\lambda^{\leq x}$ is (exp exp c, c)-conservative.

(1) We claim that $\lambda^{\leq x}$ is (exp exp c, c)-conservative. Assume that ϕ is a L′-sentence such that $PA_{\hat{c}} \vdash (\lambda^{\leq \hat{c}} \rightarrow \phi^{\leq c})$, let q be a witness for this proof. Let us proceed in $(PA_{\hat{c}} + \neg\lambda^{\leq \hat{c}})$. We prove $\phi^{\leq c}$ in this theory; thus $\phi^{\leq c}$ is provable in $PA_{\hat{c}}$. Then we have

$\hat{c} \models (\forall s,z)$(z witnesses a $PA_{\hat{c}}$-proof of $\ulcorner \overline{\lambda}^{\leq \hat{c}} \rightarrow s^{\leq c} \urcorner \rightarrow$
$\rightarrow -\Gamma(-s)$ or
$(\exists s',z' < s,z)$(z′ witnesses a $PA_{\hat{c}}$-proof of $\ulcorner \neg\overline{\lambda}^{\leq \hat{c}} \rightarrow s^{\leq c} \urcorner$ and $\neg\Gamma(s)$)) .

Substitute $\overline{\phi}$, $\overline{q}$ for s, z : then either $\neg\Gamma(\neg\overline{\phi})$, thus $\Gamma(\overline{\phi})$ and we are ready, or there are $s', z' < \overline{\phi}, \overline{q}$ such that z witnesses a $PA_{\hat{c}}$-proof of $\ulcorner \neg\lambda^{\leq \hat{c}} \rightarrow s^{\leq c} \urcorner$ and $\neg\Gamma(s)$. But the latter alternative is impossible since for each particular $\phi_1 < \phi$ and $q_1 < q$, if q_1 witnesses a (standard!) $PA_{\hat{c}}$-proof of $(\neg\lambda^{\leq \hat{c}} \rightarrow \phi_1^{\leq c})$ then we have $\phi_1^{\leq \hat{c}}$

(we are working in $PA_{\hat{c}} + \neg\lambda^{\leq\hat{c}}$!), thus $c \models \overline{\phi_1}$, thus $\Gamma(\overline{\phi_1})$.

(2) Now assume $PA_{\hat{c}} + \neg\lambda^{\leq\hat{c}} \vdash \phi^{\leq c}$, let q be a witness for this and let us proceed in $PA_{\hat{c}} + \lambda^{\leq\hat{c}}$. Then we have an s, $z < \hat{c}$ such that s is false (i.e. $\Gamma(\neg s)$) and z witnesses a $PA_{\hat{c}}$-proof of $\ulcorner\lambda^{\leq\hat{c}} \rightarrow s^{\leq c}\urcorner$. Clearly, both s and z must be <u>non-standard</u>, thus our ϕ and q , being standard, are $< s, z$. Thus the last part of the definition of λ applies: we have

$\hat{c} \models$ (if q witnesses a PA_c-proof of $\ulcorner\neg\overline{\lambda}^{\leq\hat{c}} \rightarrow \overline{\phi}^{\leq c}\urcorner$ then $\Gamma(\overline{\phi})$) , thus $\hat{c} \models \Gamma(\overline{\phi})$, thus $c \models \overline{\phi}$, i.e. $\phi^{\leq c}$. This completes the proof.

Our last task is to prove Theorem 4.

<u>Lemma 4</u>. For each recursively enumerable set $X \subseteq N$ there is an L'-formula $\rho(x)$ such that the formula $c \models \rho(x)$ numerates X in PA_c, i.e. for each $n \in N$, $n \in X$ iff $PA_c \vdash (c \models \rho(n))$.

<u>Proof</u>. Let $X \subseteq N$ be r.e. and let $\phi(y)$ be an L'-formula such that $n \in X$ iff $N \models (\exists z)(\phi(\overline{n}))^{\leq z}$. (ϕ is easy to obtain from a Σ_1-definition of X in N .)

Now if $n \in X$ then $N \models (\phi(\overline{n}))^{\leq k}$ for some k , thus $PA_c \vdash (\phi(\overline{n}))^{\leq k}$, thus $PA_c \vdash c \models (\phi(\overline{n}))^{\leq k}$, and $PA_c \vdash c \models (\exists y)(\phi(\overline{n})^{\leq y})$.

Conversely, let $PA_c \vdash c \models (\exists y)\phi(\overline{n})^{\leq y}$, then for some $k \in N$ $PA \vdash (\forall x \geq k)(x \models (\exists y)\phi(\overline{n})^{\leq y})$, thus $N \models (k \models (\exists y)\phi(n)^{\leq y})$, then $N \models (\exists y)\phi(\overline{n})^{\leq y}$, thus $n \in N$.

We have proved that the formula $c \models (\exists y)\phi(x)^{\leq y}$ numerates X in PA_c . Similarly we can numerate relations $\subseteq N^2$ instead of sets.

<u>Proof of Theorem 4</u>. (Cf. proof of Theorem 3 in [4].) Let $Y = \{k \mid (\forall m)R(k,m)\}$ be Π^0_2 , let $c \models \rho(x,y)$ numerate R in PA_c . Let for each $k \in N$ δ_k be an L'-formula such that

$$PA_{\hat{c}} \vdash \hat{c} \models \delta_k \equiv c \models (\exists s,z)(\Gamma(\neg s)) \;\&\; z \text{ witnesses a } PA_{\hat{c}}\text{-proof of } \ulcorner\hat{c} \models \delta_k \rightarrow c \models s\urcorner \text{ and } (\forall u \leq s)(c \models \rho(k,u)) .$$

Note that the function associating δ_k with k is recursive.

(1) Let $k \in Y$ and let q be a witness for a $PA_{\hat{c}}$-proof of $\ulcorner\hat{c} \models \delta_k \rightarrow c \models \phi\urcorner$; we prove $PA_{\hat{c}} \vdash c \models \phi$. Proceed in $(PA_{\hat{c}} + \hat{c} \models \neg\delta_k)$. We have (i) & (ii) → (iii), where

(i) says $\hat{c} \models \ulcorner q$ witnesses a $PA_{\hat{c}}$-proof of $\ulcorner\hat{c} \models \delta_k \rightarrow c \models \phi\urcorner\urcorner$,

(ii) says $(\forall u \leq q)(c \models \rho(\overline{k},u))$,

(iii) says $c \models \phi$.

(This is a reformulation of $\hat{c} \models \neg\delta_k$.)
Now, (i) evidently holds, (since q is standard and witnesses the desired proof.) Since $k \in Y$ we have $(\forall m)R(k,m)$, thus $(\forall m \leq q)\{PA_c \vdash \rho(k,m)\}$, i.e. $PA_c \vdash (\forall u \leq \overline{q})\rho(k,u)$ and thus (ii) holds in our theory $(PA_{\hat{c}} + \hat{c} \models \neg\delta_k)$. Thus the last theory proves $c \models \phi$ and therefore $PA_{\hat{c}} \vdash (c \models \phi)$; the formula $\delta_{\overline{k}}^{\leq x}$ is (exp, exp c, c)-conservative.

(2) Now let $k \notin Y$; take an m such that $\neg R(k,m)$. Then $PA_c \nvdash (c \models \rho(k,m))$; but we prove $(PA_{\hat{c}} + \hat{c} \models \delta_k) \vdash \{c \models \rho(k,m)\}$. Proceed in the last theory and assume $c \models \neg\rho(k,m)$. From $\hat{c} \models \delta_k$ it follows that there are s, z such that z witnesses a PA_c-proof of $(\hat{c} \models \delta_k \rightarrow c \models s)$ but in fact $c \models \neg s$ and at the same time, $(\forall u \leq z)\rho(\overline{k},u)$, thus our z is smaller than m and hence standard. Let $q_1,\ldots,q_n$ be all witnesses of $PA_{\hat{c}}$-proofs of formulas of the form $\ulcorner \hat{c} \models \delta_k \rightarrow c \models \phi_i \urcorner$ such that $q_i \leq m$. Then, on the one hand, we get $\bigvee_{i=1}^{n} (c \models \neg\phi_i)$ (since one of q_i is the z above) but, on the other hand, q_i transform to a $(PA_{\hat{c}} + \hat{c} \models \delta_k)$-proof of $\bigwedge_{i=1}^{n} (c \models \phi_i)$, thus we have got a contradiction. This shows that $(PA_{\hat{c}} + \hat{c} \models \delta_k)$ proves $c \models \rho(\overline{k},\overline{m})$, but we know that the last formula is unprovable in $PA_{\hat{c}}$. Thus $(\delta_k)^{\leq x}$ is not (exp exp c, c)-conservative. This completes the proof.

Remark.

In this paper, $\log_b x$ is defined as $\min_z (b^z \geq x)$.

References.

[1] S. Feferman: Arithmetization of metamathematics in a general setting, Fund. Math. 49 (1960), 35-92.

[2] D. Guaspari: Partially conservative extensions of arithmetic, Trans. Amer. Math. Soc. 25 (1970), 47-68.

[3] P. Hájek: On partially conservative extensions of arithmetic, in: Logic Colloquium 78, North-Holland P.C., 1979, Amsterdam.

[4] P. Lindström: Notes on partially conservative sentences and interpretability, Phil. Comm. Red Series, No.13, Göteborg 1980.

[5] K. McAloon: Completeness theorems, incompleteness theorems and models of arithmetic, Trans. A. M. S. 239 (1978) 253-277.

[6] J. Paris, C. Dimitracopoulos: Truth definitions for Δ formulas in: Logic et algoritmic, Geneve 1982.

FINITELY APPROXIMABLE SETS

Peter G. Hinman
University of Michigan
and
ETH, Zürich

Abstract. The continuous or countable functionals form a class of objects which are finitely approximable - they can be completely described by a set of (hereditarily) finite sets (approximations). We introduce and study a wider class of objects - sets, functions, and relations - all of which lay claim to a notion of finite approximability.

1. Introduction. In most contexts we are able to deal with the infinite objects of mathematics by conventions which assign them finite names. For some purposes, however, we are forced to reduce infinite objects to finite ones in a more effective way. The most obvious examples involve computation by machine. If the object reduced is "really" infinite, the reduction results in a loss of information, and we speak of the corresponding finite object as a finite approximation to the original. This process is successful in a given context if a sufficiently large part of the original information can be captured in a finite approximation. With this very rough formulation, we might tentatively call an object finitely approximable iff the process is successful in every context - that is, if the totality of the original information coincides with the sum of the information content of the finite approximations. The aim of this paper is to explore the notion of finite approximation in a general setting.

A naive first look might seem to reveal that every mathematical object is finitely approximable. If we take the usual foundational position that everything is a set, then clearly every set is the union of its finite subsets. However, a finite set whose elements are infinite is no more directly representable on a computer than is the original infinite set. In other words, we must insist that finite approximations be hereditarily finite. Thus, if HF denotes the set of hereditarily finite sets, to each finitely approximable object x we should be able to assign a set $Ap(x) \subseteq HF$ of its finite approximations in an injective way. Since HF is denumerable and has therefore continuum $(2^{\aleph_0})$ many subsets, it follows that there can be at most continuum many finitely approximable objects.

The most important model for our develpment is the class of <u>continuous</u> or <u>countable</u> functionals, introduced independently in [Kl 59] and [Kr 59] and widely studied since then. We do not assume that the reader is familiar with this theory, but the next few paragraphs may be somewhat heavier going for the uninitiated; we recommend the excellent [No 80] as reference.

A bit of notation is unavoidable here so we take this opportunity to introduce the main conventions for the whole paper. ω denotes the set $\{0,1,\dots\}$ of natural numbers and ${}^{\omega}\omega$ the set of functions from ω into ω. Letters $i,j,\dots,t,u$ will always denote natural numbers; $\alpha,\beta,\dots,\varepsilon$ belong to ${}^{\omega}\omega$. A finite sequence $(m_0,\dots,m_{n-1})$ of natural numbers has a code $\langle m_0,\dots,m_{n-1}\rangle \in \omega$; $\bar{\alpha}(n) = \langle\alpha(0),\dots,\alpha(n-1)\rangle$. We write $s \subseteq \alpha$ to mean that for some n, $s = \bar{\alpha}(n)$; n is the <u>length</u> $lg(s)$. Members of HF are denoted by letters $a,b,\dots,e$; subsets of HF by $A,B,\dots,E$. A finite sequence $\langle a_0,\dots,a_{n-1}\rangle$ of members of HF is itself in HF; by suitable coding we assume the same for subsets of HF. For any function f and set X, $f''X = \{f(x) : x \in X\}$.

The continuous functionals are arrayed in a hierarchy $\langle Ct^k : k \in \omega\rangle$. $Ct^0 = \omega$ and Ct^{k+1} is a set of functions $\varphi : Ct^k \to \omega$, which can be characterized in a number of ways. We give here basically the original definition of [Kl 59], which uses the auxiliary notion of an <u>associate</u>.

(i) $Ct^0 = \omega$;
$Ct^1 = {}^{\omega}\omega$;
a member of Ct^0 or Ct^1 is its own unique associate;

(ii) for all $k > 0$, any $\varphi : Ct^k \to \omega$, and any $\varepsilon \in {}^{\omega}\omega$,

(a) ε is an associate for φ iff for all $\psi \in Ct^k$ and all associates δ for ψ there exists $m \in \omega$ such that $\forall n < m$. $\varepsilon(\bar{\delta}(n)) = 0$ and $\forall n \geq m$. $\varepsilon(\bar{\delta}(n)) = \varphi(\psi) + 1$;

(b) $\varphi \in Ct^{k+1}$ iff φ has an associate;

(iii) $Ct = \cup\{Ct^k : k \in \omega\}$.

Each $m \in Ct^0$ is hereditarily finite and thus trivially finitely approximable. For $\alpha \in Ct^1$, the (codes of) initial segments form a natural class of approximations:

$$Ap(\alpha) = \{u : u \subseteq \alpha\}$$

Clearly α is completely determined by $Ap(\alpha)$. Consider next a function $\psi : Ct^1 \to \omega$. If ψ has an associate δ , then by (ii)(a),

$$\forall\alpha \in Ct^1 . \exists m \in \omega . \forall\beta \supseteq \bar{\alpha}(m) . \psi(\beta) = \psi(\alpha) ,$$

namely the least m such that $\delta(\bar{\alpha}(m)) = \psi(\alpha) + 1$. Therefore each value of ψ is described by some pair $\langle u,q \rangle$ such that $\forall\beta \supseteq u$. $\psi(\beta) = q$. Call such a pair <u>correct for</u> ψ and set

$$Ap(\psi) = \{b: b \text{ is a finite set of pairs correct for } \psi\} .$$

Again, ψ is determined by this set.

If we put

$$[u]^1 = \{\beta: u \subseteq \beta\} = \{\beta: u \in Ap(\beta)\} ,$$

then a pair $\langle u,q \rangle$ is correct for ψ just in case ψ has constantly the value q on $[u]^1$. Analogously, we now set

$$[b]^2 = \{\psi \in Ct^2: b \in Ap(\psi)\} .$$

Then for a function $\varphi : Ct^2 \to \omega$, we call a pair $\langle b,p \rangle$ <u>correct for</u> φ iff φ has constantly the value p on $[b]^2$ and set

$$Ap(\varphi) = \{a : a \text{ is a finite set of pairs correct for } \varphi\} .$$

Suppose now that $\varphi \in Ct^3$, $\psi \in Ct^2$, and $\varphi(\psi) = p$. Let ε be an associate for φ . For each associate δ for ψ , there is some m_δ such that $\varepsilon(\bar{\delta}(m_\delta)) = p+1$. Thus if

$$b_\delta = \{\langle u,q \rangle : u < m_\delta \text{ and } \delta(u) = q+1\} ,$$

then $a_\delta = \{\langle b_\delta ,p \rangle\} \in Ap(\varphi)$. The set

$$\{a_\delta: \delta \text{ is an associate for } \psi\}$$

is a subset of $Ap(\varphi)$ which encodes the fact that $\varphi(\psi) = p$. Note that each a_δ contains only part of this information since it is dependent not just on ψ but on a particular description δ of ψ . Thus $Ap(\varphi)$ again determines φ .

It follows from Th.3.3 that the continuous functionals are, in their context, exactly the finitely approximable ones. We next take an introductory look at the question when a <u>set</u> should be called finitely approximable. The direct analogy to the hierarchy $\langle Ct^k: k \in \omega \rangle$ would be a hierarchy $\langle Fa^k: k \in \omega \rangle$, where $Fa^0 = HF$ and Fa^{k+1} is a collection of subsets of Fa^k – those which are finitely approximable. A set $z \subseteq HF$ admits (at least) two natural notions of approximation:

$$Ap_+(z) = \{c \in HF: c \subseteq z\} ,$$

or

$$Ap_\pm(z) = \{<c^+,c^-> \in HF: c^+ \subseteq z \ \underline{\text{and}} \ c^- \cap z = \emptyset\} .$$

This dichotomy did not arise in the case of functions and causes some complication in the theory. In either case, however, every $z \subseteq HF$ is determined by its set of approximations and thus belongs to $Fa^1 = Fa^1_+ = Fa^1_\pm = \wp(HF)$.

Consider next a set $y \subseteq Fa^1$. Under the first (positive) model, a $b \in Ap_+(y)$ should code the information that certain $z \subseteq HF$ belong to y ; according to the second model b would also contain information about z which are not members of y . Since b is finite it cannot refer directly to (infinite) z but only to their approximations. For $*$ either $+$ or $\pm$, set

$$[c]^1_* = \{z \in Fa^1 : c \in Ap_*(z)\} .$$

Then the two suggested versions of approximation are:

$$Ap_+(y) = \{b \in HF: \forall c \in b.\ [c]^1_+ \subseteq y\} ,$$

or

$$Ap_\pm(y) = \{<b^+,b^-> \in HF: \forall c \in b^+.\ [c]^1_\pm \subseteq y \ \underline{\text{and}}$$
$$\forall c \in b^-.\ [c]^1_\pm \cap y = \emptyset\} .$$

Then we would put

$$y \in Fa^2_+ \leftrightarrow \forall z \in y.\ \exists c \in Ap_+(z).\ \exists b \in Ap_+(y).\ c \in b$$
$$\leftrightarrow \forall z \in y.\ \exists c.\ z \in [c]^1_+ \subseteq y ,$$

and

$$y \in Fa^2_\pm \leftrightarrow \forall z \in y.\ \exists c.\ z \in [c]^1_\pm \subseteq y \ \underline{\text{and}}$$
$$\forall z \in Fa^1 \sim y.\ \exists c.\ z \in [c]^1_\pm \subseteq Fa^1 \sim y .$$

Just as the members of Ct^2 are exactly the functions continuous in the usual Baire topology on ${}^\omega\omega$, so also both versions of Fa^2 have natural topological characterizations: Fa^2_+ consists of the <u>open</u> subsets of $\wp(HF)$ in the positive topology, while $Fa^2_\pm$ consists of the <u>closed-open</u> subsets in the Baire topology.

We claim that neither of these classes coincides with the intuitive notion of finite approximability at this level. In the positive case fix $d \in HF$ and set

$$y_d = \{z \subseteq HF: d \notin z\} .$$

Clearly $y_d \notin Fa^2_+$, but in a natural sense y_d is determined by only finitely much information so should be classed as finitely approximable. Although y_d does belong to $Fa^2_{\pm}$, this class is an even less attractive choice for Fa^2. For one thing, it follows from König's lemma (or compactness) that every $y \in Fa^2_{\pm}$ is really finitely describable as a finite union of "intervals" $[c]^1_{\pm}$. As a specific example consider

$$y^* = \{z \subseteq HF: z \cap \omega \neq \emptyset \ \underline{\text{and}} \ (\text{least } n \in (z \cap \omega)) \text{ is odd } \} .$$

$y^* \notin Fa^2_{\pm}$ since $(HF \sim \omega) \notin y^*$ but no finite amount of (positive and negative) information about $(HF \sim \omega)$ suffices to guarantee this. Another way of putting the problem with $Fa^2_{\pm}$ is that it requires for a set to be finitely approximable that its complement be so also. This seems inherently an unreasonable restriction.

The principle we wish to draw from these examples follows a middle course between incorporating all or none of the negative information about a set in its description(s) via finite approximations: a description should include all positive information and that part of the negative information which is finitely describable. For example, that $d \notin z$ above is a finite piece of negative information about z and is therefore reasonable to be included in any description of z via finite approximations. While the fact that $(HF \sim \omega) \notin y^*$ is not finitely describable, since it depends on infinitely much information about $(HF \sim \omega)$, that $z^* = \{2,3,4\} \notin y^*$ is describable by the finite information that $1 \notin z^*$ but $2 \in z^*$.

Thus, in brief, we shall use the $\pm$-version of Ap and [ ] but the +-version of Fa :

$$y \in Fa^2 \longleftrightarrow \forall z \in y \ . \ \exists c \ . \ z \in [c]^1_{\pm} \subseteq y .$$

The full definition, to be given in abstract form in §2 , involves the notion of a full description of a set, the analogue here of an associate. A full description $B \in Fd(y)$ is a subset of $Ap(y)$ which preserves the full information content of the whole set - that is, for all $z \in Fa^1$, if $z \in y$. Then

$$\exists c \in Ap(z) \ . \ \exists b \in B \ . \ c \in b^+ ,$$

and for $z \not\leq y$,
if $\exists c \in Ap(z) . \exists b \in Ap(y) . c \in b^-$,
then $\exists c \in Ap(z) . \exists b \in B . c \in b^-$.

In the general case the clause '$\exists c \in Ap(z)$' is replaced by '$\forall C \in Fd(z) . \exists c \in C$' , but for $z \in Fa^1$ this complication is not needed. (This corresponds to the fact that $\alpha \in Ct^1$ have unique associates.)

In topological terms, Fa^2 consists of the open sets in the Baire topology and the $z \not\leq y$ which satisfy the above condition are just those in the <u>exterior</u> of y . This relationship is less clear at higher levels (cf. end of §2.)

After working out these ideas we shall have a reasonable notion of finitely approximable set to go with the previous notion of finitely approximable functional. How do these notions fit together and where are we then with respect to our goal of characterizing the class of all finitely approximable mathematical objects? It is a happy accident of set theory that the simple notion of set as characterized by the hierarchy

$$V_\sigma = \cup\{\wp(V_\tau) : \tau < \sigma\}$$

suffices to define all sorts of things of interest to mathematics. When we restrict to finitely approximable sets this seems no longer to be the case. Although clearly any $\psi \in Ct^2$ can be regarded as a subset y_ψ of $\wp(HF)$, y is not an open set for at least two reasons: because it is a function and because its domain is not an open subset of $\wp(HF)$. Thus it appears that in the finitely approximable universe we must treat functions and sets differently. This leads us in §4 to a general construction involving sets, relations, and functions, which seems at least to include a wide variety of finitely approximable objects.

The work reported here has a second, more technical motivation. Kleene's version of the continuous functionals arose in connection with his work on the theory of recursion over the maximal type structure over ω :

$$Tp^o = \omega ;$$

$$Tp^{k+1} = \{\varphi : \varphi \text{ is a function } Tp^k \to \omega\} ;$$

$$Tp = \cup\{Tp^k : k \in \omega\} .$$

For a development of this theory see [Ke-Mo 77] or [Hi 78]. In [Kl 59] he showed that the schemas for recursion over Tp make perfect sense if one simply considers the variables of type k as ranging over Ct^k instead of Tp^k. In particular, Ct is closed under relative recursion in this sense. Much later, Normann [No 78] and independently Moschovakis (unpublished) introduced a generalization of Kleene's theory to one for recursion over the whole universe V of sets. Strictly speaking this <u>set recursion</u> generalizes <u>normal</u> Kleene recursion, the specialization of the general theory which assumes the equality relations over each Tp^k to be computable. The question then arises whether there is a solution to the proportion

$$Tp : Ct :: V : ?\ ,$$

that is, a subclass X of V such that the restriction (in some sense) of set recursion to X is at the same time an extension of Kleene recursion over Ct. We offer the class of finitely approximable objects as a natural choice for X, but argue in §5 that it seems doubtful that there is a natural corresponding recursion theory.

I want to thank several members of the audience at the meeting for perceptive and helpful comments, particularly Dana Scott and Sol Feferman. The hospitality of the ETH and my colleagues here made possible the latter stages of the work. The typing was done cheerfully and excellently by Rahel Boller.

2. Abstract finite approximation structures

In this section we introduce a class of abstract structures which possess many of the characteristics of the notion of finite approximability as discussed in §1. These structures are closely connected with the filter spaces of [Hy 79]; the connection will be explored in § 3.

2.1. Definition. An abstract finite approximation structure (AFAS) is a triple $\mathfrak{X} = (X, Ap, Fd)$ such that X is a set and

(1) Ap is a function from X into $\mathfrak{P}(HF) \sim \{\emptyset\}$;

(2) Fd is a function with domain X such that for all $x \in X$, $\emptyset \neq Fd(x) \subseteq \mathfrak{P}(Ap(x))$;

(3) let $[a] = \{x \in X : a \in Ap(x)\}$; then for any $x \in X$ and any $A \in Fd(x)$,

$$\forall a,b \in A.\ \exists c \in A\ .\ [c] \subseteq [a] \cap [b]\ .$$

An $a \in Ap(x)$ is to be thought of as a finite approximation to each $x \in [a]$ and $A \in Fd(x)$ as a full description of x . Of course, at this abstract level nothing is said of the context in which this approximation process is to be interpreted; that will be provided in the applications. When necessary (usually) we shall decorate Ap, Fd, and $[\cdot]$ with a superscript $\mathfrak{X}$ to distinguish various different AFAS. We call X the universe of $\mathfrak{X}$ and sometimes write $x \in \mathfrak{X}$ for $x \in X$.

A trivial but important example of an AFAS is the space $\mathfrak{HF} = (HF.Ap^{\mathfrak{HF}}, Fd^{\mathfrak{HF}})$, where for each $a \in HF$,

$$Ap^{\mathfrak{HF}}(a) = \{a\}\ ,\ \text{and}$$

$$Fd^{\mathfrak{HF}}(a) = \{\{a\}\}\ .$$

Slightly less trivial is $\mathfrak{PHF}$ with universe $\mathfrak{P}(HF)$ and for $z \subseteq HF$,

$$Ap^{\mathfrak{PHF}}(z) = \{<c^+, c^-> : c^+ \subseteq z \ \underline{\text{and}}\ c^- \cap z = \emptyset\}$$

and for $C \subseteq HF$,

$$C \in Fd^{\mathfrak{PHF}}(z) \leftrightarrow C \subseteq Ap^{\mathfrak{PHF}}(z) \ \underline{\text{and}}$$

$$\forall e \in HF[e \in z \rightarrow \exists c \in C.e \in c^+ \ \underline{\text{and}}\ e \notin z \rightarrow \exists c \in C.e \in c^-]\ .$$

Another class of examples is provided by the restriction of an AFAS to a subset. If $\mathfrak{X} = (X, Ap^{\mathfrak{X}}, Fd^{\mathfrak{X}})$ is an AFAS and $U \subseteq X$, then set

$$\mathcal{X}\upharpoonright U = \mathfrak{U} = (U, Ap^{\mathfrak{U}}, Fd^{\mathfrak{U}}) ,$$

where $Ap^{\mathfrak{U}}$ and $Fd^{\mathfrak{U}}$ are the restrictions of $Ap^{\mathcal{X}}$ and $Fd^{\mathcal{X}}$ to U . Note that

$$[a]^{\mathfrak{U}} = [a]^{\mathcal{X}} \cap U .$$

In particular we shall have cause later to refer to $\Omega = \mathfrak{H}\mathfrak{F}\upharpoonright\omega$

We first consider the question of subsets. Given an AFAS $\mathfrak{Y} = (Y, Ap^{\mathfrak{Y}}, Fd^{\mathfrak{Y}})$, we want to define a new AFAS $\mathcal{X}$ with $X \subseteq \mathfrak{P}(Y)$ and the elements of X being the finitely approximable subsets of Y . The process should be thought of (and in §4 will be used) as a step in an inductive definition: if we know how to approximate each $y \in Y$, then when and how can we approximate $x \subseteq Y$? The definition follows directly from the considerations of §1 , which provides natural examples.

2.2. <u>Definition</u>. For any AFAS $\mathfrak{Y}$, $Op(\mathfrak{Y})$ is the AFAS defined as follows. For each $x \subseteq Y$ set

$$Ap(x) = \{\langle a^+, a^-\rangle : \forall b \in a^+ . [b]^{\mathfrak{Y}} \subseteq x \text{ and } \forall b \in a^- . [b]^{\mathfrak{Y}} \cap x = \emptyset\} ,$$

and for any $A \subseteq HF$, let

$$[A]^+ = \{y \in Y : \forall B \in Fd^{\mathfrak{Y}}(y) . \exists b \in B . \exists a \in A . \ b \in a^+\}$$

and

$$[A]^- = \{y \in Y : \forall B \in Fd^{\mathfrak{Y}}(y) . \exists b \in B . \exists a \in A . \ b \in a^-\} .$$

Then $Op(\mathfrak{Y}) = \mathcal{X}$, where

(1) $X = \{x \subseteq Y : x = [Ap(x)]^+\}$;

(2) $Ap^{\mathcal{X}}$ is the restriction of Ap to X ;

(3) for each $x \in X$, $Fd^{\mathcal{X}}(x)$ is the set of all $A \subseteq Ap^{\mathcal{X}}(x)$ such that

(i) $\langle a_0^+, a_0^-\rangle, \langle a_1^+, a_1^-\rangle \in A \rightarrow \langle a_0^+ \cup a_1^+, a_0^- \cup a_1^-\rangle \in A$;

(ii) $[A]^+ = x$;

(iii) $[A]^- = [Ap^{\mathcal{X}}(x)]^-$.

First note that the conditions of 2.1 are indeed fulfilled. The only point to verify is (3), which follows from the fact that

$$[\langle a_0^+, a_0^-\rangle]^{\mathcal{X}} \cap [\langle a_1^+, a_1^-\rangle]^{\mathcal{X}} = [\langle a_0^+ \cup a_1^+, a_0^- \cup a_1^-\rangle]^{\mathcal{X}} .$$

Note that $Ap^{\mathcal{X}}(x)$ itself belongs to $Fd^{\mathcal{X}}(x)$, and in fact $Fd^{\mathcal{X}}(x)$ is closed under supersets which satisfy (3)(i). This will be true in all the examples and would be a natural condition to add to the definition - if A fully describes x and $A \subseteq A'$, then intuitively A' also describes x.

For any A, $[A]^+$ is the set of y which A guarantees to be in (some) x - for any description B of some y, an element a of A can "use" a finite piece b of B to check that y belongs, $b \in a^+$. Similarly, $[A]^-$ is the set of y which A verifies not to belong. By Lemma 2.3 below, for all $x \subseteq Y$, $[Ap(x)]^+ \subseteq x$ and thus $x \in X$ iff every $y \in x$ is there by virtue of finite approximations. A full description of x tells us as much about x and its complement as does the full set $Ap^{\mathfrak{X}}(x)$.

The designation $Op(\mathfrak{Y})$ arises, of course, from the fact that the elements of X are exactly the open subsets of Y under a natural topology. For the moment we take this as a definition; from the results of §3 we have an independent description of this topology in terms of filter spaces.

2.3. Lemma. For $\mathfrak{X}$ and $\mathfrak{Y}$ as in Definition 2.2 and any $x \subseteq Y$, each of the following propositions implies the next. If $x \in X$, then they are all equivalent.

(i) $y \in [Ap(x)]^+$;

(ii) $\forall B \in Fd^{\mathfrak{Y}}(y).\ \exists b \in B.\ [b]^{\mathfrak{Y}} \subseteq x$;

(iii) $\exists b \in Ap^{\mathfrak{Y}}(y).\ \exists a \in Ap(x).\ b \in a^+$;

(iv) $\exists b \in Ap^{\mathfrak{Y}}(y).\ [b]^{\mathfrak{Y}} \subseteq x$;

(v) $\exists b.\ y \in [b]^{\mathfrak{Y}} \subseteq x$;

(vi) $y \in x$.

Proof. Immediate from the definitions. ■

The hierarchy $\langle Fa^k : k \in \omega \rangle$, which was discussed informally in §1, now can be defined formally by:

$$\mathfrak{FA}^0 = \mathfrak{SF}, \text{ and}$$

$$\mathfrak{FA}^{k+1} = Op(\mathfrak{FA}^k) .$$

We turn next to functions. Given a pair $(\mathfrak{Y}, \mathfrak{Z})$ of AFAS, which functions $f : Y \to Z$ are finitely approximable. Consider first the case when $\mathfrak{Z}$ is a discrete space, say Ω. By analogy with Ct as discussed in §1, a finite approximation to f should be a set of pairs $\langle b,p \rangle$ such that f is constant p on $[b]^{\mathfrak{Y}}$ and f is finitely approximable iff

$\forall y \in Y.\ \forall B \in Fd^{\mathfrak{Y}}(y).\ \exists b \in B.\ f$ is constant on $[b]^{\mathfrak{Y}}$.

In the general case the finite value p must be replaced by an approximation to a value z. For each fixed full description B of an argument y, there should be some full description C of z such that values of f "arbitrarily close" to z as given by elements of C, will be guaranteed by restricting y to some $[b]^{\mathfrak{Y}}$ with $b \in B$.

2.4. Definition. For any pair $(\mathfrak{Y}, \mathfrak{Z})$ of AFAS, $Ct(\mathfrak{Y}, \mathfrak{Z})$ is the AFAS defined as follows. For each $f : Y \to Z$, set

$$Ap(f) = \{a : \forall \langle b,c\rangle \in a.\ f''[b]^{\mathfrak{Y}} \subseteq [c]^{\mathfrak{Z}}\},$$

and for any $A \subseteq HF$, set

$$[A]_f = \{y \in Y : \forall B \in Fd^{\mathfrak{Y}}(y).\ \exists C \in Fd^{\mathfrak{Z}}(z).\ \forall c \in C.\ \exists b \in B.\ \exists a \in A.\ \langle b,c\rangle \in a\}.$$

Then $Ct(\mathfrak{Y}, \mathfrak{Z}) = \mathfrak{X}$, where

(1) $X = \{f : [Ap(f)]_f = Y\}$;

(2) $Ap^{\mathfrak{X}}$ is the restriction of Ap to X;

(3) for each $f \in X$, $Fd^{\mathfrak{X}}(f)$ is the set of all $A \subseteq Ap^{\mathfrak{X}}(f)$ such that

(i) $a_0, a_1 \in A \to a_0 \cup a_1 \in A$;
(ii) $[A]_f = Y$.

It is, of course, intended that this definition should generalize that of the hierarchy $\langle Ct^k : k \in \omega\rangle$, and it will follow from the results of the next section that if we set

$$\mathfrak{Ct}^0 = \Omega, \quad \text{and}$$
$$\mathfrak{Ct}^{k+1} = Ct(\mathfrak{Ct}^k, \Omega),$$

then the universe of $\mathfrak{Ct}^k$ is exactly Ct^k.

We shall need two other similar general constructions, which we give without further comment in the hope that the intuitions are now sufficiently clear.

2.5. Definition. For any pair $(\mathfrak{Y}, \mathfrak{Z})$ of AFAS, $\mathfrak{Y} \times \mathfrak{Z}$ is the AFAS $\mathfrak{X}$ defined as follows:

(1) $X = Y \times Z$;

(2) for each $x = (y,z) \in X$, $Ap^{\mathfrak{X}}(x) = Ap^{\mathfrak{Y}}(y) \times Ap^{\mathfrak{Z}}(z)$;

(3) for each $x = (y,z) \in X$,

$Fd^{\mathscr{X}}(x) = \{B \times C : B \in Fd^{\mathfrak{Y}}(y)$ and $C \in Fd^{\mathfrak{Z}}(z)\}$.

2.6. Definition. For any HF-indexed family $<\mathfrak{Y}_c : c \in HF>$ of AFAS, $\cup <\mathfrak{Y}_c : c \in HF>$ is the AFAS $\mathscr{X}$ defined as follows:

(1) $X = \cup\{Y_c : c \in HF\}$;

(2) for each $x \in X$,
$Ap^{\mathscr{X}}(x) = \{a : \forall <b,c> \in a.\ x \in Y_c$ and $b \in Ap^{\mathfrak{Y}_c}(x)\}$;

(3) for each $x \in X$, $Fd^{\mathscr{X}}(x)$ is the set of all $A \subseteq Ap^{\mathscr{X}}(x)$ such that

(i) $a_o, a_1 \in A \rightarrow a_o \cup a_1 \in A$;
(ii) for some $\emptyset \neq d \in HF$, $\forall c \in d.\ x \in Y_c$ and $\{b : \exists a \in A <b,c> \in a\} \in Fd^{\mathfrak{Y}_c}(x)$.

We close this section with some observations and a question which are not of direct relevance to the paper. Let $\mathfrak{Y}$ be any AFAS and for any $x \subseteq Y$ define $Ap(x)$ as in Definition 2.2. By Lemma 2.3, $[Ap(x)]^+ \subseteq x$; similarly $[Ap(x)]^- \cap x = \emptyset$. If $x \notin Op(\mathfrak{Y})$, then $[Ap(x)]^+ \subsetneq x$ and the following sequence of sets seems to be non-trivial:

$x_o = x$ and for any ordinal $\sigma > o$,

$x_\sigma = [Ap(\cap\{x_\tau : \tau < \sigma\})]^+$.

The sequence is monotone decreasing and must therefore become constant at some ordinal σ_x ; let x° denote x_{σ_x} .

By definition, $x^\circ = [Ap(x^\circ)]^+$, so $x^\circ \in Op(\mathfrak{Y})$. Furthermore, for any $u \in Op(\mathfrak{Y})$, if $u \subseteq x$, then by a simple induction, for all σ , $u \subseteq x_\sigma$ and thus $u \subseteq x^\circ$. In other words, x° is the <u>interior</u> of x . Thus, in some sense, the ordinal σ_x measures how far x is from being finitely approximable. Then, for example, the ordinal

$\sigma_{\mathscr{X}} = \sup^+\{\sigma_x : (Y \sim x) \in \mathscr{X} = Op(\mathfrak{Y})\}$

is a measure of the complexity of $\mathscr{X}$. We would be interested in its value for $\mathscr{X} = \mathfrak{F}A^k$.

The second remark concerns the topologies on the AFAS $\mathfrak{F}A^k$, where of course the open subsets of Fa^k are exactly the elements of Fa^{k+1} . For $k = 0$ and 1 these are well-known and well-behaved. For larger k this is no longer true; the topology on $\mathfrak{F}A^2$ is not Hausdorff or even T_0. Consider the sets $y = \mathfrak{P}(HF) \sim \{\emptyset\}$ and $z = \mathfrak{P}(HF)$; clearly $y, z \in Fa^2$. Let x be any open neighborhood of y, so for some $b \in Fd^{\mathfrak{F}\lambda^2}(y)$. $y \in [b]^{\mathfrak{F}\lambda^2} \subseteq x$. Since for no c is $\emptyset \neq [c]^{\mathfrak{F}\lambda^1}$ disjoint from y, also $z \in [b]^{\mathfrak{F}\lambda^2}$ so $z \in x$. Similarly, any open neighborhood of z also contains x.

3. Filter Spaces

Martin Hyland in [Hy 79] provided several elegant characterizations of the classes Ct^k in terms of filter and limit spaces. It was an analysis of the filter space construction that led us to the abstract formulation of §2; this section is devoted to showing the connection. For the convenience of the casual reader we have removed some of the more technical proofs to the Appendix.

We do not want to reproduce too much of [Hy 79] here but will make this section sufficiently self-contained that the reader should have to rely on [Hy 79] only for technical background information.

The notion of a filter space is a generalization of that of a topological space; a good discussion of the motivation may be found in §X.1 of [Du 66]. A <u>filter</u> Φ on a set X is a non-empty collection of non-empty subsets of X closed under superset and finite intersection. For example, associated with each point $x \in X$ is the <u>principal filter at</u> x, Π_x, consisting of all $U \subseteq X$ such that $x \in U$. A <u>filter structure</u> on X is a function $\mathfrak{F}$ which to each $x \in X$ associates a collection $\mathfrak{F}(x)$ of filters on X. If X has a topology $\mathfrak{T}$, then for each $x \in X$ a natural filter on X is the <u>neighborhood filter at</u> x:

$Nb_x = \{U \subseteq X$: for some $O \in \mathfrak{T}$, $x \in O \subseteq U\}$. The corresponding <u>topological filter structure</u> is

$$\mathfrak{F}_{\mathfrak{T}}(x) = \{\Phi: \Phi \text{ is a filter on } X \text{ \underline{and} } Nb_x \subseteq \Phi\} .$$

This clearly satisfies the conditions of

3.1. <u>Definition</u>. A <u>filter space</u> is a pair $(X, \mathfrak{F})$ such that $\mathfrak{F}$ is a filter structure on the set X and for each $x \in X$, $\mathfrak{F}(x)$ contains Π_x and is closed under superfilter.

Intuitively, a filter $\Phi \in \mathfrak{F}(x)$ contains a certain amount of information about x ; smaller $U \in \Phi$ determine x more precisely so carry more information. As a rule, larger filters contain smaller sets and more total information. On the other hand in a sense the total information contained in $\mathfrak{F}(x)$ is the common part of the information carried by its elements so that the minimal elements of $\mathfrak{F}(x)$ are the determining factor. Clearly Π_x is a maximal element. In a topological filter structure $\mathfrak{F}(x)$ has also a smallest element Nb_x , but in many interesting cases $\mathfrak{F}(x)$ has no smallest element. Note that $Nb_x = \Pi_x$ for all $x \in X$ just in case $\mathfrak{T}$ is the discrete topology on X .

With <u>any</u> filter space $(X,\mathfrak{F})$ may be associated a topology: for any $U \subseteq X$, U is <u>open in</u> $(X,\mathfrak{F})$ iff

$$\forall x \in U.\ \forall \Phi \in \mathfrak{F}(x).\ U \in \Phi .$$

Note that by definition, for each $x \in X$ the neighborhood filter Nb_x for this topology is included in every $\Phi \in \mathfrak{F}(x)$, but unless Nb_x itself belongs to $\mathfrak{F}(x)$, $\mathfrak{F}(x)$ is strictly smaller (and thus carries <u>more</u> information) than the induced topological filter structure. In particular, this must be true if $\mathfrak{F}(x)$ has no smallest element.

Associated with each AFAS $\mathfrak{X}$ is a natural filter space $(X,\mathfrak{F}^{\mathfrak{X}})$ defined as follows. For any $A \subseteq HF$ set

$$\theta_A^{\mathfrak{X}} = \{U \subseteq X : \exists a \in A.\ [a]^{\mathfrak{X}} \subseteq U\} .$$

and for $x \in X$ let

$$\mathfrak{F}^{\mathfrak{X}}(x) = \{\Phi : \Phi \text{ is a filter on } X \text{ \underline{and} for some } A \in Fd^{\mathfrak{X}}(x) ,\ \theta_A^{\mathfrak{X}} \subseteq \Phi\} .$$

Since $a \in A \in Fd^{\mathfrak{X}}(x) \rightarrow a \in Ap^{\mathfrak{X}}(x) \rightarrow x \in [a]^{\mathfrak{X}}$, for all $A \in Fd^{\mathfrak{X}}(x)$, $\theta_A^{\mathfrak{X}} \subseteq \Pi_x$ so $\Pi_x \in \mathfrak{F}^{\mathfrak{X}}(x)$. That each $\mathfrak{F}^{\mathfrak{X}}(x)$ is closed under superfilter is obvious. Note that condition 2.1 (3) guarantees that for each $A \in Fd^{\mathfrak{X}}(x)$, $\theta_A^{\mathfrak{X}}$ is closed under finite intersection, is therefore itself a filter, and thus belongs to $\mathfrak{F}^{\mathfrak{X}}(x)$. These $\theta_A^{\mathfrak{X}}$ are the minimal elements of $\mathfrak{F}^{\mathfrak{X}}(x)$.

3.2. <u>Lemma</u>. For any AFAS $\mathfrak{Y}$, if $\mathfrak{X} = Op(\mathfrak{Y})$, then X consists exactly of the sets open in $(Y,\mathfrak{F}^{\mathfrak{Y}})$.

<u>Proof</u>. We introduce the abbreviation

$$P(B,x) \quad \text{for} \quad \exists b \in B.\ \exists a \in Ap^{\mathfrak{X}}(x).\ b \in a^+ .$$

For any $B \subseteq HF$ and any $x \subseteq Y$ we have

$$P(B,x) \leftrightarrow \exists b \in B.\ [b]^{\mathfrak{Y}} \subseteq x$$
$$\leftrightarrow x \in \theta_B^{\mathfrak{Y}} .$$

Then by definition 2.2 and Lemma 2.3, we have for any $x \subseteq Y$,

$$x \in X \leftrightarrow x \subseteq [Ap^{\mathfrak{X}}(x)]^+$$
$$\leftrightarrow \forall y \in x.\ \forall B \in Fd^{\mathfrak{Y}}(y).\ P(B,x)$$
$$\leftrightarrow \forall y \in x.\ \forall B \in Fd^{\mathfrak{Y}}(y).\ x \in \theta_B^{\mathfrak{Y}}$$
$$\leftrightarrow \forall y \in x.\ \forall \Phi \in \mathfrak{F}^{\mathfrak{Y}}(y).\ x \in \Phi$$
$$\leftrightarrow x \text{ is open in } (Y,\mathfrak{F}^{\mathfrak{Y}}) . \blacksquare$$

A similar result holds for the Ct-construction. For filter spaces $(Y,\mathfrak{G})$ and $(Z,\mathfrak{H})$, a function $f : Y \rightarrow Z$ has a natural extension to a map f^* on filters given by

$$f^*(\Psi) = \{W \subseteq Z : \exists V \in \Psi.\ f''V \subseteq W\} .$$

Then f is called <u>continuous from</u> $(Y,\mathfrak{G})$ <u>to</u> $(Z,\mathfrak{H})$ iff for all $y \in Y$ and all $\Psi \in \mathfrak{G}(y)$, $f^*(\Psi) \in \mathfrak{H}(f(y))$. It is easily checked that if both spaces are topological, this notion coincides with the usual one.

3.3. <u>Lemma</u>. For any pair $(\mathfrak{Y},\mathfrak{Z})$ of AFAS, let $\mathfrak{X} = Ct(\mathfrak{Y},\mathfrak{Z})$. Then X consists exactly of the functions continuous from $(Y,\mathfrak{F}^{\mathfrak{Y}})$ to $(Z,\mathfrak{F}^{\mathfrak{Z}})$.

<u>Proof</u>. The computation is parallel to that of Lemma 3.2. We introduce the abbreviation

$P(B,f,y)$ for $\exists C \in Fd^{\mathfrak{Z}}(f(y)).\forall c \in C.\exists b \in B.\exists a \in Ap^{\mathfrak{X}}(f).\langle b,c\rangle \in a$.

For any $B \subseteq HF$, $f : Y \rightarrow Z$. and $y \in Y$, we have

$$P(B,f,y) \leftrightarrow \exists C \in Fd^{\mathfrak{Z}}(f(y)).\ \forall c \in C.\ \exists b \in B.\ f''[b]^{\mathfrak{Y}} \subseteq [c]^{\mathfrak{Z}}$$
$$\leftrightarrow \exists C \in Fd^{\mathfrak{Z}}(f(y)).\ \theta_C^{\mathfrak{Z}} \subseteq f^*(\theta_B^{\mathfrak{Y}})$$
$$\leftrightarrow f^*(\theta_B^{\mathfrak{Y}}) \in \mathfrak{F}^{\mathfrak{Z}}(f(y)) .$$

Then for any $f : Y \rightarrow Z$,

$$f \in X \leftrightarrow \forall y \in Y.\ \forall B \in Fd^{\mathfrak{Y}}(y).\ P(B,f,y)$$
$$\leftrightarrow \forall y \in Y.\ \forall B \in Fd^{\mathfrak{Y}}(y).\ f^*(\theta_B^{\mathfrak{Y}}) \in \mathfrak{F}^{\mathfrak{Z}}(f(y))$$
$$\leftrightarrow \forall y \in Y.\ \forall \Psi \in \mathfrak{F}^{\mathfrak{Y}}(y).\ f^*(\Psi) \in \mathfrak{F}^{\mathfrak{Z}}(f(y))$$
$$\leftrightarrow f \text{ is continuous from } (Y,\mathfrak{F}^{\mathfrak{Y}}) \text{ to } (Z,\mathfrak{F}^{\mathfrak{Z}}) . \blacksquare$$

One of the main results of [Hy 79] is that for each k the associates generate a natural filter structure $\mathfrak{F}^k$ on Ct^k such that Ct^{k+1} consists exactly of the functions continuous from $(Ct^k, \mathfrak{F}^k)$ to $(Ct^o, \mathfrak{F}^o)$ <u>and</u> that $\mathfrak{F}^{k+1}$ has a natural characterization in terms of $\mathfrak{F}^k$ without reference to associates. The preceding Lemma is essentially a translation of the first of these results into our context; we want now to translate the second and then extend the analysis to the operation Op .

Let $(Y,\mathfrak{G})$ and $(Z,\mathfrak{H})$ be filter spaces and X the set of functions continuous from $(Y,\mathfrak{G})$ to $(Z,\mathfrak{H})$. For any $U \subseteq X$ and $V \subseteq Y$, set

$$U^*(V) = \cup\{g''V : g \in U\} .$$

For Φ and Ψ filters on X and Y , respectively,

$$\Phi^*(\Psi) = \{W \subseteq Z : \exists U \in \Phi. \exists V \in \Psi. U^*(V) \subseteq W\}$$

is easily seen to be a filter on Z . The filter structure $\mathfrak{F}$ on X is defined by setting, for each $f \in X$,

$$\mathfrak{F}(f) = \{\Phi: \forall y \in Y. \forall \Psi \in \mathfrak{G}(y). \Phi^*(\Psi) \in \mathfrak{H}(f(y))\} .$$

3.4. <u>Theorem</u>. For any pair $(\mathfrak{Y},\mathfrak{Z})$ of AFAS let $\mathcal{X} = Ct(\mathfrak{Y},\mathfrak{Z})$. Let $\mathfrak{F}$ be the filter structure on X constructed as above from the filter spaces $(Y,\mathfrak{F}^{\mathfrak{Y}})$ and $(Z,\mathfrak{F}^{\mathfrak{Z}})$. Then $\mathfrak{F} = \mathfrak{F}^{\mathcal{X}}$.

<u>Proof</u>. See Appendix. ∎

The corresponding constructions for cartesian product and restriction are exactly as given in [Hy 79] and we leave it to the diligent reader to make the comparisons with the definitions of §2. For the countable union of filter spaces we proceed as follows.

Let $<(Y_c,\mathfrak{G}_c): c \in HF>$ be an HF-indexed family of filter spaces. Set $X = \cup\{Y_c: c \in HF\}$. For any $d \in HF$ and any d-indexed family

$$\mathfrak{D} = <\Psi_c: c \in d>$$

of filters such that each Ψ_c is on Y_c , set

$$\Phi_{\mathfrak{D}} = \{ U \subseteq X: \text{for some } V_c \in \Psi_c \ (c \in d). \cap\{V_c: c \in d\} \subseteq U\} .$$

Call $\mathfrak{D}$ <u>worthy</u> iff $\emptyset \notin \Phi_{\mathfrak{D}}$; in this case $\Phi_{\mathfrak{D}}$ is a filter on X . Let $\mathfrak{F}$ be the filter structure on X such that for all $x \in X$ and all filters Φ on X ,

$\Phi \in \mathfrak{F}(x) \longleftrightarrow$ for some $\emptyset \neq d \in HF$, $\forall c \in d.\ x \in Y_c$ and

for some worthy family $\mathfrak{D} = \langle \Psi_c \in \mathfrak{G}_c(x) : c \in d \rangle$,

$\Phi_{\mathfrak{D}} \subseteq \Phi$.

We call the filter structure $(X, \mathfrak{F})$ the union of the family $\langle (Y_c, \mathfrak{G}_c) : c \in HF \rangle$. Note that for each $x \in X$, any family of filters in which each Ψ_c is the principal filter at x (on Y_c) is worthy; thus $\mathfrak{F}(x)$ does contain the principal filter at x (on X).

3.5. Theorem. For any HF-indexed family $\langle \mathfrak{Y}_c : c \in HF \rangle$ of AFAS, let $\mathcal{X} = \cup \langle \mathfrak{Y}_c : c \in HF \rangle$. Let $\mathfrak{F}$ be the filter structure on X constructed as above from the family $\langle (Y_c, \mathfrak{F}^{\mathfrak{Y}_c}) : c \in HF \rangle$ of filter spaces. Then $\mathfrak{F} = \mathfrak{F}^{\mathcal{X}}$.

Proof. See Appendix. ■

It remains in this section to show that if $\mathcal{X} = Op(\mathfrak{Y})$, then the filter structure $\mathfrak{F}^{\mathcal{X}}$ has a natural characterization in terms of $\mathfrak{F}^{\mathfrak{Y}}$. In fact, this construction turns out to be a natural adaptation of that above for continuous functions.

Fix a filter space $(Y, \mathfrak{G})$ and let X be the set of subsets of Y open in $(Y, \mathfrak{G})$. We seek a natural filter structure $\mathfrak{F}$ on X. Let $\hat{X}$ denote (briefly) the set of characteristic functions

$$f_x(y) = \begin{cases} 0, & \text{if } y \in x; \\ 1, & \text{if } y \notin x, \end{cases}$$

of subsets x of Y closed-open in $(Y, \mathfrak{G})$. Conversely, for each $f : Y \to \{0,1\}$, set

$$x_f^i = \{y \in Y : f(y) = i\}.$$

Let $\mathfrak{H}$ be the discrete filter structure on $\{0,1\}$. Then $\hat{X}$ is just the set of functions continuous from $(Y, \mathfrak{G})$ to $(\{0,1\}, \mathfrak{H})$. Let $\hat{\mathfrak{F}}$ denote the induced filter structure on $\hat{X}$ as described before Theorem 3.4. Then for any $f \in \hat{X}$, a filter $\hat{\Phi}$ on $\hat{X}$ belongs to $\hat{\mathfrak{F}}(f)$ iff for all $y \in Y$ and all $\Psi \in \mathfrak{G}(y)$ all of the following equivalent conditions hold:

$$\hat{\Phi}^*(\Psi) \in \mathfrak{H}(f(y));$$

$$\hat{\Phi}^*(\Psi) = \Pi_{f(y)};$$

$$\exists U \in \hat{\Phi}.\ \exists V \in \Psi.\ U^*(V) = \{f(y)\};$$

$$\exists U \in \hat{\Phi}.\ \exists V \in \Psi.\ V \subseteq \cap\{x_g^{f(y)} : g \in U\};$$

$$\exists U \in \hat{\Phi}.\ \cap\{x_g^{f(y)} : g \in U\} \in \Psi.$$

Returning now to X and thinking of an open set as X_f^o for a partially continuous function f we have as one natural condition for a filter Φ on X to belong to $\mathfrak{F}(x)$:

$$\forall y \in x.\ \forall \Psi \in \mathfrak{Y}(y).\ \exists U \in \Phi.\ \cap U \in \Psi.$$

This condition corresponds to the clause 3 (ii), $[A]^+ = x$, in Definition 2.2. To describe the analogue for clause (3)(iii) we introduce some notation.

For any filter Φ on X , set

$$\Phi^+ = \{y \in Y\colon \ \forall \Psi \in \mathfrak{G}(y).\ \exists U \in \Phi.\ \cap U \in \Psi\} ;$$

$$\Phi^- = \{y \in Y\colon \ \forall \Psi \in \mathfrak{G}(y).\ \exists U \in \Phi.(Y \sim \cup U) \in \Psi\} .$$

The above condition is thus simply $x \subseteq \Phi^+$. Suppose for a moment that $\mathfrak{G} = \mathfrak{F}^{\mathfrak{Y}}$ for an AFAS $\mathfrak{Y}$ with universe Y . Then for any $x \subseteq Y$ we have

$$\begin{aligned} y \in \Pi_x^+ &\leftrightarrow \forall \Psi \in \mathfrak{F}^{\mathfrak{Y}}(y).\ \exists U \in \Pi_x.\ \cap U \in \Psi \\ &\leftrightarrow \forall \Psi \in \mathfrak{F}^{\mathfrak{Y}}(y).\ x \in \Psi \\ &\leftrightarrow \forall B \in Fd^{\mathfrak{Y}}(y).\ \exists b \in B.\ [b]^{\mathfrak{Y}} \subseteq x \\ &\leftrightarrow y \in [Ap(x)]^+ \end{aligned}$$

and similarly $\Pi_x^- = [Ap(x)]^-$. Thus in the general case it is natural to put

$$\Phi \in \mathfrak{F}(x) \leftrightarrow x \subseteq \Phi^+ \ \underline{\text{and}} \ \ \Pi_x^- \subseteq \Phi^-.$$

3.6 Theorem. For any AFAS $\mathfrak{Y}$, let $\mathfrak{X} = Op(\mathfrak{Y})$. Let $\mathfrak{F}$ be the filter structure on X constructed as above from the filter space $(Y, \mathfrak{F}^{\mathfrak{Y}})$. Then $\mathfrak{F} = \mathfrak{F}^{\mathfrak{X}}$.

Proof. See Appendix.

4. The finitely approximable universe

We propose here a candidate for the class of all finitely approximable objects. It is designed, of course, to include all the examples we have discussed together with a great deal more. We do not at this point have much in the way of theorems about this structure, but it appears to be rich and potentially interesting.

We shall in fact define a class of AFAS - those which can be built up from $\mathfrak{SF}$ by the operations of §2. The basic principle is that if the elements of Y and Z are finitely approximable, then so are the open subsets of Y, the continuous functions from Y to Z, etc. This simple formulation is a bit too broad since, for example, if all elements of Y are finitely approximable, then the same is true of any $Y' \subseteq Y$ and Y' is an open subset of itself.

These considerations are related to the fact that an object is not finitely approximable in vacuo but in a context. What is required of a set x for it to be considered finitely approximable depends on how we view x - for example, of what Y it is seen as a subset. Thus a scheme for finitely approximating an object x must be accompanied by a description of the context $\mathscr{X}$ in which the approximation takes place. We show below that $\mathscr{X}$ can be described by a set $D_{\mathscr{X}} \subseteq HF$ so that a complete description of any finitely approximable object x is provided by the pair of sets $(Ap^{\mathscr{X}}(x), D_{\mathscr{X}})$.

4.1. Definition. Fa is the smallest collection of AFAS such that:

(1) $\mathfrak{H}\mathfrak{F} \in Fa$;

(2) for any $\mathfrak{Y} \in Fa$, $Op(\mathfrak{Y}) \in Fa$ and for each $x \in Op(\mathfrak{Y})$, $\mathfrak{Y}\upharpoonright x \in Fa$;

(3) for any $\mathfrak{Y}, \mathfrak{Z} \in Fa$, $\mathfrak{Y} \times \mathfrak{Z} \in Fa$ and $Ct(\mathfrak{Y}, \mathfrak{Z}) \in Fa$;

(4) for any family $<\mathfrak{Y}_c : c \in HF>$ of elements of Fa,

$\cup <\mathfrak{Y}_c : c \in HF> \in Fa$.

If we view Fa as being generated in a transfinite sequence of levels Fa^{σ}, then the denumerability of the families in (4) ensures that the sequence terminates at the latest at $\aleph_1$. Clearly each level has at most $2^{\aleph_0}$ members and thus so does Fa. Furthermore, for each $\mathscr{X} \in Fa$, X has power at most $2^{\aleph_0}$.

4.2. Definition. An object x is finitely approximable iff for some $\mathscr{X} \in Fa$, $x \in \mathscr{X}$.

The basic property of finite approximation is given by

4.3. Lemma. For any $\mathscr{X} \in Fa$ and any $x, x' \in \mathscr{X}$, if $Ap^{\mathscr{X}}(x) = Ap^{\mathscr{X}}(x')$, then $x = x'$.

Proof. We proceed by induction over Fa. The result is obvious for $\mathscr{X} = \mathfrak{H}\mathfrak{F}$. If $x, x' \in \mathscr{X} = Op(\mathfrak{Y})$ and $Ap^{\mathscr{X}}(x) = Ap^{\mathscr{X}}(x')$, then by Lemma 2.3, for any $y \in Y$,

$$\begin{aligned} y \in x &\longleftrightarrow \exists b \in Ap^{\mathfrak{Y}}(y) \,.\, \exists a \in Ap^{\mathscr{X}}(x) \,.\, b \in a^{+} \\ &\longleftrightarrow \exists b \in Ap^{\mathfrak{Y}}(y) \,.\, \exists a \in Ap^{\mathscr{X}}(x') \,.\, b \in a^{+} \\ &\longleftrightarrow y \in x' \end{aligned}$$

(no induction hypothesis is needed). Suppose that the result holds for $\mathfrak{Y}$ and $\mathfrak{Z}$ and let $x,x' \in \mathfrak{X} = \mathfrak{Y} \times \mathfrak{Z}$ with $Ap^{\mathfrak{X}}(x) = Ap^{\mathfrak{X}}(x')$. Then $x = \langle y,z \rangle$ and $x' = \langle y',z' \rangle$ with $y,y' \in Y$ and $z,z' \in Z$. By the definition of $Ap^{\mathfrak{X}}$, $Ap^{\mathfrak{Y}}(y) = Ap^{\mathfrak{Y}}(y')$ and $Ap^{\mathfrak{Z}}(z) = Ap^{\mathfrak{Z}}(z')$. Hence $y = y'$ and $z = z'$ so $x = x'$. The remaining clauses are treated similarly. ∎

We now define the sets $D_{\mathfrak{X}}$ mentioned above. The idea behind $D_{\mathfrak{X}}$ is that it should code the path over which $\mathfrak{X}$ is constructed in the process of building up Fa. Unfortunately this path is not unique - for example, if $x,x' \in Op(\mathfrak{Y})$ with $x \subseteq x'$, then $\mathfrak{Y}\restriction x = (\mathfrak{Y}\restriction x')\restriction x$. This is a standard phenomenon for inductively defined sets and the problem has a standard solution. We define inductively a set FA of pairs $(\mathfrak{X},A)$ such that A codes <u>a</u> path over which $\mathfrak{X}$ can be constructed. There will thus not be a unique $D_{\mathfrak{X}}$ but rather one for each way of constructing $\mathfrak{X}$.

4.4. <u>Definition</u>. FA is the smallest collection of pairs $(\mathfrak{X},A)$ such that $\mathfrak{X}$ is an AFAS, $A \subseteq HF$, and

(1) $(\mathfrak{HF}, \langle 0,HF \rangle) \in FA$;

(2) for any $(\mathfrak{Y},B) \in FA$,

  (i) $(Op(\mathfrak{Y}), \langle 1,B \rangle) \in FA$;

  (ii) for each $x \in Op(\mathfrak{Y})$, $(\mathfrak{Y}\restriction x, \langle 2,B,Ap^{Op(\mathfrak{Y})}(x) \rangle \in FA$;

(3) for all $(\mathfrak{Y},B)$ and $(\mathfrak{Z},C) \in FA$,

  (i) $(\mathfrak{Y} \times \mathfrak{Z}, \langle 3,B,C \rangle) \in FA$;

(4) for any family $\langle (\mathfrak{Y}_c,B_c) : c \in HF \rangle$ of elements of FA,

  $(\cup \langle \mathfrak{Y}_c : c \in HF \rangle, \langle 5, \{\langle b,c \rangle : b \in B_c, c \in HF\} \rangle) \in FA$.

4.5. <u>Lemma</u>. For any $(\mathfrak{X},A)$, $(\mathfrak{X}',A') \in FA$, if $A = A'$, then $\mathfrak{X} = \mathfrak{X}'$.

<u>Proof</u>. Straightforward. ∎

4.6. <u>Theorem</u>. For any $A,B \subseteq HF$, there is at most one object x such that for some $\mathfrak{X}$, $(\mathfrak{X},A) \in FA$ and $B = Ap^{\mathfrak{X}}(x)$.

<u>Proof</u>. Immediate from 4.3 and 4.5. ∎

5. Recursion over Fa

As we intimated at the end of §1, this will be a short section. We have considered a great many possible definitions for a natural recursion theory over Fa and found them all lacking. What we present here is one basic flaw that seems insurmountable.

From the proof that Kleene recursion restricted from Tp to Ct is a viable theory one may extract a corollary that is, in fact, the heart of the proof. Roughly, it reads: if φ is Kleene recursive in $\psi_o,\ldots,\psi_n$, then there is a recursive function f such that for all associates $\delta_o,\ldots,\delta_n$ for $\psi_o,\ldots,\psi_n$, respectively, $f(\delta_o,\ldots,\delta_n)$ is an associate for φ. Let us assume that an analogous result should hold for any recursion theory over Fa. It seems also reasonable to assume for such a theory that the function which from two sets produces their intersection is recursive - in other words, that $x \cap y$ should be recursive in x,y. Note that the class of finitely approximable sets is closed under intersection. Under these two assumptions, the following theorem has as a consequence that there is no reasonable recursion theory over Fa.

5.1. Theorem. There is no recursive function f such that for all $\mathfrak{X} \in Fa$, all sets $x,y \in \mathfrak{X}$, and all $A \in Fd^{\mathfrak{X}}(x)$ and $B \in Fd^{\mathfrak{X}}(y)$, $f(A,B) \in Fd^{\mathfrak{X}}(x \cap y)$.

Proof. It suffices to consider Fa^2. Let (as in §1)

$$y^* = \{z \subseteq HF: z \cap \omega \neq \emptyset \text{ and } (\text{least } n.\ n \in (z \cap \omega)) \text{ is odd}\}$$

and

$$x^* = \{z \subseteq HF: z \cap \omega \neq \emptyset \text{ and } (\text{least } n.\ n \in (z \cap \omega) \text{ is even}\}.$$

Then $x^*,y^* \in Fa^2$ and $x^* \cap y^* = \emptyset$, in particular $\emptyset \notin x^* \cap y^*$.

In the rest of the proof we write superscript k in place of $\mathfrak{F}A^k$. For any $z \subseteq HF$, $\langle\emptyset,\emptyset\rangle \in Ap^1(z)$ - that is $\emptyset \in [\langle\emptyset,\emptyset\rangle]^1 = \mathfrak{P}(HF)$, so $\langle\emptyset,\{\langle\emptyset,\emptyset\rangle\}\rangle \in Ap^2(x^* \cap y^*)$ and thus $\emptyset \in [Ap^2(x^* \cap y^*)]^-$. Hence for any $C \in Fd^2(x^* \cap y^*)$, also $\emptyset \in [C]^-$.

Suppose now that $A^* \in Fd^2(x^*)$, $B^* \in Fd^2(y^*)$ and f were as in the statement so that $C = f(A^*,B^*) \in Fd^2(x^* \cap y^*)$. Since $\emptyset \in [C]^-$, there is some $c \in C$ and some $d \in Ap^1(\emptyset)$ such that $d \in c^-$. Then by the continuity of recursive functions there exist finite $a \subseteq A^*$ and $b \subseteq B^*$ such that $c \in f(a,b)$. But a and b are consistent with $\emptyset$ being a member of $x^* \cap y^*$ - i.e. a and b can be extended to full descriptions A and B for sets x and y, respectively, such that $\emptyset \in x \cap y$. However, $c \in f(A,B)$ so $\emptyset \in [f(A,B)]^-$ and thus $f(A,B) \notin Fd^2(x \cap y)$. ∎

6. Appendix

We give here the proofs of the three technical results, Theorems 3.4 - 3.6.

Proof of Theorem 3.4. We show first that $\mathfrak{F}^{\mathfrak{X}} \subseteq \mathfrak{F}$. It clearly suffices to show that for each $f \in X$ and each $A \in Fd^{\mathfrak{X}}(f)$, $\theta_A^{\mathfrak{X}} \in \mathfrak{F}(f)$. We introduce the abbreviation

$$P(A,B,f,y) \quad \text{for} \quad \exists C \in Fd^{\mathfrak{Z}}(f(y)).\ \forall c \in C.\ \exists b \in B.\ \exists a \in A.\ \langle b,c\rangle \in a\ .$$

Then from the definitions we have easily

$$\begin{aligned} P(A,B,f,y) &\leftrightarrow \exists C \in Fd^{\mathfrak{Z}}(f(y)).\ \forall c \in C.\ \exists b \in B.\ \exists a \in A.\ ([a]^{\mathfrak{X}})*([b]^{\mathfrak{Y}}) \subseteq [c]^{\mathfrak{Z}} \\ &\leftrightarrow \exists C \in Fd^{\mathfrak{Z}}(f(y)).\ \theta_C^{\mathfrak{Z}} \subseteq (\theta_A^{\mathfrak{X}})*(\theta_B^{\mathfrak{Y}}) \\ &\leftrightarrow (\theta_A^{\mathfrak{X}})*(\theta_B^{\mathfrak{Y}}) \in \mathfrak{F}^{\mathfrak{Z}}(f(y))\ . \end{aligned}$$

Then, since $A \in Fd^{\mathfrak{X}}(f)$,

$$\begin{aligned} &\forall y \in Y.\ \forall B \in Fd^{\mathfrak{Y}}(y).\ P(A,B,f,y) \rightarrow \\ &\rightarrow \forall y \in Y.\ \forall B \in Fd^{\mathfrak{Y}}(y).(\theta_A^{\mathfrak{X}})*(\theta_B^{\mathfrak{Y}}) \in \mathfrak{F}^{\mathfrak{Z}}(f(y)) \\ &\rightarrow \forall y \in Y. \forall \Psi \in \mathfrak{F}^{\mathfrak{Y}}(y).\ (\theta_A^{\mathfrak{X}})*(\Psi) \in \mathfrak{F}^{\mathfrak{Z}}(f(y)) \\ &\rightarrow \theta_A^{\mathfrak{X}} \in \mathfrak{F}\ . \end{aligned}$$

Toward the inclusion $\mathfrak{F} \subseteq \mathfrak{F}^{\mathfrak{X}}$ fix $f \in X$ and $\Phi \in \mathfrak{F}(f)$. We define $A \in Fd^{\mathfrak{X}}(f)$ such that $\theta_A^{\mathfrak{X}} \subseteq \Phi$. Let

$$A = \{a \in Ap^{\mathfrak{X}}(f) : \forall \langle b,c\rangle \in a.\ \exists U \in \Phi.\ U*([b]^{\mathfrak{Y}}) \subseteq [c]^{\mathfrak{Z}}\}\ .$$

For any $y \in Y$ and $B \in Fd^{\mathfrak{Y}}(y)$,

$$\begin{aligned} \Phi*(\theta_B^{\mathfrak{Y}}) \in \mathfrak{F}^{\mathfrak{Z}}(f(y)) &\leftrightarrow \exists C \in Fd^{\mathfrak{Z}}(f(y)).\theta_C^{\mathfrak{Z}} \subseteq \Phi*(\theta_B^{\mathfrak{Y}}) \\ &\leftrightarrow \exists C \in Fd^{\mathfrak{Z}}(f(y)).\ \forall c \in C.\ \exists U \in \Phi.\ \exists b \in B.\ U*([b]^{\mathfrak{Y}}) \subseteq [c]^{\mathfrak{Z}} \\ &\leftrightarrow P(A,B,f,y)\ . \end{aligned}$$

Since $\Phi \in \mathfrak{F}(f)$, we have

$$\forall y \in Y.\ \forall B \in Fd^{\mathfrak{Y}}(y).\Phi*(\theta_B^{\mathfrak{Y}}) \in \mathfrak{F}^{\mathfrak{Z}}(f(y))\ ,$$

and therefore

$$\forall y \in Y.\ \forall B \in Fd^{\mathfrak{Y}}(y).\ P(A,B,f,y)\ ,$$

from which it follows that $A \in Fd^{\mathcal{X}}(f)$.

To prove that $\theta^{\mathcal{X}}_A \subseteq \Phi$, it suffices to show that for any $a \in A$, $[a]^{\mathcal{X}} \in \Phi$. But for each $<b,c> \in a$, there is by definition $U_{b,c} \in \Phi$ such that $U_{b,c} \subseteq [\{<b,c>\}]^{\mathcal{X}}$. Thus

$$U = \cap\{U_{b,c}: <b,c> \in a\} \subseteq \cap\{[\{<b,c>\}]^{\mathcal{X}}: <b,c> \in a\} = [a]^{\mathcal{X}}\ ,$$

and U , being a finite intersection of members of Φ, belongs to Φ .■

<u>Proof of Theorem 3.5</u>. We show first that $\mathfrak{F}^{\mathcal{X}} \subseteq \mathfrak{F}$. As above fix $x \in X$ and $A \in Fd^{\mathcal{X}}(x)$: we show $\theta^{\mathcal{X}}_A \in \mathfrak{F}(x)$. By the definition of $Fd^{\mathcal{X}}$, there is a $d \in HF$ such that

$$\forall c \in d.\ x \in Y_c\ \underline{\text{and}}\ B_c = \{b: \exists a \in A. <b,c> \in a\} \in Fd^{\mathfrak{Y}_c}(x)\ .$$

Let $\Psi_c = \theta^{\mathfrak{Y}_c}_{B_c}$; then $\Psi_c \in \mathfrak{F}^{\mathfrak{Y}_c}(x)$. Set

$$\mathfrak{D} = <\Psi_c: c \in d>\ .$$

We claim that $\Phi_{\mathfrak{D}} \subseteq \theta^{\mathcal{X}}_A$. Fix $U \in \Phi_{\mathfrak{D}}$ and the corresponding $V_c \in \Psi_c$ such that $\cap\{V_c: c \in d\} \subseteq U$. For each $c \in d$ there exists $b_c \in B_c$ with $[b_c]^{\mathfrak{Y}} \subseteq V_c$, and $a_c \in A$ with $<b_c,c> \in a_c$. Choose $a \in A$ such that $[a]^{\mathcal{X}} \subseteq \cap\{[a_c]^{\mathcal{X}}: c \in d\}$. Then for any $x' \in X$,

$$\begin{aligned} x' \in [a]^{\mathcal{X}} &\rightarrow \forall c \in d.\ a_c \in Ap^{\mathcal{X}}(x') \\ &\rightarrow \forall c \in d.\ x' \in Y_c\ \underline{\text{and}}\ b_c \in Ap^{\mathfrak{Y}_c}(x') \\ &\rightarrow \forall c \in d.\ x' \in [b_c]^{\mathfrak{Y}_c} \\ &\rightarrow x' \in \cap\{V_c: c \in d\} \subseteq U\ . \end{aligned}$$

Then $[a]^{\mathcal{X}} \subseteq U$ so $U \in \theta^{\mathcal{X}}_A$.

Toward the inclusion $\mathfrak{F} \subseteq \mathfrak{F}^{\mathcal{X}}$, fix $x \in X$ and $\Phi \in \mathfrak{F}(x)$. Fix also d such that $\forall c \in d.\ x \in Y_c$, a worthy family $\mathfrak{D} = <\Psi_c \in \mathfrak{F}^{\mathfrak{Y}_c}(x): c \in d>$ such that $\Phi_{\mathfrak{D}} \subseteq \Phi$, and $B_c \in Fd^{\mathfrak{Y}_c}(x)$ such that $\theta^{\mathfrak{Y}_c}_{B_c} \subseteq \Psi_c$. We define $A \in Fd^{\mathcal{X}}(x)$ such that $\theta^{\mathcal{X}}_A \subseteq \Phi$. Let

$$A = \{a \in Ap^{\mathcal{X}}(x): \forall <b,c> \in a.\ c \in d\ \underline{\text{and}}\ b \in B_c\}\ .$$

It suffices to show for each $a \in A$ that $[a]^{\mathcal{X}} \in \Phi$. For each $c \in d$,

let
$$V_c = \begin{cases} \cap\{[b]^{\mathfrak{Y}_c} : \langle b,c\rangle \in a\}, & \text{if } \exists b.\ \langle b,c\rangle \in a\ ; \\ Y_c\ , & \text{otherwise.} \end{cases}$$

Then $V_c \in \Psi_c$ and easily
$$\cap\{V_c : c \in d\} \subseteq [a]^{\mathfrak{X}}\ .$$

Hence, by the definition of $\Phi_{\mathfrak{D}}, [a]^{\mathfrak{X}} \in \Phi_{\mathfrak{D}} \subseteq \Phi$. ∎

<u>Proof of Theorem 3.6</u>. The argument is parallel to that for Theorem 3.4 but somewhat easier. Note first that for any a and b,

$$b \in a^{+} \rightarrow [b]^{\mathfrak{Y}} \subseteq \cap[a]^{\mathfrak{X}}\quad , \text{ and}$$
$$b \in a^{-} \rightarrow [b]^{\mathfrak{Y}} \cap \cup[a]^{\mathfrak{X}} = \emptyset\ .$$

We introduce the abbreviations
$$P^{\pm}(A,B) \quad \text{for} \quad \exists b \in B.\ \exists a \in A.\ b \in a^{\pm}\ .$$

Then
$$\begin{aligned} P^{+}(A,B) &\rightarrow \exists b \in B.\ \exists a \in A.\ [b]^{\mathfrak{Y}} \subseteq \cap[a]^{\mathfrak{X}} \\ &\rightarrow \exists a \in A.\ \cap[a]^{\mathfrak{X}} \in \theta_B^{\mathfrak{Y}} \\ &\rightarrow \exists U \in \theta_A^{\mathfrak{X}}.\ \cap U \in \theta_B^{\mathfrak{Y}}\ , \end{aligned}$$

and similarly
$$P^{-}(A,B) \rightarrow \exists U \in \theta_A^{\mathfrak{X}}.\ (Y \sim \cup U) \in \theta_B^{\mathfrak{Y}}\ .$$

For the inclusion $\mathfrak{F}^{\mathfrak{X}} \subseteq \mathfrak{F}$, fix $x \in X$ and $A \in Fd^{\mathfrak{X}}(x)$. For any $y \in Y$, we have

$$\begin{aligned} y \in x &\rightarrow \forall B \in Fd^{\mathfrak{Y}}(y).\ P^{+}(A,B) \\ &\rightarrow \forall B \in Fd^{\mathfrak{Y}}(y).\ \exists U \in \theta_A^{\mathfrak{X}}.\ \cap U \in \theta_B^{\mathfrak{Y}} \\ &\rightarrow \forall \Psi \in \mathfrak{F}^{\mathfrak{Y}}(y).\ \exists U \in \theta_A^{\mathfrak{X}}.\ \cap U \in \Psi \\ &\rightarrow y \in (\theta_A^{\mathfrak{X}})^{+} \end{aligned}$$

so that the first condition, $x \subseteq (\theta_A^{\mathfrak{X}})^{+}$, is satisfied. A similar proof shows $[A]^{-} \subseteq (\theta_A^{\mathfrak{X}})^{-}$. We observed just before Theorem 3.6 that $\Pi_x^{-} = [Ap(x)]^{-}$ and since $A \in Fd^{\mathfrak{X}}(x)$, $[A]^{-} = [Ap(x)]^{-}$ so we have also $\Pi_x^{-} \subseteq (\theta_A^{\mathfrak{X}})^{-}$ and thus $\theta_A^{\mathfrak{X}} \in \mathfrak{F}(x)$.

Towards the inclusion $\mathfrak{F} \subseteq \mathfrak{F}^{\mathfrak{X}}$, fix $x \in X$ and $\Phi \in \mathfrak{F}(x)$. We define $A \in Fd^{\mathfrak{X}}(x)$ such that $\theta_A^{\mathfrak{X}} \subseteq \Phi$. Set

$$A = \{a \in Ap^{\mathfrak{X}}(x) : \forall b \in a^+ .\ \exists U \in \Phi .\ [b]^{\mathfrak{Y}} \subseteq \cap U \ \underline{\text{and}}$$
$$\forall b \in a^- .\ \exists U \in \Phi .\ [b]^{\mathfrak{Y}} \cap \cup U = \emptyset\}.$$

First, for any $y \in Y$,

$$y \in \Phi^+ \leftrightarrow \forall B \in Fd^{\mathfrak{Y}}(y) .\ \exists U \in \Phi .\ \cap U \in \Theta_B^{\mathfrak{Y}}$$
$$\leftrightarrow \forall B \in Fd^{\mathfrak{Y}}(y) .\ \exists b \in B .\ \exists U \in \Phi .\ [b]^{\mathfrak{Y}} \subseteq \cap U$$
$$\leftrightarrow y \in [A]^+ .$$

Similarly, $\Phi^- = [A]^-$.

Thus $x \subseteq [A]^+$ and $[Ap^{\mathfrak{X}}(x)]^- = \Pi_x^- \subseteq \Phi^- = [A]^-$, so $A \in Fd^{\mathfrak{X}}(x)$.

Finally we show that for each $a \in A$, $[a]^{\mathfrak{X}} \in \Phi$. For each $b \in a^+$ choose $U_b \in \Phi$ such that $[b]^{\mathfrak{Y}} \subseteq \cap U_b$, and for each $b \in a^-$ choose $U_b \in \Phi$ such that $[b]^{\mathfrak{Y}} \cap \cup U_b = \emptyset$. Set

$$U = \cap\{U_b : b \in a^+ \cup a^-\} .$$

Then $U \in \Phi$ and for any $x \in U$ and any b,

$$b \in a^+ \to [b]^{\mathfrak{Y}} \subseteq x \ \underline{\text{and}} \ b \in a^- \to [b]^{\mathfrak{Y}} \cap x = \emptyset, \text{ so that } x \in [a]^{\mathfrak{X}} .$$

Thus $U \subseteq [a]^{\mathfrak{X}}$ so $[a]^{\mathfrak{X}} \in \Phi$. ■

References

[Du 66] Dugundji, J., Topology, Allyn and Bacon, Boston, 1966, xvi + 447 pp.

[Hi 78] Hinman, P.G., Recursion-Theoretic Hierarchies, Perspectives in Mathematical Logic, Springer-Verlag, Berlin-Heidelberg-New York, 1978, xii + 480 pp.

[Hy 79] Hyland, J.M.E., Filter spaces and continuous functionals, Annals of Mathematical Logic 16 (1979) 101-143.

[Ke-Mo 77] Kechris, A.S. and Moschovakis, Y.N., Recursion in higher types, Handbook of Mathematical Logic, ed. K.J. Barwise, North-Holland, Amsterdam, 1977, 681-737.

[Kl 59] Kleene, S.C., Countable functionals, Constructivity in Mathematics, ed. A. Heyting. North-Holland, Amsterdam, 1959,81-100.

[Kr 59] Kreisel, G., Interpretation of analysis by means of functionals of finite type, Constructivity in Mathematics, ed. A. Heyting, North-Holland, Amsterdam, 1959, 101-128.

[No 78] Normann, D., Set recursion, Generalized Recursion Theory II, ed. J.E. Fenstad, R.O. Gandy, and G.E. Sacks, North-Holland, Amsterdam, 1978, 303-320.

[No 80] Normann, D., Recursion on the countable functionals, Lecture Notes in Mathematics 811, Springer-Verlag, Berlin-Heidelberg-New York, 1980, viii + 191 pp.

A Unified Approach to Constructive and Recursive Analysis

by

Christoph Kreitz and Klaus Weihrauch

1. Introduction

Many mathematicians familiar with the constructivistic objections to classical mathematics concede their validity but remain unconvinced that there ist any satisfactory alternative.

Among others Bishop [2] and Bridges [3] showed that large parts of classical analysis and functional analysis can be formulated in a constructive way. But it does not seem to be likely that their concept of constructivity will be generally accepted.

The previous attempts to formulate and study effectivity in analysis can roughly be devided into three classes. The constructivists only study "constructive" objects and only use "constructive" proofs e.g. by using intuitionistic logic (Brouwer [4], Lorenzen [15], Bishop [2], Bridges [3], et al.). The other two attempts are based on recursion theory. The "Russian school" (Ceitin [5], Kushner [14], Aberth [1], et al.) starts with an "effective" partial numbering of the set of computable real numbers $\mathbb{R}_c$ by which computability on $\mathbb{N}$ is transferred to $\mathbb{R}_c$. The "Polish school" (Grzegorczyk [8], Klaua [9], et al.) starts (essentially) with an "effective" representation of all real numbers by $\mathbb{F} := \mathbb{N}^{\mathbb{N}}$, by which computability of operators on $\mathbb{F}$ is transferred to computability on $\mathbb{R}$. The approach presented here is a consequent continuation of that of the Polish school. It is formulated as a theory of numberings $\nu: \mathbb{N} \dashrightarrow S$ (Ershov [7]) and of representations $\delta: \mathbb{F} \dashrightarrow M$ (Kreitz and Weihrauch [12]) and admits to study continuity, computability and computational complexity (Ko [10]) which can be considered as different degrees of constructivity. We will only outline basic definitions and properties and show by examples how analysis can be developed in this context.

We shall consider the set $\mathbb{N}$ of natural numbers as the (w.l.g. single) concrete set of finite objects and the set $\mathbb{F} = \mathbb{N}^{\mathbb{N}}$ of sequences of natural numbers as the (w.l.g. single) concrete set of infinite objects on which "constructions" can be performed (e.g. by a computer or by men with paper and pencil). Constructions on $\mathbb{F}$ are governed by Baire's topology, thus continuity of functions $\mathbb{F} \dashrightarrow \mathbb{F}$ or $\mathbb{F} \dashrightarrow \mathbb{N}$ ($\mathbb{N}$ with the discrete topology) is the weakest form of constructivity. Computability is the next additional stronger requirement for functions $\mathbb{F} \dashrightarrow \mathbb{F}$, $\mathbb{F} \dashrightarrow \mathbb{N}$, or $\mathbb{N} \dashrightarrow \mathbb{N}$. Further conditions on the computational complexity (e.g. primitive recursive or polynomial) yield more restricted kinds of constructivity. As a basis for the studies in Chapter 2 a unified concise Type 2 theory of continuity and computability on $\mathbb{F}$ is outlined, which is formally similar to ordinary Type 1 recursion theory on $\mathbb{N}$. More details can be found in a forthcoming paper (Weihrauch [23]).

For all other objects the elements of $\mathbb{N}$ or $\mathbb{F}$ are used as names. Let S be a set to be named by numbers. Then any $s \in S$ must have a name and any number is name of at most one $s \in S$. Thus a naming of a set S by numbers is a possibly partial surjective function $\nu: \mathbb{N} \dashrightarrow S$, which we call a numbering. The theory of numberings is studied in detail by Ershov [7] (see also Mal'cev [16]). Similarly a naming of a set M by elements of $\mathbb{F}$ is a possibly partial surjective function $\delta: \mathbb{F} \dashrightarrow M$ which we call a representation. Constructivity on S or M is defined via constructivity on concrete objects, namely the names w.r.t. given numberings or representations.

Chapter 3 gives an outline of a general theory of representations. An essential point is the definition of admissible representations of separable T_o-spaces. Again in this theory topological (t-) and computational (c-) aspects are considered simultaneously (Kreitz and Weihrauch [12]).

In Chapter 4 as an example representations of the real numbers are studied. It is shown that the significant differences between previously defined representations are topological ones. As a further application compactness on $\mathbb{R}$ will be studied in Chapter 5.

2. Type 2 Recursion Theory

As we already have outlined computability and continuity on $\mathbb{N}$ and $\mathbb{F}$ are the basis of our approach to constructive and recursive analysis. We assume the reader is familiar with ordinary recursion theory on $\mathbb{N}$ and with basic poperties of numberings (Mal'cev [16], Rogers [2o], Ershov [7]). Let φ be a standard numbering of the unary partial recursive functions $P^{(1)}$, let $< >: \mathbb{N}^n \to \mathbb{N}$ be Cantor's n-tupling function. By f: A ----→ B (with dotted arrow) we denote a possibly partial function from A to B. Unlike to ordinary (Type 1) recursion theory for Type 2 recursion theory there is no generally accepted formalism. Below we outline a unified approach which is formally similar to the Type 1 formalism. More details can be found in Weihrauch's paper [23].

We start with some topological preliminaries. Let $\mathbb{F} := \mathbb{N}^{\mathbb{N}}$, $\mathbb{B} := \mathbb{F} \cup W(\mathbb{N})$ where $W(\mathbb{N})$ is the set of words (i.e. finite sequences) over $\mathbb{N}$. On $\mathbb{B}$ a partial order is defined by $b \sqsubseteq c :\Longleftrightarrow b$ is a prefix of c. On $\mathbb{B}$ we shall assume the topology corresponding to the cpo $(\mathbb{B}, \sqsubseteq, \varepsilon)$ (Egli and Constable [6], Scott [21]). The topology induced on the subset $\mathbb{F}$ is Baire's topology. On $\mathbb{N}$ we consider the discrete topology.

First a standard representation ψ of $[\mathbb{F} \to \mathbb{B}]$, the set of continuous functions from $\mathbb{F}$ to $\mathbb{B}$ is defined. From ψ we derive representations of certain continuous functions from $\mathbb{F}$ to $\mathbb{F}$ and from $\mathbb{F}$ to $\mathbb{N}$. The construction of ψ rests on the following property. Let $\gamma := W(\mathbb{N}) \to W(\mathbb{N})$ be isotone (w.r.t. $\sqsubseteq$). Then the function $\overline{\gamma}: \mathbb{F} \to \mathbb{B}$, defined by

$\overline{\gamma}(p) := \sup\{\gamma(w) \mid w \sqsubseteq p\}$, ist continuous. And for any continuous function $\Gamma: \mathbb{F} \to \mathbb{B}$, $\Gamma = \overline{\gamma}$ for some isotone $\gamma: W(\mathbb{N}) \to W(\mathbb{N})$. The function γ specifies, how from prefixes of $p \in \mathbb{F}$ sufficiently many prefixes of $\Gamma(p) = \overline{\gamma}(p)$ can be determined. A function $\Gamma: \mathbb{F} \to \mathbb{B}$ is called computable, iff $\Gamma = \overline{\gamma}$ for some computable function γ. The computable functions $\Gamma: \mathbb{F} \to \mathbb{B}$ can easily be characterized by oracle Turing machines which on input $p \in \mathbb{F}$ from time to time read a value $p(i)$ and from time to time write one of the values $q(0), q(1), \ldots$ (in this order) of the result $q \in \mathbb{B}$. For transforming n-ary functions on $\mathbb{F}$ to unary ones, the following tupling functions $\Pi^{(n)}: \mathbb{F} \to \mathbb{F}$ are used:

$\Pi(p,q)(i) := (p(x)$ if $i = 2x, q(x)$ if $i = 2x+1)$, $\Pi^{(1)}(p) := p$

$\Pi^{(n+1)}(p_1, \ldots, p_{n+1}) := \Pi(\Pi^{(n)}(p_1, \ldots, p_n), p_{n+1})$,

notation: $\langle p_1, \ldots, p_n \rangle := \Pi^{(n)}(p_1, \ldots, p_n)$.

Also ω-ary tupling is possible: $\Pi^{(\omega)}(p_0, p_1, \ldots)\langle i,j \rangle := p_i(j)$. The functions $\Pi^{(n)}$ and $\Pi^{(\omega)}$ are homeomorphisms w.r.t. the product topologies. The projections of their inverses are computable.

The definition of $\overline{\gamma}$ is effective in the following sense. There is a computable (by an oracle Turing machine) operator $\Gamma_u: \mathbb{F} \to \mathbb{B}$ with the following property. On input p,q it determines $\overline{\gamma}(q)$ if $\gamma := \nu_N p \nu_N^{-1}$ (ν_N is a bijective standard numbering of $W(\mathbb{N})$) is isotone, $\Gamma(q)$ for some continuous $\Gamma: \mathbb{F} \to \mathbb{B}$ otherwise. Then by $\psi_p(q) := \psi(p)(q) := \Gamma_u\langle p,q \rangle$ a representation $\psi: \mathbb{F} \to [\mathbb{F} \to \mathbb{B}]$ of the continuous functions from $\mathbb{F}$ to $\mathbb{B}$ is defined, which satisfies the "universal Turing machine theorem" and the "smn-theorem".

Theorem:

(1) $\psi_p(q) = \Gamma_u\langle p,q\rangle$ for some computable $\Gamma_u \in [\mathbb{F} \to \mathbb{B}]$

(2) $\psi\langle q,r\rangle = \psi_{\Sigma\langle p,q\rangle}(r)$ for some computable $\Sigma \in [\mathbb{F} \to \mathbb{B}]$ with range $(\Sigma) \subseteq \mathbb{F}$.

The proof is similar to that in ordinary recursion theory.

Notice that Γ_u and Σ are not only continuous but even computable. Similar to Type 1 recursion theory the utm-theorem and the snm-theorem characterize the representation ψ uniquely up to (computable) equivalence (see Chapter 3). More interesting than ψ itself are two representations derived from ψ.

Definition:

(1) Define a set $[\mathbb{F} \longrightarrow \mathbb{N}]$ of partial functions from $\mathbb{F}$ to $\mathbb{N}$ and a representation $\chi: \mathbb{F} \longrightarrow [\mathbb{F} \longrightarrow \mathbb{N}]$ by: $\chi_p(q) := \chi(p)(q) :=$ (div if $\psi_p(q) = \varepsilon \in \mathbb{B}$, the first number of the sequence $\psi_p(q)$ otherwise).

(2) Define a set $[\mathbb{F} \longrightarrow \mathbb{F}]$ of partial functions from $\mathbb{F}$ to $\mathbb{F}$ and a representation $\tilde{\psi}: \mathbb{F} \longrightarrow [\mathbb{F} \longrightarrow \mathbb{F}]$ by $\tilde{\psi}_p(q) := \tilde{\psi}(p)(q) :=$ ($\psi_p(q)$ if $\psi_p(q) \in \mathbb{F}$, div otherwise).

This definition extends well known concepts of computable operators and functionals to a uniform topological description, where the elements computable w.r.t. a given representation are those with computable names. The functions from $[\mathbb{F} \longrightarrow \mathbb{N}]$ and from $[\mathbb{F} \longrightarrow \mathbb{F}]$ have natural domains (c.f. domains of partial recursive functions). But the set of domains is sufficiently rich such that any continuous function is essentially considered.

Theorem:

(1) $[\mathbb{F} \longrightarrow \mathbb{N}]$ is the set of all continuous functions $\Sigma: \mathbb{F} \dashrightarrow \mathbb{N}$ such that $\mathrm{dom}(\Sigma)$ is open. For any continuous function $\Gamma: \mathbb{F} \dashrightarrow \mathbb{N}$ there is some $\Sigma \in [\mathbb{F} \longrightarrow \mathbb{N}]$ which extends Γ.

(2) A valid statement is obtained by substituting "$\mathbb{N}$" by "$\mathbb{F}$" and "open" by "G_δ-subset" in (1).

Also the representations χ and $\tilde{\psi}$ satisfy the utm- and the smn-theorem. This leads to a rich theory of continuity and of computability which is formally similar to Type 1 recursion theory. From the above theorem we conclude that by $\omega'(p) := \mathrm{dom}(\chi_p)$ a representation ω' of the open subsets of $\mathbb{F}$ is defined, which corresponds to the numbering $i \longmapsto \mathrm{dom}(\varphi_i)$ of the r.e. subsets of $\mathbb{N}$. We call a subset $A \subseteq \mathbb{F}$ t-open (c-open) iff $A = \omega'(p)$ for some (computable) p. A is t-clopen (c-clopen), iff A and $\mathbb{F} \setminus A$ are t-open (c-open). The t-open (c-open) sets are exactly the projections of the t-clopen (c-clopen) sets. The self applicability and the halting problem of χ can be formulated. They are c-open, not t-clopen, c-complete and c-productive. Also effective inseparability can be defined. The sets $\{p \mid \chi_p(p) = 0\}$ and $\{p \mid \chi_p(p) = 1\}$ are c-effectively inseperable. This property can be used in the study of precomplete representations. Many other properties can be proved easily but more questions are still unsolved in this theory of continuity and computability on $\mathbb{F}$.

3. Theory of representations

In order to define computability and constructivity on a set M with cardinality not greater than that of the continuum, we represent M by a surjective mapping $\delta: \mathbb{F} \dashrightarrow M$, called representation of M. Some examples for representations are the enumeration representation $\mathbb{M}: \mathbb{F} \longrightarrow P_\omega$ with $\mathbb{M}(p) := \{i \mid i+1 \in \text{range } p\}$, the representation δ_{cf} of P_ω by characteristic functions with $\delta_{cf}(p) := \{i \mid p(i) = 0\}$, and the representations $\psi: \mathbb{F} \longrightarrow [\mathbb{F} \longrightarrow \mathbb{B}]$, $\tilde{\psi}: \mathbb{F} \longrightarrow [\mathbb{F} \longrightarrow \mathbb{F}]$, $\chi: \mathbb{F} \longrightarrow \mathbb{F} \longrightarrow \mathbb{N}$, $\omega': \mathbb{F} \longrightarrow \{x \subseteq \mathbb{F} \mid x \text{ is open}\}$ introduced in chapter 2.

Effectivity properties of theorems, functions, sets, predicates etc. can be expressed by effectivity of correspondences (i.e. multivalued functions) which are triples $f = (M, M', P)$ where $P \subseteq M \times M'$.

Definition:

Let δ,δ' be representations of M resp. M' and let $f = (M,M',P)$ be a correspondence. f is called weakly (δ,δ)-t-(c-) effective iff there is some (computable) $\Gamma \in [\mathbb{F} \longrightarrow \mathbb{F}]$ such that

$$(\delta q,\delta'\Gamma q) \in P \quad \text{for all} \quad q \in \delta^{-1}\,\mathrm{dom}(f).$$

f is called (δ,δ')-t-(c-) effective, iff in addition $\Gamma(q)$ is undefined for all $q \in \delta^{-1}(M\setminus \mathrm{dom}\, f)$.

(δ,ν)-effectivity of a correspondence $f = (M,S,P)$ where ν is a numbering of S is defined accordingly using $[\mathbb{F} \longrightarrow \mathbb{N}]$ instead of $[\mathbb{F} \longrightarrow \mathbb{F}]$. For convenience we shall say "continuous" instead of "t-effective" and "computable" instead of "c-effective".

Since a partial function is a single valued correspondence the above definition is applicable to functions. A subset $A \subseteq \mathbb{M}$ can either be characterized as the domain of a partial function or by its characteristic function.

A set $A \subseteq M$ is called δ-(c-) open iff $d_A := (M, \mathbb{N}, A \times \mathbb{N})$ is $(\delta,id_{\mathbb{N}})$t-(c-)effective. A is called δ-t-(c-) clopen iff $c_A := M, \mathbb{N}, \{(x,0) \mid x \in A\} \cup \{(y,1) \mid y \in M\setminus A\})$ is $(\delta,id_{\mathbb{N}})$ t-(c-) effective.

Usually we say "provable" instead of "c-open" and "decidable" instead of "c-open".

The δ-effectivity on M strongly depends on the representation δ. Consider the two questions whether complementation on P_ω is effective and whether countable union on P_ω is effective. There is no absolute answer but only one relative to the considered representation: Complementation is $(\delta_{cf},\delta_{cf})$-computable but not even $(\mathbb{M},\mathbb{M})$-continuous. countable union is computable w.r.t. $\mathbb{M}$ but not even weakly continuous w.r.t. δ_{cf} (use $\Pi^{(\omega)}$ for formalization). This difference can be explained using the intuitive concept of finitely (or continuously) accessible (f.a.) information. Every true information $n \in \mathbb{M}(p)$ is f.a. from p, no true information $n \notin \mathbb{M}_p$ is f.a. from p. But every true information $n \in \delta_{cf}(p)$ or $m \notin \delta_{cf}(p)$ is f.a. from p. Representations may be changed in a certain way without changing the induced effectivity.

For any two representations δ, δ' of M resp. M' define

$\delta \leq_t \delta'$: <==> $M \subseteq M'$ and $id_{M,M'}$ is (δ,δ') t-effective,

$\delta \equiv_t \delta'$: <==> $\delta \leq_t \delta'$ and $\delta' \leq_t \delta$.

c-reducibility ($\leq_c$) and c-equivalence ($\equiv_c$) are defined accordingly. It is easy to show that $\delta_{cf} \leq_c \mathbb{M}$ and that $\mathbb{M}$ and δ_{cf} are not t-equivalent.

Since effective functions are closed under composition two representations are t- (c-) equivalent if and only if they define the same continuity (computability theory).

<u>Theorem:</u>

Let δ, δ' be representations of M. Then (1), (2) and (3) are equivalent.

(1) $\delta \leq_t \delta'$

(2) For any representation $\delta_1 : \mathbb{F} \dashrightarrow M_1$ and any correspondence $f = (M, M_1, P)$:
f (weakly)(δ_1, δ) t-effective ==> f (weakly)(δ_1, δ') t-effective

(3) For any representation $\delta_2 : \mathbb{F} \dashrightarrow M_2$ and any correspondence $g = (M_2, M, P)$:
g (weakly)(δ', δ_2) t-effective ==> g (weakly)(δ, δ_2) t-effective.

<u>Proof:</u>

(1) ⇒ (2) ; (1) ⇒ (3): Immediately from "$\delta \leq_t \delta' \Rightarrow id_M$ is (δ, δ')-continuous"

(2) ⇒ (1) : Choose $\delta_1 := \delta$ and $f := id_M$

(3) ⇒ (1) : Choose $\delta_2 := \delta'$ and $g := id_M$

□

Every representation $\delta : \mathbb{F} \dashrightarrow M$ induces a topology τ_δ on M by $x \in \tau_\delta$: <==> $\delta^{-1}x = A \cap \text{dom}\,\delta$ for some open subset $A \subseteq \mathbb{F}$.

τ_δ is called the final-topology of δ and it consists exactly of all the δ-open subsets of M. For example $\tau_{\mathbb{M}}$, the final topology of the enumeration representation of P_ω is determined by the basis $\{O_e | e \subseteq \mathbb{N}, \text{finite}\}$ where $O_e := \{x \subseteq \mathbb{N} | e \subseteq x\}$. Clearly t-equivalent representations have the same final topologies but the converse does not hold in general (a counterexample is presented in chapter 4). If on M already a topology τ is given then $\tau = \tau_\delta$ should hold for any "reasonable" representation of M. (In some special cases there might be reasons for choosing $\tau_\delta \neq \tau$.) For separable T_o-spaces representations equivalent to a standard-representation defined as follows seem to be the most natural ones.

__Definition:__

Let (M,τ) be a separable T_o-space and let U be a numbering of some basis of τ. For $x \in M$ let $\varepsilon_u(x) := \{i | x \in U_i\}$.
A standard-representation δ_u of (M,τ) is defined by

$$\text{dom}\delta_u := \mathbb{M}^{-1}\varepsilon_u(M) \quad \text{and} \quad \delta_u(p) := \varepsilon_u^{-1}\,\mathbb{M}(p) \text{ whenever } p\ \text{dom}\delta_u.$$

A standard-representation δ_u of a seperable T_o-space has some remarkable properties:

__Lemma:__

(1) δ is continuous and open, expecially $\tau = \tau_{\delta_u}$,

(2) For any space (M',τ') and any $H: M \dashrightarrow M'$
$H \circ \delta_u$ is continuous $\iff$ H is continuous,

(3) $\zeta \leq_t \delta_u$ for any continuous $\zeta: \mathbb{F} \dashrightarrow M$.

__Proof:__

(1) ε_u is (τ,τ_M)-continuous and open and $\mathbb{M}$ is open and continuous w.r.t. τ_M.

(2) Follows from (1).

(3) Define $\Delta: \mathbb{F} \to \mathbb{F}$ by
$\Delta(p)\langle n,k\rangle := (n + d \text{ if } \zeta[p^{[k]}] \subset U_n,\ 0 \text{ otherwise})$
Then $\mathbb{M}_{\Delta(p)} = \{n | \zeta_{(p)} \in U_n\} = \varepsilon_u(\zeta(p))$ for every $p \in \text{dom}\zeta$
Therefore $\zeta \leq_t \delta_u$ by Δ.

□

An immediate consequence ist that all the standard-representations of the same space are t-equivalent and therefore the equivalence class $\{\delta \mid \delta \equiv_t \delta_u\}$ does not depend on the numbering U. Since every representation equivalent to δ_u induces the some continuity-theory we call a representation δ of a seperable T_o-space t-effective (or admissible) iff $\delta \equiv_t \delta_u$ for some standard-representation δ_u. The representations $\mathbb{M}$ and δ_{cf} of P_ω are admissible. The decimal-representation of the real numbers is not (see next chapter).

For admissible representations of a space (M,τ) the final topology is identical with τ. Furthermore topological continuity and continuity w.r.t. these representations are closely related.

<u>Theorem:</u>
Let (M_i,τ_i) be separable T_o-spaces and let $\delta_i : \mathbb{F} \dashrightarrow M_i$ be admissible representations $(i = 1,2)$. Let $F: M_1 \dashrightarrow M_2$, then:

(1) $F(\tau_1,\tau_2)$-continuous $\Longleftrightarrow$ F weakly (δ_1,δ_2)-continuous,

(2) $F(\tau_1,\tau_2)$-continuous $\wedge$ $\text{dom}F \in G_\delta(\tau_1) \Rightarrow F\ (\delta_1,\delta_2)$-continuous.

<u>Proof:</u>
W.l.g. we may assume δ_1 and δ_2 to be standard representations.

(1) Let $F: M_1 \dashrightarrow M_2$ be (τ_1,τ_2)-continuous and let $\delta' := F \circ \delta_1$. Then $\delta': F \dashrightarrow M$ is continuous and therefore $\delta' \leq_t \delta_2$. i.e. $F\delta_1(p) = \delta_2\Gamma(p)$ for all $p \in \text{dom}F\delta_1$ with some continuous Γ. Conversely let F be weakly (δ_1,δ_2)-continuous. i.e. $F\delta_1 = \delta_2\Gamma$ for some continuous Γ. Since δ_2 is continuous the same holds for $F\delta_1$ and hence also F is continuous.

(2) Let F be continuous and $\text{dom}F \in G_\delta(\tau_1)$ i.e. $\text{dom}F = \bigcap_{i\in\mathbb{N}} O_i$ where $O_i \in \tau_1$ for $i \in \mathbb{N}$. Since δ_1 is continuous there are sets O_i' open in $\mathbb{F}$ such that $\text{dom}F\delta_1 = \delta_1^{-1}\,\text{dom}F = \bigcap_{i\in\mathbb{N}} O_i' \cap \text{dom}\delta_1$.

By (1) there is some $\Gamma \in [\mathbb{F} \to \mathbb{F}]$ with $F\delta_1(p) = \delta_2\Gamma(p)$ for all $p \in \text{dom}F\delta_1$ Now let Γ_1 be the restriction of Γ to the G_δ-set $\cap\, O_i$. Then $\text{dom}\Gamma_1 = \text{dom}\Gamma \cap \cap\, O_i$ is a G_δ-set and hence $\Gamma_1 \in [\mathbb{F} \to \mathbb{F}]$ and for every $p \in \text{dom}\delta_1$

$$p \in \mathrm{dom}F\delta_1 \Rightarrow \delta_2\Gamma_1(p) = F\delta_1(p)$$

and $p \notin \mathrm{dom}F\delta_1 \Rightarrow p \notin \mathrm{dom}\Gamma_1$.

This means F is strongly (δ_1,δ_2)-continuous.

□

For some representations the converse of (2) also holds (e.g. for the representation ρ of $\mathbb{R}$ by normed Cauchy-sequences-see next chapter).

There are many other aspects of representations which should be studied, for example recursion-theoretic properties, computable elements, the structure of equivalence degrees, closure properties etc. There are also natural representations the final topologies of which are not separable. See Kreitz & Weihrauch [12] for further discussion.

4. Representations of the Real Numbers

An excellent recursion theoretic comparison of many representations of $\mathbb{R}$ has been given by T. Deil [25]. In this chapter we show by examples that the essential differences between the representations which are mainly discussed in constructive and computable analysis already are of topological nature. More details can be found in a paper by the authors [24].

Let $\mathbb{R}$ be the set of real numbers, let $\tau_R \subseteq 2^{\mathbb{R}}$ be the set of open subsets $A \subseteq \mathbb{R}$, where A is open iff it is the union of open intervals $(x;y) := \{z \in R \mid x < z < y\}$. The space $(\mathbb{R},\tau_R)$ is a seperable T_o-space. From Chapter 3 we know that there are admissible representations δ of $\mathbb{R}$ for which expecially $\tau_\delta = \tau_R$. The authors have shown [12] that for any complete separable metric space an admissible representation can be defined via normed Cauchy sequences. In the case of $\mathbb{R}$ a useful representation of this type is as follows:

Definition:

Let $Q_n := \{m \cdot 2^{-n} | m \in \mathbb{Z}\}$, $Q_D := \bigcup Q_n$, let ν_D be a standard numbering of Q_D. Then the standard representation ρ of the real numbers is defined by $dom(\rho) := \{p \in \mathbb{F} | (\forall k)(\nu_D p(k) \in Q_k \wedge | \nu_D p(k+1)| < 2^{-k})\}$,

$$\rho(p) := \lim \nu_D p(n) \quad \text{for all} \quad p \in dom\rho.$$

This representation is admissible and its final topology is τ_R. As a product of admissible representations ρ^n, defined by $\rho^n\langle p_1,\ldots,p_n\rangle := (\rho(p_1),\ldots,\rho(p_n))$, is admissible for any $n > 0$. Therefore ρ^n satisfies Rice's theorem which states that $\rho^{-1}X$ is not t-clopen if $X \in R^n$ is not trivial. Especially relations like $x < y$, $x = y, \ldots$ are not ρ^2-decidable.

The definition of the real numbers as cuts of rational numbers induces several representations.

Definition:

Define a representation $\rho_<$ of $\mathbb{R}$ as follows:

$$dom\rho_< := \{p \in \mathbb{F} \mid \nu_D \mathbb{M}_p \text{ has an upper bound}\}$$

$$\rho_<(p) := \sup \nu_D \mathbb{M}_p \quad \text{for} \quad p \in dom(\rho_<).$$

A representation $\rho_>$ is defined correspondingly by $\rho_>(p) = \inf \nu_D \mathbb{M}_p$.

The representations $\rho_<$ and $\rho_>$ are admissible w.r.t. their final topologies $\tau_<$ and $\tau_>$, respectively, where

$$\tau_< = \{(x;\infty)\ x \in \mathbb{R} \cup \{-\infty\}\},\ \tau_> = \{(-\infty;x)\ x \in \mathbb{R} \cup \{\infty\}\}.$$

It is easy to show that $\rho \in \inf_c(\rho_<,\rho_>)$ (c.f. [24]) and by the characterization of the final topologies

$\rho_< \not\leq_t \rho, \rho_> \not\leq_t \rho, \rho_< \not\leq_t \rho_>$ and $\rho_> \not\leq_t \rho_<$. The characteristic properties of ρ, $\rho_<$ and $\rho_>$ are given by the finitely accessible information in each case. For any $i \in \mathbb{N}$ and $p \in \mathbb{F}$ the property

$\nu_D(i) < \rho_<(p)$ can be proved in finitely many steps iff it is true, more precisely $\{(d,x) \in Q_D \times \mathbb{R} \mid d < x\}$ is $[\nu_D,\rho_<]$-provable. $\rho_<$ is the greatest representation (w.r.t. $\leq_c$ or $\leq_t$) with this property (similarly for $\rho_>$). Finally ρ is the greatest representation δ of $\mathbb{R}$ such that

$$\{(d,e,x) \in Q_D \times Q_D \times \mathbb{R} \mid d < x < e\} \text{ is } [\nu_D,\nu_D,\delta]\text{-provable.}$$

Classical analysis suggests the following representation δ_c by unrestricted Cauchy sequences: $\mathrm{dom}(\delta_c) = \{p \mid \nu_D p$ is a Cauchy sequence$\}$, $\delta_c(p) := \lim \nu_D p(n)$ for $p \in \mathrm{dom}(\delta_c)$. But in this case no prefix $p^{[n]}$ of p gives any information of $\delta_c(p)$, no information is finitely accessible. This is formally expressed by the characterization of the final topology τ_c of δ_c: $\tau_c = \{\emptyset, \mathbb{R}\}$, i.e. the indiscrete topology on $\mathbb{R}$. Therefore, δ_c is not useful in constructive analysis for purely topological reasons.

The most commonly used representations are the r-adic representations (r=2,8,1o,...). The finite r-adic fractions are dense in $\mathbb{R}$, the infinite r-adic fractions, however, are not appropriate for representing the real numbers. Again, this has topological reasons. For simplicity we consider the case r = 1o.
Define δ_{DEZ} by $\mathrm{dom}(\delta_{DEZ}) = \{p \in \mathbb{F} \mid (\forall n > 0)p(n) < 1o\}$, $\delta_{DEZ}(p) := (\pi_1 p(0) - \pi_2 p(0)) + \Sigma\ \{p(i) \cdot 1o^{-i} \mid i \in \mathbb{N}\}$. It can be shown that δ_{DEZ} has the final topology τ_R, especially $\delta_{DEZ} \leq_c \rho$. But δ_{DEZ} is not admissible since $\rho \not\leq_t \delta_{DEZ}$. The names w.r.t. δ_{DEZ} contain more finitely accessible information than those w.r.t. ρ. A very undesirable consequence is that many trivial functions on $\mathbb{R}$ are not even $(\delta_{DEZ},\delta_{DEZ})$-continuous.
Example: The function $x \longmapsto 3x$ is not $(\delta_{DEZ},\delta_{DEZ})$-continuous:

Consider the sequences (p_n) and (p_n) on $\mathbb{F}$ defined by $p_n = (0\,0\,3\ldots3\,0\ldots)$, $q_n = (0\,0\,3\ldots3\,4\,0\ldots)$ (each with n-times "3"). Suppose $3\cdot\delta_{DEZ}(r) = \delta_{DEZ}\Gamma(r)$ $(r\in \mathrm{dom}(\delta_{DEZ}))$ for some $\Gamma:\mathbb{F}\dashrightarrow\mathbb{F}$. Then $\lim p_n = \lim q_n = (0\,0\,3\,3\ldots)$, but $\lim\Gamma p(n) \neq \lim\Gamma q(n)$, hence Γ is not continuous. Therefore the decimal representation is not very useful for analysis, again for topological reasons.

Finally we consider the representations by characteristic functions of left cuts (or right cuts) which is also often considered. Let ν be a total bijective numbering of some dense (w.r.t. τ_R) subset of $\mathbb{R}$. Define $\delta_{L\nu}$ by $\mathrm{dom}(\delta_{L\nu}) := \{p\in\mathbb{F} \mid (\exists x\in R)(\forall i)(q(i) = 0 \iff \nu(i) < x\}$, $\delta_{L\nu}(p) := \sup \nu p^{-1}\{0\}$ for $p\in\mathrm{dom}(\delta_{L\nu})$. Then $\delta_{L\nu}$ is admissible and its final topology $\tau_{L\nu}$ is generated by the basis $\{(x;y] \mid x,y\in S, x<y\}$. Especially, one easily shows $\delta_{L\nu}\leq_c\rho$ and $\tau_R\subseteq\tau_{L\nu}$. Notice that the final topology $\tau_{L\nu}$ depends on the dense subset $S\subseteq\mathbb{R}$ which is numbered by ν. Therefore no left cut representation $\delta_{L\nu}$ of $\mathbb{R}$ can be called "natural" for topological reasons. Right cut representations $\delta_{R\nu}$ are defined accordingly. It should be noted that the representations defined by continued fractions is the infimum of $\delta_{L\nu}$ and $\delta_{R\nu}$ if ν is some standard numbering of $\mathbb{Q}$.

Among all the representations of $\mathbb{R}$ we have considered, ρ is the only one which has the appropriate topological properties for the study of analysis. The special definition of ρ guarantees some positive computability properties which become important in complexity theory (Ko [10], Kreitz and Weihrauch [11]).

5. Compactness on $\mathbb{R}$

As an application of the theory of representations to constructive analysis we consider compactness on $\mathbb{R}$. We show that (at least) two different reasonable kinds of compactness can be formulated in our theory. Let I, defined by $I_{<j,k>} := (\nu_D(j)-2^{-k};\nu_D(j)+2^{-k})$, be a standard numbering of a basis of τ_R. A standard representation ω of τ_R is defined by $\omega(p) := \cup\{I_k \mid k \in \mathbb{M}_p\}$. The corresponding representation α of the closed subsets is $\alpha(p) := \mathbb{R}\setminus\omega(p)$. Another way specifiying a closed set A is to list all the open sets I_k such that $A \cap I_k \neq \emptyset$. Let $\mathrm{dom}(\alpha_c) := \{p \mid (\exists A,\text{closed})\ \mathbb{M}_p = \{k \mid A \cap I_k \neq \emptyset\}\}$ $\alpha_c(p) := \bigcap_k \bigcup_j \{I_{<j,k>} \mid <j,k> \in \mathbb{M}_p\}$. The representations α and α_c are incomparable w.r.t. topological reducibility because the finitely accessible informations are too different. Both representations are admissible w.r.t. their final topologies. Let α_1 be the standard infimum of α and α_c. This representation corresponds to the concept of locatedness in constructive analysis (Bishop [2]). For a closed set $A, A=\{x \mid d(x,A)=0\}$, therefore the closed sets can be characterized by distance functions which are (ρ,ρ)-continuous. The $\tilde{\psi}$-names of the corresponding operators $\mathbb{F}$ yield a representation α' of the nonempty closed subsets of $\mathbb{R}$ which directly expresses locatedness. The representation α_1 (restricted to the nonempfty sets) is equivalent to α'. A subset of $\mathbb{R}$ is compact, iff it is closed and bounded. Therefore, the restriction $\overline{\alpha}(\overline{\alpha}_c,\overline{\alpha}_1)$ of $\alpha(\alpha_c,\alpha_1)$ to the bounded subsets yields a representation of the compact subsets $K(\mathbb{R})$. Unfortunately sup: $K(\mathbb{R}) \longrightarrow \mathbb{R}$ is not $(\overline{\alpha},\rho)$ - $((\overline{\alpha}_c,\rho)$ -, (α_1,ρ) -) continuous, that is, a ρ-name for sup (S) cannot be continuously obtained from an $\overline{\alpha}$-name of S (etc.). However the following holds:

Theorem:

There are computable functions Σ,Γ,Δ: $\mathbb{F} \dashrightarrow \mathbb{F}$ such that:

$\rho_< \Sigma(p) = \sup \alpha_c(p)$ if $\alpha_c(p)$ is bounded,

$\rho_> \Gamma\langle i,p\rangle = \sup \alpha(p)$ if $\alpha(p) \subseteq [-i;i]$,

$\rho \; \Delta\langle i,p\rangle = \sup \alpha_1(p)$ if $\alpha_1(p) \subseteq [-i;i]$.

Proof:

Since $\mathbb{M}_p = \{j \mid I_j \cap \alpha_c(p) \neq \emptyset\}$ for every $p \in \mathrm{dom}\,\alpha_c$, there is some computable $\Sigma \in [\mathbb{F} \to \mathbb{F}]$ such that $\mathbb{M}_{\Sigma(p)} = \{i \mid \nu_D(i) < \sup \alpha_c(p)\}$ and therefore $\rho_< \Sigma(p) = \sup \alpha_c(p)$ whenever $\alpha_c p$ is bounded.

Now let $p \in \mathrm{dom}\,\alpha$, $i \in \mathbb{N}$ such that $\alpha(p) \subseteq [-i;i]$. Then

$\nu_D(j) > \sup \alpha(p) \iff (\nu_D(j) > i$ or $[\nu_D(j);i] \subseteq \mathbb{R} \setminus \alpha(p))$

Since $\mathbb{R} \setminus \alpha(p) = \cup \{I_k \mid k \in \mathbb{M}_p\}$ and $[\nu_D(j);i]$ is compact in $\mathbb{R}$, there is some computable $\Gamma \in [\mathbb{F} \to \mathbb{F}]$ such that

$M_{\Gamma(p)} = \{j \mid \nu_D(j) > \sup \alpha(p)$ - i.e. $\rho_> \Gamma(p) = \sup \alpha(p)$ -

whenever

$\alpha(p) \subseteq [-i;i]$.

The last claim follows immediately from $\alpha_e \in \mathrm{Inf}_c\{\alpha_c, \alpha\}$.

□

Therefore, for determining the supremum (continuously) it is not sufficient to know that $\alpha_1(p)$ is bounded but a bound must be known (c.f. Bishop's concept of constructive compactness). The information of a bound can be inserted into the name. Define representations α^b and α_1^b of the compact subsets of $\mathbb{R}$ by $\mathrm{com}(\alpha^b) = \{\langle i,p\rangle \mid \alpha(p) \subseteq [-i;i]\}$, $\alpha^b\langle i,p\rangle := \alpha(p)$, and α_1^b correspondingly with α_1. Then sup is $(\alpha^b, \rho_>)$-computable and (α_1^b, ρ)-computable (but not (α^b, ρ)-continuous). The compact subsets of $\mathbb{R}$ can also be characterized by their Heine-Borel property which leads to (at least) two different characterizations. Let $C_p := \{I_j \mid j \in \mathbb{M}_p\}$, $C_{p,n} := \{I_j \mid (\exists i \in D_n) j+1 = p(i)\}$. Let us say "$\Omega$: $\mathbb{F} \dashrightarrow \mathbb{N}$ proves compactness of $S \subseteq \mathbb{R}$" iff

$(\forall p)[S \subseteq \cup C_p \iff p \in \mathrm{dom}(\Omega)$ and $S \subseteq \cup C_p \Longrightarrow S \subseteq \cup C_{p,\Omega(p)}]$.

This induces a representation $\varkappa_w$ of the compact sets by $\mathrm{dom}(\varkappa_w) = \{p \mid \chi_p$ proves compactness of some $S \subseteq R\}$, $\varkappa_w(p)$ = the set $S \subseteq \mathbb{R}$ for which χ_p proves compactness. Ourfirst constructive version of the Heine-Borel theorem for $\mathbb{R}$ is as follows:

<u>Theorem</u> (Heine-Borel, weak)

$$\alpha^b \equiv_c \varkappa_w$$

<u>Proof:</u>

Let $\langle i,p\rangle \in \mathrm{dom}\,\alpha^b$. Then $\alpha^b\langle i,p\rangle = [-i;i] \setminus \cup \{I_j \mid j \in \mathbb{M}_p\}$.
By compactness of $[-i;i]$ we have

$\alpha^b\langle i,p\rangle \subseteq \cup C_p \iff (\exists n)\ [-i;i] \subseteq \cup (C_{q,n} \cup C_{p,n})$
$\qquad \Rightarrow (\exists n)\ \alpha^b\langle i,p\rangle \subseteq \cup C_{q,n}$.

Since $\langle -i;i\rangle \subseteq \cup (C_{q,n} \cup C_{p,n})$ is decidable there is some computable $\Sigma \in [\mathbb{F} \to \mathbb{F}]$ with $\alpha^b\langle i,p\rangle = \varkappa_w\, \Sigma\langle i,p\rangle$ vor every $\langle i,p\rangle \in \mathrm{dom}\,\alpha^b$.

Conversely $\varkappa_w(p) \subseteq C_{id}$ holds for every $p \in \mathrm{dom}\,\varkappa_w$ (where $id(n) = n$) and since $\cup C_{id,n} \subseteq [-i;i]$ is decidable, there is some computable $\Delta \in [\mathbb{F} \to \mathbb{N}]$ with $\varkappa_w(p) \subseteq [-\Delta(p), \Delta(p)]$ if $p \in \mathrm{dom}\,\varkappa_w$.

Futhermore $\varkappa_w(p) = \mathbb{R} \setminus \cup \{I_j \mid \overline{I}_j \cap \varkappa_w(p) = \emptyset\}$ and

$\overline{I}_j \cap \varkappa_w(p) = \emptyset \iff (\exists p \in \mathrm{dom}\,\chi_p)\ \overline{I}_j \cap \cup C_{q,\chi_p(q)} = \emptyset$

(where $\overline{I}_j$ denates the closure of the interval I_j). Using the projection theorem (see [23]) one can construct some computable $\Gamma \in [\mathbb{F} \to \mathbb{F}]$ with $\mathbb{M}_{\Gamma(p)} = \{j \mid \bar{I}_j \subseteq \mathbb{R} \setminus \varkappa_w(p)\}$ - especially $\alpha\Gamma(p) = \mathbb{R} \setminus \omega\Gamma(p) = \varkappa_w(p)$ - whenever $p \in \mathrm{dom}\,\varkappa_w$.

□

Since sup, therefore, is not $(\varkappa_w, \rho)$-continuous, $\varkappa_w$ is called the weak Heine-Borel representation. As we already know, the names p w.r.t. α_1^b contain more finitely accessible information. It can be shown that from such a name p for any covering C_p a minimal covering can be obtained. Let $\varkappa$ be the restriction of $\varkappa_w$ to those $p \in \mathbb{F}$ such that $C_{q,\chi_p(q)}$ is a minimal covering of $\varkappa_w(p)$, i.e. $\varkappa_w(p) \not\subseteq \cup E$ for any proper subset E of $C_{q,\chi_p(p)}$. Then the (strong) Heine-Borel theorem can be formulated as follows:

<u>Theorem</u> (Heine-Borel, strong)

$$a_1^b \equiv_c \varkappa$$

The proof is similar to that of the weak Heine-Borel theorem. More details will be presented in a forthcoming paper [13].

6. Conclusion

In this contribution we have presented an approach to constructive analysis which is based on Type 1 and Type 2 recursion theory and on the theory of numberings and of representations. For the Type 2 theory, topology plays an important role. The approach extends the approach of the Polish school. Since for any representation δ there is a canonical numbering $\nu_\delta := \delta\varphi$ of the computable elements it includes the approach of the Russian school, and also **essential ideas of the approaches** like Bishop's can be expressed and studied without change of logic. We claim that this is the adequate way for investigating constructivity, computability and computational complexity in analysis (and other areas of mathematics).

The properties "continuous", "computable", "easily (e.g. polynomially) computable" yield a basic hierarchy of constructivity. It has turned out that in most cases a property which is not "effective" is not even continuous and a property which is "effective" is easily computable. This means that most of the negative results have topological reasons and are independent of Church's thesis. The fundamental role of topology in recursive analysis has already been pointed out by Nerode [18] and others.

The approach presented here does not depend on specific representations but different representations may be chosen from case to case. First of all a representation has to be topologically sound depending on the intended application. Then computability and computational complexity have to be considered. For any representation δ the corresponding numbering ν_δ of the δ-computable elements brings up the question how ν_δ-computability and δ-computability are related. Complete or partial answers are known only for very few cases (Myhill and Shepherdson [17]. Ceitin [5], Spreen [22]). Seemingly the clear topological properties of δ are concealed by the composition with the numbering φ the topological properties of which are not easily to understand.

In this contribution we only gave some examples of application. It should be mentioned that the kind of computability introduced by Pour-El and Richards [19] on the L^p-spaces can very naturally be defined in our framework by use of representations. A general theory of constructive normed spaces can be developed. Even measure theory can easily be approached. By $d(X,Y) := \mu(X \Delta Y)$ a metric on the open sets $J_k := \cup \{I_n \mid n \in D_k\}$ can be defined. The completion of this space is (essentially) the set of Borel sets factorized by the null sets.

Fachbereich Mathematik und Informatik
Fernuniversität Hagen
D-5800 Hagen

References

[1] ABERTH, O.: Computable analysis, Mc Graw-Hill, New York, 1980.

[2] BISHOP, E.: Foundations of constructive analysis, Mc Graw-Hill, New York, 1967.

[3] BRIDGES, D.S.: Constructive functional analysis, Pitman, London, 1979.

[4] BROUWER, L.E.J.: Zur Begründung der intuitionistischen Mathematik I,II,III, Math. Annalen 93 (1924), 244-258, 95 (1925), 453-473, 96 (1926), 451-489.

[5] CEITIN, G.S.: Algorithmic operators in constructive metric spaces. Trudy Mat.Inst.Steklov 67 (1962), 295-361. English translation: Amer. Math. Soc. Trans. (2) 64 (1967), 1-80.

[6] EGLI,H.; CONSTABLE, R.L.: Computability concepts for programming languages semantics, TCS 2 (1976) 133-145.

[7] ERSHOV, JU.L.: Theorie der Numerierungen I, Z.f. math. Logic 19 (1973), 289-388.

[8] GRZEGORCZYK, A.: On the definition of computable real continuous functions, Fund. Math. 44 (1957), 61-71.

[9] KLAUA, D.: Konstruktive Analysis, Deutscher Verlag der Wissenschaften, Berlin, 1961.

[10] KO, K.; FRIEDMANN, H.: Computational complexity of real functions, TCS 20 (1982) 323-352.

[11] KREITZ, C.; WEIHRAUCH, K.: Complexity theory on real numbers and functions, Proc. 6th GI-Conf., Lecture Notes on comp. Sci. 145 (1982) 165-174.

[12] KREITZ, C.; WEIHRAUCH, K.: Theory of representations (to appear) Towards a theory of representations, Informatik Berichte 40, Fernuniversität Hagen (1983).

[13] KREITZ, C.; WEIHRAUCH, K.: Compactness in constructive analysis revisited (to appear).

[14] KUSHNER, B.A.: Lectures on constructive mathematical analysis, Monographs in mathematical logic and foundations of mathematics, Izdat. "Nauka", Moskau, 1973.

[15] LORENZEN, P.: Differential und Integral - eine konstruktive Einführung in die klassische Analysis, Akademische Verlagsgesellschaft, Frankfurt a.M. 1965.

[16] MAL'CEV, A.I.: Algorithms and recursive functions, Wolters-Noordhoff, Groningen, 1970

[17] MYHILL, J.; SHEPHERDSON, J.C.: Effective operations on partial recursive functions, Zeitschrift f. math. Logik 1 (1955), 310-317.

[18] NERODE, A.: Lecture on "Constructive analysis", 1982 Summer Institute on recursion theory, Cornell University.

[19] POUR-EL, M.B.: RICHARDS, I.: L^p-computability in recursive analysis, Technical Report University of Minnesota (1983).

[20] ROGERS, H. jr.: Theory of recursive functions and effective computability, Mc Graw-Hill, New York, 1967.

[21] SCOTT, D.: Data Types as lattices, SIAM J.Comp. 5 (1976), 522-587.

[22] SPREEN, D.: Effective operations in a topological setting, technical report, RWTH Aachen, 1983.

[23] WEIHRAUCH, K.: Type 2 recursion theory (to be published).

[24] WEIHRAUCH, K.; KREITZ, C.: Representations of the real numbers and of the open subsets of the set of real numbers (to appear).

[25] DEIL, T.: Darstellungen und Berechenbarkeit reeller Zahlen, Dissertation, Fernuniversität Hagen, 1983.

ON FAITHFUL INTERPRETABILITY

Per Lindström
Dept of Philosophy
University of Göteborg
Göteborg, Sweden

First some notation and terminology. In the following S, T, A, B, C are consistent primitive recursive theories (sets of sentences).(By Craig's theorem, each r.e. theory is deductively equivalent to a primitive recursive theory; in fact to every Σ_1^0 formula $\gamma(x)$ we can effectively find a PR formula $\delta(x)$ s. t. $P \vdash Pr_\delta(x) \leftrightarrow Pr_\gamma(x)$ and so $P \vdash Con_\delta \leftrightarrow Con_\gamma$, where P is Peano arithmetic (cf. [1]).) For simplicity we assume that our theories contain only finitely many non-logical constants. We write $S \leq T$ to mean that S is (relatively) interpretable in T. $\{S \leq T\}$ is the set of interpretations of S in T. S is <u>faithfully interpretable</u> in T, $S \trianglelefteq T$, if there is an $f \in \{S \leq T\}$ which is faithful in the sense that for every sentence φ of S, if $T \vdash f\varphi$, then $S \vdash \varphi$. Let $\{S \trianglelefteq T\}$ be the set of faithful interpretations of S in T. These concepts were introduced by Feferman, Kreisel, and Orey [2]. More generally we shall say that S is <u>X-faithfully interpretable</u> in T, $S \trianglelefteq_X T$, where X is a set of sentences, if there is an $f \in \{S \leq T\}$ which is X-faithful, i.e. for every $\varphi \in X$, if $T \vdash f\varphi$, then $S \vdash \varphi$. We write $S \vdash X$ or $X \dashv S$ to mean that $S \vdash \varphi$ for every $\varphi \in X$. Thus $S \dashv T$ iff T is an extension of S. T is Σ_1^0-sound if all Σ_1^0 sentences provable in T are true. Notation and terminology not explained here are standard (cf. [1]).

The following results are proved in [2].

THEOREM A. Suppose $P \dashv T$, T is Σ_1^0-sound, and $\sigma(x)$ is a Σ_1^0 numeration of S. Then $S \trianglelefteq T + Con_\sigma$.

THEOREM B. Suppose $P \dashv T$ and T is reflexive and Σ_1^0-sound. If $S \leq T$, then $S \trianglelefteq T$.

It is also shown in [2] that Theorems A and B become false if the condition that T be Σ_1^0-sound is omitted. Indeed if $Q \dashv S$, $P \dashv T$, and T is reflexive, then for any $f \in \{S \leq T\}$ and any Σ_1^0 sentence φ, if $T \vdash \varphi$, then $T \vdash f\varphi$. Thus, in particular, if $S \trianglelefteq T$ and S is Σ_1^0-sound, then so is T. This result and Theorems A and B naturally suggest the problem of finding conditions (in terms of provability and interpretability) that are necessary and sufficient in order that $S \trianglelefteq T$ or, more generally,

$S \trianglelefteq_X T$ under the additional assumption that T is, say, an essentially reflexive extension of P. The solution of this problem is the main contribution of this paper (see Theorem 3 below).

Let

$T^{[S]} = \{\varphi\colon \exists m\ T \vdash Pr_{S\restriction m}(\bar{\varphi})\}$,

$T^{\{S\}} = \{\varphi\colon \text{if } \varphi \in T^{[S]}, \text{ then } S \vdash \varphi\}$.

Our principal result is the following

THEOREM 1. If $P \dashv T$ and there is a formula $\sigma(x)$ binumerating S in T and s. t. $T \vdash Con_\sigma$, then $S \trianglelefteq_{T^{\{S\}}} T$.

From Theorem 1 we can, using standard methods, easily derive

THEOREM 2. If $P \dashv T$, T is reflexive, and $S \leq T$, then $S \trianglelefteq_{T^{\{S\}}} T$.

PROOF. To each m, there is an n s. t. $S\restriction m \leq T\restriction n$. It follows that $P \vdash Con_{T\restriction n} \rightarrow Con_{S\restriction m}$. Since T is reflexive, this implies that for every m, $T \vdash Con_{S\restriction m}$. Now let $\sigma_0(x)$ be a PR binumeration of S and let $\sigma(x)$ be the formula

$$\sigma_0(x) \wedge Con_{\sigma_0}(y) \wedge y \leq x.$$

Then $P \vdash Con_\sigma$ and $\sigma(x)$ binumerates S in T (cf. [1, 7]). Hence, by Theorem 1, $S \trianglelefteq_{T^{\{S\}}} T$, as was to be shown.

If T is Σ_1^0-sound, then $T^{\{S\}}$ is the set of sentences of S. Thus Theorem B follows at once from Theorem 2. To obtain Theorem A from Theorem 1 use Craig's theorem in combination with the fact that if T is Σ_1^0-sound and $\sigma(x)$ is a Σ_1^0 numeration of S, then $T + Con_\sigma$ is Σ_1^0-sound. In fact Theorem A is an immediate consequence of Lemma 4 below.

To prove Theorem 1 we need the following result due to Guaspari [3] (cf. also [6]). In the following Γ is either Σ_{n+1}^0 or Π_{n+1}^0 and $\breve{\Gamma}$ is the dual of Γ. φ is Γ-conservative over T, $\varphi \in Cons(\Gamma, T)$, if for every Γ sentence ψ, if $T + \varphi \vdash \psi$, then $T \vdash \psi$.

LEMMA 1. Suppose $P \dashv T$ and let X be any r.e. set. There is then a Γ formula $\gamma(x)$ s. t.

(i) if $k \in X$, then $T \vdash \gamma(\bar{k})$,

(ii) if $k \notin X$, then $\neg\gamma(\bar{k}) \in Cons(\breve{\Gamma}, T)$.

Suppose $P \dashv T$ and $\sigma(x)$ numerates S in T. If Y_σ is the set numerated by $Pr_\sigma(x)$ in T, then clearly $T^{[S]} \subseteq Y_\sigma$. The content of the following lemma is that there are formulas $\sigma(x)$ (bi)numerating S in T s. t. $Y_\sigma = T^{[S]}$.

LEMMA 2. Suppose $P \dashv T$ and $\sigma(x)$ (is Σ_{n+1}^0 and) binumerates S in T. Then there is a (Σ_{n+1}^0) formula $\sigma^*(x)$ s. t.

(i) $\vdash \sigma^*(x) \to \sigma(x)$ and so $\vdash \mathrm{Con}_\sigma \to \mathrm{Con}_{\sigma^*}$,

(ii) $\sigma^*(x)$ binumerates S in T,

(iii) $\mathrm{Pr}_{\sigma^*}(x)$ numerates $T^{[S]}$ in T.

PROOF. By Lemma 1, there is a Σ_1^0 formula $\gamma(x)$ s. t.

(1) if $\varphi \in T^{[S]}$, then $T \vdash \gamma(\bar\varphi)$,

(2) if $\varphi \notin T^{[S]}$, then $\neg\gamma(\bar\varphi) \in \mathrm{Cons}(\Sigma_1^0, T)$.

Let $\tau(x)$ be a PR binumeration of T and let PrS(x,y,z) be a PR formula s. t. for all $\alpha(x)$ and φ,

$$P \vdash \mathrm{PrS}(\bar\alpha, \bar\varphi, z) \leftrightarrow z = \overline{\mathrm{Pr}_\alpha(\bar\varphi)}.$$

Finally, applying self-reference, let $\sigma^*(x)$ be s. t.

$$P \vdash \sigma^*(x) \leftrightarrow \sigma(x) \wedge \forall yz \le x(\exists u(\mathrm{PrS}(\overline{\sigma^*}, y, u) \wedge \mathrm{Prf}_\tau(u,z)) \to \gamma(y)).$$

Then $\sigma^*(x)$ is Σ_{n+1}^0 if $\sigma(x)$ is Σ_{n+1}^0. Also (i) is obvious. To prove (ii) and (iii) we first prove that

(3) if $T \vdash \mathrm{Pr}_{\sigma^*}(\bar\varphi)$, then $\varphi \in T^{[S]}$.

Suppose $T \vdash \mathrm{Pr}_{\sigma^*}(\bar\varphi)$, let p be a proof of $\mathrm{Pr}_{\sigma^*}(\bar\varphi)$ in T, and set q = max{p, φ}. Then

$$P \vdash \neg\gamma(\bar\varphi) \to (\sigma^*(x) \to \sigma(x) \wedge x \le \bar q),$$

whence

$$P \vdash \neg\gamma(\bar\varphi) \to (\mathrm{Pr}_{\sigma^*}(\bar\varphi) \to \mathrm{Pr}_{\sigma(x) \wedge x \le \bar q}(\bar\varphi)),$$

whence, since $\sigma(x)$ binumerates S in T and $P \dashv T$,

(4) $T \vdash \neg\gamma(\bar\varphi) \to \mathrm{Pr}_{S \restriction q}(\bar\varphi)$.

Suppose $\varphi \notin T^{[S]}$. Then, by (2) and (4), $T \vdash \mathrm{Pr}_{S\restriction q}(\bar\varphi)$ and so $\varphi \in T^{[S]}$. Thus $\varphi \in T^{[S]}$ and (3) is proved.

To prove (ii) it suffices to show that for all φ and p,

(5) $T \vdash \mathrm{Prf}_\tau(\overline{\mathrm{Pr}_{\sigma^*}(\bar\varphi)}, \bar p) \to \gamma(\bar\varphi)$.

But this is trivial unless p is a proof of $\mathrm{Pr}_{\sigma^*}(\bar\varphi)$ in T. So suppose it is. Then, by (3), $\varphi \in T^{[S]}$, whence, by (1), $T \vdash \gamma(\bar\varphi)$. Thus (5) holds in this case too and (ii) is proved. Finally (iii) follows at once from (ii) and (3).

The main proof in [2] uses a theorem of Mostowski [8] on completely independent formulas. In place of this result we shall apply the following improvement of Mostowski's theorem due to Scott [10]. φ^i is φ if i = 0 and $\neg\varphi$ if i = 1.

LEMMA 3. Suppose $P \dashv T$. To any (Σ_n^0) formula $\eta(x)$, there is a (Σ_{n+1}^0) formula $\xi(x)$ s. t. for all $g, h \in {}^\omega 2$, if $T_g = T \cup \{\eta(\bar n)^{g(n)}: n \in \omega\}$ is consistent, then so is $T_g \cup \{\xi(\bar n)^{h(n)}: n \in \omega\}$.

LEMMA 4. Suppose $P \dashv T$, $\sigma(x)$ numerates S in T, and $T \vdash \mathrm{Con}_\sigma$. There is then an $f \in \{S \le T\}$ s. t. for all φ, if $T \vdash f\varphi$, then $T \vdash \mathrm{Pr}_\sigma(\bar\varphi)$.

PROOF. The following proof is a modification of the standard proof that if S and T are as assumed, then $S \leq T$ (cf. [1]). The $f \in \{S \leq T\}$ constructed in that proof need, however, not necessarily have the additional property that $T \vdash f\varphi$ implies $T \vdash Pr_\sigma(\bar{\varphi})$. (Indeed if $T \vdash \neg Pr_\sigma(\bar{\varphi})$ and $T \vdash \neg Pr_\sigma(\overline{\neg\varphi})$, then either $T \vdash f\varphi$ or $T \vdash f\neg\varphi$.) The idea of ensuring this by introducing the set X (see (3) below) and letting X be formally represented by a sufficiently independent formula is taken from [2].

Let c_n, $n \in \omega$, be new individual constants and let L be the set of sentences obtained from formulas of S by replacing variables by constants c_n. Let $\exists v_n \alpha_n(v_n)$, $n \in \omega$, be a primitive recursive enumeration of all members of L of the form indicated. As is well-known there is then a primitive recursive set

$$D = \{\exists v_n \alpha_n(v_n) \to \alpha_n(c_{j_n}) : n \in \omega\}$$

and a PR binumeration $\delta(x)$ of D s. t. for every sentence φ of S,

(1) $P \vdash Pr_\gamma(\bar{\varphi}) \to Pr_\sigma(\bar{\varphi})$,

where $\gamma(x)$ is $\sigma(x) \vee \delta(x)$. Since $T \vdash Con_\sigma$, it follows that

(2) $T \vdash Con_\gamma$.

Now let θ_n, $n \in \omega$, be a primitive recursive enumeration of L. Consider the following recursive definition, where X is any subset of ω.

$$(3) \quad \varphi_n = \begin{cases} \theta_n \text{ if } S \cup D \vdash \bigwedge_{m<n} \varphi_m \to \theta_n \text{ or} (S \cup D \nvdash \bigwedge_{m<n} \varphi_m \to \neg\theta_n \ \& \ n \in X), \\ \neg\theta_n \text{ otherwise.} \end{cases}$$

(Here $\bigwedge_{m<0}\varphi_m$ is $0 = 0$). Let $\xi(x)$ be the formula given by Lemma 3 for $\eta(x) = Pr_\gamma(x)$. Next let $\chi(x,y)$ be the formalization of the result of converting (3) into an explicit definition in the usual way using $\gamma(x)$ and $\xi(x)$ to represent $S \cup D$ and X, respectively. Let $\beta(x)$ be $\exists y \chi(x,y)$ and let Scm_β be the sentence saying that the set defined by $\beta(x)$ is strongly complete, i.e. is complete and consistent and contains $\alpha(c_k)$ for some k whenever it contains $\exists v \alpha(v)$. Then, by a standard argument, $P \vdash Con_\gamma \to Scm_\beta$ (for more details see [1,2]). So, by (2), $T \vdash Scm_\beta$. But from this and since $P \vdash \sigma(x) \to \beta(x)$ it follows, as is well-known (cf. [1]) that there is an $f \in \{S \leq T\}$ s. t.

(4) $T \vdash f\varphi \leftrightarrow \beta(\bar{\varphi})$.

Suppose now $T \nvdash Pr_\sigma(\bar{\varphi})$, where φ is a sentence of S. To complete the proof we must show that $T \nvdash f\varphi$. By (1), $T \nvdash Pr_\gamma(\bar{\varphi})$. Let $g \in {}^{\omega}2$ be s. t. $g(\varphi) = 1$ and

(5) $T \cup Y_g$ is consistent,

where $Y_g = \{Pr_\gamma(\bar{n})^{g(n)} : n \in \omega\}$. Next define ψ_n as follows (compare (3)).

$$\psi_n = \begin{cases} \theta_n \text{ if } Pr_\gamma(\overline{\bigwedge_{m<n}\psi_m \to \theta_n}) \in Y_g \text{ or } (Pr_\gamma(\overline{\bigwedge_{m<n}\psi_m \to \neg\theta_n}) \notin Y_g \ \& \\ \qquad Pr_\gamma(\overline{\bigwedge_{m<n}\psi_m \wedge \theta_n \to \varphi}) \notin Y_g, \\ \neg\, \theta_n \text{ otherwise.} \end{cases}$$

Let $h \in {}^{\omega}2$ be s. t.

$$h(n) = 0 \text{ iff } Pr_\gamma(\overline{\bigwedge_{m<n}\psi_m \wedge \theta_n \to \varphi}) \notin Y_g$$

and set $Y_{g,h} = Y_g \cup \{\xi(\bar{n})^{h(n)}: n \in \omega\}$. Then, by (5) and the choice of $\xi(x)$,

(6) $T \cup Y_{g,h}$ is consistent.

Recalling the definition of $\chi(x,y)$, we can now show, by induction, that for every n, $P \cup Y_{g,h} \vdash \chi(\overline{\psi_n},\bar{n})$, whence

(7) $P \cup Y_{g,h} \vdash \beta(\overline{\psi_n})$.

We now show that for every n,

(8) $Pr_\gamma(\overline{\bigwedge_{m<n}\psi_m \to \varphi}) \notin Y_g$.

Note that, by (5), $\{\psi: Pr_\gamma(\bar{\psi}) \in Y_g\}$ is closed under logical deduction. Since $Pr_\gamma(\bar{\varphi}) \notin Y_g$, (8) holds for $n = 0$. Suppose it holds for $n = k$.
CASE 1. $\psi_k = \theta_k$. Then either $Pr_\gamma(\overline{\bigwedge_{m<k}\psi_m \to \theta_k}) \in Y_g$ or $Pr_\gamma(\overline{\bigwedge_{m<k+1}\psi_m \to \varphi}) \notin Y_g$. Hence, by the inductive assumption, (8) holds for $n = k+1$.
CASE 2. $\psi_k = \neg\theta_k$. Then $Pr_\gamma(\overline{\bigwedge_{m<k}\psi_m \wedge \theta_k \to \varphi}) \in Y_g$, since otherwise $Pr_\gamma(\overline{\bigwedge_{m<k}\psi_m \to \neg\theta_k}) \notin Y_g$ and so $\psi_k = \theta_k$. But then, by the inductive assumption, (8) holds for $n = k+1$. This proves (8).

Finally, in view of (8), it follows that for some k, $\psi_k = \neg\varphi$. Hence, by (7), $P \cup Y_{g,h} \vdash \beta(\overline{\neg\varphi})$, whence, by (4) and (6), $T \nvdash f\varphi$. Thus we have shown that f is as claimed and the proof is complete.

Note that Theorem A is an immediate consequence of Lemma 4.

PROOF OF THEOREM 1. By Lemma 2, there is a formula $\sigma^*(x)$ binumerating S in T s. t. $T \vdash Con_{\sigma^*}$ and $Pr_{\sigma^*}(x)$ numerates $T^{[S]}$ in T. Next, by Lemma 4 with $\sigma(x)$ replaced by $\sigma^*(x)$, there is an $f \in \{S \le T\}$ s. t. for every sentence φ of S, if $T \vdash f\varphi$, then $T \vdash Pr_{\sigma^*}(\bar{\varphi})$. It follows that f is $T^{[S]}$-faithful and so the proof is complete.

Note that if $P \dashv T$ and $T^{[\emptyset]} \dashv S$, then $T^{[S]} \dashv S$. Thus from Theorems 1 and 2 we get the following

COROLLARY 1. (i) If $P \dashv T$, $T^{[\emptyset]} \dashv S$, and there is a formula $\sigma(x)$ binumerating S in T s. t. $T \vdash Con_\sigma$, then $S \trianglelefteq T$.
(ii) If $P \dashv T$, T is reflexive, and $T^{[\emptyset]} \dashv S \le T$, then $S \trianglelefteq T$.

In the following A, B, C are essentially reflexive extensions of P.

THEOREM 3. $S \trianglelefteq_X A$ iff $S \le A$ and $X \subseteq A^{\{S\}}$.

PROOF. "If" is just a repetition of Theorem 2. To prove "only if" suppose $S \trianglelefteq_X A$. Then $S \le A$. To show that $X \subseteq A^{\{S\}}$ suppose $\varphi \in X$ and $A \vdash Pr_{S\restriction m}(\bar{\varphi})$. Let ψ be the sentence $\bigwedge S\restriction m \to \varphi$. Then $A \vdash Pr_\emptyset(\bar{\psi})$. Suppose

$f \in \{S \leq A\}$ is X-faithful and let λ_f be the sentence saying that $f \in \{L^S \leq A\}$, where L^S is the set of logically valid sentences of S. Then

$$P \vdash Pr_{\emptyset}(\bar{\psi}) \to Pr_{\emptyset}(\overline{\lambda_f \to f\psi}).$$

But then $A \vdash Pr_{\emptyset}(\overline{\lambda_f \to f\psi})$. Since A is essentially reflexive, it follows that $A \vdash \lambda_f \to f\psi$. But $A \vdash \lambda_f$ and $A \vdash f \bigwedge S\restriction m$. Hence $A \vdash f\varphi$. But f is X-faithful and $\varphi \in X$. Therefore $S \vdash \varphi$. Thus $\varphi \in A^{\{S\}}$ and the proof is complete.

COROLLARY 2. (i) $S \trianglelefteq A$ iff $A^{[\emptyset]} \dashv S \leq A$.

(ii) If $S \leq A$, then $S \trianglelefteq_X A$, where $X = \{\varphi : S \trianglelefteq_{\{\varphi\}} A\}$.

(iii) If $S \trianglelefteq_X A$ and $S \leq B \dashv A$, then $S \trianglelefteq_X B$.

(iv) If $S \trianglelefteq A$ and $S \dashv T \leq B \dashv A$, then $T \trianglelefteq B$.

$\trianglelefteq$ cannot be replaced by $\trianglelefteq_X$ in Corollary 2 (iv). In fact there are sentences ψ, θ, χ s. t. $Q \trianglelefteq_{\{\theta\}} P + \chi$, $Q + \psi \dashv P$, and $Q + \psi \not\trianglelefteq_{\{\theta\}} P+\chi$. Let ψ be any sentence s. t. $Q \nvdash \psi$ and $P \vdash \psi$. Let

$$Y = \{\varphi : P \vdash Pr_{Q+\psi}(\bar{\varphi}) \to Pr_Q(\bar{\psi})\}.$$

Then Y and Th(Q + ψ) are disjoint. It follows that there is a sentence $\varphi \notin Y \cup Th(Q + \psi)$. Let $\theta = \varphi \wedge \psi$ and $\chi = Pr_{Q+\psi}(\bar{\theta})$. Then, by Theorem 3, ψ, θ, χ are as desired.

Let T^X, where X is a set of sentences, be the set of members of X provable in T. We write $S \dashv_X T$ to mean that $S^X \subseteq T^X$.

The following lemma is well-known (cf. [3,4,5]).

LEMMA 5. $A \leq B$ iff $A \dashv_{\Pi_1^0} B$.

THEOREM 4. $A \trianglelefteq_X B$ iff $A \cup B^{\Sigma_1^0} \dashv_X A \dashv_{\Pi_1^0} B$.

PROOF. By Theorem 3 and Lemma 5, it suffices to show that

$$A \cup B^{\Sigma_1^0} \dashv_X A \text{ iff } X \subseteq B^{\{A\}}.$$

Suppose first $A \cup B^{\Sigma_1^0} \dashv_X A$. Let $\varphi \in X$. To prove that $\varphi \in B^{\{A\}}$, suppose $\varphi \in B^{[A]}$, i.e. for some m, $B \vdash Pr_{A\restriction m}(\bar{\varphi})$. Then $Pr_{A\restriction m}(\bar{\varphi}) \in B^{\Sigma_1^0}$. Moreover, A being essentially reflexive, $A \vdash Pr_{A\restriction m}(\bar{\varphi}) \to \varphi$. But then $A \cup B^{\Sigma_1^0} \vdash \varphi$, whence $A \vdash \varphi$. It follows that $\varphi \in B^{\{A\}}$. Thus $X \subseteq B^{\{A\}}$. Next suppose $X \subseteq B^{\{A\}}$. Let $\varphi \in X$ and suppose $A \cup B^{\Sigma_1^0} \vdash \varphi$. There is then a $\psi \in B^{\Sigma_1^0}$ s. t. $A \vdash \psi \to \varphi$. Since ψ is Σ_1^0, there is a k s. t. $P \vdash \psi \to Pr_{A\restriction k}(\bar{\psi})$. It follows that there is an n s. t. $P \vdash \psi \to Pr_{A\restriction n}(\bar{\varphi})$. But then $B \vdash Pr_{A\restriction n}(\bar{\varphi})$, whence $\varphi \in B^{[A]}$. But $\varphi \in B^{\{A\}}$ and so $A \vdash \varphi$. Thus we have shown that $A \cup B^{\Sigma_1^0} \dashv_X A$ and the proof is complete.

COROLLARY 3. $A \trianglelefteq B$ iff $A \dashv_{\Pi_1^0} B \dashv_{\Sigma_1^0} A$ iff $A \trianglelefteq_{\Sigma_1^0} B$ iff $A \trianglelefteq_{\{\varphi\}} B$ for every Σ_1^0 sentence φ.

Let us write $A \equiv B$ and $A \trianglelefteq\!\trianglerighteq B$ to mean that $A \leq B \leq A$ and $A \trianglelefteq B \trianglelefteq A$, respectively. It is known that if $A \dashv B$, then there is a θ s. t. $A + \theta$

$\equiv$ B (cf. [5,6]). This can be improved as follows.

COROLLARY 4. If $A \dashv B$, then there is a (Σ_2^0, Π_2^0) sentence θ s. t. $A + \theta \equiv B$.

PROOF. By Theorem 4 [6], there is a Γ sentence θ s. t. $B \dashv_\Gamma A + \theta \dashv_{\check{\Gamma}} B$. Now let Γ be either Σ_2^0 or Π_2^0 and apply Corollary 3.

The above characterizations of $\trianglelefteq$ can be applied to evaluate certain sets of sentences defined in terms of $\trianglelefteq$ as follows.

THEOREM 5. (i) If $A \leq B$, then $X = \{\varphi \in \Sigma_1^0 : A + \varphi \trianglelefteq B\}$ is a complete Π_2^0 set.
(ii) If $Q \dashv T \trianglelefteq A$, then $X = \{\varphi : T \trianglelefteq A + \varphi\}$ is a complete Π_2^0 set.

PROOF. The relation $\{\langle\varphi,\psi\rangle : S + \varphi \leq C + \psi\}$ is Π_2^0 (cf. [4,5]). Hence, by Corollary 2 (i), the sets X mentioned in (i) and (ii) are Π_2^0.

(i) By Theorem 4 [6], there is a Σ_1^0 sentence θ s. t. $B \dashv_{\Sigma_1^0} A + \theta \dashv_{\Pi_1^0} A \cup B^{\Sigma_1^0}$. By Lemma 5, $A + \theta \leq B$. Hence, by Corollary 3, $A + \theta \trianglelefteq B$. Let $Y = \{\varphi \in \Sigma_1^0 : A + \theta + \varphi \leq B\}$. Then Y is a complete Π_2^0 set (cf. [5,6, 11]). Moreover $\varphi \in Y$ iff $\theta \wedge \varphi \in X$. Thus X is a complete Π_2^0 set.

(ii) Let Y be any Π_2^0 set. By the proof of Theorem 6 [6], there is a formula $\delta(x)$ s. t.
(1) if $k \in Y$, then $\delta(\bar{k}) \in \mathrm{Cons}(\Sigma_1^0, A)$,
(2) if $k \notin Y$, then there is a Σ_1^0 sentence θ s. t. $A + \delta(\bar{k}) \vdash \theta$ and $T \nvdash \theta$.
It follows that
(3) $Y = \{k : \delta(\bar{k}) \in X\}$.
To see this suppose first $k \in Y$. Let $\psi \in (A + \delta(\bar{k}))^{[\emptyset]}$, i.e. $A + \delta(\bar{k}) \vdash Pr_\emptyset(\bar{\psi})$. Then, by (1), $A \vdash Pr_\emptyset(\bar{\psi})$. But $T \trianglelefteq A$ and so, by Corollary 2 (i), $A^{[\emptyset]} \dashv T$. It follows that $T \vdash \psi$. Thus we have shown that $(A + \delta(\bar{k}))^{[\emptyset]} \dashv T$ and so, again by Corollary 2 (i), $\delta(\bar{k}) \in X$. Suppose next $k \notin Y$. Let θ be as in (2). Since θ is Σ_1^0, we have $P \vdash \theta \rightarrow Pr_Q(\bar{\theta})$, whence $P \vdash \theta \rightarrow Pr_\emptyset(\overline{\bigwedge Q \rightarrow \theta})$, whence $A + \delta(\bar{k}) \vdash Pr_\emptyset(\overline{\bigwedge Q \rightarrow \theta})$. Also clearly $T \nvdash \bigwedge Q \rightarrow \theta$. Hence $(A + \delta(\bar{k}))^{[\emptyset]} \not\dashv T$ and so $\delta(\bar{k}) \notin X$. This proves (3) and so concludes the proof of (ii).

Since our theories contain only finitely many non-logical constants, interpretations can be identified with natural numbers in an obvious way. In [5] it was (in fact) shown that if $A \leq T$, then $\{A \leq T\}$ is $\Pi_2^0 - \Sigma_2^0$. Part (ii) of our next result is a partial improvement of this.

THEOREM 6. (i) Suppose $Q \dashv T \trianglelefteq A$ and either T is finite or T = B. Then $\{T \trianglelefteq A\}$ is a complete Π_2^0 set.
(ii) If $A \leq B$, then $\{A \leq B\}$ is a complete Π_2^0 set.

PROOF. (i) We first note that there is a sentence θ s. t. $T \nvdash \theta$ and $T + \theta \leq A$. If T = B, this follows from the result of Orey [9] (cf. also

[5,6]) that there is a sentence ψ s. t. $B + \psi, B + \neg\psi \leq B$. Suppose $Q \dashv T$ and T is finite. Let θ be a Rosser sentence for T. Then $T \nvdash \theta$. Moreover $P \vdash Con_T \rightarrow Con_{T+\theta}$. Since $T \leq A$ and T is finite, we have $A \vdash Con_T$. Hence $A \vdash Con_{T+\theta}$ and so $T + \theta \leq A$.

Now let $X = \{k: \forall m R(k,m)\}$, where $R(k,m)$ is r.e., be any Π^0_2 set. Let $\rho(x,y)$ be a formula numerating $R(k,m)$ in $T + \neg\theta$ and let

$$S_k = T \cup \{\theta \vee \rho(\bar{k},\bar{m}): m \in \omega\}.$$

Since $T \trianglelefteq A$, we have $A^{[\emptyset]} \dashv T$, whence $A^{[\emptyset]} \dashv S_k$, whence $A^{[S_k]} \dashv S_k$. Since $S_k \leq A$, we can effectively find formulas $\sigma_k(x)$ binumerating S_k in A and s. t. $P \vdash Con_{\sigma_k}$ (cf. the proof of Theorem 2). But then it follows from the proofs of Lemmas 2 and 4 that to each k we can effectively find an $f_k \in \{S_k \trianglelefteq A\}$. But then $k \in X$ iff $T \vdash \theta \vee \rho(\bar{k},\bar{m})$ for every m iff $f_k \in \{T \trianglelefteq A\}$. Thus

$$X = \{k: f_k \in \{T \trianglelefteq A\}\}$$

and so the proof is complete.

(ii) Let X and $R(k,m)$ be as above and let $\rho(x,y)$ be a formula numerating $R(k,m)$ in B. Let $\alpha(x)$ be a formula binumerating A in B and let $\sigma(x,y)$ be the formula

$$\alpha(x) \wedge Con_{\alpha(z) \wedge z \leq x} \wedge \forall z \leq x \rho(y,z).$$

Then

(1) if $k \notin X$, then there is an m s. t. $B \nvdash \exists x(\bar{m} \leq x \wedge \sigma(x,\bar{k}))$.

$P \vdash Con_{\sigma(x,\bar{k})}$ for every k and $B \vdash Con_{A \upharpoonright n}$ for every n (cf. the proof of Theorem 2). Hence

(2) if $k \in X$, then $\sigma(x,\bar{k})$ binumerates A in B.

Moreover, by the proof of Lemma 4, to each k we can effectively find an f_k s. t.

(3) $f_k \in \{\{\varphi: B \vdash \sigma(\bar{\varphi},\bar{k})\} \leq B\}$,

(4) if $B \vdash f_k\varphi$, then $B \vdash Pr_{\sigma(x,\bar{k})}(\bar{\varphi})$.

To complete the proof it suffices to show that

(5) $X = \{k: f_k \in \{A \leq B\}\}$.

If $k \in X$, then, by (2) and (3), $f_k \in \{A \leq B\}$. Next suppose $k \notin X$. Let m be as in (1) and let θ be s. t. $P \vdash \theta \leftrightarrow \neg Pr_{A \upharpoonright m}(\bar{\theta})$. Then, by a standard argument,

(6) $A \vdash \theta$.

Since $A \leq B$, it follows, by Lemma 5, that $B \vdash \theta$ and so $B \vdash \neg Pr_{A \upharpoonright m}(\bar{\theta})$. Hence, by the definition of $\sigma(x,y)$,

$$B \vdash Pr_{\sigma(x,\bar{k})}(\bar{\theta}) \rightarrow \exists x(\bar{m} \leq x \wedge \sigma(x,\bar{k})).$$

But then, by (1), $B \nvdash Pr_{\sigma(x,\bar{k})}(\bar{\theta})$, whence, by (4), $B \nvdash f_k\theta$, whence, by (6), $f_k \notin \{A \leq B\}$. This completes the proof of (5) and thereby the proof of (ii).

An interpretation in T will be said to be Γ w. r. t. T if there is a Γ formula $\beta(x)$ s. t. $T \vdash f\varphi \leftrightarrow \beta(\bar{\varphi})$. f is Δ_k^0 w. r. t. T if it is Π_k^0 and Σ_k^0 w. r. t. T. Let $f \sim_T g$ mean that

$\{\varphi: T \vdash f\varphi\} = \{\varphi: T \vdash g\varphi\}$.

Now let g be any interpretation in A. We can than find a PR formula $\sigma_0(x)$ binumerating a theory $S \dashv\vdash \{\varphi: A \vdash g\varphi\}$. Let $\sigma(x)$ be the formula $\sigma_0(x) \wedge \mathrm{Con}_{\sigma_0}(y) \wedge y \leq x$. Then $\sigma(x)$ is Π_1^0 and $P \vdash \mathrm{Con}_\sigma$. Next let $\sigma^*(x)$ be as in Lemma 2. Then $\sigma^*(x)$ is Σ_2^0, whence $\mathrm{Pr}_\gamma(x)$, where $\gamma(x)$ is as in the proof of Lemma 4 with $\sigma(x)$ replaced by $\sigma^*(x)$, is Σ_2^0. It follows that the formulas $\xi(x)$, $\chi(x,y)$, $\beta(x)$ mentioned in that proof are Σ_3^0, Σ_4^0, and Σ_4^0, respectively. Also $P \vdash \mathrm{Con}_{\sigma^*}$. Thus the f defined in the proof of Theorem 1, with T replaced by A, is Σ_4^0 w. r. t. A. Since moreover $A \vdash \mathrm{Scm}_\beta$ and so $A \vdash \neg\beta(\bar{\varphi}) \leftrightarrow \beta(\overline{\neg\varphi})$, it follows that f is Π_4^0 w. r. t. A. Clearly $S \trianglelefteq A$. Hence, by Theorem 3, $A^{[S]} \dashv S$, whence $f \in \{S \trianglelefteq A\}$ and so $f \sim_A g$. Thus we have proved the following

THEOREM 7. To any interpretation g in A we can effectively find an interpretation $f \sim_A g$ s. t. f is Δ_4^0 w. r. t. A.

In [2] it is shown that if A is Σ_1^0-sound, Δ_4^0 can be replaced by Δ_3^0. The problem if these estimates are optimal remains open.

References

[1] S. Feferman, Arithmetization of metamathematics in a general setting, Fund. Math. 49 (1960), 35-92.

[2] S. Feferman, G. Kreisel, and S. Orey, 1-consistency and faithful interpretations, Archiv f. math. Logik u. Grundlf. 6 (1960), 52-63.

[3] D. Guaspari, Partially conservative extensions of arithmetic, Trans. of the Amer. Math. Soc. 254 (1979), 47-68.

[4] M. Hájková and P. Hájek, On interpretability in theories containing arithmetic, Fund. Math. 76 (1972), 131-137.

[5] P. Lindström, Some results on interpretability, Proc. of the 5th Scand. Logic Symp. 1979, Aalborg 1979, 329-361.

[6] P. Lindström, On partially conservative sentences and interpretability, to appear.

[7] P. Lindström, On certain lattices of degrees of interpretability, to appear.

[8] A. Mostowski, A generalization of the incompleteness theorem, Fund. Math. 49 (1961), 205-232.

[9] S. Orey, Relative interpretations, Zeitschr. f. math. Logik u. Grundl. d. Math. 7 (1961), 146-153.

[10] D. Scott, Algebras of sets binumerable in complete extensions of arithmetic, Recursive function theory, Providence 1962, 117-121.

[11] R. Solovay, On interpretability in set theories, to appear.

Abstract Recursion
as a Foundation for the Theory of Algorithms

by

Yiannis N. Moschovakis*
Department of Mathematics
University of California
Los Angeles, CA 90024

Contents

*During the preparation of this paper the author was partially supported by NSF Grant #MCS 83-02555.

Introduction

The main object of this paper is to describe an abstract, (axiomatic) theory of recursion and its connection with some of the basic, foundational questions of computer science.

The models of this theory include (of course) ordinary recursion on the natural numbers, recursion in higher types (Kleene [1959], [1963]), positive elementary induction (Moschovakis [1974]) and most of the generalized recursion theories studies by logicians. They also include the basic computational models of theoretical computer science, and they provide natural structures in which higher level programming languages can be interpreted.

One of the main tools of the theory is a formal language of recursion, REC, which yields natural systems for the specification and mathematical analysis of algorithms, particularly concurrent algorithms.

From the technical point of view, one can take this work as the theory of many-sorted, concurrent and (more significantly) second order recursion schemes. At the same time, logicians will recognize the present theory as a simultaneous refinement and generalization of first order definability, in the spirit of abstract model theory.

Most of the results of ordinary recursion theory are usually formulated as properties of the collection of recursive partial functions - closure properties. structure properties, existence results for recursively enumerable sets which satisfy certain conditions, etc. The proofs of these results, however, typically involve more subtle aspects of the definitions of recursive partial functions, for example the length of computations or the properties of delicate (typically concurrent) combinations of recursive definitions. Because of this, the notion of recursive partial function is not an adequate primitive for axiomatizing the more significant aspects of recursion, although it has been used to construct some interesting mathematical theories, see Wagner [1969], Strong [1968], Friedman [1971].

It is tempting to take computations and their lengths as primitives, as in Moschovakis [1971] and Fenstad [1980]. One can also view the study of abstract complexity measures of Blum as attempts in this direction. The main problem with the theory developed on these primitives is that it is rather weak and does not have natural finite models.

Our present approach is to simply take as primitive the very notion of a recursive definition or recursion, represented semantically by an operative system of partial, monotone functionals. A very abstract version of this was described in Moschovakis [1976] and then in Kechris-Moschovakis [1977] the special case of recursion in higher types was studied in this framework. The refined (and simplified) version which we will describe here allows naturally recursion on finite universes among its models and appears to provide the correct context for studying some of the important notions of computer science, for example complexity and concurrency.

Similar approaches to abstract recursion theory have been suggested by others, in particular Feferman [1965] and Kleene [1981].

If, by Church's Thesis the precise, mathematical notion of recursive function captures the intuitive notion of computable function, then the precise, mathematical notion of recursion (an operative system of explicit functionals, as we will define it) should model adequately the mathematical properties of the intuitive notion of algorithm. We will make precise this refined version of Church's Thesis and we will strive to discuss it intelligently; because of the intensional nature of the concepts involved, it is both more subtle and more problematical than the classical Church's Thesis which it generalizes.

This paper is aimed at mathematicians, logicians and computer scientists who share an interest in the foundations of the theory of algorithms. My aim here is to motivate and describe the theory of recursion rather than to develop it rigorously - so there will be many examples, some very detailed definitions and almost no proofs. There are good reasons for choosing this rather dreary style of exposition, but it is sure to drive the mathematically minded reader up the wall; he or she is advised to relieve the monotony by trying to separate from the many claimed facts those few which cannot be proved directly by induction on something obvious, but do require some ingenuity.

In the words of Landin [1964], which are more true of this paper than of his, "most of the ideas here are to be found in the literature." It will be easy to recognize the influence of the work of Feferman, Gandy, Kleene, Spector, Backus, McCarthy and Scott - but I am also indebted deeply to many others, some of whom do not even appear in the references. What appears to be new in this work is the attempt to explicate the connection between recursions and algorithms - and in fact to bring the notion (a particular notion) of algorithm center stage.

On a more personal level, I would like to thank my student Michel de Rougemont, and also Marina Chen, Carver Mead and Carolyn Talcott, who took the time to talk to me. I learned a great deal from them, but perhaps I did not get it quite right, so they bear no responsibility for the ideas expressed here.

Part 1. Examples

In this first part, we will give a detailed analysis of five well-known algorithms, aimed primarily at the readers who are not familiar with abstract recursion or the fixed-point theory of programs. The analysis has been shaped to explain our approach to the theory of recursion and our choice of basic notions, so that the experts too may find it useful to peruse it.

1A. The Euclidean algorithm; fixed points. We are given positive integers $x \geq y \geq 1$ and we wish to compute their greatest common divisor,

(1A.1) $\gcd(x,y)$ = the largest integer z which divides both x and y;

more precisely, we want to define an algorithm which will compute $\gcd(x,y)$, for arbitrary $x \geq y \geq 1$.

If we assume that the remainder function

(1A.2) $\mathrm{rem}(x,y)$ = the largest r such that $0 \leq r < y$ and for some q, $x = y \cdot q + r$

is available to us, then we can define the Euclidean algorithm for the gcd succintly, by the single equation

$$(1A.3) \qquad \gcd(x,y) = \begin{cases} y & \text{if } \mathrm{rem}(x,y) = 0, \\ \gcd(y,\mathrm{rem}(x,y)) & \text{otherwise.} \end{cases}$$

Using (1A.3) and trivial facts about the remainder function, it is easy to establish by induction on x the

Basic Fact about the Euclidean Algorithm. For each $x \geq y \geq 1$, the Euclidean algorithm will compute $\gcd(x,y)$. ⊣

Now the mathematically conscientious reader who stopped for a minute to think through the proof of this basic fact, must have used an imaginative, non-literal reading of equation (1A.3); because at it stands, formally, (1A.3) is not a definition, but only an equation with the symbol "gcd" occurring on both sides of the "=" sign.

There is a familiar, non-literal understanding of (1A.3) which is very well understood and we need not take much space to belabor its analysis (de Bakker [1980] gives both an exposition and many references to the fixed point theory of programs). Briefly, one interprets (1A.3) as a condition on an arbitrary partial function gcd, one proves that there exists a least partial function which satisfies this condition and then one takes (1A.3) as an implicit definition of this least solution, which is of course the gcd function.

Still, there must be more than this in (1A.3) to justify the Basic Fact, since that makes an assertion about the Euclidean algorithm and not about the gcd function, which was after all defined by (1A.1).

The problem then becomes how to read (1A.3) directly as the definition of an "algorithm," so that the intuitive proof of the Basic Fact becomes meaningful and correct.

One solution is to interpret (1A.3) as a recipe for an iterative computation scheme which ultimately will compute gcd(x,y). Our approach here is mathematically simpler and more widely applicable, although in this case it will essentially amount to the same thing: since the difficulty comes from the fact that the symbol "gcd" occurs on both sides of the "=", it must be that these two occurrences of "gcd" stand for different objects; and it is natural to assume, following usual mathematical practice, that the object on the left is defined in terms of the object on the right.

Consider then the following re-write of (1A.3), where we have embellished the notation, so that different objects are named by distinct symbols.

$$\text{(1A.4)} \qquad \underset{\sim}{\mathrm{gcd}}(x,y,\underline{\mathrm{gcd}}) \simeq \begin{cases} y & \text{if } \mathrm{rem}(x,y) = 0, \\ \underline{\mathrm{gcd}}(y,\mathrm{rem}(x,y)) & \text{otherwise.} \end{cases}$$

Here the straight-underlined variable "$\underline{\mathrm{gcd}}$" varies over binary, partial functions on the integers and the value $\underset{\sim}{\mathrm{gcd}}(x,y,\underline{\mathrm{gcd}})$ which we are defining depends both on the given integers x, y and on the given partial function $\underline{\mathrm{gcd}}$. There is no doubt now that (1A.4) defines directly the operation $\underset{\sim}{\mathrm{gcd}}$, assuming as before that we are taking the remainder function rem as given.

This operation $\underset{\sim}{\mathrm{gcd}}$ is a typical example of the kind of mathematical object which we will use to model algorithms. It has some important basic properties.

(1) $\underset{\sim}{\mathrm{gcd}}$ is a <u>second-order operation</u> - we will call it a <u>functional</u> - whose arguments include not only the integers x and y, but also the partial function $\underline{\mathrm{gcd}}$.

It will become clear in a moment why we take "$\underline{\mathrm{gcd}}$" to range over partial and not just total (completely defined) functions. The symbol "$\simeq$" stands for the so-called <u>strong equality</u> between "values" of partial functions, i.e.

$$f(x) \simeq g(y) \Leftrightarrow \text{either both } f(x) \text{ and } g(y) \text{ are undefined, or they are both defined and } f(x) = g(y).$$

In dealing with partial functions, we will also use the notations:

$$f(x)\downarrow \Leftrightarrow f(x) \text{ is defined,}$$
$$f \subseteq g \Leftrightarrow \text{for all } x, w,\ f(x) \simeq w \Rightarrow g(x) \simeq w.$$

(2) The functional $\underset{\sim}{\mathrm{gcd}}$ is <u>explicitly defined</u> from the given function rem.

Later on we will make precise just what kinds of explicit definitions we will use. It is obvious, however, that if we use functionals to model algorithms, then we must restrict ourselves to functionals which are obviously (and directly) "computable" in terms of the "givens."

(3) gcd is an operative functional, i.e. its partial function argument gcd varies over binary partial functions and for each gcd, the partial function

$$(x,y) \mapsto \underset{\sim}{gcd}(x,y,\text{gcd})$$

is also binary.

We will see in a minute how this makes it possible to iterate the functional gcd.

(4) gcd is monotone in its partial function argument, i.e. if

$$\text{gcd} \subseteq \text{gcd}',$$

then

$$\underset{\sim}{gcd}(x,y,\text{gcd}) \simeq d \Rightarrow \underset{\sim}{gcd}(x,y,\text{gcd}') \simeq d.$$

(5) gcd is compact (or continuous), i.e. the value of gcd(x,y,gcd) depends only on finitely many values of gcd (in fact on at most one value, gcd(y,rem(x,y)), if rem(x,y) > 0).

The twin properties of monotonicity and operativeness make it possible to iterate gcd in the following way. For each integer $m = 0,1,2,\ldots$, the partial function gcd^m of two variables is defined by the recursion

(1A.5) $$\text{gcd}^0(x,y) \simeq \underset{\sim}{gcd}(x,y,\emptyset)$$

($\emptyset$ is the totally undefined, empty function),

(1A.6) $$\text{gcd}^{m+1}(x,y) \simeq \underset{\sim}{gcd}(x,y,\text{gcd}^m),$$

so that (for example),

$$\text{gcd}^0(x,y)\downarrow \Leftrightarrow \text{rem}(x,y) = 0,$$
$$\text{gcd}^0(x,y)\downarrow \Rightarrow \text{gcd}^0(x,y) = y.$$

It is easy to verify by induction (using monotonicity) that these partial functions form a non-decreasing sequence

(1A.7) $$\text{gcd}^0 \subseteq \text{gcd}^1 \subseteq \cdots$$

and then (by the compactness) the limit partial function

(1A.8) $$\text{gcd}^\infty = \bigcup_{m=0}^{\infty} \text{gcd}^m$$

is the least fixed point of the functional gcd, i.e.

(1A.9) for all x, y, $\underset{\sim}{gcd}(x,y,\text{gcd}^\infty) \simeq \text{gcd}^\infty(x,y)$,

(1A.10) if for all x, y,

$$\underset{\sim}{gcd}(x,y,p) \simeq w \Rightarrow p(x,y) \simeq w,$$

then

$$\underset{\sim}{gcd}^{\infty} \subseteq p.$$

The partial functions $\underset{\sim}{gcd}^0, \underset{\sim}{gcd}^1, \ldots$ are the stages or approximations to the fixed point $\underset{\sim}{gcd}^{\infty}$.

This iteration process which will be associated with each (monotone, operative) functional plays the role of computation in the fixed-point analysis of algorithms.

Now the precise version of the basic fact about the Euclidean algorithm is the assertion

(1A.11) $$\underset{\sim}{gcd}^{\infty}(x,y) = \text{the greatest common divisor of } x \text{ and } y,$$

which follows directly from (1A.9) by induction on x.

How does this compare with the proof of the basic fact produced by the mathematically conscientious reader just after reading the beginning of this section? More likely than not, he or she set up come specific implementation - a computation scheme - for the algorithm; whatever the precise details of that proof, it is a safe bet that its mathematical content - its idea - is very close to the simple argument outlined above.

Following the drift of the discussion, we might he expected at this point to simply identify the Euclidean algorithm with the functional $\underset{\sim}{gcd}$. We will not go quite that far, because the time-honored intuitive concept of algorithm carries many linguistic and intensional connotations (some of them tied up with implementations) with which we have not concerned ourselves. Instead we will make the weaker (and almost trivial) claim that the functional $\underset{\sim}{gcd}$ embodies all the essential mathematical properties of the Euclidean algorithm.

1B. Linear representation of the gcd; simultaneous recursion. A familiar property of gcd(x,y) is that it is a linear combination of x and y; i.e. for all $x \geq y \geq 1$, there exist whole numbers A(x,y), B(x,y) (positive, negative or zero) such that

(1B.1) $$gcd(x,y) = A(x,y) \cdot x + B(x,y) \cdot y.$$

It is quite trivial to verify that (1B.1) follows from the following two equations:

(1B.2) $$A(x,y) = \begin{cases} 0 & \text{if } rem(x,y) = 0, \\ B(y,rem(x,y)) & \text{if } rem(x,y) > 0, \end{cases}$$

(1B.3) $$B(x,y) = \begin{cases} 1 & \text{if } rem(x,y) = 0, \\ A(y,rem(x,y)) - B(y,rem(x,y)) \cdot quot(x,y) & \text{if } rem(x,y) > 0. \end{cases}$$

Here, of course, quot(x,y) is the (non-negative, integer) quotient of x by y.

We can now repeat the analysis of the preceding section to get out of (1B.2), (1B.3) a mathematical object which represents an algorithm for computing A(x,y) and B(x,y). Instead of a single, operative functional, in this case we come up with an operative system of two functionals. Here it is:

$$(1B.4) \qquad \underset{\sim}{A}(x,y,A,B) \simeq \begin{cases} 0 & \text{if } \operatorname{rem}(x,y) = 0 \\ B(y,\operatorname{rem}(x,y)) & \text{if } \operatorname{rem}(x,y) > 0, \end{cases}$$

$$(1B.5) \quad \underset{\sim}{B}(x,y,A,B) \simeq \begin{cases} 1 & \text{if } \operatorname{rem}(x,y) = 0, \\ A(y,\operatorname{rem}(x,y)) - B(y,\operatorname{rem}(x,y)) \cdot \operatorname{quot}(x,y) & \text{if } \operatorname{rem}(x,y) > 0. \end{cases}$$

Now both "A" and "B" vary over binary, partial functions on the whole numbers and the system is operative in the sense that for all A, B, both

$$(x,y) \mapsto \underset{\sim}{A}(x,y,A,B),$$
$$(x,y) \mapsto \underset{\sim}{B}(x,y,A,B)$$

are binary, like A and B. The functionals $\underset{\sim}{A}$ and $\underset{\sim}{B}$ are obviously both monotone and compact. They are also explicitly defined in terms of rem, quot, and $-$.

The (simultaneous) iterates of the system $[\underset{\sim}{A},\underset{\sim}{B}]$ are defined in the obvious way:

$$(1B.6) \qquad \begin{cases} \underset{\sim}{A}^0(x,y) \simeq \underset{\sim}{A}(x,y,\emptyset,\emptyset), \\ \underset{\sim}{B}^0(x,y) \simeq \underset{\sim}{B}(x,y,\emptyset,\emptyset) \end{cases}$$

and then, inductively,

$$(1B.7) \qquad \begin{cases} \underset{\sim}{A}^{m+1}(x,y) \simeq \underset{\sim}{A}(x,y,\underset{\sim}{A}^m,\underset{\sim}{B}^m) \\ \underset{\sim}{B}^{m+1}(x,y) \simeq \underset{\sim}{B}(x,y,\underset{\sim}{A}^m,\underset{\sim}{B}^m). \end{cases}$$

In the limit, using compactness, nen can easily check that the unions

$$\underset{\sim}{A}^\infty = \cup_m \underset{\sim}{A}^m; \quad \underset{\sim}{B}^\infty = \cup_m \underset{\sim}{B}^m$$

are precisely the simultaneous fixed points of the system $[\underset{\sim}{A},\underset{\sim}{B}]$, so that in particular, for all x, y

$$(1B.8) \qquad \begin{cases} \underset{\sim}{A}^\infty(x,y) \simeq \underset{\sim}{A}(x,y,\underset{\sim}{A}^\infty,\underset{\sim}{B}^\infty), \\ \underset{\sim}{B}^\infty(x,y) \simeq \underset{\sim}{B}(x,y,\underset{\sim}{A}^\infty,\underset{\sim}{B}^\infty), \end{cases}$$

i.e. $A = \underset{\sim}{A}^\infty$ and $B = \underset{\sim}{B}^\infty$ satisfy (1B.2) and (1B.3).

In this example, the mathematical object needed to represent the algorithm turned out to be an operative system of functionals rather than a single functional.

1C. The mergesort; parameters and implementation-independent properties. For a third example, consider the sorting problem. We are given a set X, an ordering $\leq$ or X and a positive integer d; we must define an algorithm which will sort an arbitrary sequence

$$a = (a(0), a(1), \ldots, a(n-1)) \qquad (n \leq d)$$

of members of X in increasing (non-decreasing) order, i.e. compute the function

$$\text{(1C.1)} \qquad \text{sort}(a) = (a(k_0), a(k_1), \ldots, a(k_{n-1})),$$

where $k_0, k_1, \ldots, k_{n-1}$ is a permutation of $0, 1, \ldots, n-1$ such that

$$\text{(1C.2)} \qquad a(k_0) \leq a(k_1) \leq \cdots \leq a(k_{n-1}).$$

Among the many known sorting algorithms, the mergesort is best for illustrating the notions with which we are concerned here. It is determined by the equations

$$\text{(1C.3)} \quad \text{sort}(a) = \begin{cases} a & \text{if } \text{length}(a) = n \leq 1, \\ \text{merge}(\text{sort}(\text{1st}(a)), \text{sort}(\text{2nd}(a))), & \text{otherwise,} \end{cases}$$

where

$$\text{(1C.4)} \quad \text{merge}(b,c) = \begin{cases} b & \text{if } c = \emptyset \text{ (i.e. } c \text{ is the empty sequence),} \\ c & \text{otherwise, if } b = \emptyset, \\ b(0)\frown\text{merge}(\text{tail}(b),c) & \text{otherwise, if } b(0) \leq c(0), \\ c(0)\frown\text{merge}(b,\text{tail}(c)) & \text{otherwise.} \end{cases}$$

Here 1st(a) and 2nd(a) give the first and second half of a (appropriately adjusted when length(a) is odd), $x\frown c$ is the sequence $(x, c(0), \ldots, c(n-1))$ and

$$\text{tail}(c(0), c(1), \ldots, c(n-1)) = (c(1), c(2), \ldots, c(n-1)).$$

We assume that these functions are given.

An intuitive reading of (1C.3) and (1C.4) (like our first reading of the corresponding equation (1A.3) for the gcd) leads easily to at least a heuristic proof of the

Basic Fact about the mergesort. The mergesort algorithm will work and if length(a) = n, then no more than $n \cdot \log_2(n)$ comparisons of members of X will be required to sort the sequence a. ⊣

Our task is to turn this heuristic argument into a precise theorem about the properties of the algorithm implicitly described by (1C.3) and (1C.4).

According to our analysis in the preceding two sections, (1C.3) and (1C.4) are shorthand for the following operative system of two functionals which embodies the mathematical properties of the mergesort.

$$\text{(1C.5)} \quad \underset{\sim}{\text{sort}}(a, \underline{\text{sort}}, \underline{\text{merge}}) \simeq \begin{cases} a & \text{if } \text{length}(a) \leq 1, \\ \underline{\text{merge}}(\underline{\text{sort}}(\text{1st}(a)), \underline{\text{sort}}(\text{2nd}(a))), & \text{otherwise,} \end{cases}$$

$$\text{(1C.6)}\quad \underline{\text{merge}}(b,c,\underline{\text{sort}},\underline{\text{merge}}) \simeq \begin{cases} b & \text{if } c = \emptyset,\\ c & \text{if } b = \emptyset,\\ b(0)\hat{\ }\underline{\text{merge}}(\text{tail}(b),c), & \text{if } b(0) \leq c(0),\\ c(0)\hat{\ }\underline{\text{merge}}(b,\text{tail}(c)). \end{cases}$$

(From here on, we will omit repeating "otherwise" in putting down definitions, by cases; they are always meant to be read sequentially, with an "else" or "otherwise" between them.)

The simultaneous inerates of the system $[\underline{\text{sort}},\underline{\text{merge}}]$ are defined again by the recursion

$$\text{(1C.7)}\qquad \underline{\text{sort}}^m(a) \simeq \underline{\text{sort}}(a,\underline{\text{sort}}^{m-1},\underline{\text{merge}}^{m-1}),$$

$$\text{(1C.8)}\qquad \underline{\text{merge}}^m(b,c) \simeq \underline{\text{merge}}(b,c,\underline{\text{sort}}^{m-1},\underline{\text{merge}}^{m-1}),$$

where we have combined the basis and the successor steps using the convention that for any functional f,

$$\text{(1C.9)}\qquad f^{-1} = \emptyset.$$

The simultaneous fixed points of $[\underline{\text{sort}},\underline{\text{merge}}]$ are defined again by

$$\text{(1C.10)}\qquad \underline{\text{sort}}^\infty = \cup_m \underline{\text{sort}}^m;\ \underline{\text{merge}}^\infty = \cup_m \underline{\text{merge}}^m.$$

Lemma 1. For any two sequences b, c, if $m \geq \text{length}(b) + \text{length}(c)$, then $\text{merge}^m(b,c)$ is defined and if b, c are sorted sequences, then $\text{merge}^m(b,c)$ is the sort of their concatenation.

In particular, $\text{merge}^\infty(b,c)$ is defined for all sequences b, c and if b, c are sorted, then $\text{merge}^\infty(b,c)$ is the sort of their concatenation.

Proof is by induction on $\text{length}(b) + \text{length}(c)$. ⊣

Lemma 2. For each sequence a, $\underline{\text{sort}}^\infty(a)$ is defined and gives the sort of a, i.e.

$$\underline{\text{sort}}^\infty(a) = \text{sort}(a).$$

Proof is by induction on $\text{length}(a)$, using Lemma 1 and (for $\text{length}(a) \geq 2$) the fixed-point equation

$$\underline{\text{sort}}^\infty(a) \simeq \underline{\text{merge}}^\infty(\underline{\text{sort}}^\infty(\text{1st}(a)),\underline{\text{sort}}^\infty(\text{2nd}(a))). \quad ⊣$$

This proves the first part of the Basic Fact about the mergesort, that the system $[\underline{\text{sort}},\underline{\text{merge}}]$ computes (as its principal fixed point) the sorting function.

To prove the second assertion in the basic fact about the number of comparisons required by the mergesort, we must give a definition of "required" which refers only to the functionals sort, merge and does not depend on any concepts of implementation.

The value of merge (b, c, sort, merge) does not depend only on the given values of b, c, sort, merge, but also on the ordering $\leq$ of X, which we have taken as fixed for the discussion. It is now useful to think of $\leq$ as a variable over partial, binary functions on X into the Boolean set $\{\underset{\sim}{0},\underset{\sim}{1}\}$,

$$(1C.11) \qquad \leq : X \times X \to \{\underset{\sim}{0},\underset{\sim}{1}\},$$

and redefine merge,

$$(1C.12) \quad \text{merge}(b,c,\text{sort},\text{merge};\leq) \simeq \begin{cases} b & \text{if } c = \emptyset, \\ c & \text{if } b = \emptyset, \\ b(0)^\frown\text{merge}(\text{tail}(b),c) & \text{if } \leq(b(0),c(0)) \simeq \underset{\sim}{1}, \\ c(0)^\frown\text{merge}(b,\text{tail}(c)) & \text{if } \leq(b(0),c(0)) \simeq \underset{\sim}{0}. \end{cases}$$

We can also think of sort as depending trivially on the new variable $\leq$,

$$(1C.13) \quad \text{sort}(a,\text{sort},\text{merge};\leq) \simeq \begin{cases} a & \text{if length}(a) \leq 1, \\ \text{merge}(\text{sort}(\text{1st}(a)),\text{sort}(\text{2nd}(a))). \end{cases}$$

This new pair [merge,sort] is an operative system with parameters, namely $\leq$ here, which we have separated from the other variables by a ";" to indicate that it will not be involved in the recursion process.

For each fixed $\leq$, we can go back and define

$$(1C.14) \qquad \text{sort}_{\leq}(a,\text{sort},\text{merge}) \simeq \text{sort}(a,\text{sort},\text{merge};\leq),$$

$$(1C.15) \qquad \text{merge}_{\leq}(b,c,\text{sort},\text{merge}) \simeq \text{merge}(b,c,\text{sort},\text{merge};\leq),$$

so that whenever $\leq$ is the characteristic function of a total ordering, $[\text{sort}_{\leq},\text{merge}_{\leq}]$ is just the system we had before.

The iterates of the system [sort,merge] are computed by fixing the value of the parameter $\leq$ and iterating the system $[\text{sort}_{\leq},\text{merge}_{\leq}]$. Thus we get for each m fixed points

$$(1C.18) \qquad \text{sort}^{\infty}(a;\leq) \simeq \text{sort}^{\infty}_{\leq}(a),$$

$$(1C.19) \qquad \text{merge}^{\infty}(b,c;\leq) \simeq \text{merge}^{\infty}_{\leq}(b,c).$$

This is just a lot of notation, but it can be very useful. The use of parameters, as in this example allows us to define complicated operations (e.g. functionals, with partial function arguments) using relatively simple systems whose recursion variables are few and easy to deal with. In terms of implementations, one can think of parameters as read-only variables or functions which can only be called by value during the recursion process. Here, we will make a much more direct, mathematical use of the parameter $\leq$.

Suppose $f(x,\leq)$ is any functional whose variables include some list x and $\leq$ varying over partial functions on $X \times X$ to $\{0,1\}$. For each fixed x and $\leq$, we will say that the value $f(x,\leq)$ depends on $\leq k$ values of $\leq$ if

(1C.20) $f(x,\leq) \simeq w \Rightarrow$ there exists some partial function $\leq' \subseteq \leq$ defined on no more than k pairs, such that $f(x,\leq') \simeq w$.

Intuitively, this means that we need only know k values of $\leq$ to compute $f(x,\leq)$.

Notice that if $f(x,\leq)$ is not defined, then the value $f(x,\leq)$ depends on ≤ 0 values of $\leq$; also, if the value of $f(x,\leq)$ depends on $\leq k$ values of $\leq$, then it depends on $\leq k'$ values of $\leq$, for any $k' \geq k$.

With this definition, the following facts are very simple to check.

Lemma 3. The value of $\underset{\sim}{\text{merge}}(b,c,\underset{\sim}{\text{sort}},\underset{\sim}{\text{merge}};\leq)$ depends on ≤ 1 values of $\leq$.

Proof is by inspection of (1C.12). ⊣

Lemma 4. For each m, the value of $\underset{\sim}{\text{merge}}^m(b,c;\leq)$ depends on $\leq m$ values of $\leq$; hence if $\text{length}(b) + \text{length}(c) = n$, then the value of $\underset{\sim}{\text{merge}}^\infty(b,c;\leq)$ depends on $\leq n$ values of $\leq$.

Proof is by induction on m and an application of Lemma 1. ⊣

Lemma 5. If a is a sequence of length n, then the value of $\underset{\sim}{\text{sort}}^\infty(a;\leq)$ depends on $\leq n \cdot \log_2(n)$ values of $\leq$.

Proof is by induction on n, using Lemma 4 and the equation

$$\underset{\sim}{\text{sort}}^\infty(a;\leq) \simeq \underset{\sim}{\text{merge}}^\infty(\underset{\sim}{\text{sort}}^\infty(\text{1st}(a);\leq),\underset{\sim}{\text{sort}}^\infty(\text{2nd}(a);\leq);\leq)$$

when $n \geq 2$. ⊣

In the case when $\leq$ is (the characteristic function of) a total ordering on X, Lemma 5 gives one precise version of the assertion that "if a has length n, then the mergesort will use no more than $n \cdot \log_2 n$ comparisons to sort a."

How does the heuristic proof of the Basic Fact about the mergesort to which we alluded above compare with this precise version? The mathematically conscientious reader who stopped to make that precise, most likely came up with a more direct argument, which makes some assumptions about implementations, perhaps even setting up a specific "implementation scheme" for the mergesort. Nevertheless the "idea" of any such proof is undoubtedly very close to the argument we outlined above.

For our purposes, it is very important to emphasize that the precise version and the proof of the Basic Fact about the mergesort which we gave in Lemmas 1-5 depend in no way on any assumptions about implementations. Perhaps it will clarify matters if we point out here that there are, in fact, at least two fundamentally different classes of implementations of this alogrithm.

Sequential implementations. It is easy to transform the basic equations (1C.3) and (1C.4) into a formal program in a language which allows recursive calls, like Pascal, to be compiled and executed by a von Neumann processor. Whatever the details of space allocation, stack management and the like, any such compiled program will interpret substitution in the basic calls sequentially, i.e. to sort a sequence $a = (a(0),a(1),a(2),a(3))$, it will first sort $(a(0),a(1))$, then it will sort $(a(2),a(3))$ and only then will it merge the two results.

Concurrent implementations. Consider a binary tree of merger processors M,

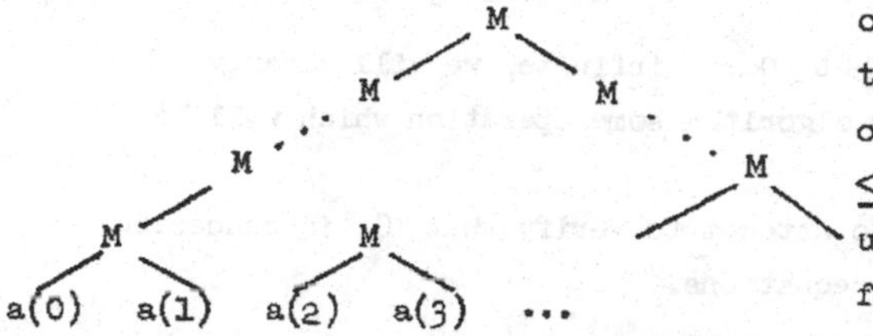

of depth k, where we assume that the upper bound d on the length of the sequences to be sorted is $\leq 2^k$. Each M processor waits until it receives two sequences from its children, and then it merges them and passes them to its parent - or the output at the root. We imagine that the input a is fed to the bottom line of mergers, as a sequence of one-element sequences.

This system of processors implements the simplified mergesort algorithm, where we are given the merge function along with 1st, 2nd, tail, etc. To define a system which implements the mergesort as we defined it, we replace each M-

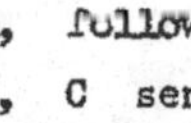

processor above by a pair of processors A (for accumulator) and C (for comparator). When C receives two sequences b and c, it checks if one of them is $\emptyset$, and if so it sends the other one to A, followed by an end of file (EOF) signal; if neither b, nor c are empty, C sends the lesser of $b(0)$, $c(0)$ to A and then sends to itself the pair $(\text{tail}(b),c)$ or $(b,\text{tail}(c))$ accordingly, as $b(0) \leq c(0)$ or not. The processor A accumulates into a single sequence all the terms or sequences it receives, and when it receives the EOF signal from C it sends the accumulated total to its parent. (At the root these instructions are slightly modified.)

It is clear that this concurrent system of processors implements the merge-sort (given 1st, 2nd, tail, etc.). In fact the timing of this implementation is identical to the staging determined by the approximations $\underset{\sim}{\text{sort}}^0, \underset{\sim}{\text{merge}}^0, \underset{\sim}{\text{sort}}^1, \underset{\sim}{\text{merge}}^1, \ldots$ in our recursive analysis of the algorithm.

1D. Connectedness of a graph; discontinuous functionals. Suppose we are given an undirected graph

$$\mathcal{G} = \langle G, R\rangle, \tag{1D.1}$$

i.e. a structure where G is an arbitrary set and R is any binary relation on G such that

$$R(x,y) \Rightarrow R(y,x).$$

We wish to describe an "algorithm" which will verify the connectedness of $\mathcal{G}$, if $\mathcal{G}$ is connected; i.e. we wish to compute the partial (constant) function

(1D.2) $\quad \text{conn} \simeq \underset{\sim}{1} \Leftrightarrow$ for all x, y in G either $x = y$ or there exists a path $x_1,\ldots,x_n$ such that $R(x,x_1), R(x_1,x_2),\ldots,R(x_n,y)$.

Since we are allowing the possibility that G is infinite, we will clearly have to take among the "givens" for our algorithm some operation which will be infinitary, when G is infinite.

There is an obvious, natural way to attempt to verify that $\mathcal{G}$ is connected, which is described by the following two equations.

$$\text{conn} \simeq \underset{\sim}{1} \Leftrightarrow \text{for all } x, y, \ \text{transclos}(x,y) \simeq \underset{\sim}{1}, \tag{1D.3}$$

$$\text{transclos}(x,y) \simeq \underset{\sim}{1} \Leftrightarrow x = y \text{ or there exists some } z \text{ such that } [\text{transclos}(x,z) \simeq \underset{\sim}{1} \text{ and } R(z,y)]. \tag{1D.4}$$

If we read these two equations as the shorthand definition of an operative system in the style of the last three sections, we get the following:

$$\underset{\sim}{\text{conn}}(\text{conn},\text{transclos}) \simeq \underset{\sim}{1} \Leftrightarrow (\forall x)(\forall y)\ \text{transclos}(x,y) \simeq \underset{\sim}{1}, \tag{1D.5}$$

$$\underset{\sim}{\text{transclos}}(x,y,\text{conn},\text{transclos}) \simeq \underset{\sim}{1} \Leftrightarrow x = y \vee (\exists z)[\text{transclos}(x,z) \simeq \underset{\sim}{1} \ \&\ R(z,y)]. \tag{1D.6}$$

The system $[\underset{\sim}{\text{conn}}, \underset{\sim}{\text{transclos}}]$ is (formally) operative, because conn varies over 0-ary partial functions on G to the Boolean set $\{\underset{\sim}{0},\underset{\sim}{1}\}$ (i.e. the partial constants in $\{\underset{\sim}{0},\underset{\sim}{1}\}$) and $\underset{\sim}{\text{conn}}$ has no variables over G, while translos varies over binary partial functions on G to $\{\underset{\sim}{0},\underset{\sim}{1}\}$ and $\underset{\sim}{\text{transclos}}$ has two variables over G. Also, both conn and transclos are clearly monotone in their partial function arguments, so the system can be iterated as before:

(1D.7) $$\underset{\sim}{\text{conn}}^0 \simeq \underset{\sim}{\text{conn}}(\emptyset,\emptyset)$$

(1D.8) $$\underset{\sim}{\text{transclos}}^0(x,y) \simeq \underset{\sim}{\text{transclos}}(x,y,\emptyset,\emptyset),$$

(1D.9) $$\underset{\sim}{\text{conn}}^{m+1} \simeq \underset{\sim}{\text{conn}}(\underset{\sim}{\text{conn}}^m,\underset{\sim}{\text{transclos}}^m),$$

(1D.10) $$\underset{\sim}{\text{transclos}}^{m+1}(x,y) \simeq \underset{\sim}{\text{transclos}}(x,y,\underset{\sim}{\text{conn}}^m,\underset{\sim}{\text{transclos}}^m).$$

Using these definitions, it is trivial to check that

Lemma 1. For each graph $\mathcal{G} = \langle G,R\rangle$ and each $m \geq 0$, x, y in G,

$\underset{\sim}{\text{transclos}}^m(x,y) \simeq \underset{\sim}{1} \Leftrightarrow$ there is a path of length $\leq m$ joining x to y,

$\underset{\sim}{\text{conn}}^m \simeq \underset{\sim}{1} \Leftrightarrow$ each two points in G can be connected by a path of length $< m$. ⊣

Since there are connected graphs where the distance (shortest path) between arbitrary points is not bounded, the lemma implies that, in general, the unions

$$\bigcup_{m=0}^{\infty} \underset{\sim}{\text{conn}}^m, \quad \bigcup_{m=0}^{\infty} \underset{\sim}{\text{transclos}}^m$$

will not define the pair of the least simultaneous fixed points of the system [conn, transclos]. The obvious reason for this is that the functional conn is not continuous - the value of conn (conn, transclos) clearly depends on infinitely many values of transclos (x,y), when G is infinite.

It is well-known that the existence and the basic properties of fixed points for operative systems can be established without any assumptions of continuity; one extends the iteration of the systmm into the transfinite and uses simple facts about ordinal numbers to argue that the iteration eventually will stop and will produce the simultaneous least fixed points, which we will again denote by

$$\underset{\sim}{\text{conn}}^{\infty}, \ \underset{\sim}{\text{transclos}}^{\infty}.$$

These partial functions satisfy the fixed-point equations (1D.3) and (1D.4), which imply in particular

$$\underset{\sim}{\text{conn}}^{\infty} \simeq \underset{\sim}{1} \Leftrightarrow \mathcal{G} \text{ is connected,}$$

so that we may say that the "algorithm" defined by the system [conn, transclos] does what we wanted it to do.

The question is, how can be accept the operative system [conn, transclos] as an "algorithm," since its iteration may extend into the transfinite and not be "implementable" in any natural way. In the theory to be developed here, the key test whether a system represents an algorithm will not be "implementability" (which is in fact very difficult to make precise) but explicit or recursive definability in terms of "the givens."

Consider the functional $E_G^{\#}$ which is defined on (some) unary partial functions

$$p : G \to \{0,1\},$$

takes values in $\{0,1\}$ and embodies quantification on the set G:

(1D.11)
$$E_G^{\#}(p) \simeq \begin{cases} 1 & \text{if } (\exists x \in G)[p(x) \simeq 1], \\ 0 & \text{if } (\forall x \in G)[p(x) \simeq 0]. \end{cases}$$

(The notation with the "$\#$" is a historical relic which we will follow but not bother to explain here.) Once we make things precise in Part 2, it will be obvious that both conn and transclos are explicitly definable in terms of $=$, R and $E_G^{\#}$ on G. We will then say that the system [conn,transclos] represents an explicit algorithm in the givens $=$, R, $E_G^{\#}$, or in the structure

(1D.12)
$$\mathcal{G}^{\#} = \langle G,=,R,E_G^{\#}\rangle.$$

Such recursion structures with (possibly discontinuous) functionals in them will be the basic objects of our study.

One of our main aims is to develop a theory of recursion which covers equally naturally both the "implementable" algorithms of computer science and the discontinuous structures of interest to logicians, such as (1D.12).

1E. Computation schemes. We have not discussed yet the most familiar computation model, which (under various names and with minor variations) is often taken as the natural, mathematical definition of "algorithm."

A computation scheme for a partial function

$$f : X \to W$$

is a quintuple

(1E.1)
$$C = \langle S,\text{input},\text{test},\ \text{transition},\text{output}\rangle,$$

where the following conditions hold.

(1) S is a set and input, test, transition, output are all total functions on the following domains and ranges:

$$\text{input} : X \to S,$$

$$\text{test} : S \to \{0,1\},$$

$$\text{transition} : S \to S,$$

$$\text{output} : S \to W.$$

(2) For each $x \in X$,

(1E.2) $$f(x) \simeq \text{output}(s^n),$$

where

(1E.3) $$s^0 = \text{input}(x), \quad s^{i+1} = \text{transition}(s^i)$$

and

(1E.4) $$n \simeq \text{least } i \text{ such that } \text{test}(s^i) = \underset{\sim}{0}.$$

The idea is that we compute $f(x)$ by starting with the initial state $s^0 = \text{input}(x)$, and then iterating the transition function to produce $s^1, s^2, \ldots$, until (and if ever) we find some first s^n such that $\text{test}(s^n) = \underset{\sim}{0}$; we then apply the output function to that s^n to get the value $f(x)$.

There is a plausible approach to a mathematical theory of computation which simply identifies "algorithms" with "computation schemes." Part of the appeal here is that implementable algorithms clearly correspond to computation schemes, and quite special ones at that.

One of the reasons we are not taking this route is that we want a theory which will cover - simply and naturally - even the non-implementable algorithms of 1D. But even for implementable recursive algorithms, their reduction to computation schemes often brings in a great deal of seemingly irrelevant detail, (stacks and the like) which does not always help in the mathematical analysis of the algorithm.

Later on we will discuss the role of computation schemes in the present theory. Here we will just put down the familiar fixed-point analysis of computation schemes.

Given $C = (S, \text{input}, \text{test}, \text{transition}, \text{output})$ which computes $f : X \to Y$, let

(1E.5) $$\underset{\sim}{\text{value}}(x, \underset{\sim}{\text{value}}, \underset{\sim}{\text{loop}}) \simeq \text{output}(\underset{\sim}{\text{loop}}(\text{input}(x))),$$

(1E.6) $$\underset{\sim}{\text{loop}}(s, \underset{\sim}{\text{value}}, \underset{\sim}{\text{loop}}) \simeq \begin{cases} s & \text{if } \text{test}(s) = \underset{\sim}{0}, \\ \underset{\sim}{\text{loop}}(\text{transition}(s)) & \text{if } \text{test}(s) = \underset{\sim}{1}. \end{cases}$$

Clearly $[\underset{\sim}{\text{value}}, \underset{\sim}{\text{loop}}]$ is an operative, continuous system and for all x in X,

$$\underset{\sim}{\text{loop}}^{\infty}(x) \simeq f(x);$$

moreover, the most direct attempt to set up a computation scheme which will implement the recursion determined by $[\underset{\sim}{\text{value}}, \underset{\sim}{\text{loop}}]$ will obviously lead us back to C.

Part 2. Introduction to the Theory of Recursion

This is the main part of the paper, in which we will define precisely the basic notions of the theory of recursion, we will establish a few basic facts and (mostly) we will discuss how algorithms can be represented and studied in this framework.

2A. Universes and functionals. Contrary to the usual practice in first order logic, it is convenient here to deal right from the start with many-sorted structures.

A (many-sorted) universe is a (possibly finite) sequence of basic sets

$$\Sigma = \{\underset{\sim}{0},\underset{\sim}{1}\}, S_1, S_2, \ldots ,$$

starting with the fixed Boolean set $S_0 = \{\underset{\sim}{0},\underset{\sim}{1}\}$.

Thus the simplest (trivial) universe has $\{\underset{\sim}{0},\underset{\sim}{1}\}$ as its only basic set. In a more typical case, Σ will have just one more basic set, for example the set

$$N = \{0,1,2,\ldots\}$$

of (non-negative) integers, the set of vertices of a graph or the set

(2A.1) $S^* =$ all finite sequences from S,

where S is some given set.

Sometimes the basic sets of a universe are defined inductively and can become quite complicated, as in the following example of the universe Σ_{BP} (BP stands for "baby programming").

First we define by induction the simple basic sets of Σ_{BP}.

(1) $\{\underset{\sim}{0},\underset{\sim}{1}\}$ in a simple, basic set of Σ_{BP}; for some fixed positive integer maxint, the set int of integers between -maxint and maxint is a simple basic set of Σ_{BP}; some fixed finite set char is a simple basic set of Σ_{BP}.

(2) For each simple basic set T of Σ_{BP}, the set

(2A.2) array-of-T = all finite sequences from T of length < maxint

is a simple basic set of Σ_{BP}.

Now the basic sets of Σ_{BP} are the simple basic sets of Σ_{BP} together with all sets of the form

(2A.3) file-of-T = $\{(f,i) : f$ is a finite sequence from T and $i \leq \text{length}(f)\}$

where T is any simple basic set of Σ_{BP}.

Each time, when we give a recursive definition of a universe like this, we will assume that we adopt some fixed enumeration of the basic sets, to fit the formal definition of a universe as a <u>sequence</u> of sets. This particular example Σ_{BP} will be used to illustrate how we will deal with the <u>types</u> of programming languages - called <u>sorts</u> by logicians and by the neutral term <u>basic sets</u> here to minimize confusion.

A <u>recursion space</u> of a given universe Σ is any cartesian product

$$U = U_1 \times \cdots \times U_n \tag{2A.4}$$

of basic sets of Σ; a <u>partial function space</u> of Σ is any set

$$P(U,V) = \text{all partial functions } p : U \to V, \tag{2A.5}$$

where U and V are recursion spaces of Σ. Finally a <u>product space</u> or simply <u>space</u> of Σ is any cartesian product

$$X = X_1 \times \cdots \times X_n, \tag{2A.6}$$

where each X_i is a basic set or a partial function space of Σ.

If $X = X_1 \times \cdots \times X_n$ and $Y = Y_1 \times \cdots \times Y_m$ are spaces of Σ, then (by definition)

$$X \times Y = X_1 \times \cdots \times X_n \times Y_1 \times \cdots \times Y_m. \tag{2A.7}$$

Similarly, if $x = (x_1,\ldots,x_n)$, $y = (y_1,\ldots,y_m)$ are members of X and Y respectively, then (by definition)

$$(x,y) = x,y = (x_1,\ldots,x_n,y_1,\ldots,y_m). \tag{2A.8}$$

As usual, we will allow caretsian products of length 1, i.e. we will associate with each set A (and usually identify with A) the set of all one-term sequences from A. What may be less usual, we will also admit <u>the cartesian product of no factors</u>, $\mathrm{I\!I}$; formally

$$\mathrm{I\!I} = \{\emptyset\} \tag{2A.9}$$

has only the empty tuple as a member and satisfies

$$X \times \mathrm{I\!I} = \mathrm{I\!I} \times X = X \tag{2A.10}$$

for any product space X.

A (partial, monotone) <u>functional</u> on a universe Σ is any partial function

$$f : X_1 \times \cdots \times X_n \to W, \tag{2A.11}$$

where $X = X_1 \times \cdots \times X_n$ is a product space of Σ, W is a recursion space of Σ

and f is monotone in its partial function arguments: i.e. if $x = (y,p,z) \in X$ and p varies over $P(U,V)$, then

(2A.12) $$[f(y,p,z) \simeq w \ \& \ p \subseteq p'] \Rightarrow f(y,p',z) \simeq w.$$

In addition, f is <u>compact</u> or <u>continuous</u> if whenever $x = (y,p,z) \in X$ and p varies over $P(U,V)$, then

(2A.13) $f(y,p,z) \simeq w \Rightarrow$ there is some $p_0 \subseteq p$, p_0 defined only on finitely many points, such as $f(y,p_0,z) \simeq w$.

Every partial function

$$f : U_1 \times \cdots \times U_n \to W$$

is a functional, by the definition. This includes the case when $U_1 \times \cdots \times U_n = I$ is the empty product, so that the <u>partial constants</u>

$$f : I \to W$$

are also among the functionals.

For each two recursion spaces U, W in a universe Σ, we define

$$\underline{ap}: P(U,W) \times U \to W$$

by

(2A.14) $$\underline{ap}(p,u) \simeq p(u).$$

There is one such <u>application functional</u> for each U, W - we will denote all of them by the same symbol "<u>ap</u>".

Suppose in a given universe Σ we have basic sets S, W and the corresponding sets of finite strings S^*, W^*. Define

$$\underline{aptoall}: S^* \times P(S,W) \to W^*$$

by

(2A.15) $$\underline{aptoall}((s_0,\ldots,s_{n-1}),p) \simeq (p(s_0),\ldots,p(s_{n-1})).$$

In the same context, with $S = W$, define the <u>binary iteration functional</u>

$$\angle : S^* \times P(S \times S,S) \to S$$

by

(2A.16) $$\angle((s_0,s_1,\ldots,s_{n-1}),p) \simeq p(s_0,p(s_1,(\ldots\ p(s_{n-2},s_{n-1}))\ldots).$$

(We leave $\angle(s,p)$ undefined when s is a string of length less than 2.)

We also saw many examples of functionals in Part 1, including <u>the quantifier</u>

$E_G^{\#}$, defined by (1D.11) on any universe which has the basic set G. A related functional is the half quantifier $E_G^{1/2}$ which is often used to introduce nondeterminacy:

$$E_G^{1/2}(p) \simeq \begin{cases} \underset{\sim}{1} & \text{if } (\exists x \in G)p(x) \simeq \underset{\sim}{1} \\ \text{undefined} & \text{otherwise.} \end{cases} \tag{2A.17}$$

For the complete logic pendant, we should point out that these definitions of product spaces, partial function spaces, functionals etc. are meant to introduce new (typed, labeled, free or whatever) objects. For example, we may well wish to have the set

$$\mathcal{P} = \text{all partial functions on } N \text{ to } N$$

among our basic sets, in a universe Σ; now the partial function space $P(N,N)$ is considered formally as a different object (space) from $\mathcal{P}$, even though as sets, $\mathcal{P}$ and $P(N,N)$ are equal. Going one step further, in these cricumstances we may well want to study the functional of Σ

$$\underset{\sim}{c} : P(N,N) \to \mathcal{P}$$

which is simply the identity on the partial functions on N to N,

$$\underset{\sim}{c}(p) = p.$$

One may happily live a full mathematical life without ever worrying about this kind of logical detail, and certainly for reading this paper, this is exactly what one should do.

The signature of a universe Σ is $k+1$ if Σ has k basic sets or 0 if Σ has infitely many basic sets; the signature of a basic set S_i in Σ is the integer i; the signature of a recursion space $U_1 \times \cdots \times U_n$ in Σ is the tuple $\langle u_1,\ldots,u_n\rangle$, where each u_i is the signature of U_i; the signature of a partial function space $P(U,V)$ is the sequence $\langle u,v\rangle$, where u and v are the signatures of U and V; finally, the signature of a functional $f : X_1 \times \cdots \times X_n \to W$ is the tuple $\langle x_1,\ldots,x_n,w\rangle$, where $x_1,\ldots,x_n,w$ are the signatures of $X_1,\ldots,X_n,W$, respectively. We assume some fixed coding of tuples of integers, so that all signatures are integers.

A functional

$$f : X_1 \times \cdots \times X_n \to W$$

is operative (with parameters) if for some recursion space U, $X_1 = U$ and $X_2 = P(U,W)$, so that in fact

(2A.18) $$f : U \times P(U,W) \times X_3 \times \cdots \times X_n \to W.$$

The functional f has no parameters if $n = 2$.

Here we allow U to be the empty product $\mathbb{I}$, in which case we do not even include it in the notation, so that each functional of the form

$$f : P(I,W) \times X_2 \times \cdots \times X_n \to W$$

is operative.

Similarly, a system of functionals

$$f = [f_1,\ldots,f_n]$$

is operative in Σ, if the domains and ranges of $f_1,\ldots,f_n$ match properly, i.e. of for some recursion spaces $U_1,W_1,U_2,W_2,\ldots,U_n,W_n$,

(2A.19) $$\begin{cases} f_1 : U_1 \times P(U_1,W_1) \times \cdots \times P(U_n,W_n) \times Y_1 \times \cdots \times Y_m \to W_1, \\ f_2 : U_2 \times P(U_1,W_1) \times \cdots \times P(U_n,W_n) \times Y_1 \times \cdots \times Y_m \to W_2, \\ \cdots \\ f_n : U_n \times P(U_1,W_1) \times \cdots \times P(U_n,W_n) \times Y_1 \times \cdots \times Y_m \to W_n. \end{cases}$$

Here the parameters (if any) are the tuples in $Y = Y_1 \times \cdots \times Y_m$. Again, any one of the spaces $U_1,\ldots,U_n$ may be the empty product $\mathbb{I}$, so that for example, the system [conn, transclos] of 1D is operative.

Sometimes it is useful notationally to separate the recursion variables from the parameters by a ";", so that (for example) the value of one of the functionals above will be denoted by

$$f_i(u_i,p_1,\ldots,p_n;y_1,\ldots,y_m).$$

It is also useful to think of a system with parameters as an operation, which assigns to each value of the parameters the corresponding parameterless system. This just means that for the system $f = [f_1,\ldots,f_n]$ above and each

$$y = y_1,\ldots,y_m,$$

we will write

(2A.20) $$f(y) = [f_{1,y},f_{2,y},\ldots,f_{n,y}],$$

where

$$f_{i,y}(u_i,p_1,\ldots,p_n) \simeq f_i(u_i,p_1,\ldots,p_n;y) \qquad (i = 1,\ldots,n).$$

The iterates of a system are defined as in the examples of Part 1. First we define for each fixed value y of the parameters the iterates of the system

$f(y)$, by recursion on the integer k (or the ordinal k if the system is not compact):

(2A.21) $$f^k_{i,y}(u_i) \simeq f_{i,y}(u_i, f^{<k}_{1,y}, \ldots, f^{<k}_{n,y}) \qquad (i = 1,\ldots,n),$$

where

(2A.22) $$f^{<k}_{i,y} = \bigcup_{\ell<k} f^{\ell}_{i,y} \qquad (i = 1,\ldots,n).$$

From these we get the iterates of the original system f,

(2A.23) $$f^k_i(u_i;y) \simeq f^k_{i,y}(u_i) \quad (i = 1,\ldots,n)$$

and in the limit the simultaneous fixed points of f,

(2A.24) $$f^\infty_i = \bigcup_k f^k_i \qquad (i = 1,\ldots,n),$$

so that

$$f^\infty_i(u_i;y) \simeq w \Leftrightarrow \text{for some } k,\ f^k_i(u_i;y) \simeq w.$$

The smallest ordinal κ such that

(2A.25) $$f^\kappa_i = f^{<\kappa}_i \qquad (\text{for } i = 1,\ldots,n)$$

is the closure ordinal of f, and for compact f, easily $\kappa \leq \omega$ = the least infinite ordinal.

It is also convenient to fix the notation

(2A.26) $$\overline{f}(u_1;y) \simeq f^\infty_1(u_1;y),$$

so that $\overline{f}$ is the principal fixed point of the system f, the functional which f is usually expected to compute. Typically u_1 ranges over the empty product Π, in which case for each y,

(2A.27) $$\overline{f}(y) \simeq f^\infty_1(y)$$

is just the constant computed by the system $f(y)$ of (2A.20).

For the final two examples of this section we will use recursion to define the basic while and ancestral functionals. These embody deterministic and non-deterministic iteration, respectively.

On any universe Σ which contains the recursion spaces U and W, let

(2A.28) $$\text{whileloop}_{U,W}(u,\text{loop};\text{test},\text{action},\text{value}) \simeq \begin{cases} \text{value}(u), & \text{if } \text{test}(u) \simeq 0, \\ \text{loop}(\text{action}(u)), & \text{if } \text{test}(u) \simeq 1, \end{cases}$$

where $u \in V$, test $: U \to \{0,1\}$, action $: U \to U$ and value $: V \to W$, and set

(2A.29) $\underset{\sim}{\text{while}}_{U,W}(\underline{\text{test}}, u, \underline{\text{action}}, \underline{\text{value}}) \simeq \underset{\sim}{\text{whileloop}}^{\infty}_{U,W}(u; \underline{\text{test}}, \underline{\text{action}}, \underline{\text{value}})$.

Clearly,

$$\underset{\sim}{\text{while}}_{U,W}(\underline{\text{test}}, u, \underline{\text{action}}, \underline{\text{value}}) \simeq \underline{\text{value}}(\underline{\text{action}}^{n}(u)),$$

where

$$\underline{\text{action}}^{0}(u) \simeq u, \quad \underline{\text{action}}^{i+1}(u) \simeq \underline{\text{action}}(\underline{\text{action}}^{i}(u))$$

and

$$n \simeq \text{least } i \text{ such that for all } j < i, \ \underline{\text{test}}(\underline{\text{action}}^{j}(u)) \simeq \underset{\sim}{1} \text{ and } \underline{\text{test}}(\underline{\text{action}}^{i}(u)) \simeq \underset{\sim}{0}.$$

Similarly, let

(2A.30) $\underset{\sim}{\text{ancloop}}_U(u, \underline{\text{loop}}; v, r) \simeq \underset{\sim}{1} \Leftrightarrow r(u,v) \simeq \underset{\sim}{1} \vee (\exists u')[r(u,u') \simeq \underset{\sim}{1} \ \& \ \underline{\text{loop}}(u') \simeq \underset{\sim}{1}]$,

where $u, v \in U$, $r : U \times U \to \{\underset{\sim}{0}, \underset{\sim}{1}\}$, $\underline{\text{loop}} : U \to \{\underset{\sim}{0}, \underset{\sim}{1}\}$ and set

(2A.31) $\underset{\sim}{\text{ancestral}}_U(u,v,r) \simeq \underset{\sim}{\text{ancloop}}^{\infty}_U(u; v, r)$.

Now

$$\underset{\sim}{\text{ancestral}}_U(u,v,r) \simeq \underset{\sim}{1} \Leftrightarrow \text{there exists a sequence } u_0, \ldots, u_{n-1} \text{ such that } r(u,u_0) \simeq r(u_0,u_1) \simeq \ldots \simeq r(u_{n-1},v) \simeq \underset{\sim}{1}.$$

We will use the ambiguous "$\underset{\sim}{\text{while}}$" and "$\underset{\sim}{\text{ancestral}}$" to denote any particular instances of these functionals for specific U and W, when no ambiguity can occur. As one might expect, $\underset{\sim}{\text{while}}$ turns out to be recursive on every structure, while $\underset{\sim}{\text{ancestral}}$ is recursive in the half-quantifier $E^{1/2}$ defined by (2A.17), the basic "non-deterministic" functional.

2B. Recursion structures. A recursion structure (or simply structure) is a pair

$$\mathfrak{A} = \langle \Sigma, \Phi \rangle,$$

where Σ is a universe and

$$\Phi = \varphi_0, \varphi_1, \ldots$$

is a (finite or infinite, possibly empty) sequence of functionals on Σ. The signature of $\mathfrak{A}$ is the sequence

(2B.1) $$\text{sign}(\mathfrak{A}) = k, e_0, e_1, \ldots,$$

where each e_i is the signature of φ_i and either $k = 0$ and $\mathfrak{A}$ has infinitely many basic sets or $k > 0$ and $\mathfrak{A}$ has exactly $k - 1$ basic sets.

In the next two sections, we will define and study the basic properties of explicit and recursive functionals on an arbitrary structure $\mathfrak{A}$. Before going into this, it will be useful to set up some notational conventions and to look at a few of the most important structures with which we will be concerned.

As we have pointed out, partial functions are (somewhat degenerate cases of) functionals, so when we know the recursive functionals of $\mathfrak{A}$, we will also know the recursive partial functions of $\mathfrak{A}$. The situation is a bit more complicated with relations, i.e. subsets $R \subseteq U$ of the recursion spaces of $\mathfrak{A}$.

We will represent each relation $R \subseteq U$ by two partial functions, the characteristic (total) function of R,

(2B.2)
$$\chi_R(u) = \begin{cases} \underset{\sim}{1} & \text{if } R(u), \\ \underset{\sim}{0} & \text{if } \neg R(u) \end{cases}$$

and the semicharacteristic (partial) function of R,

(2B.3)
$$s\chi_R(u) \simeq \underset{\sim}{1} \Leftrightarrow R(u).$$

For example, in ordinary recursion theory, R is recursive if χ_R is recursive, while R is semirecursive (or recursively enumerable) if $s\chi_R$ is recursive.

If Γ is a collection of functionals on a structure $\mathfrak{A}$, we let

(2B.4)
$$\begin{aligned} \text{Sec } \Gamma &= \{R : R \text{ is a relation on } \mathfrak{A} \text{ and } \chi_R \in \Gamma\} \\ &= \text{the section of } \Gamma, \end{aligned}$$

(2B.5)
$$\begin{aligned} \text{Env } \Gamma &= \{R : R \text{ is a relation on } \mathfrak{A} \text{ and } s\chi_R \in \Gamma\} \\ &= \text{the envelope of } \Gamma. \end{aligned}$$

The terminology comes from abstract recursion theory and is quite useful.

We will also study uniform or global definability on a class $\mathcal{K}$ of structures of the same signature.

A global functional or $\mathcal{K}$ (or simply $\mathcal{K}$-functional) is an operation f which assigns a functional $f^{\mathfrak{A}}$ on each $\mathfrak{A}$ in $\mathcal{K}$, with all $f^{\mathfrak{A}}$'s having the same, fixed signature. For example, on the class of all graphs, we can consider the global functional $E^{\#}$ which assigns to each graph $\mathfrak{G}$ the existential quantifier $E^{\#}_G$ defined by (1D.11). (Gurevich [1983] uses the same term, at least for the case where each $f^{\mathfrak{A}}$ is a partial function.)

A relational query R on a class of structures $\mathcal{K}$ of the same signature is an operation which assigns a relation

$$R^{\mathfrak{A}} \subseteq U^{\mathfrak{A}}$$

to each $\mathfrak{A} \in \mathcal{K}$, where $U^{\mathfrak{A}}$ is the recursion space determined by the fixed signature of the query R. (This convenient terminology was introduced by Chandra-Harel [1980]).

We will assign to each relational query R on $\mathcal{K}$ two global functionals on $\mathcal{K}$, $\chi(R) = \chi_R$ coming from the characteristic function of $R^{\mathfrak{A}}$

$$\chi_R(u) = \begin{cases} \underset{\sim}{1} & \text{if } R^{\mathfrak{A}}(u), \\ \underset{\sim}{0} & \text{if } \neg R^{\mathfrak{A}}(u) \end{cases} \tag{2B.6}$$

and $s\chi(R) = s\chi_R$ coming from the semicharacteristic partial function,

$$s\chi_R^{\mathfrak{A}}(u) \simeq \underset{\sim}{1} \Leftrightarrow R^{\mathfrak{A}}(u). \tag{2B.7}$$

In the special case where $U_{\mathfrak{A}} = \mathbb{I}$ (the empty product), $R^{\mathfrak{A}}$ is simply true or false; we call R a <u>Boolean query</u> and we often identify it with the class of structures it determines,

$$R \sim \{\mathfrak{A} \in \mathcal{K} : R^{\mathfrak{A}} \text{ is true in } \mathfrak{A}\}. \tag{2B.8}$$

Now $\chi(R)$ and $s\chi(R)$ are constant global functionals on $\mathcal{K}$.

For example, if

$$\mathfrak{Conn} = \text{the class of connected graphs,}$$

then for each graph $\mathcal{G}$,

$$\chi(\mathfrak{Conn})^{\mathcal{G}} \simeq \begin{cases} \underset{\sim}{1} & \text{if } \mathcal{G} \text{ is connected,} \\ \underset{\sim}{0} & \text{if } \mathcal{G} \text{ is disconnected} \end{cases}$$

and

$$s\chi(\text{conn})^{\mathcal{G}} \simeq \underset{\sim}{1} \Leftrightarrow \mathcal{G} \text{ is connected.}$$

If Γ is a collection of global functionals on a class $\mathcal{K}$, we naturally put

$$\text{Sec}\,\Gamma = \{R : R \text{ is a relational query on } \mathcal{K} \text{ and } \chi(R) \in \Gamma\}, \tag{2B.9}$$

$$\text{Env}\,\Gamma = \{R : R \text{ is a relational query on } \mathcal{K} \text{ and } s\chi(R) \in \Gamma\}. \tag{2B.10}$$

After these preliminary remarks, we now look at some of the examples.

(2B.11) <u>Recursion on the natural numbers.</u> Let

$$\mathfrak{N} = \langle N, 0, \text{succ}, \text{pred}, \chi_{=0} \rangle,$$

where $N = \{0,1,2,\ldots\}$, $\text{succ}(x) = x + 1$, $\text{pred}(0) = 0$, $\text{pred}(x + 1) = x$ and $\chi_{=0}$ is the characteristic function of equality to 0, i.e.

$$\chi_{=0}(x) = \begin{cases} \underset{\sim}{1} & \text{if } x = 0, \\ \underset{\sim}{0} & \text{if } x \neq 0. \end{cases}$$

The basic sets of $\mathfrak{n}$ are N and $\{\underset{\sim}{0},\underset{\sim}{1}\}$, which we will not bother to list. It will turn out of course that the $\mathfrak{n}$-recursive functionals are precisely the ordinary, familiar recursive partial functionals on the natural numbers.

Notice that here 0 is formally considered as a functional, with domain the empty product $\mathbb{I}$ and constant value 0.

This is an important special case of the next example.

(2B.12) First order recursion. Let

$$\mathfrak{G} = \langle A_1,\ldots,A_n,R_1,\ldots,R_m,f_1,\ldots,f_k,c_1,\ldots,c_\ell\rangle$$

be a (many-sorted) structure in the sense of model theory, with basic sets $A_1,\ldots,$ relations $R_1,\ldots,$ functions $f_1,\ldots$ and distinguished elements $c_1,\ldots$. Most commonly there will be just one basic set, as in the case of a graph $\mathcal{G} = \langle G,R\rangle$ or a field $\mathfrak{F} = \langle F,+,\cdot,0,1\rangle$. We will view $\mathfrak{G}$ as a recursion structure, with basic sets $\{\underset{\sim}{0},\underset{\sim}{1}\}$, $A_1,\ldots,$ and given functionals the functions $f_1,\ldots,$ the constants $c_1,\ldots$ (viewed as functions on $\mathbb{I}$, as above) and the characteristic functions $\chi_1,\ldots$ of $R_1,\ldots,$

$$\chi_i(x) = \begin{cases} \underset{\sim}{1} & \text{if } R_i(x), \\ \underset{\sim}{0} & \text{if not } R_i(x). \end{cases}$$

We will let

$$\mathfrak{FO}(\sigma) = \text{all first order structures of signature } \sigma$$

and often refer to "$\mathfrak{FO}$" when σ is irrelevant or understood.

These first order structures with no "true" functionals carry an interesting and non-trivial recursion theory, even when they are finite. Among them we can find quite faithful representations of the familiar, restricted computation models, like automata, flowcharts, recursion schemes and Turing machines. Here we will list just one more example of first order recursion in an infinite structure.

(2B.13) Pure LISP. Let A (the atoms) be the set of all words from a fixed, finite alphabet excluding the symbols ".", "(" and ")". The set S of S-expressions is the smallest set of words which contains the atoms and is closed under the function

$$\text{cons}(x,y) = (x.y),$$

where $(x.y)$ is the concatenation of "(", x, ".", etc. The structure of pure

LISP is

$$\mathfrak{PLISP} = \langle S, \text{atom}, \text{eq}, \text{cons}, \text{car}, \text{cdr}, \{c_y : y \text{ a letter}\}\rangle,$$

where atom is the (Boolean-valued) characteristic function of A, eq is the characteristic function of equality of A (undefined on S - A), car and cdr are defined on S - A so that

$$\text{car}((x.y)) = x,$$
$$\text{cdr}((x.y)) = y$$

and for each letter y in the given fixed alphabet, c_y is the constant function with no arguments and value the S-expression y.

It would be more accurate to call $\mathfrak{PLISP}$ the structure of concurrent pure LISP, since our definition of functional iteration implicitly assumes parallel computations, while the evaluation algorithm of McCarthy [1960] is sequential; i.e. our semantics (to be described in 2D, 2E) will compute cons(f(x),g(y)) by computing simultaneously f(x) and g(y) and then taking the pair, while McCarthy [1960] computes f(x) first, then g(y) and then takes the pair. This kind of subtle distinction becomes much more important when we consider the extensions of LISP by assignments, set's and setq's.

Recursion in $\mathfrak{PLISP}$ is also a formulation in the present theory of the prime computability theory of Moschovakis [1969].

Many times we give recursive definitions which are meant to apply not to just one, fixed structure but to a whole class of structures. The following class of finite, initial segments of $\mathfrak{N}$ is very natural.

(2B.14) Recursion on initial segments of N. Let $\mathfrak{IN}$ ("$\mathfrak{I}$" for "initial") be the collection of all structures of the form

$$\mathfrak{A} = \langle \{0,1,\ldots,n-1\}, 0, \text{succ}, \text{pred}, \chi_{=0}\rangle$$

where n is a positive integer and succ is the usual successor on N, with succ(n - 1) = 0.

A related collection of structures in $\mathfrak{BSTR}$ (for "binary strings") which consists of all structures of the form

$$\mathfrak{A}_\sigma = \langle \{0,\ldots,n-1\}, 0, \text{succ}, \text{pred}, \chi_{=0}, \sigma\rangle,$$

where σ is a unary relation on $\{0,1,\ldots,n-1\}$, so that its characteristic function (which we will also call σ in this case) is a binary string of length n.

For each $k \geq 1$, let $\mathfrak{BSTR}^{(k)}$ be the collection of all (k + 1-sorted) structures of the form

$$\mathfrak{A}_\sigma^{(k)} = \langle\{0,1,\ldots,n-1\},F_1,\ldots,F_k,0,\text{succ},\text{pred},\chi_{=0},\sigma,$$
$$\text{ap}_1,\ldots,\text{ap}_k,\text{assign}_1,\ldots,\text{assign}_k\rangle,$$

where for $i = 1,\ldots,k$,

F_i = all total functions of i arguments on $\{0,1,\ldots,n-1\}$ to $\{\underset{\sim}{0},\underset{\sim}{1}\}$

and the total functions ap_i, assign_i,

$$\text{ap}_i : F_i \times \{0,1,\ldots,n-1\}^i \to \{\underset{\sim}{0},\underset{\sim}{1}\},$$
$$\text{assign}_i : F_i \times \{0,1,\ldots,n-1\}^i \times \{\underset{\sim}{0},\underset{\sim}{1}\} \to F_i$$

are defined as follows:

$$\text{ap}_i(f,x_1,\ldots,x_i) = f(x_1,\ldots,x_i),$$
$$\text{assign}_i(f,x_1,\ldots,x_i,t) = g,$$

where

$$g(x'_1,\ldots,x'_i) = \begin{cases} f(x'_1,\ldots,x'_i) & \text{if } (x'_1,\ldots,x'_i) \neq (x_1,\ldots,x_i),\\ t & \text{if } (x'_1,\ldots,x'_i) = (x_1,\ldots,x_i).\end{cases}$$

It will turn out that uniform (or global) recursion on these classes of structures gives simple characterizations of many interesting complexity classes of languages (sets of strings), e.g. recursion on $\mathfrak{BSTR}$ is precisely polynomial time computability. We will discuss these facts (and give references) in the next two sections.

(2B.15) Expansions. If $\mathfrak{A} = \langle\Sigma,\Phi\rangle$ is a structure and $\Psi = \psi_1,\psi_2,\ldots$ is a sequence of functionals on the universe Σ, we will let

$$(\mathfrak{A},\Psi) = \langle\Sigma,\Phi \cup \Psi\rangle$$

be the expansion of $\mathfrak{A}$ which has the additional functionals in Ψ - and similarly when we wish to add more basic sets to $\mathfrak{A}$. We assume a standard interweaving of the sequences in $\mathfrak{A}$ and the new sequences of objects that we add.

For example, we may wish to study the expansion $(\mathfrak{A},=_S)$, where $=_S$ is the equality relation on some basic set of $\mathfrak{A}$, or the larger expansion $(\mathfrak{A},=)$ where we add the equality relation on each basic set of $\mathfrak{A}$.

Another very natural expansion is

(2B.16) $$(\mathfrak{A},\underset{\sim}{\text{while}}),$$

where we add to $\mathfrak{A}$ all the while functionals defined by (2A.29), one for each pair U, W of recursion spaces of $\mathfrak{A}$. The while functionals are recursive on every structure and hence (as one would expect) $\mathfrak{A}$ and $(\mathfrak{A},\text{while})$ have exactly the same recursive functionals. Nevertheless, it will be useful to consider recursive expansions like $(\mathfrak{A},\text{while})$ which have a richer collection of algorithms and a much stronger language for specifying algorithms.

Other recursive functionals which can be profitably added to structures include the aptoall and $/$ of (2A.15) and (2A.16), when the structure has the appropriate basic sets of strings, as in the example below.

We will often expand all the structures in a class, e.g.

$$(\mathfrak{FO},\text{while}) = \{(\mathfrak{A},\text{while}) : \mathfrak{A} \text{ is a first order structure}\},$$

$$(\mathfrak{M},\text{while}) = \{(\mathfrak{A},\text{while}) : \mathfrak{A} \in \mathfrak{M}\},$$

etc.

(2B.17) Baby programming. Let Σ_{BP} be the universe for baby programming defined in 2A and consider the structure

$$\mathfrak{BP} = \langle \Sigma_{BP}, \Phi_{BP} \rangle,$$

where in Φ_{BP} we put the following partial functions and functionals.

(1) The constant 0 in int and some fixed, finite set of total functions with variables in $\{0,1\}$, int and char, including the identity functions, (the restricted) succ, pred, perhaps +, * and =, etc.

(2) The following partial functions, where i, j range over int, x ranges over some simple basic set T and a, b range over array-of-T.

$$\text{length}(a) = \text{the length of } a < \text{maxint}$$

$$\text{ap}(a,i) \simeq a(i) \qquad (\text{defined when } i < \text{length}(a))$$

$$\text{assign}(x,a,i) \simeq b \qquad (\text{defined when } i < \text{length}(a))$$

where

$$b(j) = a(j) \text{ if } j \neq i,\ b(i) = x.$$

(3) For each basic set

$$F = \text{file-of-T} = \{(f,i) : f \text{ is a finite sequence from T and } i \leq \text{length}(f)\},$$

the partial functions

$$\text{read}(f,i) \simeq (f(i), i+1) \qquad (f \in F,\ i+1 \leq \text{length}(f))$$

$$\text{write}((f,i),x) \simeq (f \cup \langle i,x\rangle, i+1) \qquad (f \in F,\ x \in T,\ i+1 \leq \text{length}(f))$$

$$\text{restore}(f,i) = (f,0) \qquad (f \in F,\ i \leq \text{length}(f)).$$

(4) The functionals while, aptoall and $\not\downarrow$ which make sense for the simple basic sets.

This example is obviously meant to illustrate the kind of recursion structures which we can use to develop semantics for programming languages, and the particular intrinsics chosen have no special significance. It is a severely limited example and it might suggest that we will not be able to model satisfactorily but the simplest applicative languages, like pure LISP. We will duscuss this question briefly in Part 3, after we have introduced all the basic notions that we need.

(2B.18) Positive elementary induction. For each first order structure

$$\mathfrak{A} = \langle A_1,\ldots,A_n,R_1,\ldots,R_m,f_1,\ldots,f_k,c_1,\ldots,c_\ell\rangle,$$

let

$$(\mathfrak{A},=,E^{\#}) = (\mathfrak{A},=_1,\ldots,=_n,E_0^{\#},E_1^{\#},\ldots,E_n^{\#})$$

be the expansion of $\mathfrak{A}$ by the equality relations and the existential quantifiers on the basic sets, i.e. the functionals

$$E_i^{\#}(p) \simeq \begin{cases} \underset{\sim}{1} & \text{if } (\exists x \in A_i)p(x) \simeq \underset{\sim}{1}, \\ 0 & \text{if } (\forall x \in A_i)p(x) \simeq \underset{\sim}{0}, \end{cases}$$

with $A_0 = \{\underset{\sim}{0},\underset{\sim}{1}\}$.

Recursion in $(\mathfrak{A},=,E^{\#})$ gives a natural and elegant treatment of the theory of positive, elementary induction (Moschovakis [1974]) and the various fix-point extensions of lower predicate calculus, cf. Aho-Ullman [1979], Chandra-Harel [1982], Immerman [1982], Harel-Kozen [1982].

For example, on the natural numbers, $(\mathfrak{N},=,E^{\#})$-recursion is precisely the classical theory of hyperarithmetic computability of Kleene.

The global version of positive elementary induction is also interesting on finite structures; for example, recursion on the class $(\mathfrak{FSTR},=,E^{\#})$ gives another characterization of polynomial time computability.

The last example in this section is quite different and suggests some of the connections of recursion theory with other areas of logic.

(2B.19) Set recursion. Let V be a transitive collection of sets with sufficiently strong closure properties (this can be made precise), e.g. V could be the class of all sets; we set

$$\mathcal{V} = \langle V,\emptyset,\in,\{,\},\cup,\underset{\sim}{rep}\rangle,$$

where $\emptyset$ is the empty set (as a constant), $\in$ is the membership relation, $\{,\}$ is the (unordered) pair, $\cup$ is the union and rep is the functional which embodies the replacement operation, i.e.

$$\underline{rep}(x,p)\downarrow \Leftrightarrow (\forall t \in x)[p(t)\downarrow],$$
$$\underline{rep}(x,p) \simeq \{p(t) : t \in x\}.$$

Recursion in $\mathcal{V}$ was called E-recursion by Normann [1978] who introduced it (in a different way) and it provides a very natural notion of "refined definability" in set theory. It is intimately related to recursion in higher types of Kleene [1959] , [1963] and to the theory of admissible sets, and it has been developed quite extensively, e.g. see Sacks [1980], [????], Slaman [????].
One of its most interesting relativizations is power set recursion, i.e. recursion in the expansion

$$\mathcal{V}(p) = (\mathcal{V}, \omega, \text{pow}),$$

where ω is the least infinite ordinal and pow is the power function,

$$\text{pow}(x) = \{y : y \subseteq x\},$$

See Moss [1984].

2C. Explicit definability. The explicit functionals of a structure $\mathfrak{A} = \langle\Sigma,\Phi\rangle$ are just those which can be defined from the given functionals in Φ and the application functional ap, using composition, definition by cases and λ-substitution into the partial function arguments of the functionals in Φ (if and of them are "true" functionals). To make this precise, we will associate with each signature

$$\sigma = k, e_0, e_1, \ldots$$

of a structure a formal language $\text{EXP}(\sigma)$ of terms, which will define the explicit functionals on every structure with signature σ.

The basic sorts of $\text{EXP}(\sigma)$ are $0,1,\ldots,k-1$ if $k > 0$ and all the nonnegative integers $0,1,2,\ldots$ if $k = 0$. The recursion space sorts of $\text{EXP}(\sigma)$ are the (codes of) tuples

$$\langle \bar{b}_1, \ldots, \bar{b}_n \rangle,$$

where each $\bar{b}_i$ in a basic sort, including the empty tuple $\langle\ \rangle$; these are precisely the signatures of recursion spaces in the structures of signature σ. Similarly, the partial function sorts of $\text{EXP}(\sigma)$ are all pairs $\langle \bar{u}, \bar{v} \rangle$, where $\bar{u}$ and $\bar{v}$ are recursion space sorts.

$\text{EXP}(\sigma)$ will use the fixed, distinct, reserved symbols and strings

$$(\quad) \quad , \quad \text{if} \quad \text{then} \quad \text{else}$$
$$\underset{\sim}{0} \quad \underset{\sim}{1} \quad \text{Booleq} \quad \text{first} \quad \text{rest},$$

an infinite list of distinct <u>basic</u> <u>variables</u>

$$v_0^b, v_1^b, \ldots$$

for each basic sort b, an infinite list of <u>partial</u> <u>function</u> <u>variables</u>

$$p_0^{\langle\bar{u},\bar{v}\rangle}, p_1^{\langle\bar{u},\bar{v}\rangle}, \ldots$$

for each partial function space sort $\langle\bar{u},\bar{v}\rangle$ and a <u>constant</u> <u>functional</u> <u>symbol</u> $\underset{\sim}{\varphi}_i$ of formal signature e_i, for each e_i in σ (if any).

If $u_1,\ldots,u_n$ are variables of basic sorts $\bar{b}_1,\ldots,\bar{b}_n$, we will call the tuple

$$u = u_1,\ldots,u_n$$

a <u>variable</u> <u>of</u> <u>sort</u> $\langle\bar{b}_1,\ldots,\bar{b}_n\rangle$. This allows us to avoid introducing separate symbols for variables over recursion spaces.

This completes the listing of the alphabet of EXP(σ). It goes without saying that in specific cases we will use variables and names for functionals which are less cumbersome and naturally associated with the intended interpretation, e.g. i,j,... for integers, +, · for specific functions, etc. Variables of partial function sort (<u>partial</u> <u>function</u> <u>variables</u>) will be denoted usually by p, q, r or descriptive words like "<u>loop</u>".

If G has signature σ, we will also call EXP(σ) the language of G,

$$\text{EXP}(G) = \text{EXP}(\sigma).$$

The terms of EXP(σ) <u>do</u> <u>not</u> <u>include</u> <u>all</u> <u>the</u> <u>variables</u>; they are defined by the following induction which also assigns a <u>recursion</u> <u>space</u> <u>sort</u> and a <u>set</u> <u>of</u> <u>the</u> <u>variables</u> to each of them.

If $\bar{u}_1,\ldots,\bar{u}_n$ are (codes of) tuples, we will let below

$$\bar{u}_1 * \cdots * \bar{u}_n$$

be the code of their concatenation.

(1) The constants $\underset{\sim}{0}$ and $\underset{\sim}{1}$ and all the Boolean variables are terms of (Boolean) sort $\langle 0\rangle$; $\underset{\sim}{0}$ and $\underset{\sim}{1}$ have no free variables, while each v_i^0 is free in the term v_i^0.

(2) If $t_1,\ldots,t_n$ are terms of sorts $\bar{u}_1,\ldots,\bar{u}_n$ respectively, then

$$(t_1,\ldots,t_n)$$

is a term of sort $\bar{u}_1 * \cdots * \bar{u}_n$ whose free variables are those which are free in one of $t_1,\ldots,t_n$; if s and t are terms of Boolean sort, so is

$$\underline{\text{Booleq}}(s,t),$$

and its free variables are those which are free in s or t; if t is a term of sort $\langle \bar{u}_1,\ldots,\bar{u}_n\rangle$ with $n \geq 2$, then

$$\text{first}(t) \qquad \text{rest}(t)$$

are terms of respective sorts $\langle \bar{u}_1\rangle$, $\langle \bar{u}_2,\ldots,\bar{u}_n\rangle$ and free variables those of t.

(3) If $t_1,\ldots,t_n$ are terms or variables of sorts $\bar{u}_1,\ldots,\bar{u}_n$ and p is a partial function space variable of sort $\langle \bar{u},\bar{v}\rangle$, with $\bar{u} = \bar{u}_1 * \cdots * \bar{u}_n$, then

$$p(t_1,\ldots,t_n)$$

is a term of sort $\bar{v}$ with free variables p, the variables among $t_1,\ldots,t_n$ and the variables which occur free in some t_i.

(4) If s is a Boolean term and t_1, t_2 are terms of the same sort $\bar{u}$, then

$$(\text{if } s \text{ then } t_1 \text{ else } t_2)$$

is a term of sort $\bar{u}$ with free variables those of s, t_1 and t_2.

For the last clause of λ-substitution we need the auxiliary notion of a λ-term.

(5a) If t is a term of sort $\bar{v}$ and u is a variable of sort $\bar{u}$, then

$$(\lambda u t)$$

is a λ-term of sort $\langle \bar{u},\bar{v}\rangle$ and free variables those of t except for the basic variables in the list u.

(5b) If $\underset{\sim}{\varphi}$ is a constant functional symbol of formal signature $\langle \bar{x}_1,\ldots,\bar{x}_n,\bar{w}\rangle$ and $t_1,\ldots,t_n$ are terms, basic variables, partial function variables or λ-terms of sorts $\bar{x}_1,\ldots,\bar{x}_n$ respectively, then $\underset{\sim}{\varphi}(t_1,\ldots,t_n)$ is a term of sort $\bar{w}$ and free variables the variables among $t_1,\ldots,t_n$ and the variables which occur free in some t_i.

A variable $u = (u_1,\ldots,u_n)$ of recursion space sort is free in t if $u_1,\ldots,u_n$ are all free in t.

The variables which occur in a term t but are not free in t are <u>bound</u> in t; notice that in EXP(σ) we have no bound partial function variables. Terms with no free variables are <u>closed</u>.

The only minor subtlety (or peculiarity) of this definition is that we have not included the basic variables among the terms. We will comment on the reason for this after explaining the obvious semantics of $EXP(\sigma)$.

Let $\mathfrak{G} = \langle \Sigma, \Phi \rangle$ be a structure of signature σ. A <u>valuation in</u> $\mathfrak{G}$ is a function α defined on the variables of $EXP(\sigma)$, such that if u is a variable of sort $\bar{u}$, then $\alpha(u)$ is an object of signature $\bar{u}$ in $\mathfrak{G}$ - e.g. if u is Boolean then $\alpha(u) \in \{\underset{\sim}{0}, \underset{\sim}{1}\}$ (or, more pedantically, $\alpha(u)$ is a one-tuple from $\{\underset{\sim}{0}, \underset{\sim}{1}\}$), and if p is a partial function variable of sort $\langle \bar{u}, \bar{v} \rangle$, then $\alpha(p) : U \to V$, where U, V are the recursion spaces of signatures $\bar{u}$ and $\bar{v}$.

We define a partial function

$$\mathrm{Val}(\mathfrak{G}, \alpha, t) \simeq \mathrm{Val}(\alpha, t)$$

on the variables, terms and λ-terms of $EXP(\sigma)$ by the following induction, corresponding to (1)-(5) above.

(1)' If t is a variable, then $\mathrm{Val}(\alpha,t) \simeq \alpha(t)$; $\mathrm{Val}(\alpha,\underset{\sim}{0}) \simeq (\underset{\sim}{0})$; $\mathrm{Val}(\alpha,\underset{\sim}{1}) \simeq (\underset{\sim}{1})$.

(2)' $\mathrm{Val}(\alpha,(t_1,\ldots,t_n)) \simeq (\mathrm{Val}(\alpha,t_1),\ldots,\mathrm{Val}(\alpha,t_n))$;

$$\mathrm{Val}(\alpha,\underline{\mathrm{Booleq}}(s,t)) \simeq \begin{cases} \underset{\sim}{1} & \text{if } \mathrm{Val}(\alpha,s) \simeq \mathrm{Val}(\alpha,t), \\ \underset{\sim}{0} & \text{if } \mathrm{Val}(\alpha,s),\ \mathrm{Val}(\alpha,t) \text{ are defined} \\ & \text{and } \mathrm{Val}(\alpha,s) \neq \mathrm{Val}(\alpha,t); \end{cases}$$

if $\mathrm{Val}(\alpha,t) \simeq (u_1,\ldots,u_n), n \geq 2$, then

$$\mathrm{Val}(\alpha,\underline{\mathrm{first}}(t)) \simeq (u_1);\ \mathrm{Val}(\alpha,\mathrm{rest}(t)) \simeq (u_2,\ldots,u_n).$$

(3)' $\mathrm{Val}(\alpha,p(t)) \simeq \alpha(p)(\mathrm{Val}(\alpha,t))$.

(4)' $$\mathrm{Val}(\alpha,(\text{if } s \text{ then } t_1 \text{ else } t_2)) \simeq \begin{cases} \mathrm{Val}(\alpha,t_1) & \text{if } \mathrm{Val}(\alpha,s) \simeq \underset{\sim}{1}, \\ \mathrm{Val}(\alpha,t_2) & \text{if } \mathrm{Val}(\alpha,s) \simeq \underset{\sim}{0}. \end{cases}$$

(5a)' $$\mathrm{Val}(\alpha,(\lambda ut)) = p$$

where $p : U \to V$ is the partial function on the space U of signature the sort of u, defined by

$$p(u') \simeq \mathrm{Val}(\mathrm{subst}(\alpha,u/u'),t).$$

Here, of course, $\mathrm{subst}(\alpha,u/u')$ is the valuation which agrees with α on all the variables, except that

$$\mathrm{subst}(\alpha,u/u')(u) = u'.$$

(5b)' $$\mathrm{Val}(\alpha,\underset{\sim}{\varphi}_i(t_1,\ldots,t_n)) \simeq f_i(\mathrm{Val}(\alpha,t_1),\ldots,\mathrm{Val}(\alpha,t_n)),$$

where f_i is the i'th functional in the structure $\mathfrak{G}$.

Using this routine definition, one can verify routinely the usual properties of semantics one would expect - that $\mathrm{Val}(\alpha,t)$ depends only on the values of α on the free variables of t, replacement and substitution properties, etc.

A functional

$$f : X_1 \times \cdots \times X_n \to W$$

on the universe of a structure $\mathfrak{A}$ of signature σ is explicit in $\mathfrak{A}$ if there is a term t of $\mathrm{EXP}(\sigma)$ whose free variables are among $x_1,\ldots,x_n$, these being of the appropriate sorts, so that

$$\alpha(x_1') = x_1 \ \& \cdots \& \ \alpha(x_n') = x_n \Rightarrow f(x_1',\ldots,x_n') \simeq \mathrm{Val}(\alpha,t).$$

The collection

$$\text{(2C.1)} \qquad \underset{\sim}{\mathrm{EXP}}(\mathfrak{A}) = \text{all explicit functionals on } \mathfrak{A}$$

contains all the functionals in $\mathfrak{A}$ and the application functionals $\underset{\sim}{ap}$ and it has all the right closure properties, including closure under composition, definition by cases and λ-substitution - e.g. if

$$f(x,y) \simeq g(x,\lambda u h(y,u))$$

and g, h are explicit, so is f.

Since we did not admit all basic variables as terms, the identity functions $id : S_i \to S_i$ on basic sets,

$$\text{(2C.2)} \qquad id(x) = x$$

are not necessarily explicit; neither are the basic equality relations on the basic sets. This is obviously a matter of choice, and in many cases it is useful to include id and $=$ among the givens of a structure. On the other hand, there are cases where id and $=$ are recursive but not explicit and where this is an accurate modelling of the situation, in terms of the natural algorithms that we wish to consider as given.

To see that the identity function, which is not explicit on $\mathfrak{N}$ is explicit on $(\mathfrak{N},\underset{\sim}{\text{while}})$, define first the iteration functional using the $\underset{\sim}{\text{while}}$,

$$\text{(2C.3)} \qquad \underset{\sim}{\text{iter}}(n,u,p) \simeq \underset{\sim}{\text{while}}(\text{test},n,u,\lambda iv(\text{pred}(i),p(v)),\lambda ivp(v))$$

where

$$\text{test}(i,v) = (\text{if } i = 0 \text{ then } \underset{\sim}{0} \text{ else } \underset{\sim}{1}).$$

It is clear that for $p : U \to U$ with $U = \mathbb{N} \times \cdots \times \mathbb{N}$ any product of copies of the integers,

(2C.4) $$\underline{iter}(n,u,p) \simeq p^{n+1}(u).$$

Thus we have

(2C.5) $$id(n) = (\text{if } n = 0 \text{ then } 0 \text{ else } iter(pred(n),0,succ)).$$

If we think of the identity function as <u>copying the input to output</u>, then the present analysis is not as silly as it might appear to be at first sight: there are cases where one may grant this copying for free, while in other situations one may wish to model the fact that the copying takes "time".

In the "baby programming" example (2B.17), we put in the structure the identities on the simple basic sets <u>int</u> and <u>char</u>, but we purposely left out the more complicated identity functions on arrays and files. These too, of course, are easily explicit in $\mathfrak{BP}$, which has the <u>while</u> functionals on its simple basic sets, but their "computation" uses the <u>while</u>.

The same remarks apply to the equality relations on basic sets.

The notion of explicit definability extends immediately to classes of structures: a global functional for a class $\mathcal{K}$ is <u>explict on</u> $\mathcal{K}$ if there is a fixed term t of EXP which defines $f^{\mathfrak{A}}$ on each $\mathfrak{A}$ in $\mathcal{K}$. We set

(2C.6) $$\underline{Exp}(\mathcal{K}) = \text{all explicit global functionals on } \mathcal{K}.$$

Using the notation of (2B.9) and (2B.10) then, we can talk of the <u>$\mathcal{K}$-explicit relational queries</u> in Sec $\underline{Exp}(\mathcal{K})$ and the <u>$\mathcal{K}$-semiexplicit relational queries</u> in Env $\underline{Exp}(\mathcal{K})$.

On the class $\mathfrak{FO}$ of first order structures, the explicit global functionals are quite trivial - they are all defined by cases from the usual (lower predicate calculus) terms of the language. On structures with true functionals, on the other hand, explicit definability can be quite strong as the next few results attest.

(2C.7) <u>Fact.</u> On the expansions ($\mathfrak{N}$,<u>while</u>) and ($\mathfrak{PLISP}$,<u>while</u>) of the structures defined in (2B.11) and (2B.13), the explicit functionals are exactly the (classical) recursive functionals. ⊣

This is just the identification of general recursive functions with μ-recursive functions, e.g. see Kleene [1952].

(2C.8) <u>Fact.</u> On the class $\mathfrak{BSTR}$ of finite structures which represent binary strings,

Sec $\underline{Exp}(\mathfrak{BSTR},\underline{while})$ = Env $\underline{Exp}(\mathfrak{BSTR},\underline{while})$ = the class of relational queries on $\mathfrak{BSTR}$ which are deterministically Turing computable in logspace.

Similarly, for $k \geq 1$,

$$\text{Sec } \underset{\sim}{\text{Exp}}(\mathcal{BSTR}^{(k)}, \underset{\sim}{\text{while}}) = \text{Env } \underset{\sim}{\text{Exp}}(\mathcal{BSTR}^{(k)}, \underset{\sim}{\text{while}}) = \text{the class of relational queries on strings which are deterministically Turing computable in space}(n^k). \dashv$$

These characterizations follow immediately from results of Immerman [1983] and Gurevich [1983], or they can be proved directly, quite easily.

If we use the ancestral functional of (2A.31), we can also formulate (and easily establish directly) in our context many of the characterizations of non-deterministic classes in Immerman [1983].

(2C.9) Fact. Env $\underset{\sim}{\text{Exp}}(\mathcal{BSTR}, \underset{\sim}{\text{ancestral}})$ = nondeterministic logspace. Also, for each $k \geq 1$,

$$\text{Env } \underset{\sim}{\text{Exp}}(\mathcal{BSTR}^{(k)}, \underset{\sim}{\text{ancestral}}) = \text{nondeterministic space}(n^k).$$

(Immerman [1983].) $\dashv$

We should also mention two totally trivial facts which do not involve recursion, but point out how first and higher order definability can be represented in EXP, presumably to be compared with other notions of definability that do involve recursion.

(2C.10) Fact. If $\mathcal{FO}$ is the class of first order structures, then

$$\text{Env } \underset{\sim}{\text{Exp}}(\mathcal{FO}, =, E^{\#}) = \text{Sec } \underset{\sim}{\text{Exp}}(\mathcal{FO}, =, E^{\#}) = \text{the elementary (first-order definable) relational queries in } \mathcal{FO}. \dashv$$

Explicit definability on $(\mathcal{FO}, =, E^{\#})$ is better understood if we introduce in EXP the following obvious abbreviations

$$(2C.11) \quad \begin{cases} \dot\neg(t) \equiv (\text{if } t \text{ then } \underset{\sim}{0} \text{ else } \underset{\sim}{1}) \\ (t \mathbin{\dot\&} s) \equiv (\text{if } t \text{ then } s \text{ else } \underset{\sim}{0}) \\ (t \mathbin{\dot\vee} s) \equiv (\text{if } t \text{ then } \underset{\sim}{1} \text{ else } s) \\ (\exists^{\#}x)t \equiv E^{\#}(\lambda x t) \\ (\forall^{\#}x)t \equiv \dot\neg(\exists^{\#}x)\dot\neg t \\ t = s \equiv \chi_{=}(t,s) \end{cases}$$

Now the Boolean terms look just like formulas of first order logic and their interpretation (on defined arguments) is precisely that of logic. The "#" is needed on the quantifiers because there are other functionals which can represent quantification; e.g. using the half-quantifier $E^{1/2}$, we might also abbreviate

$$(\exists^{1/2}x)t \equiv E^{1/2}(\lambda x t). \tag{2C.12}$$

(2C.13) Fact. Let

$$\mathfrak{BSJR}^{(\infty)} = \bigcup_k \mathfrak{BSJR}^{(k)}$$

be the class of structures of the form

$$\mathfrak{A} = \langle \{0,1,\ldots,n-1\}, F_1, F_2, \ldots, ap_1, ap_2, \ldots, assign_1, assign_2, \ldots, 0, succ, pred, \chi_{=0}, \sigma \rangle$$

as in (2B.14), and let $(\mathfrak{BSJR}^{(\infty)}, E^{1/2})$ be the class obtained by adjoining $E_S^{1/2}$ for each basic set to all the structures in $\mathfrak{BSJR}^{\infty}$. Then

$$\text{Exp } \underline{\text{Env}}(\mathfrak{BSJR}^{\infty}, E^{1/2}) = \text{all } \Sigma_1^1 \text{ relational queries on strings}$$
$$= \text{nondeterministic polynomial time.} \quad \dashv$$

This is just a translation in our notation of the basic characterization of NP in Fagin [1974].

2D. Recursion. The recursive functionals on a structure $\mathfrak{A}$ are those which can be constructed from the given functional of $\mathfrak{A}$ using explicit definitions and the (simultaneous)-fixed-point operation. To make this precise and deal effectively with the global theory of recursive definability, we will introduce here the language of recursion REC.

For each signature σ, the alphabet of REC(σ) is that of EXP(σ) (defined in 2C) augmented by the following additional reserved strings and symbols:

rec : [ ]

The terms of REC(σ) are defined inductively, by clauses (1)-(5) in the definition of the terms of EXP(σ) and the following additional clause.

(6) If $t_1,\ldots,t_n$ are terms of sorts $\bar{v}_1,\ldots,\bar{v}_n$ and $u_1,\ldots,u_n$ are variables of sorts $\bar{u}_1,\ldots,\bar{u}_n$ and $p_1,\ldots,p_n$ are partial function variables of sorts $\langle\bar{u}_1,\bar{v}_1\rangle,\ldots,\langle\bar{u}_n,\bar{v}_n\rangle$ and u_1^* is a term or variable of sort $\bar{u}_1$, then the string

$$(\text{rec } u_1,p_1,\ldots,u_n,p_n : u_1^*)[t_1,\ldots,t_n] \tag{2D.1}$$

is a term of sort $\bar{v}_1$. The free variables of this term are the free variables of u_1^* and the free variables of each t_i other than $u_i,p_1,\ldots,p_n$.

Consider for example the term

```
t ≡ (rec s,loop : s*)[if test(s) then loop (action(s)) else value(s)]
         |  |      |         |    |            |        |      |             |    |
         b  b      f         f    b            b        f      b             f    b
```

where we have identified the free and bound occurrences of the variables. In the interpretation, this will take the value of

$$\text{while}(\text{test}, s^*, \text{action}, \text{value})$$

which depends of course on the values of the free variables test, action, value and s^*, which may be a term or a variable.

For each structure $\mathfrak{A}$ of signature σ, we extend the definition of the partial function Val to the terms of $\mathrm{REC}(\sigma)$ by adding the following clause to (1)'-(5)' of 2C.

(6)' If t is a recursive term as in (2D.1), let $U_1,\ldots,U_n$, $V_1,\ldots,V_n$ be thr recursion spaces in $\mathfrak{A}$ with signatures $\bar{u}_1,\ldots,\bar{u}_n$, $\bar{v}_1,\ldots,\bar{v}_n$ and for the given valuation α consider the operative system of functionals

$$f_1(u_1', p_1',\ldots,p_n') \simeq \mathrm{Val}(\mathrm{subst}(\alpha, u_1/u_1', p_1/p_1',\ldots,p_n/p_n')),$$
$$\ldots$$
$$f_n(u_n', p_1',\ldots,p_n') \simeq \mathrm{Val}(\mathrm{subst}(\alpha, u_n/u_n', p_1/p_1',\ldots,p_n/p_n')),$$

where

$$\beta_i = \mathrm{subst}(\alpha, u_i/u_i', p_1/p_1',\ldots,p_n/p_n')$$

is the valuation which agrees with α on all the variables, except that

$$\beta(u_i) = u_i', \beta(p_1) = p_1',\ldots,\beta(p_n) = p_n'.$$

we set

$$\mathrm{Val}(\alpha,t) \simeq f_1^\infty(\mathrm{Val}(\alpha,u_1^*)).$$

The notation is a bit dense, but the idea is clear: the conditions on the formation rule (6) insure that the terms $t_1,\ldots,t_n$ define an operative system in the variables $u_1,\ldots,u_n$, $p_1,\ldots,p_n$ and the value of t is the value of the principal fixed point of that operative system at u_1^*.

Consider, for example the following closed term in the language for graphs, expanded by the quantifier $E^\#$ and $=$,

(2D.2) $\quad \mathrm{conn} \equiv (\mathrm{rec}\ \mathrm{conn}, x, y, \mathrm{transclos}:)$
$\quad\quad [\text{if } \forall^\# x \forall^\# y\ \mathrm{transclos}(x,y) \text{ then } \underset{\sim}{1} \text{ else } \mathrm{conn}(\ ),$
$\quad\quad \text{if } x = y \text{ then } \underset{\sim}{1} \text{ else } \exists^\# z(\mathrm{transclos}(x,z)\ \&\ R(z,y))].$

Here we have used the abbreviations introduced in (2C.11) as well as other obvious misspellings, e.g. "x = y" for $\chi_{=}(x,y)$, omitting parentheses and the like. Notice that <u>conn</u> here varies over partial functions on $\mathbb{I}$, i.e. with just the empty tuple for argument. From the analysis in 1D, the term conn is defined on a graph $\mathcal{G}$ (and is then = $\underset{\sim}{1}$) exactly when $\mathcal{G}$ is connected.

We could also define the transitive closure first, by a simple recursion, and then get conn explicitly from it:

$$\text{(2D.3)}\quad \text{transclos}(x,y) \equiv (\text{rec } x,y,\underline{\text{transclos}}{:}x,y)\ [\text{if } x = y \text{ then } \underset{\sim}{1} \text{ else } \exists^{\#}z(\underline{\text{transclos}}(x,z)\ \&\ (z,y))],$$

$$\text{(2D.4)}\quad \text{conn} \equiv \forall^{\#}x\forall^{\#}y\ \text{transclos}(x,y).$$

Notice that x and y have both free and bound occurrences in transclos(x,y).

The linear representation of the gcd discussed in 1B gives a good example of simultaneous recursion:

$$\text{(2D.5)}\quad \begin{aligned} A(x,y) \equiv (&\text{rec } x,y,\underline{A},x,y,\underline{B}{:}x,y)\\ &[\text{if rem}(x,y) = 0 \text{ then } 0 \text{ else } \underline{B}(y,\text{rem}(x,y)),\\ &\ \text{if rem}(x,y) = 0 \text{ then } 1\\ &\qquad \text{else } \underline{A}(y,\text{rem}(x,y)) - \underline{B}(y,\text{rem}(x,y)) * \text{quot}(x,y)]. \end{aligned}$$

This is a formal term in a suitable language for the integers (positive and negative) where rem, quot, etc. are either taken as primitives or have been already defined.

A functional

$$f : X_1 \times \cdots \times X_n \to W$$

on the universe of $\mathcal{G}$ is <u>recursive</u>, if there is a term t of REC(σ) with free variables $x_1,\ldots,x_n$ of the appropriate sorts, such that for each valuation α,

$$\alpha(x_1) = x_1' \ \&\ \cdots\ \&\ \alpha(x_n) = x_n' \Rightarrow f(x_1,\ldots,x_n) \simeq \text{Val}(\alpha,t).$$

We let

$$\text{(2D.6)}\quad \underset{\sim}{\text{Rec}}(\mathcal{G}) = \text{all recursive functionals on } \mathcal{G}$$

and for each class $\mathcal{K}$ of structures of the same signature we let

$$\text{(2D.7)}\quad \underset{\sim}{\text{Rec}}(\mathcal{K}) = \text{all } \mathcal{K}\text{-recursive global functionals},$$

where a global functional is $\mathcal{K}$-recursive if it can be defined uniformly on all the structures of $\mathcal{K}$ by a fixed term of REC(σ).

A relation $R \subseteq U$ is <u>recursive</u> on $\mathcal{G}$ if its characteristic function χ_R is recursive and <u>semirecursive</u> on $\mathcal{G}$ if its semicharacteristic (partial) function

sX_R is recursive - see (2B.6) and (2B.7) for the definitions of X_R, sX_R. In the notation of sections and envelopes introduced in (2B.9), (2B.10)

(2D.8) $$\text{Sec } \underline{\text{Rec}}(\mathfrak{A}) = \text{all recursive relations on } \mathfrak{A}$$

(2D.9) $$\text{Env } \underline{\text{Rec}}(\mathfrak{A}) = \text{all semirecursive relations on } \mathfrak{A}$$

and similarly for the global objects, Sec $\underline{\text{Rec}}(\mathcal{K})$ and Env $\underline{\text{Rec}}(\mathcal{K})$.

$\underline{\text{Rec}}(\mathfrak{A})$ is closed under explicit definability and the taking of simultanous fixed points, Sec $\underline{\text{Rec}}(\mathfrak{A})$ is closed under &, $\vee$, $\neg$, substitutions of total recursive functions, etc. and Env $\underline{\text{Rec}}(\mathfrak{A})$ is closed under & and substitution of recursive partial functions - these are all trivial facts.

On the other hand:

(2D.10) Fact. There exists a recursion structure $\mathfrak{A}$ where Env $\underline{\text{Rec}}(\mathfrak{A})$ is not closed under $\vee$. There exists an expansion $(\mathfrak{N}, f)$ of the integers by a single partial function, whose recursive sets are exactly those of $\mathfrak{N}$, but where there are semirecursive sets with semirecursive complements which are not recursive. ⊣

These counterexamples stem from the fact that the present theory is completely deterministic - any nondeterminism which is desired must be deliberately put in the structures, for example by adjoining the functional $E^{1/2}$ of (2A.17).

We have already mentioned that the recursive functionals on $\mathfrak{N}$ and PLJSP are precisely the classical recursive functionals.

For each first order structure $\mathfrak{A}$, let

(2D.11) $$\text{Ind}(\mathfrak{A}) = \text{Env } \underline{\text{Rec}}(\mathfrak{A}, =, E^{\#})$$

and for a class of structures $\mathcal{K}$,

(2D.12) $$\text{Ind}(\mathcal{K}) = \text{Env } \underline{\text{Rec}}(\mathcal{K}, =, E^{\#}).$$

(2D.13) Fact. Ind($\mathfrak{A}$) consists precisely of the (absolutely) positive elementary inductive relations on $\mathfrak{A}$, so in particular,

$$\text{Ind}(\mathfrak{N}) = \text{the classical } \Pi^1_1 \text{ relations on } N. \quad ⊣$$

This result holds globally, with the natural global definition of positive elementary induction. On finite structures, this class of queries has turned out to be quite interesting, primarily because of the following two results.

(2D.14) Fact. The negation of every inductive relational query on the class of finite first order structures is also inductive. (Immerman [1982]). ⊣

We will outline a proof of this result in (2F.18). Of course it fails on infinite structures, in general, e.g. it fails on $\mathfrak{N}$.

(2D.15) Fact. For the class $\mathcal{BSTR}$ of finite structures which represent binary strings,

$$\begin{aligned} \text{Sec } \underset{\sim}{\text{Rec}}(\mathcal{BSTR}) &= \text{Env } \underset{\sim}{\text{Rec}}(\mathcal{BSTR}) \\ &= \text{Ind}(\mathcal{BSTR}) \\ &= \text{the deterministically polynomial} \\ &\quad \text{time computable queries.} \end{aligned}$$ ⊣

Thss result, which exhibits the known "robustness" of PTIME in the present context has been realized independently by many persons. Immerman [1982] proves $\underset{\sim}{\text{Ind}}(\mathcal{BSTR})$ = PTIME and Gurevich [1983] has a variant of Sec $\underset{\sim}{\text{Rec}}(\mathcal{BSTR})$ = PTIME. The present version was established jointly with M. deRougemont, see deRougemont [1983]. (Cf. also the earlier paper Sazonov [1980]).

Like the similar characterizations (2C.8) and (2C.9), this characterization of PTIME has a simple proof and sheds no light on the P = NP question. It gives some evidence that the abstract theory of recursion gives simple and elegant "presentations" of natural notions of computability.

There are also many interesting specific results about inductive definability on classes of finite structures, e.g.:

(2D.15) Fact. On the class of all finite graphs, the Boolean query

$$\text{hamilt}^{\mathcal{G}} \Leftrightarrow \mathcal{G} \text{ is Hamiltonian}$$

is neither inductive nor coinductive (deRougemont [1983]). ⊣

Two terms t and s of REC(σ) are denotationally equivalent if for every structure $\mathfrak{A}$ and every valuation α in $\mathfrak{A}$,

$$\text{Val}(\alpha,t) \simeq \text{Val}(\alpha,s).$$

(2D.17) REDUCTION THEOREM. Every term t of the language of recursion REC can be reduced effectively to a denotationally equivalent term t^* of the form

$$(*) \qquad (\text{rec } \underline{\text{out}}, u_1, p_1, \ldots, u_n, p_n :)[t_0, t_1, \ldots, t_n]$$

where $t_0, t_1, \ldots, t_n$ are explicit and $\underline{\text{out}} : I \to W$ for some W. ⊣

Terms of the form (*) with explicit $t_0, t_1, \ldots, t_n$ are called simple and the main import of the reduction theorem is to establish the closure properties of the class of simple recursive functionals, i.e. functionals definable by simple terms. After the fact, all recursive functionals are simple; sometimes the theory is developed by taking the notion of "simple recursive" as basic and then proving the closure properties of this class, which ultimately establish that every term of REC defines a simple recursive functional.

As it stands, the reduction theorem can be viewed as a normal form theorem for recursive functionals. Of considerably more interest is the proof of (2D.17) the specific reductions which are used and which appear to preserve not only the denotation of a term but its intension, i.e. the "algorithm" described by the term. We will come back to this important point in the next section.

The reduction theorem is a bit easier to understand - and accept as almost trivial - for the case of first order structures, where (for example) it includes as a special case the classical First Recursion Theorem of Kleene [1952] on $\mathfrak{N}$. The general case, for structures with functionals which may be discontinuous was apparently announced first in Moschovakis [1976]. (Cf. also the Induction Completeness Theorem of Moschovakis [1974] and Theorem 12 in Moschovakis [1969].)

There are many interesting subclasses of the recursive functionals which deserve study. Here we will restrict ourselves to a few facts concerning the while functional and fixed points.

Since while is nothing but the abstract version of the least number operator, it is natural to think of explicit definability in while as a limited notion of recursion, the abstract version of μ-recursion on $\mathfrak{N}$. By (2C.7),

$$\underline{\text{Exp}}(\mathfrak{N},\underline{\text{while}}) = \underline{\text{Rec}}(\mathfrak{N})$$

and the same is true for $\mathfrak{PLISP}$. On the other hand, by (2C.8) and (2D.15), on $\mathfrak{BSTR}$

$$\text{Env } \underline{\text{Exp}}(\mathfrak{BSTR},\underline{\text{while}}) = \text{DET LOGSPACE}$$

$$\text{Env Rec}(\mathfrak{BSTR}) = \text{DET PTIME}$$

and the equality of these two classes is quite unlikely.

It is easy to find specific examples where explicitness in while can be proved to be distinct from resursiveness.

(2D.18) Fact. Env $\underline{\text{Exp}}(\mathfrak{N},=,E^{\#},\underline{\text{while}})$ consists precisely of all arithmetical relations, while Env $\underline{\text{Rec}}(\mathfrak{N},=,E^{\#}) = \Pi^1_1$. (Kleene [1959]). ⊣

(2D.19) Fact. Let $\mathcal{K}$ be the collection of all finite structures of the form

$$\mathfrak{G} = \langle T,\ell,r,\text{root},\text{terminal},R\rangle,$$

where $\langle T,\ell,r,\text{root},\text{terminal}\rangle$ is a binary tree on the modes T, with a root,

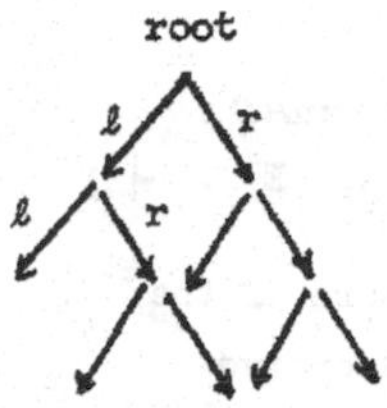

the functions ℓ and r give the children and terminal (x) is true on the terminal nodes (when $\ell(x) = r(x) = x$), and where R is an arbitrary unary relation on the nodes. The subclass

$$\mathcal{K}' = \{\mathfrak{A} \in \mathcal{K} : \mathfrak{A} \models \forall x R(x)\}$$

is recursive on $\mathcal{K}$ but not even semiexplicit on $(\mathcal{K}, \underline{\text{while}})$, so that

$$\text{Sec}\ \underline{\text{Exp}}(\mathcal{K}, \underline{\text{while}}) \subsetneq \text{Sec}\ \underline{\text{Rec}}(\mathcal{K}). \qquad \dashv$$

A partial function

$$f : U \to V$$

on recursion spaces U, V of a structure $\mathfrak{A}$ is a <u>fixed point</u> of $\mathfrak{A}$ if there exists an explicit operative functional $\varphi(u,p)$ such that

$$f(u) \simeq \varphi^{\infty}(u).$$

These are precisely the partial functions defined by terms of REC of the form

$$(\text{rec}\ u, p{:}u)t$$

where t is an explicit term with no free variables other than u and p. We let

(2D.20) $$\underline{\text{Fix}}(\mathfrak{A}) = \text{all fixed points of } \mathfrak{A}$$

and we define the global version $\underline{\text{Fix}}(\mathcal{K})$ in the obvious way.

The fixed points of many structures are a complicated and little understood class. In some of the partial results below there is a hint that these partial functions may have interesting algebraic properties.

(2D.21) <u>Fact.</u> There is a recursive, total function in $(\mathbb{N}, =, E^{\#})$, (i.e. a classical hyperarithmetic function on N) which is not a fixed point of $(\mathbb{N}, =, E^{\#})$. (Feferman [1965]). $\dashv$

This old result of Feferman's uses Cohen forcing in the language of arithmetic for its proof. In Exercises 8.13, 8.14 of Moschovakis [1974] the result was claimed for all "acceptable" structures, but the hint given these does not work and the question is still open.

(2D.22) <u>Fact.</u> On the class of finite graphs, the relational query

$$\text{transclos}^{\mathcal{G}}(x,y) \Leftrightarrow \text{there is a path in } \mathcal{G} \text{ joining } x \text{ to } y$$

is inductive, its negation $\neg$ transclos(x,y) is also inductive by Immerman's (2D.14), but $\neg$ transclos(x,y) is not a fixed point of (Finitegraphs,=,$E^{\#}$). ⊣

(2D.23) Fact. For every expansion $(\mathfrak{N}, f_1,\ldots,f_n)$ of the structure of the integers by total recursive functions $f_1,\ldots,f_n$, there are toral recursive functions on N which are not fixed points of $(\mathfrak{N}, f_1,\ldots,f_n)$. ⊣

(2D.24) Fact. The function $x \mapsto x^2$ is not a fixed point of $\mathfrak{N}$ or even $(\mathfrak{N},\text{id},=)$; in fact, if $f(x_1,\ldots,x_n)$ is an n-ary, totally defined fixed point of $(\mathfrak{N},\text{id},=)$, then there is a number M and n-tuples $(x_1,\ldots,x_n)$ with arbitrarily large $\min(x_1,\ldots,x_n)$ such that

$$f(x_1,\ldots,x_n) \leq M \cdot (\max(x_1,\ldots,x_n))^n.$$

In addition, there exist total, recursive functions with values ≤ 1 which are not fixed points of $(\mathfrak{N},\text{id},=)$ (McColm [1985]). ⊣

2E. Intensional semantics for REC. The denotational semantics for REC defined in the preceding section assigns to each term t of REC(σ) and each structure $\mathfrak{A}$ of signature σ a recursive functional

$$f^{\mathfrak{A}} : X_1 \times \cdots \times X_n \to W;$$

here W is the recursion space with signature the sort of t and $X_1,\ldots,X_n$ have the signatures $\bar{x}_1,\ldots,\bar{x}_n$ of the free variables of t, enumerated in some fixed order. We call

$$f^{\mathfrak{A}} = \text{den}(\mathfrak{A},t) \tag{2E1o}$$

the denotation of t on $\mathfrak{A}$, and for each class $\mathcal{K}$ of structures of signature σ, we will call the corresponding global functional the $\mathcal{K}$-denotation of t, den($\mathcal{K}$,t).

Terms with the same denotation may describe essentially different algorithms: e.g. both the mergesort and the bubblesort algorithms on arrays of integers are easily defined in REC($\mathcal{BP}$) by terms which have the same denotation, the sorting function on arrays.

The more refined intensional semantics for REC assigns to each term t of REC(σ) and each structure $\mathfrak{A}$ of signature σ an operative system of functionals

$$\text{int}(\mathfrak{A},t) = [f_1,\ldots,f_n]^{\mathfrak{A}} \tag{2E.2}$$

on $\mathfrak{A}$, <u>the intension of</u> t <u>on</u> $\mathfrak{A}$, which computes the denotation of t on $\mathfrak{A}$: in the notation of (2A.26)

$$\text{(2E.3)} \qquad \operatorname{den}(\mathfrak{A},t) = \overline{\operatorname{int}(\mathfrak{A},t)} = f_1^{\infty}.$$

The claim will be that the operative system int($\mathfrak{A}$,t) <u>embodies all the mathematical properties of the algorithm described by</u> t <u>on</u> $\mathfrak{A}$.

The definition of intensional semantics for REC(σ) comes down to a careful proof of the Reduction Theorem (2D.17). We assign to each term t of REC(σ) a term nf(t) of the form

$$\text{(2E.4)} \qquad (\text{rec } \underline{\text{out}}, u_1, p_1, \ldots, u_n, p_n:)[t_0, t_1, \ldots, t_n],$$

<u>the normal form of</u> t, and we take int($\mathfrak{A}$,t) to be the operative system of functionals (with parameters) defined by $t_0, t_1, \ldots, t_n$. Similarly, for each class $\mathcal{K}$ of structures of signature σ, the <u>$\mathcal{K}$-global intension</u> int($\mathcal{K}$,t) of t or $\mathcal{K}$ is the operative system of $\mathcal{K}$-global functionals defined by $t_0, t_1, \ldots, t_n$.

The terms $t_0, t_1, \ldots, t_n$ are explicit and of a special <u>simple basic</u> form, each of them expressing directly one of the givens of the structure. Moreover, the system $[t_0, t_1, \ldots, t_n]$ has some simple properties, e.g. the variable <u>out</u> does not actually occur in $t_0, t_1, \ldots, t_n$.

The reduction

$$\text{(2E.5)} \qquad t \mapsto \operatorname{nf}(t)$$

is defined by induction on the construction of the term t, and is naturally quite complicated to describe in full detail. Here we will concentrate on a few of the cases and on some examples which illustrate the basic idea. Most of the reductions we will use are already included in the <u>normal form for explicit functionals</u> of Kolaitis [1984].

(1) If t is $\underset{\sim}{0}$, $\underset{\sim}{1}$ or a Boolean variable, then

$$\text{(2E.6)} \qquad t \to (\text{rec } \underline{\text{out}}:)t.$$

Here <u>out</u> : $I \to \{\underset{\sim}{0},\underset{\sim}{1}\}$ and the trivial operative system defined by the constant or parameter t closes off in exactly one step.

(2) Suppose

$$t \equiv \operatorname{first}(s)$$

and by induction hypothesis

$$s \mapsto (\text{rec } \underline{\text{out}}_0, u_1, p_1, \ldots, u_n, p_n:)[s_0, s_1, \ldots, s_n].$$

We set

(2E.7) $t \mapsto (\text{rec } \underline{out}, \underline{out}_0, u_1, p_1, \ldots, u_n, p_n:)[\text{first}(\underline{out}_0(\ )), s_0, s_1, \ldots, s_n]$.

To get just a bit behind the symbolism in this trivial case, suppose $s \equiv s(x)$ has just one free variable, so that $\text{int}(\mathfrak{A}, s)$ defined by $s_0, s_1, \ldots, s_n$ looks like

$$f_0(\text{out}_0, p_1, \ldots, p_n; x)$$
$$f_1(u_1, \text{out}_0, p_1, \ldots, p_n; x)$$
$$\ldots$$
$$f_n(u_n, \text{out}_0, p_1, \ldots, p_n; x),$$

where by hypothesis, "out_0" is a dummy variable, such that none of $f_0, f_1, \ldots, f_n$ depends on its value. Now the value of $s(x)$ on $\mathfrak{A}$ is given by

$$f_0^\infty(x)$$

and the value of $t(x) \equiv \text{first}(s(x))$ is first $(f_0^\infty(x))$, which is precisely the principal fixed point of the system determined by the term $\text{nf}(t)$ in (2E.7), i.e. the value of $\text{int}(\mathfrak{A}, t)$. Moreover, the recursion determined by $\text{int}(\mathfrak{A}, t)$ closes just one stage after the recursion determined by $\text{int}(\mathfrak{A}, s)$, and computes t in the obvious way - by computing first the value of s and then applying "first" to it.

A similar trivial construction of $\text{nf}(t)$ can be given for all the other cases under (2) in the definition of terms and under (3). We put down the formula for case (4), just to have one more example for perusal.

(4) If

$$t \equiv (\text{if } s \text{ then } z \text{ else } w)$$

and

$$s \mapsto (\text{rec } \underline{out}_s, u_1^s, p_1^s, \ldots, u_n^s, p_n^s:)[s_0, s_1, \ldots, s_n]$$

and similarly with the terms z and w, with principal variables $\underline{out}_z$, $\underline{out}_w$, we set

(2E.8) $t \mapsto (\text{rec } \underline{out}, \underline{out}_s, \ldots, \underline{out}_z, \ldots, \underline{out}_w, \ldots:)$
$$[(\text{if } \underline{out}_s(\ ) \text{ then } \underline{out}_z(\ ) \text{ else } \underline{out}_w(\ )), s_0, \ldots, z_0, \ldots, w_0, \ldots].$$

(5) To take a special case, suppose

$$t \equiv \underline{\varphi}(x, \lambda v s(v)),$$

where x is a variable and $s(v)$ (which may have other free variables) has been reduced,

$$s(v) \mapsto (\text{rec } \underline{out}_0, u_1, p_1, \ldots, u_n, p_n :)$$
$$[s_0(v,p_1,\ldots,p_n), s_1(v,u_1,p_1,\ldots,p_n), \ldots, s_n(v,u_n,p_1,\ldots,p_n)].$$

We replace the partial function variables $out_0, p_1, \ldots, p_n$ by similarly named variables which also take v as argument and set

(2E.9) $$t \mapsto (\text{rec } \underline{out}, v, \underline{out}_0, v, u_1, p_1, \ldots, v, u_n, p_n :)$$
$$[\varphi(x, \underline{out}_0), s_0(v, \lambda u_1 p_1(v,u_1), \ldots, \lambda u_n p_n(v,u_n),$$
$$\ldots, s_n(v, \lambda u_1 p_1(v,u_1), \ldots, \lambda u_n p_n(v,u_n))].$$

The verification that this $nf(t)$ is denotationally equivalent to t and in fact that $int(G,t)$ computes the value of t on each G in the obvious, natural way is trivial, as in case (2) above.

(6) This is the interesting case of the construction, when

(2E.10) $$t \equiv (\text{rec } u_1, p_1, u_2, p_2, \ldots, u_n, p_n : u_1^*)[t_1, \ldots, t_n]$$

is defined by the recursion operator.

The general computation of the normal form for t from those of $u_1^*, t_1, \ldots, t_n$ is quite complicated here. We will confine ourselves to a very special and simple case, which contains the mathematically interesting part of the argument.

(2E.11) The Recursion Theorem. Suppose $g(u,q;v,p)$ is an operative functional on a universe Σ, where u and v vary over the recursion spaces U and V, let $u^* \in U$, $v^* \in V$ and suppose that the functional

$$f(v,p) \simeq g^\infty(u^*;v,p)$$

is also operative. Then

$$f^\infty(v^*) \simeq h\,(u^*,v^*),$$

where h is the operative functional defined by

$$h(u,v,r) \simeq g(u, \lambda u'\, r(u',v); v, \lambda v'\, r(u^*,v')). \qquad \dashv$$

This is the local version of the result and it is quite easy to verify for the general case, where g may be discontinuous.

Moreover, if one makes any reasonable assumptions about implementations and assumes that a procedure G for computing $g(u,q,v,p)$ is available (G operating on arbitrary values of u, v and procedures for computing q and p passed to it), then one can argue quite convincingly that the computation of $h^\infty(u^*,v^*)$ described by h is precisely the natural way to go about computing $f^\infty(v^*)$, using G.

The global version of the Recursion Theorem is quite elegant, if a bit opague on first sight.

(2E.12) Fact. Suppose $t(u,q,v,p)$ is a term of REC (perhaps with additional free variables) where u and v are distinct and do not occur free in the terms u^*, v^*, and assume that the sorts of u, q, v, p, u^*, v^* are such that the terms below make sense; then

$$(\text{rec } v,p{:}v^*)(\text{rec } u,q{:}u^*)t(u,q,v,p) \tag{2E.13}$$

is denotationally equivalent with

$$(\text{rec } u,v,r{:}u^*,v^*)t(u,\lambda u' r(u',v),v,\lambda v' r(u^*,v')). \tag{2E.14}$$ ⊣

To define the normal form of a general recursive term (2E.10), we use reductions which are somewhat messier but have the same general form as (2E.13) ↦ (2E.14). We then prove that $nf(t)$ has the same denotation as t and we argue that it describes the natural way that we would go about computing the value of t, making only direct appeals to assumed procedures that compute the givens of the structure.

A perusal of the formulas (2E.6), (2E.7), (2E.8), (2E.9) and (2E.14) gives a general idea of the very special simple basic terms which appear in normal forms. It is not hard to identify combinatorically these terms and give a precise definition of what it means for a term to be in normal form.

To illustrate the notions, let

$$\mathfrak{N}^{+\cdot} = (\mathfrak{N},+,\cdot)$$

be the expansion of the structure of the integers by the usual $+$ and $\cdot$.

Notice first that the two terms

$$x + y \quad \text{and} \quad y + x$$

have normal forms

$$(\text{rec } \underline{\text{out}}{:})x + y \quad \text{and} \quad (\text{rec } \underline{\text{out}}{:})y + x$$

and hence have the same intension, by the commutativity of $+$, even though they are syntactically distinct. This kind of "semantic input" into the definition of intensions becomes much more involved for complicated terms, and there is no way to identify intensions with syntactic objects.

On the other hand, it is easy to compute that

$$x \cdot (y + z) \mapsto (\text{rec } \underline{\text{out}}, P_+{:})[x \cdot P_+(\ ), y + z] \tag{2E.13}$$

and

$$(2E.14)\quad (x\cdot y)+(x\cdot z)\mapsto(\text{rec } \underline{out}, P_y, P_z:)[P_y(\ )+P_z(\ ), x\cdot y, x\cdot z],$$

so that $x\cdot(y+z)$ and $(x\cdot y)+(x\cdot z)$ have different intensions, as they should. Notice that both operative systems in (2E.13) and (2E.14) have closure ordinal 2; the first computes $y+z$ in the first stage and then adds the result to x in the second stage, while the second (concurrently) computes $x\cdot y$ and $x\cdot z$ in the first stage and then adds the results in the second.

For a more complicated example which involves functionals, consider the expansion

$$\mathfrak{N}^{+\#}=(\mathfrak{N},+,\cdot,E^{\#},\forall^{\#},\dot{\&},\dot{\vee},\dot{\neg},=)$$

of $\mathfrak{N}$, (where the additional primitives are defined in (2C.8)) and let

$$t\equiv(\exists^{\#}x)(\forall^{\#}y)(x+y=\mathrm{succ}(y))$$

be the term which takes value $\underset{\sim}{1}$ if 1 exists. Computing normal forms successively by the rules outlined above, we get:

$$x+y\mapsto(\text{rec } P:)x+y,$$

$$\mathrm{succ}(y)\mapsto(\text{rec } S:)\mathrm{succ}(y),$$

$$x+y=\mathrm{succ}(y)\mapsto(\text{rec } E,P,S:)[X_{=}(P(\ ),S(\ )),x+y,\mathrm{succ}(y)],$$

$$(\forall^{\#}y)x+y=\mathrm{succ}(y)\mapsto(\text{rec } A,y,E,y,P,y,S:)$$
$$[\forall^{\#}(E),X_{=}(P(y),S(y)),x+y,\mathrm{succ}(y)]$$

and finally

$$(\exists^{\#}x)(\forall^{\#}y)(x+y=\mathrm{succ}(y))$$
$$\mapsto(\text{rec } \underline{out},x,A,x,y,E,x,y,P,x,y,S:)$$
$$[\exists^{\#}(A),\forall^{\#}(\lambda yE(x,y)),X_{=}(P(x,y),S(x,y)),x+y,\mathrm{succ}(y)].$$

It should be clear that the operative system defined by the normal form of t describes precisely the (concurrent) algorithm which we would naturally use to check the value of t, if we did not know it.

This of course is a general fact: <u>for any relational first order structure $\mathfrak{A}$, the intension of an explicit closed term t of $(\mathfrak{A},E^{\#},\forall^{\#},\dot{\&},\dot{\vee},\dot{\neg},=)$ describes the natural algorithm for computing the truth value of the sentence expressed by t</u>.

These simple computations of the intensions of explicit temss point out what may appear to be a paradoxical feature of our approach: <u>we are assigning meaning to explicit computation by reducing it to recursion</u>. Of course the relevant recursions are trivial, and they only serve to mark semantically the

the order in which subcomputations must be performed, and which can be done concurrently.

On the other hand, the intension of a recursive term t certainly combines all the nested recursions in t into just one (an "outside loop" in the first order case) but then appeals to the mathematical, analysis of that recursion by iteration, as in 2A, and fixes no specific, operational, implementable computation for actually carrying out that recursion.

We are forced to leave this final step open, partly because the theory covers structures with arbitrary, discontinuous functionals, where the "computation" of fixed points by iteration is all that we can do. One can also argue that even in the implementable case, the abstract "computation" by functional iteration and the consequent ramification of the simultaneous fixed points into stages already encodes all the important (implementation independent) properties of the algorithm, and that it provides a good tool for studying these properties.

2F. Concurrency and normality. It is quite obvious that our modelling of algorithms by operative systems of functionals handles easily and faithfully the most usual (deterministic) concurrent combinations of algorithms.

For example, it is easy to construct from given operative systems f_1 and g_1 which compute

$$f, g : X \to W,$$

an operative system h_1 which computes the pair functional

$$h(x) \simeq (f(x), g(x)),$$

and in such a way that the stages match, i.e. the intended computation is parallel.

The situation is more interesting when we do not need both values $f(x)$ and $g(y)$, but would be happy with one of them. Let us call a functional

$$ch : X \times Y \to \{\underset{\sim}{0}, \underset{\sim}{1}\}$$

a choice functional for f and g as above, if

(2F.1) $$ch(x,y) \simeq \underset{\sim}{1} \Rightarrow f(x)\downarrow,$$

(2F.2) $$ch(x,y) \simeq \underset{\sim}{0} \Rightarrow g(y)\downarrow,$$

(2F.3) $$f(x)\downarrow \vee g(y)\downarrow \Rightarrow ch(x,y)\downarrow;$$

now the functional

(2F.4) $$h(x,y) \simeq (\text{if } ch(x,y) \text{ then } f(x) \text{ else } g(y))$$

is defined precisely when at least one of $f(x)$ or $g(y)$ is defined and returns one of their values.

As it turns out, in most interesting structures we can always find a recursive partial choice function ch, whenever f and g are recursive partial functions, by analysing the natural staging determined by a recursive definition.

Suppose $f = [f_1,\ldots,f_n]$ is an operative system of functionals on the universe of some structure $\mathfrak{G}$,

$$f_i : U_i \times P(U_1,W_1) \times \cdots \times P(U_n,W_n) \times X \to W_i \qquad (i = 1,\ldots,n),$$

and for $i = 1,\ldots,n$ let

$$(2F.5) \qquad \mathrm{stage}_{f,i}(u_i;x) \simeq \text{least } \xi \text{ such that } f_i^{\xi}(u_i;x)\downarrow.$$

Now

$$(2F.6) \qquad \mathrm{stage}_{f,i}(u_i;x)\downarrow \Leftrightarrow f_i^{\infty}(u_i;x)\downarrow$$

and $\mathrm{stage}_{f,i}$ takes ordinal values - integers if f is a continuous system. The closure ordinal of f is

$$(2F.7) \qquad \kappa(f) = \sup\{\mathrm{stage}_{f,i}(u_i;x) : i = 1,\ldots,n,\ u_i \in U_i,\ x \in X\}.$$

Typically we are interested in the stages of the principal fixed point of f, and we use the simplified notation

$$(2F.8) \qquad |u;x|_f \simeq \mathrm{stage}_{f,1}(u;x).$$

Now the partial function $|\ |_f$ assigns ordinal stages to the domain of the functional

$$\overline{f} = f_1^{\infty}$$

computed by f.

If the set of integers N is one of the basic sets of $\mathfrak{G}$ and if f is a continuous system, we can ask directly whether the functionals $\mathrm{stage}_{i,f}$ are recursive. In the general case, we must compare stages indirectly.

For given systems $f = [f_1,\ldots,f_n]$, $g = [g_1,\ldots,g_m]$, put

$$(2F.9) \quad \mathrm{comp}_{f,g}(u;x;v;y) \simeq \begin{cases} \underset{\sim}{1} & \text{if } \overline{f}(u;x)\downarrow \;\&\; [\overline{g}(v;y)\uparrow \vee |u;x|_f \le |v;y|_g], \\ \underset{\sim}{0} & \text{if } \overline{g}(v;y)\downarrow \;\&\; [\overline{f}(u;x)\uparrow \vee |v;y|_g < |u;x|_f]. \end{cases}$$

It is clear that the stage comparison functional $\mathrm{comp}_{f,g}$ is a choice functional for $\overline{f}$ and $\overline{g}$ in the sense of (2F.1)-(2F.3). The problem is to find reasonable and useful hypotheses on f and g which insure that $\mathrm{comp}_{f,g}$ is recursive.

To see first that even in very simple circumstances, comp need not be recursive, suppose

$$f : N \times P(N,N) \to N$$

is a single operative functional on the integers, with no parameters, and assume that the stage comparison partial function

$$\mathrm{comp} = \mathrm{comp}_{f,f}$$

is in fact recursive in $\mathfrak{N}$. If it so happens that $f(0,\emptyset)\downarrow$, so that $|0|_f = 0$, then

$$\begin{aligned} f(x,\emptyset)\downarrow &\Leftrightarrow f^0(x)\downarrow \\ &\Leftrightarrow \mathrm{comp}(x,0) = \underset{\sim}{1}, \end{aligned}$$

and $\lambda x\mathrm{comp}(x,0)$ is total, so that the relation

$$R(x) \Leftrightarrow f(x,\emptyset)\downarrow \qquad (2F.10)$$

is recursive in $\mathfrak{N}$. It is of course easy to define recursive functionals f for which the relation R of (2F.10) is complete semirecursive on $\mathfrak{N}$, so that comp cannot be recursive in general.

On the other hand, it is quite easy to prove that in the case of $\mathfrak{N}$, $\mathrm{comp}_{f,g}$ is always recursive when f and g are explicit operative systems with no partial function parameters. This easy proof extends directly to first order structures which satisfy some minimal conditions. For the general case of structures with functionals with which we are concerned here, we need a more sophisticated approach.

We will first identify a large class of special, normal structures in which we can establish a very general stage comparison theorem for explicit systems with no partial function parameters. Not all interesting structures are normal - but they are at least recursive expansions of normal structures, and for those we will show that every recursive partial function has a recursive choice partial function.

Consider first the case of a functional with just one partial function argument, i.e.

$$f : V \times P(U,W) \to W',$$

where V, U, W, W' are recursion spaces of some structure $\mathfrak{A}$. A normalizing functional for f is any functional

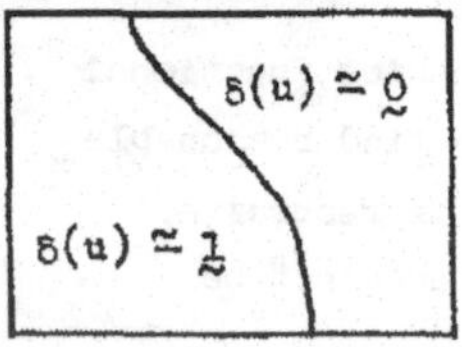

$$\Delta : V \times P(U,W) \times P(U,\{\underset{\sim}{0},\underset{\sim}{1}\}) \to \{\underset{\sim}{0},\underset{\sim}{1}\}$$

with the following two properties, where we use the variable

$$\delta : U \to \{\underset{\sim}{0},\underset{\sim}{1}\}$$

to "cut out" subsets of U.

(2F.11) $\qquad f(v,p\restriction\{u : \delta(u) \simeq \underset{\sim}{1}\})\downarrow \Rightarrow \Delta(v,p,\delta) \simeq \underset{\sim}{1}.$

(2F.12) $\qquad$ If $\delta : U \to \{\underset{\sim}{0},\underset{\sim}{1}\}$ is a total function and
$$\forall u[\delta(u) = \underset{\sim}{1} \Rightarrow p(u)\downarrow],$$

then
$$f(v,p\restriction\{u : \delta(u) = \underset{\sim}{1}\})\uparrow \Rightarrow \Delta(v,p,\delta) \simeq \underset{\sim}{0}.$$

This definition can be extended directly to functionals with more than one partial function arguments; in general, a normalizing functional for $f(v,p_1,p_2,\ldots,p_n)$ has the form $\Delta(v,p_1,\delta_1,p_2,\delta_2,\ldots,p_n,\delta_n)$.

A functional f is <u>normal in a structure</u> $\mathfrak{A}$ if f is $\mathfrak{A}$-recursive and has an $\mathfrak{A}$-recursive normalizing functional. This means - roughly speaking - that we can compute recursively in the structure the domain of definition of the functional, <u>insofar as this is possible</u>.

Consider, for example, the application functional
$$\underset{\sim}{ap}(p,u) \simeq p(u).$$

It is obvious that the relation
$$R(p,u) \Leftrightarrow p(u)\downarrow$$

cannot be recursive, since its characteristic function is not monotone in p. On the other hand, the functional
$$\Delta(p,\delta,u) \simeq \delta(u)$$

is easily seen to be a normalizing functional for $\underset{\sim}{ap}$, so $\underset{\sim}{ap}$ <u>is normal</u> in <u>every structure</u>.

The next few simple facts will illustrate a bit the notion of normality.

(2F.13) <u>Fact</u>. A partial function (i.e. a functional with no partial function arguments) on the univers of a structure $\mathfrak{A}$ is normal in $\mathfrak{A}$ if and only if its domain of definition is a recursive set of $\mathfrak{A}$. ⊣

This is true essentially by definition, since when the variable "p" is absent, (2F.10) and (2F.11) imply that Δ is the characteristic function of the domain of f. It implies, in particular, that every total function is normal.

(2F.14) <u>Fact</u>. The functionals $E^{\#}_{G}$ (1D.11), $\underset{\sim}{aptoall}$ (2A.15), $\angle$ (2A.16), $\underset{\sim}{iter}$ (2C.3), $\underset{\sim}{rep}$ (2B.19), are normal in every structure where they are defined and recursive - and hence in every structure where they are among the givens. ⊣

For example, using the abbreviations of (2C.11), it is easy to check that

$$\Delta(p,\delta) \simeq (\exists^{\#}t)[\delta(t) \mathbin{\dot{\&}} p(t)] \mathbin{\dot{\vee}} (\forall^{\#}t)[\delta(t) \mathbin{\dot{\&}} \dot{\neg}\, p(t)],$$

is a normalizing functional for $E^{\#}$, and it is in fact explicit in $E^{\#}$.

(2F.15) Fact. If all the given functionals in a structure $\mathcal{A} = \langle \Sigma, \Phi \rangle$ are normal in $\mathcal{A}$, then every $\mathcal{A}$-explicit functional is normal in $\mathcal{A}$. ⊣

This is proved by assigning directly to each term t of EXP a term t^* which defines a normalizing functional for the functional defined by t.

Now (2F.13)-(2F.15) imply immediately that most of the interesting functionals we study are in fact normal in the structures where they occur; important exceptions are the while and the ancenstral which we will discuss shortly.

(2F.16) The Stage Comparison Theorem. If f and g are normal, operative systems without any partial function parameters in a structure $\mathcal{A}$, then the associated stage comparison partial function $comp_{f,g}$ of (2F.9) is recursive in $\mathcal{A}$.

Proof. For the simple (but typical) case where we are given operative functionals $f(u,p)$ and $g(v,q)$ with no parameters and normalizing functionals Δ_f and Δ_g, let

$$h(u,v,r) \simeq \Delta_f(u,f^{\infty},\lambda u' \dot{\neg}\, \Delta_g(v,g^{\infty},\lambda v' \dot{\neg}\, v(u',v')))$$

and verify that

$$comp_{f,g}(u,v) \simeq h^{\infty}(u,v). \qquad ⊣$$

A structure $\mathcal{A}'$ is normal if all the given functionals of $\mathcal{A}$ (and hence all $\mathcal{A}$ explicit functionals) are normal in A.

(2F.17) Fact. If $\mathcal{A}$ is normal, then Env Rec($\mathcal{A}$) is closed under disjunction and the recursive relations of $\mathcal{A}$ are precisely those which are semirecursive and have semirecursive complements. ⊣

All first order structures are normal by (2F.13), and in that case many of the consequences of normality are easily proved directly. For structures with functionals, however, simple structural properties like (2F.17) can become quite difficult to establish and the approach through a stage comparison theorem is often the simplest.

Gandy [1967] formulated and proved the first abstract stage comparison theorem for Kleene (normal) recursion in type 2, and he used it to establish all the basic properties of that structure. His results were extended to higher types by Moschovakis [1967] and Platek [1966] , and then to (what we call now) positive elementary induction in Moschovakis [1969] , [1974] . The general version (2F.16), based on the hypothesis of normality was announced in Moschovakis [1976] and proved (for the special case of recursion in higher types, with a generally valid argument) in Kechris-Moschovakis [1977] . Harel-Kozen [1982] gives a version of the stage comparison theorem in a command-language formulation of positive elementary induction.

In the infintary structures studied by the logicians, the stage comparison theorem is the first (and simplest) of a sequence of <u>selection theorems</u> which lead to the mathematically most interesting parts of the subject (see Kechris-Moschovakis [1977] for references). Instead of discussing these developments here, we will make just a few comments on the relevance of these notions for the structures which arise naturally in computer science.

Notice first that all the definitions extend directly to the global case, so we can talk about a global functional on a class of structures $\mathcal{K}$ being <u>normal</u> on $\mathcal{K}$, and of a <u>normal class</u> $\mathcal{K}$ of structures. The class $\mathcal{FO}$ of all first order structures is normal, as is the class

$$\mathcal{FO}^{\#} = \{(\mathfrak{A},=,E^{\#}) : \mathfrak{A} \text{ is first order}\},$$

whose semirecursive relational queries are the positive, elementary inductive queries on $\mathcal{FO}$.

A good illustration if the uses of the stage comparison theorem is the following simple version of Immerman's proof of (2D.14).

(2F.18) <u>Fact.</u> Let $\mathcal{K}$ be a normal class of finite structures and let

$$\mathcal{K}^{\#} = (\mathcal{K},E^{\#})$$

be the class of expansions, where we add the existential quantifier functionals for all the basic sets in the structures of $\mathcal{K}$; the collection of global semi-recursive queries on $\mathcal{K}^{\#}$ is closed under negation. (Essentially Immerman [1982].)

<u>Proof.</u> It is easy to reduce the theorem to the following simpler proposition: suppose $f(u,p)$ is an explicit and operative (global) functional on $\mathcal{K}^{\#}$, assume that for each $\mathfrak{A}$ in $\mathcal{K}^{\#}$,

$$\mathfrak{A} \models (\exists^{\#}u)f^{\infty}(u)\downarrow$$

and prove that

$$\{u : f^{\infty}(u)\uparrow\}$$

is a semirecursive query in $\mathcal{K}^{\#}$.

Define $g(i,u,q)$ (on each G in $\mathcal{K}^{\#}$) with i ranging over $\{\underset{\sim}{0},\underset{\sim}{1}\}$ and q taking arguments in $\{\underset{\sim}{0},\underset{\sim}{1}\} \times U$, by

$$g(i,u,q) \simeq \begin{cases} f(u,\lambda u' q(\underset{\sim}{0},u')) & \text{if } i = \underset{\sim}{0}, \\ q(0,u) & \text{if } i = \underset{\sim}{1}. \end{cases}$$

Clearly, for all u

$$g^{\infty}(\underset{\sim}{0},u) \simeq g^{\infty}(\underset{\sim}{1},u) \simeq f^{\infty}(u)$$

and for the stages,

$$|\underset{\sim}{1},u|_g \simeq |\underset{\sim}{0},u|_g + 1 \simeq |u|_f + 1.$$

Since the structures of $\mathcal{K}$ are finite, on each of them there is a maximum stage $|u|_f$, for $f^{\infty}(u)\downarrow$; moreover, this is defined by

$$\text{Max}(u) \Leftrightarrow (\forall^{\#}u')\{\text{comp}(\underset{\sim}{0},u',\underset{\sim}{0},u) \simeq \underset{\sim}{1} \vee \text{comp}(\underset{\sim}{0},u',\underset{\sim}{1},u) \simeq \underset{\sim}{0}\},$$

so Max is a $\mathcal{K}^{\#}$-semirecursive query by (2F.14)-(2F.15). But

$$f^{\infty}(u)\uparrow \Leftrightarrow (\exists^{\#}u')[\text{Max}(u') \ \& \ \text{comp}(\underset{\sim}{0},u,\underset{\sim}{0},u') \simeq \underset{\sim}{0})],$$

so $\{u : f^{\infty}(u)\uparrow\}$ is also semirecursive in $\mathcal{K}^{\#}$. ⊣

In connection with these ideas, we should mention still another functional representation of the existential quantifier, that originally chosen by Kleene [1959], [1963] in his work on higher types. For each set G and $p : G \to \{\underset{\sim}{0},\underset{\sim}{1}\}$, let

$$\text{(2F.19)} \qquad E_G(p) \simeq \begin{cases} \underset{\sim}{1} & \text{if } p \text{ is total and } (\exists t \in G)p(t) = \underset{\sim}{1}, \\ \underset{\sim}{0} & \text{if } p \text{ is total and } (\forall t \in G)p(t) = \underset{\sim}{0}; \end{cases}$$

thus E_G is just the restriction of $E_G^{\#}$ to <u>total</u> functions.

Recursion in E is very well understood in the infinite models, but there is very little known about it in the finite case. For example, if we take

$$(\text{Fin Graphs},=,E) = \{\langle G,R,=,E_G\rangle : G \text{ finite}, R \text{ is binary on } G\},$$

it is not known whether the class

$$\Gamma = \text{Env}\ \underset{\sim}{\text{Rec}}(\text{Fin Graphs},=,E)$$

coincides with Ind(Fin Graphs), whether it is closed under negation, or even whether it contains the Boolean query "$\mathcal{G}$ is connected". (It should not.) The algorithms of recursion in E appear to be completely deterministic and much more restricted than those of recursion in $E^{\#}$, but this too is not well understood in the finite case.

The while functional is not always normal, because of its strong expressive power. Notice that by (2F.13) and (2F.15), the domain of every explicit partial function in a normal structure is recursive; thus while cannot be normal where we have non-trivial partial functions which are explicit in while.

(2F.20) Fact. (a) while is not normal in $\mathfrak{N}$.

(b) while is not globally normal on the class of finite structures with one basic set and a single unary function, i.e. the structures of the form

$$\mathfrak{A} = \langle A, \pi, = \rangle$$

where $\pi : A \to A$ is total and A is finite. ⊣

The upshot is that we do not always have recursive computations of staging when we take the while among our primitives. On the other hand, while of course is recursive, so we can always compute the staging relative to the simpler structure, without the while. The relevant mathematical fact is contained in the next simple result.

(2F.21) Fact. If $\mathfrak{A}$ is a recursive expansion of a normal structure $\mathfrak{A}_0$, then for every two $\mathfrak{A}$-recursive partial functions f and g, we can find an $\mathfrak{A}$-recursive choice partial function ch satisfying (2F.1)-(2F.3). ⊣

It is quite obvious that all structures which model implementable computations are in fact recursive expansions of normal structures. In 3 we will discuss briefly the uses of the Stage Comparison Theorem (2F.16) and its corollary (2F.21) in the study of REC and its explicit sublanguages as systems for specifying concurrent algorithms.

2G. Foundational considerations; Church's Thesis for algorithms. In adopting the present approach to the theory of algorithms, we have made several foundational assumptions, some of them involving choices among competing, plausible alternatives. Here we will identify the most important of these assumptions and we will explain briefly the reasons for our choices, without attempting a full justification.

In the discussion, we will refer to the intensions of a structure $\mathfrak{A}$, i.e. the intensions of the terms of REC($\mathfrak{A}$). These represent the basic algorithms of $\mathfrak{A}$. The recursive and the explicit algorithms of $\mathfrak{A}$ are defined by operative systems of recursive or explicit functionals on $\mathfrak{A}$, respectively.

A. The set-theoretic context. It is sometimes suggested that to understand algorithms, one must work within constructive mathematics, or that one cannot define algorithms in terms of known concepts, so that they must be taken as primitives and axioms about them should be formulated. Here, we have accepted right from the start that an adequate foundation for the theory of algorithms can be developed within the familiar context of classical mathematics; moreover, algorithms will be represented (or modelled) by set-theoretic objects which "act" on "data" and are defined relative to "givens".

Our choice of many-sorted universes to represent data appears non-controversial and we will discuss shortly the decision to model the "givens" by functionals.

B. <u>What is always given</u>? Our formal defintions of explicit definability and recursion assume implicitly that if we are given the basic sets in some universe Σ, then we are automatically also given all finite cartesian products of basic sets, as well as the functions needed to construct tuples (of each fixed length) and to take them apart. There are several other primitives and operations which could be considered <u>absolutely computable</u>, to be included automatically among the givens in every situation.

B1. <u>The identity functions and equality relations on basic sets</u>. This is a minor point, which we have already discussed in 2C circa (2C.2)-(2C.3). The theory of structures with id and = among the givens is quite similar to the general theory, except that it excludes certain interesting models and it forces some technicalities in others.

For example, it is convenient to have all recursive partial functions take Boolean values, when the given functionals are all Boolean.

In all interesting structures, the equality relations are either recursive, semirecursive (typically in algebraic situations, e.g. finitely generated groups) or cosemirecursive (e.g. on the real numbers).

B2. <u>Stacks</u>. It can be argued that in order to develop a reasonable theory of recursion, one should always include the natural numbers (with 0, succ, pred and $\chi_{=0}$, at least) among the givens. A reasonable extension of this argument would grant along with each basic set S, automatically, the set

$$S^* = \text{all finite strings from } S,$$

together with natural tuple-forming and tuple-decoding functions, i.e. (minimally) the structure

$$\text{Stack}(S) = \langle S^*, \text{push}_S, \text{pull}_S \rangle$$

defined in the obvious way.

The idea that each set S brings with it its own stack is very natural, in situations where one is attempting to define the most general notion of recursion on S, with a view towards formulating and justifying an abstract form of Church's Thesis for S. Mc Carthy [1960] took this as the basic principle in defining symbolic computation and Kleene [1959] also accepted it (implicitly) in his definition of recursion in higher types, the first fully developed theory of recursion which is fundamentally different from ordinary recursion on the integers. Many attempts to axiomatize recursion theory are also based on this idea, e.g. Wagner [1969] and Moschovakis [1969], [1971].

On the other hand, Stack(S) is always infinite (when $S \neq \emptyset$) and has in it a copy of $\mathfrak{N}$; a theory of recursion which automatically admits a stack for each basic set will have no finite models and will exclude many interesting examples, e.g. recursion on $\mathfrak{BSJR}$ (polynomial time computability by (2D.15)).

Sometimes it is desirable to admit stacks over some basic sets, but not on others. For example, in the baby programming structure $\mathfrak{BP}$ (2B.17), we have files of integers and characters but no files of files - these can be introduced and manipulated recursively, but the theory keeps track of the complexity of these manipulations.

Thus it appears that our choice to leave stacks out (except when they are deliberately put in) can be justified on the mathematical grounds, that we want a theory of recursion which has non-trivial finite models and which covers naturally all the interesting examples.

Foundationally, however, the thesis for the representation of algorithms which we will formulate at the end of this section goes further than this. In effect, we are claiming that <u>the notion of algorithm has nothing to do with the natural numbers</u>. Algorithms are iterative processes, defined by combining the givens in some simple ways; for some givens, the iterations thus defined may all close off at some finite stage - or they may extend into the transfinite. It is not part and parcel of the notion of algorithm that these processes must always include ordinary recursion on the natural numbers.

B3. <u>Finiteness; computations as data.</u> Kreisel [1961] argued that every reasonable generalization of recursion theory should also generalize the notion of <u>finiteness</u>, that the "computations" of a theory should be "finite" objects (in the sense appropriate to the theory) and that the natural "domain" of the theory should include these finite objects. He also gave a reformulation in this spirit of classical hyperarithmetic theory, i.e. recursion in $(\mathfrak{N}, E^{\#})$ in our terminology. Later, Kreisel-Sacks [1965] gave a more complete development of this <u>metarecursion theory</u> on the Church-Kleene ordinal ω_1^{CK}; metarecursion has all the nice properties demanded by Kreisel, and the metarecursive functions <u>on the integers</u> are precisely the functions recursive in $(\mathfrak{N}, E^{\#})$.

This in turn led to the Kripke-Platek theory of <u>recursion on admissible ordinals</u> which has been developed very extensively, cf. Shore [1977].

The idea that relative finiteness should be a basic notion of recursion theory was elaborated in Kreisel [1965] and was very influential in the further development of abstract recursion theory, including the line which led to the present approach; cf. Moschovakis [1971], [1977].

Although we will not pursue the matter in this paper, the idea is sufficiently simple that we can give the basic definitions here. A subset $F \subseteq U$ of a recursion space in some structure $\mathfrak{A}$ is absolutely, strongly $\mathfrak{A}$-finite if the functional $E_F^{\#}$ (1D.11) which embodies quantification on F is $\mathfrak{A}$-recursive; in effect, F is absolutely, strongly $\mathfrak{A}$-finite if we can search through F recursively. A better relative notion is obtained if we allow parameters from $\mathfrak{A}$ in the recursive definition of $E_F^{\#}$, and it leads to a decent theory of finiteness, for structures which satisfy some reasonable hypotheses. For such "nice" structures (which include all familiar examples) we can also define "computations," and they turn out to be "finite" objects.

On the other hand, we do not see why the finite sets must always be among the data, or even "codable" by data in a canonical way. The obvious counter-examples to this are the finite structures which we wish to admit and which simply have more finite subsets than elements.

The Kreisel-Sacks connection between recursion in $(\mathfrak{N}, E^{\#})$ and metarecursion was the first of several interesting companion theorems, where one (generally) relates a structure $\mathfrak{A}$ with simple universe and complicated functionals (typically second-order) to another structure $\mathfrak{A}'$ on a more complicated universe but with simpler functionals (often just functions); cf. Barwise-Grandy-Moschovakis [1971], Chapter 9 in Moschovakis [1974], Fenstad [1980]. Even if one were to grant that only structures with the Kreisel regularities are nice (at least in the infinitary case), one must have the less nice structures around in order to be able to express the connection precisely.

Actually, our position is a bit stronger than the preceding discussion might suggest. We do not insist that computations always be included among the data, simply because we do not see that the concept of algorithm carries that implication.

C. Algorithms acting on algorithms; reflection. In many programming languages, it is possible to define "functions" which take other "functions" as arguments.

Consider for example a program P which computes the maximum of an array $a = \langle a_0, \ldots, a_n \rangle$ of integers, where, however, the integers are to be ordered using an "external" (as yet unknown) function ord. According to the present theory, we would expect that the algorithm described by P can be represented by an operative system of functionals

$$f = [f_1, \ldots, f_n]$$

in the parameters a, ord, so that for each a and each ord,

max of a relative to ord = $\overline{f}$(a,ord)
= the principal fixed point of the system f(a,ord)

in the notation of (2A.27).

Now this mathematical modelling of the situation is not completely faithful, because the formal variable "ord" of the program P was never really meant to range over arbitrary partial (or even total) functions; the intent is to link P (later) with some other program C which computes an ordering function and then run the combined, linked program. One may be justified in saying that the algorithm described by P is meant to apply to an array a and an arbitrary algorithm, among those which can be described by programs of the language at hand.

Nevertheless, the modelling we described will describe the situation faithfully, most of the time: it will be accurate if the programming language calls functions by value.

In programming terminology, "calling ord by value" simply means that you can occassionally demand (of the program computing ord) specific values of ord(x,y), for arguments that you provide, but you may not get into and inspect the very program which computes ord. Thus the overall algorithm depends on ord extensionally, and it defines a functional in our sense.

In contrast to this, there is call by name, where the calling program has access to the actual definition of the called function. Suppose, for example, f is defined by

(*) f(x,y) = (if x = 0 then 1 else y)

and some program Q needs to call f to compute f(0,g(0)), for some g which Q knows. In the usual call by value, Q will first compute g(0), then ask for the value of f at 0, g(0) - so in particular the computation will diverge if the computation of g(0) does not terminate. In one version of call by name, on the other hand, Q might call f before computing g(0), realize from (*) that f(0,y) does not depend on y and give output 1 without ever computing g(0).

In our approach, to model faithfully the algorithms of a language which uses call by name, we would introduce an additional basic set of program descriptions and the appropriate givens which express the "reflective power" of the language. The variable "ord" then in the algorithm described by P will not range over functions, but over these program descriptions.

To look at a different example, recall that an effective operation (Rogers [1967] , 11.5) is a partial function

$$F : \text{Total Rec} \to N$$

defined on total recursive functions on N, such that for some recursive partial function $f : N \to N$,

(*) if e is a Gödel number of α, then $F(\alpha) \simeq f(e)$.

Stretching things a bit, one can think of effective operations as partial functions on Total Rec which call their arguments by name, the Gödel number is this instance.

It does not appear promising to search for a recursion structure $\mathfrak{A}$ which has Total Rec among its basic sets, and where the givens are chosen so that the effective operations turn out naturally to be the $\mathfrak{A}$-recursive partial functions. Put another way: it may be that effective operations can be viewed directly as "algorithms" on Total Rec, but these "algorithms" are (apparently) not among the iterative processes which we are studying here. From our simple-minded point of view, the algorithm which computes the values of F above is the algorithm which computes the f that satisfies (*) - and it operates on natural numbers, not general recursive functions.

This is not to say that effective operations (and their generalizations to countable functionals of higher types) are not important, or that the rich mathematical theory of these structures cannot be developed within the theory of recursion, as we conceive it here. It simply means that these are derived notions which do not correspond directly to the alogirithms that we take as our basic objects of study.

The study of higher level programming languages involves many subtleties other than "call by name" (which is in fact quite rare) and it has led to a substantial theory of computation, founded mathematically on Scott's theory of domains.

This is not the place to attempt a ten-line summary of denotational semantics and the theory of domains, see e.g. Stoy [1977] , Tennent [1981] , de Bakker [1980] , the original Scott-Strackey [1971] , the most recent Scott [1982] and for quick and pleasant reading, Gordon [1979] and Stoy's Chapter 3 in Bjørner-Jones [1982] . What is relevant for our discussion is the implicit assumption in Scott's approach to denotational semantics, that to understand the algorithms which can be described by the programs of a higher-level language on certain data, you must imbed that data in a structured domain, which will in fact "contain" all the algorithms among its points; in effect, the algorithms become part of the data, and they can be applied to themselves.

In the present approach, not only do we not automatically include the algorithms among the data, but we draw a very sharp line between these two categories of objects. The data are represented by the basic sets of some structure $\mathfrak{A}$ and the algorithms are represented by the intensions of $\mathfrak{A}$; algorithms on $\mathfrak{A}$ cannot directly take algorithms on $\mathfrak{A}$ as arguments.

On the other hand, algorithms are represented by functionals, these functionals can take partial functions as arguments and among these partial functions are *the denotations* of other algorithms - or at least the partial functions obtained from such denotations by fixing some of their arguments; in this way, *algorithms can act extensionally on algorithms,* in effect calling them by value.

To elucidate this a bit, in the case of the example above, suppose $t(\dot{a},\dot{\mathrm{ord}})$ is a term of REC with a formal array variable $\dot{a}$ and a formal partial function variable $\dot{\mathrm{ord}}$ which defines the given maximum algorithm f, i.e. (in the relevant structure $\mathfrak{A}$)

(2G.1) $$f = \mathrm{int}\ t(\dot{a},\dot{\mathrm{ord}}),$$

let $s(\dot{x},\dot{y})$ define some ordering algorithm ord in the same way,

(2G.2) $$\mathrm{ord} = \mathrm{int}\ s(\dot{x},\dot{y})$$

and put

(2G.3) $$h = \mathrm{int}\ t(\dot{a},\lambda\dot{x}\dot{y}s(\dot{x},\dot{y})).$$

It is obvious that *the denotation* of the term $t(\dot{a},\lambda\dot{x}\dot{y}s(\dot{x},\dot{y}))$ for $\dot{a} = a$ is just the maximum of a relative to the ordering function $\overline{\mathrm{ord}}$ defined by ord, i.e. in the notation of (2A.26),

(2G.4) $$\overline{h}(a) \simeq \overline{f}(a,\lambda xy\overline{\mathrm{ord}}(x,y)).$$

It is then natural to abbreviate the definition (2G.1)-(2G.3) by

(2G.5) $$h(a) = f(a,\lambda xy\mathrm{ord}(x,y))$$

and take this as *the definition of the* λ*-substitution of one algorithm in another.* (It is easy to check that h is independent of the particular terms chosen with intensions f and ord.)

This restricted action by "calling" provides a limited amount of reflection, which is all that we can expect in the general setting, with the finite models in the picture. Of course on rich structures with stacks, one can code algorithms by objects in the structure and study reflection phenomena indirectly, as we do in arithmetic.

D. The question of identity of algorithms. It is easy to ssociate with each operative system $f = [f_1, f_2, f_3]$ a new system $f' = [f'_1, f'_3, f'_2]$, where f'_1, f'_2, f'_3 are essentially the same functionals as f_1, f_2, f_3, except that their variables are "shuffled" so that $[f'_1, f'_3, f'_2]$ is operative; one can then argue quite convincingly that the system f' represents the same algorithm as the system f.

Thus, there may be many operative systems which represent the same algorithm and our modelling does not capture directly the relation of identity among algorithms.

Actually this notion of identity of algorithms appears to be quite slippery and we do not claim to have elucidated it here.

After these cautionary remarks and explanations, we can now formulate the refined and generalized version of Church's Thesis which is the foundational basis of our approach to the theory of algorithms.

Church's Thesis for algorithms. Every algorithm P, on given data and relative to given operations, can be represented faithfully in all its essential mathematical features by an operative system of functionals f_P, which is explicitly definable in terms of the givens.

More precisely, the given data and operations can be represented faithfully by a recursion structure $\mathfrak{G}$ and the algorithm can be represented faithfully by the intension of a term t in the language REC($\mathfrak{G}$); in particular, the principal fixed point den($\mathfrak{G}$,t) is the constant, function or functional computed by the algorithm P.

There is an obvious global version of this thesis for algorithms which are meant to act on multiple data and relative to different operations of a certain kind.

The classical Church's Thesis identifies the intuitive notion of computable function on the integers with the precise, mathematical notion of recursive function. It is a serious mistake to understand it as just a definition of computability.

In the present case, however, it would not be too far off the mark to take the refined version for algorithms above as just a definition of algorithm. The non-trivial (modest) content of the thesis is the claim that among the many available competing possibilities, there is a most natural and useful notion of abstract algorithm; and that this notion models equally naturally both concurrent and sequential, finitary and infinitary algorithms and can be defined with no reference to implementations or concepts of storage.

Part 3. Directions

Here we will describe very briefly some interesting directions for research which are suggested by the notions just introduced and are relevant to computer science.

3A. The logic of recursion. If $\mathcal{K}$ is a class of recursion structures of signature σ and t is a Boolean term of $\mathrm{REC}(\sigma)$, put

$$\text{(3A.1)} \qquad \mathcal{K} \models t \Leftrightarrow \text{for all } \mathfrak{A} \text{ in } \mathcal{K} \text{ and all valuations } \alpha \text{ in } \mathfrak{A},\ \mathrm{Val}(\mathfrak{A},\alpha,t) \simeq \underset{\sim}{1}.$$

We would like to find natural axioms and rules of inference which will prove $\mathcal{K} \models t$ when this holds, at least for special cases of $\mathcal{K}$ and t.

Notice first that our implicit identification of the formulas of REC with Boolean terms is not as restrictive as it might appear. For example, if t and s have the same sort and = on that sort is given (or recursive), we can set

$$\text{(3A.2)} \qquad \mathcal{K} \models t \simeq s \Leftrightarrow \mathcal{K} \models \chi_{=}(t,s).$$

The expressive power of the formulas of $\mathrm{REC}(\mathcal{K})$ depends on just what global functionals are recursive on $\mathcal{K}$.

It is well-known that we cannot in general develop a complete proof theory for $\mathrm{REC}(\mathcal{K})$, since in most interesting cases the set of semantic consequences of $\mathcal{K}$,

$$\mathrm{Cons}(\mathcal{K}) = \{t : \mathcal{K} \models t\}$$

is not recursively enumerable - e.g. this is the case when $\mathcal{K}$ is the class of all finite first-order structures. Despite this, the identification of a natural, strong logic for EXP and REC could be very useful and would help clarify the concepts. See de Bakker [1980] for references to the considerable body of work in this direction for languages with similar aims as REC.

There is one theorem in this connection which follows easily from the main result in Barwise-Moschovakis [1978].

For lack of a better name, let us call a class of structures $\mathcal{K}$ close to first order (CFO) if

$$\text{(3A.3)} \qquad \mathcal{K} = (\mathcal{K}_0, \Phi),$$

where the following conditions hold.

(1) $\mathcal{K}_0$ is a classical pseudoelementary class of first-order structures, i.e. for some set T of sentences of lower predicate calculus,

$$\mathcal{K}_0 = \{\mathfrak{A}_0 : \text{for some relations } R_1,\ldots,R_n \text{ on the basic sets of } \mathfrak{A}_0,\ (\mathfrak{A}_0,R_1,\ldots,R_n) \models T\}.$$

(2) Each global functional in Φ is recursive on the class of structures

$$\mathcal{K}_0^{\#} = \{(\mathfrak{A}_0,=,E^{\#}) : \mathfrak{A}_0 \in \mathcal{K}_0\}.$$

Each pseudoelementary class of first order structures is CFO, and so is, obviously

$$\mathcal{FO}^{\#} = \{(\mathfrak{A},=,E^{\#}) : \mathfrak{A} \text{ is first order}\}.$$

For each pseudoelementary $\mathcal{K}_0$, the following expansions (for example) are also CFO:

$$(\mathcal{K}_0,\underline{\text{while}}),\ (\mathcal{K}_0,=,\underline{\text{ancestral}}),\ (\mathcal{K}_0,=,E).$$

The class of finite graphs is not CFO and neither is any collection of structures with basic sets S, S^* (say), where S^* must consist of precisely all finite strings from S.

(3A.4) Weak Completeness Theorem for CFO classes. If $\mathcal{K}$ is a CFO collection of structures, then its set of semantic consequences

$$\mathrm{Cons}(\mathcal{K}) = \{t : t \text{ is a term of } \mathrm{REC}(\mathcal{K}) \text{ and } \mathcal{K} \models t\}$$

is recursively enumerable.

Proof. Using the reduction theorem (2D.17), the characterization of recursion in $E^{\#}$ given in (2D.13) and the completeness theorem for lower predicate calculus, we can reduce this result to the following

Lemma. If $\mathcal{K}_0$ is an elementary class of first order structures, $\varphi(\bar{x},S)$ is a formula of the language of $\mathcal{K}_0$ with S occurring only positively and for each $\mathfrak{A}$ in $\mathcal{K}_0$,

$$\mathfrak{A} \models \forall x\, \varphi^{\infty}(\bar{x}),$$

then there is a fixed natural number k such that for each $\mathfrak{A}$ in $\mathcal{K}_0$,

$$\mathfrak{A} \models \forall x\, \varphi^{k}(\bar{x}).$$

Here the iterates of $\varphi(\bar{x},S)$ are defined on the ordinals by

$$\varphi^{\xi}(\bar{x}) \Leftrightarrow \varphi(\bar{x}, \bigcup_{\eta<\xi} \varphi^{\eta})$$

and the fixed point

$$\varphi^{\infty} = \bigcup_{\xi} \varphi^{\xi}$$

may be attained at an infinite ordinal on some infinite structures.

This lemma is a special case of the main Theorem B of Barwise-Moschovakis [1978], but it is worth including here a brief outline of its proof.

If the lemma is false and $K_0 = \{\mathfrak{G} : \mathfrak{G} \models T\}$ for a set of sentences T, then

$$S = T \cup \{\exists\bar{x} \neg \varphi^0(\bar{x}), \exists\bar{x} \neg \varphi^1(\bar{x}), \ldots\}$$

is consistent, and it only has infinite models, so it has a countable model $\mathfrak{G}$. By an old result of Keisler [1965] (which in the present context is reproved in Exercises 4.6, 4.7 of Moschovakis [1974]), $\mathfrak{G}$ has an elementary extension $\mathfrak{G}'$ with closure ordinal ω, i.e. such that on $\mathfrak{G}'$,

$$\varphi^{\infty} = \bigcup_{k=0}^{\infty} \varphi^k;$$

this immediately contradicts $\mathfrak{G}' \models S$. ⊣

An analysis of the proof of this theorem leads to a reasonable logic for REC: briefly, we must have enough axioms and rules to prove formally the denotational equivalence of a term with its reduced normal form and we must build in both the lower predicates calculus and the characteristic properties of fixed points. One would expect to get a more natural logic for REC by attacking the problem directly and avoiding the somewhat unnatural introduction of ideas and results from predicate logic.

3B. REC <u>as a programming language</u>. The language of recursion was designed for theoretical reasons, to aid in the formulation of a solid mathematical foundation for the theory of algorithms. Once it is there, however, and since our extended Church's Thesis claims that we can describe in it naturally all algorithms, there is the natural temptation to consider implementing it.

In programming terminology, REC is a "functional" (applicative) language (like pure LISP), with no assignments and no "commands" directly controlling "storage." The good thing about such languages is the absence of side effects, which facilitates considerably the verification of programs and allows for many powerful combinations of algorithms. The bad thing about these languages is the absence of side effects, which necessitates an unfamiliar style of programming, far removed from our basic understanding of the working of von Neumann machines. Functional languages also have a reputation of being "slow," particularly if they depend excessively on recursion.

A great deal of research has been done on applicative languages, but here we will just cite Backus [1978]. This seminal paper contains (among many other things) both references and an eloquent discussion of the important role that the study of applicative languages can play in understanding and advancing the art of language design.

Some of the characteristic features of REC among applicative languages are that it is typed (unlike the λ-calculus), truly second-order (unlike Lisp, so that the programs of an implementable version of REC will define global functionals, in our technical sense of this term) and that it allows only controlled and preannounced recursion (i.e. a function cannot call itself, unless its definition is preceded by the "rec" quantifier which announces that recursion will take place and binds and initializes the appropriate variables). The language also enforces a clear and sharp distinction between recursion (local) variables and parameters. There are some indications that these features of REC may make possible reasonably efficient implementations.

In practice, one would expect that most programs would not utilize recursion at all, but would be written in rich, explicit sublanguages of REC which have among their primitives a variety of powerful functionals like while, aptoall, and other forms of iteration which require no stack for their implementation. This is one of the key suggestions in Backus [1978], and there are good reasons to think that REC provides the proper comtext for carrying it out.

3C. The specification problem for concurrent algorithms. In order to design efficiently VLSI circuits which may involve hundreds of thousands of transistors, it is necessary to develop tools for describing faithfully and succintly just what it is that a VLSI circuit does.

The commonly suggested methodology is to start with very small circuits which do simple things, combine these into larger circuits which do more complicated things and continue with this process until you complete the design for the circuit which does the thing you want. At each stage you should be able to specify - and if possible prove - that indeed the circuit just designed does the thing it is supposed to do. The crux of the problem is to find a precise mathematical model of "what a VLSI circuit does," which is faithful and not too complicated, since the complete description of chips as "circuits" or automata quickly becomes unmanageably large.

The problem has obvious practical applications, and of course there are many solutions to it, since in fact people do design successfully terribly complicated chips. Still, there is a lot of research effort devoted to it, both from the practical and the theoretical point of view.

Now VLSI circuits, of course, implement concurrent algorithms and we have claimed that all concurrent algorithms can be represented faithfully by the intensions of terms of REC. The question is whether the representations in REC of the special concurrent algorithms which are implementable on a chip are faithful and succint enough, so that we can develop within REC a practical system for describing and analysing them. Here we will make just a few observations in connection with this very interesting project.

(1) The relevant structures will all be finite, but not first order - i.e. they will have true functionals among their givens. Typical basic sets for these structures are the m-bit numbers, but we can also introduce easily arrays, records and the like.

(2) As a first approximation, we can consider a simple notion of what it means for a <u>circuit to implement a functional</u>. For example, suppose we have a functional $f(x,p)$, where x varies over X, $p : U \to V$ and $f(x,p)$ takes values (when defined) in W; think of X, U, V and W as sets of m-bit numbers for simplicity. A circuit implementing f will have <u>ports</u> as in the drawing,

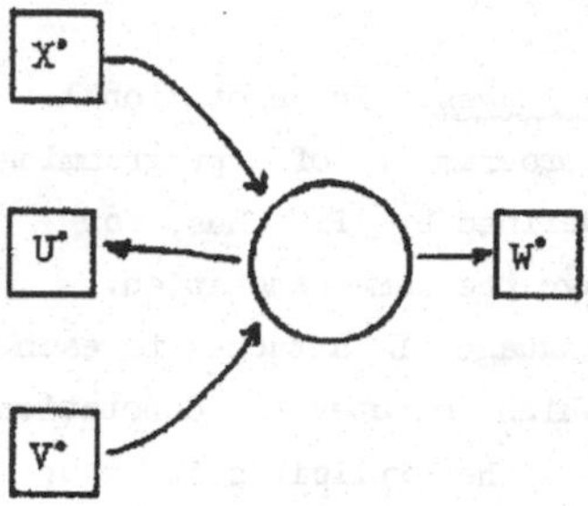

in which members of X, U, V, W can be stored. It will start working when it receives a message from X^* that the input x is ready and when it needs to compute $p(u)$ for some u, it will serve u in U^* and wait until a value v is placed in V^* before it continues; if and when it computes an output, it will place it in W^* and send a message that it has finished.

This kind of simple implementation - for <u>functions</u> - is described in the Ph.D. Thesis of Marina Chen [1983], who developed similar recursive representations for concurrent algorithms independently. It is quite easy to design circuits that implement in this way (e.g.) the "systolic" algorithms for matrix multiplication and the like described in Mead-Conway [1980] and Chen [1983].

It is also easy to check that the basic functionals we have been studying, <u>while</u>, <u>aptoall</u>, $\angle$, <u>iter</u>, $E^{\#}_{G}$ (on finite G) and many others can be implemented, and that the collection of implementable functionals is closed under <u>explicit definability</u>.

(3) Full recursion is not directly implementable, since in general it will need stacks. Thus the appropriate language seems to be $EXP(\sigma)$, for a signature σ which includes at least the functionals above. It should be obvious that there are many special forms of recursion which can be implemented without a stack, e.g. primitive recursion, various forms of "nested" recursion, tail recursion, etc.; these can all be represented directly by functionals which may be taken among the primitives.

One would expect that a very rich, explicit language can be constructed along these lines, whose programs can be "compiled" for the chip essentially automatically and in which one can specify compactly and faithfully a large collection of concurrent algorithms.

(4) The stage comparison algorithm described in 2F is relevant to this project. It can be used (at least in a limited way, since we do not have full recursion) to enrich the explicit language we will be using by commands of the form "do both P and Q and return when one of them becomes defined." It also gives directly a simulator for concurrent algorithms which counts the correct "timing" and can be run in any implementation of REC on a von Neumann machine.

These remarks should be taken with a grain of salt, since we have over-simplified many aspects of the problem and have completely by-passed others, in particular the critical notion of implementations which are faithful for the timing. It would appear, nevertheless, that the project of constructing rich, explicit languages for programming algorithms for the chip is feasible and worth-while.

3D. Intensional semantics for assignment-based languages. In denotational semantics, the "meaning" (denotation) assigned to each program P of a programming language L is (basically) the input-output function defined by P. Thus, for example, all sorting programs for arrays of integers have the same denotation. At the other extreme, the operational semantics of a language L attaches to each program P a computation scheme (in the sense of 1E) which computes the denotation of P. In the case of a recursive program, this involves the explicit setting up of stacks and other details of implementation which complicate the picture substantially and do not seem to help us understand better the "meaning" of the program.

The theory of algorithms presented here suggests an intermediate intensional semantics, where we take the "meaning" of a program P to be the algorithm described by P, represented by an operative system of functionals on a suitable recursion structure. For applicative languages which can be viewed directly as sublanguages of REC, we have already considered this in 2E. It is not hard to see (at least in principle) that the same can be done for the more traditional, assignment-based languages.

The basic techniques for developing denotational semantics for an assignment-based language L, ultimately involve translating (interpretive) L into an applicative language L', typically the untyped λ-calculus. One can use minor variations of these techniques to translate L instead into REC, and then pull the intensional semantics of REC back to L.

We will omit a detailed discussion of this project here, since we are planning to publish fairly soon an exposition of this work.

References

Aho, A.V. and Ullman, J.D. [1979]
Universality of data retrieval languages, in: Proc. of 6th ACM Symp. on Principles of Programming Languages, 1979, 110-117.

Backus, J. [1978]
Can programming be liberated from the von Neumann style? A functional style and its algebra of programs, Comm. of the ACM, 21 (1978), 613-641.

Barwise, J. and Moschovakis, Y.N. [1978]
Global inductive definability, J. of Symbolic Logic, 43 (1978), 521-534.

Barwise, J., Gandy, R.O. and Moschovakis, Y.N. [1971]
The next admissible set, J. of Symbolic Logic, 36 (1971), 108-120.

Bjoerner, D. and Jones, C.B. [1982]
Formal specification and software development, Prentice-Hall, 1982.

Chandra, A. and Harel, D. [1982]
Structure and complexity of computable queries, J. of Computer and System Sciences, 25 (1982), 99-128.

Chandra, A.K. and Harel, D. [1980]
Computable queries for relational data bases, J. of Computer and System Sciences, 21 (1980), 156-178.

Chen, M. [1983]
Space-time algorithms: semantics and methodology, Ph.D. Thesis, Cal. Inst. of Technology, 1983.

de Bakker, J. [1980]
Mathematical theory of program correctness, Prentice-Hall, 1980.

de Rougemont, M. [1983]
Second order and inductive definability on finite structures, Ph.D. Thesis, UCLA, 1983.

Fagin, R. [1974]
Generalized first-order spectra and polynomial-time recognizible sets, in: Complexity of computation, R. Karp, ed., SIAM-AMS Proc. 7, 1974, 43-73.

Feferman, S. [1965]
Some applications of forcing and generic sets, Fund. Math. 56 (1965), 325-345.

Feferman, S. [1977]
Inductive schemata and recursively continuous functionals, in: Colloquium '76, R.O. Gandy, J.M.E. Hyland eds., Studies in Logic, North Holland, Amsterdam, 1977, 373-392.

Fenstad, J.-E. [1980]
General recursion theory, Perspectives in Mathematical Logic, Springer, Berlin, 1980.

Friedman, H. [1971]

Aiiomatic recursive function theory, in: Logic Colloquium '69, R.O. Gandy, C.E.M. Yates eds., Studies in Logic, North Holland, Amsterdam, 1971, 113-137.

Gandy, R.O. [1967]

General recursive functionals of finite type and hierarchies of functionals, Ann. Fac. Sci. Univ. Clermont-Ferrand, 35, 1967, 5-24.

Gordon, M.J.C. [1979]

The denotational description of programming languages, Springer, 1979.

Gurevich, Y. [1983]

Algebras of feasible functions, Proc. of 24th IEEE Symp. on Foundations of Computer Science, 1983, 210-213.

Harel, D. and Kozen, D. [1982]

A programming language for the inductive sets, and applications, Proc. of the 9th ICALP, Springer, 1982.

Immerman, N. [1982]

Relational queries computable in polynomial time, Proc. of the 14th ACM Symp. on Theory of Computing, 1982, 147-152.

Immerman, N. [1983]

Languages which capture complexity classes, Proc. of 15th ACM Symp. on the theory of computing, 1983, 347-354.

Kechris, A.S. and Moschovakis, Y.N. [1977]

Recursion in higher types, in: Handbook of Logic, J. Barwise ed., Studies in Logic, North Holland, Amsterdam, 1976, 681-737.

Keisler, H.J. [1965]

Finite approximations of infinitely long formulas, in: The Theory of Models, J.W. Addison et al, eds., Studies in Logic, North Holland, Amsterdam, 1965, 158-169.

Kleene, S.C. [1952]

Introduction to metamathematics, van Nostrand, Princeton, 1952.

Kleene, S.C. [1959]

Recursive functionals and quantifiers of finite type I, Trans. Amer. Math. Soc. 91 (1959), 1-52.

Kleene, S.C. [1963]

Recursive functionals and quantifiers of finite type II, Trans. Amer. Math. Soc. 108 (1963), 106-142.

Kleene, S.C. [1981]

The theory of recursive functions approaching its centennial, Bull. Amer. Math. Soc. (new series) 5 (1981), 43-61.

Kolaitis, Ph. [????]

Canonical forms and hierarchies in generalized recursion theory, to appear.

Kreisel, G. [1961]

Set theoretic notions suggested by the notion of potential totality, in: Infinitistic Methods, Pergamon, Oxford, 1961, 103-140.

Kreisel, G. [1965]

Model theoretic invariants: applications to recursive and hyperarithmetic operations, in: The Theory of Models, J.W. Addison et al, eds., Studies in Logic, North Holland, Amsterdam, 1965.

Kreisel, G. and Sacks, G.E. [1965]

Metarecursive sets, J. of Symbolic Logic, 30 (1965), 318-338.

Landin, P.J. [1964]

The mechanical evaluation of expressions, Computer J. 6 (1964), 308-320.

Mc Carthy, J. [1960]

Recursive functions of symbolic expressions and their computation by machine, Part I, Comm. of the ACM, 3 (1960), 184-195.

Mc Colm, G. [????]

Ph.D. Thesis, UCLA, in preparation.

Mead, C. and Conway, L. [1980]

Introduction to VLSI Systems, Addison-Wesley, Reading Mass., 1980.

Moschovakis, Y.N. [1967]

Hyperanalytic predicates, Trans. Amer. Math. Soc. 129 (1967), 249-282.

Moschovakis, Y.N. [1969]

Abstract first order computability I and II, Trans. Amer. Math. Soc., 138 (1969), 427-504.

Moschovakis, Y.N. [1971]

Axioms for computation theories - first draft, in: Logic Colloquium '69, R.O. Gandy and C.E.M. Yates eds., Studies in Logic, North Holland, Amsterdam, 1971, 199-255.

Moschovakis, Y.N. [1974]

Elementary Induction on Abstract Structures, Studies in Logic, North Holland, Amsterdam, 1974.

Moschovakis, Y.N. [1977]

On the basic notions on the theory of induction, in: Logic, Foundations of Mathematics and Computability, Butts, Hintikka, eds., Reidel, 1977, 207-236.

Moss, L. [1984]

Power set recursion, Ph.D. Thesis, UCLA, 1984.

Normann, D. [1978]

Set recursion, in: Generalized Recursion Theory II, J.-E. Fenstad, R.O. Gandy, G.E. Sacks, eds., Studies in Logic, North Holland, Amsterdam, 1978.

Platek, R. [1966]

Foundations of recursion theory, Ph.D. Thesis, Stanford Univ., 1966.

Rogers, H. Jr. [1967]

Theory of recursive functions and effective computability, McGraw-Hill, New York, 1967.

Sacks, G.E. [1980]

Three aspects of recursive enumerability, in: Recursion Theory: Its Generalizations and Applications, F.R. Drake and S.S. Wainer, eds., Cambridge Univ. Press, 1980, 184-214.

Sacks, G.E. [????]

On the limits of recursive enumerability, to appear.

Sazonov, V.Y. [1980]

A logical approach to the problem P = NP, Math. Found. of Comp. Science, Springer CS notes #88, 1980, 562-575.

Scott, D.S. [1982]

Domains for denotational semantics, ICALP '82, Aarhus, 1982.

Scott, D.S. and Strachey, C. [1971]

Towards a mathematical semantics for computer languages, Proc. of the Symposium on Computers and Automata, in: J. Fox ed., Polytechnic Institute of Brooklyn Press, New York, 1971, 19-46.

Shore, R.A. [1977]

α-recursion theory, in: The Handbook of Logic, J. Barwise, ed., Studies in Logic, North Holland, Amsterdam, 1977, 653-680.

Slaman, T.A. [????]

Reflection and forcing in E-recursion theory, to appear.

Stoy, J.E. [1977]

Denotational semantics: The Scott-Strachey approach, MIT Press, Cambridge, Mass., 1977.

Strong, R. [1968]

Algebraically generalized recursive function theory, IBM J. Res. and Dev. 12 (1968), 415-475.

Tennent, R.D. [1981]

Principles of programming languages, Prentice-Hall, 1981.

Wagner, E.G. [1969]

Uniform reflexive structures: on the nature of Godelizations and relative computability, Trans. Amer. Math. Soc., 144 (1969), 1-41.

Some Logical Problems Connected with a Modular Decomposition Theory of Automata

Dieter Rödding †
Institut für Mathematische Logik
und Grundlagenforschung
Einsteinstraße 62
D-4400 Münster

0. Introduction:

This article deals with connections between Normed Network Theory and logic. Normed Network Theory ist the theory of modular decomposition of (sequential) automata by networks over special basis automata and we want to illustrate the simplicity and expressiveness of this conception by some investigations in terms of classical logic. Firstly, we give some fundamental definitions and discuss the so-called basis question, i.e. the problem to decide whether or not a given finite set of sequential automata forms a basis with respect to simulation of arbitrary sequential automata. Next, we indicate that we needn't consider arbitrary network constructions but can constrain ourselves to concatenation (chains) of automata. The standard-interpreted first-order language with constants and variables for automata, a function symbol for concatenation and a predicate symbol for the relation between automata A and B, describing that A is equivalent to a sub-automaton of B, has the full arithmetical complexity (and is therefore undecidable). The axiomatization problem for the equivalence of chains has been solved in a special sense by specifying some simple axioms and rules. Another universal algebraic construction principle is presented, for which the corresponding decision and axiomatization problems are open. On the other hand we can use networks over sequential automata and register components to represent functionals of finite type. For type 0 resp. $0 \to 0$ we obtain by this method of course exactly $\mathbb{N}_0$ (the set of natural numbers) resp. $F_{\mu,\text{part}}$ (the set of (partial) recursive functions), and we can represent some important functionals like the combinators K and S, recursors and μ-operators by such networks. Finally, we describe loop-problems of signals running through a network in terms of P-NP-PSPACE problems and discuss the result with respect to a special basis theorem.

1. Fundamental definitions and the basis question

For an extended motivation of the Normed Network Theory and especially of the following definitions we refer to [Rö 83] and [BrPrRöSchä 84].

Definition 1.1:

1. A sequential automaton is a tuple $A=(I_A,O_A,S_A,\delta_A,\lambda_A)$, consisting of finite, non-empty sets I_A (input lines), O_A (output lines), S_A (states), and two partial functions $\delta_A, I_A \times S_A \to S_A$ (transition function) and $\lambda_A, I_A \times S_A \to O_A$ (output function) with dom δ_A=dom λ_A.

2. A sequential A is totally defined iff dom δ_A = dom λ_A = $I_A \times S_A$.

3. For a sequential automaton A, $i \in I_A$, $s,s' \in S_A$, $o \in O_A$ we write "$i\ s \longrightarrow_A s'o$ or simply "$i\ s \longrightarrow s'o$ instead of "$\delta_A(i,s) \cong s' \wedge \lambda_A(i,s) \cong o$."

4. Next we give some special automata which play a crucial role in this article:

Definition (1.2):

$H = (\{t,s_u,s_d\}, \{t_u,t_d,s'_u,s'_d\}, \{u,d\}, \to_H)$,

$E = (\{t,s\}, \{t_u,t_d,s'\}, \{u,d\}, \to_E)$,

$T = (\{t,s\}, \{t_u,t_d\}, \{u,d\}, \to_T)$,

$F = (\{i\}, \{o_1,o_2\}, \{1,2\}, \to_F)$,

$K = (\{i_1,i_2\}, \{o\}, \{0\}, \to_K)$ with

$t\ u \to_H u\ t_u$	$t\ u \to_E u\ t_u$
$t\ d \to_H d\ t_d$	$t\ d \to_E d\ t_d$
$s_u\ u \to_H u\ s'_u$	$s\ u \to_E d\ s'$
$s_u\ d \to_H u\ s'_u$	$s\ d \to_E u\ s'$
$s_d\ u \to_H d\ s'_d$	
$s_d\ d \to_H d\ s'_d$	

$t\ u \rightarrow_T u\ t_u$ $\qquad$ $i\ 1 \rightarrow_F 2\ o_1$

$t\ d \rightarrow_T u\ t_d$ $\qquad$ $i\ 2 \rightarrow_F 1\ o_2$

$s\ u \rightarrow_T d\ t_d$

$i_1\ 0 \rightarrow_K 0\ o$

$i_2\ 0 \rightarrow_K 0\ o.$

($\rightarrow_T$ has no entry (s,d). This means that $\delta_T(s,d)$ and $\lambda_T(s,d)$ are undefined).

These automata are graphically represented by

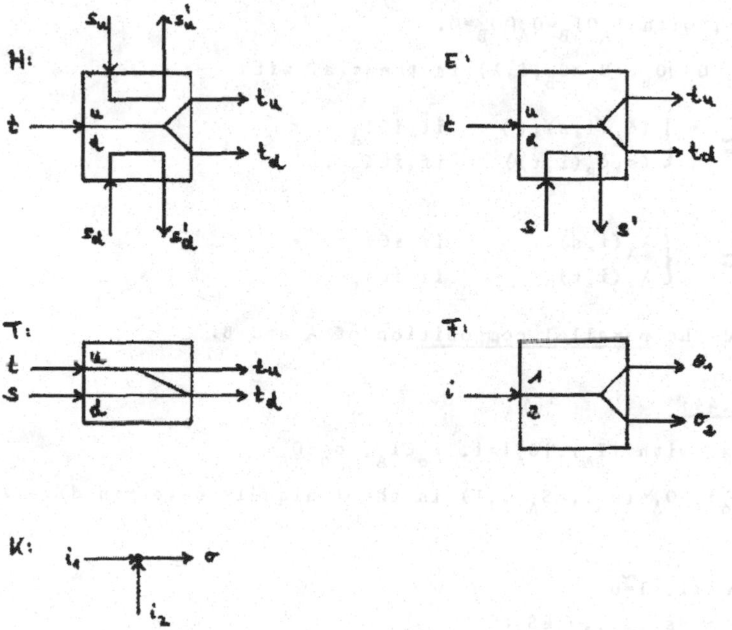

Sequential automata interact with their environment in a sequential operation modus:

The automaton receives an (unit) input signal on some input line from the environment, processes it according to its actual internal state and its transition and output function, and the environment removes the output signal before it sends another input to the automaton.

The 1-out-of-n-code of input and output signals for sequential automata gives rise to combining sequential automata to networks in a canonical way, simply by identifying some output lines with some input lines of the components of the network. This can be done by iterating the processes parallel composition (to collect the components of the network) and feedback (to make the interconnections between them).

Abbreviation: In the following, we shall abbreviate the term "sequential automaton" resp. "sequential automata" simply by "s.a.".

Definition (1.3):

Let A,B be s.a. with $I_A \cap I_B = O_A \cap O_B = \emptyset$.

$A \| B := (I_A \cup I_B,\ O_A \cup O_B,\ S_A \times S_B, \delta, \lambda)$ is the s.a. with

$$\delta(i,(s,t)) \simeq \begin{cases} (\delta_A(i,s),t) & \text{if } i \in I_A \\ (s,\delta_B(i,t)) & \text{if } i \in I_B, \end{cases}$$

$$\lambda(i,(s,t)) \simeq \begin{cases} \lambda_A(i,s) & \text{if } i \in I_A \\ \lambda_B(i,t) & \text{if } i \in I_B \end{cases}$$

$A \| B$ is called the parallel composition of A and B.

Definition (1.4):

Let A be a s.a. with $|I_A|, |O_A| > 1$, $i_o \in I_A$, $o_o \in O_A$.

$A_{i_o}^{o_o} := (I_A \setminus \{i_o\},\ O_A \setminus \{o_o\},\ S_A, \delta, \lambda)$ is the (uniquely determined) s.a. with

$$\begin{aligned} & \delta(i,s) \simeq s' \wedge \lambda(i,s) \simeq o \\ & \Longleftrightarrow \exists n \in \mathbb{N}_+ : \exists s_o, \ldots, s_n \in S_A : \\ & \qquad [s_o = s \wedge s_n = s' \\ & \qquad \wedge\ \delta_A(i,s) \simeq s_1 \wedge \forall 1 \le \nu \le n-1 : \delta_A(i_o, s_\nu) \simeq s_{\nu+1} \wedge n=1 \rightarrow \lambda_A(i,s_o) \simeq o) \\ & \qquad \wedge\ (n>1 \rightarrow \lambda_A(i,s_o) = o_o \wedge \forall 1 \le \nu < n-1 : \lambda_A(i_o,s_\nu) = o_o \wedge \lambda_A(i_o, s_{n-1} = o)) \end{aligned}$$

$A_{i_o}^{o_o}$ is called the feedback of A concerning i_o and o_o.

Parallel composition and feedback are graphically represented as

$A \# B$

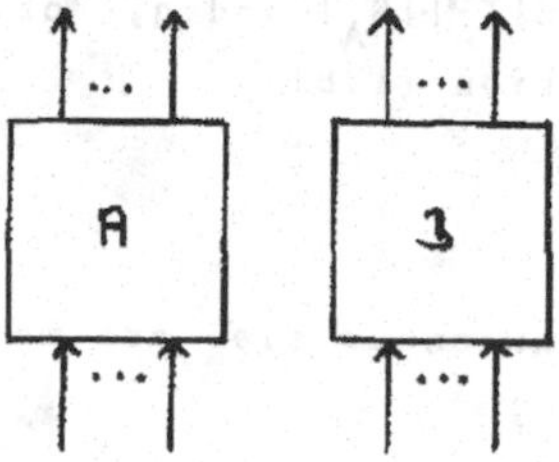

$A^{o_0}_{i_0}$

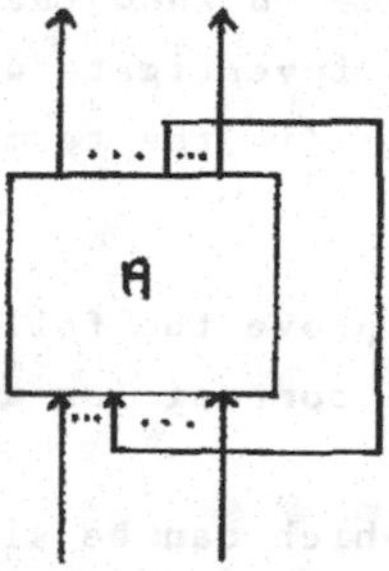

<u>Definition (1.5):</u>

Let $\mathcal{B}$ be a class of s.a. The class NN($\mathcal{B}$) of Normed Networks over $\mathcal{B}$ is the smallest class of s.a. which contains $\mathcal{B}$ and is closed under parallel composition, feedback and isomorphisms.
We call such networks "normed", because only one signal runs through the net at any time.

Our aim is now to decompose s.a. by networks over some simple components. "Decomposition" means that the resulting network (which is a s.a. again) <u>simulates</u> the decomposed automaton.

<u>Definition (1.6.):</u>

1. A s.a. B <u>simulates</u> a s.a. A <u>(isomorphically)</u> iff:
   a) $I_A \leftrightarrow I_B$, $O_A \leftrightarrow O_B$, $S_A \leftrightarrow S_B$.
   b) $\forall i \in I_A : \forall s \in S_A : (\lambda_A(i,s) \cong \lambda_B(i,s) \wedge \delta_A(i,s) \cong \delta_B(i,s)$
2. A class $\mathcal{B}$ of s.a. is called a <u>basis</u> for s.a. iff for all s.a. C there exists a N$\in$NN($\mathcal{B}$) which simulates C.

We illustrate the above definitions by simulating H by a Normed Network N_H over E and K:

Proof:

1. It is routine to show that for the monotony test of a s.a. A one only has to investigate words p of length $\leq|I_A|\cdot|S_A|$ and q,r of length $\leq|S_A|$ (in the terminology of Definition (1.8)).
2. trivial
3. One has to prove the following:
   a) Parallel composition and feedback of monotonous s.a. are monotonous.
   b) A s.a. which can be simulated by a monotonous s.a. is monotonous itself.
   c) H is antimonotonous.

This can be done straightforward.

Theorem (1.10):

Let A be a totally defined s.a. with deg(A)$\leq$1. Then there are equivalent:

1. A is antimonotonous.
2. {A,F,K} is a basis for s.a.

Proof:

"2. ⇒ 1." follows from Lemma (1.9) (note that F,K are monotonuous).

"1. ⇒ 2.": As the difficulties of the theorem do not lay in the proof but in finding the monotony-property, the following hints will be sufficient for the reader to find the proof himself:

1. A has exactly two input lines, output lines and states.
2. Make a list of nine antimonotonous s.a. $D_o,\ldots,D_8$ with degree 1 and show for each D_i that $\{D_i,F,K\}$ forms a basis.
3. Show that each antimonotonous automaton of degree 1 is isomorphic to some D_i.

[Vo 80] has generalized Theorem (1.10) for totally defined s.a. with two internal states, (Theorem (1.12)). For his proof he needs Theorem (1.10) and the following result of [Ott 78]:

Lemma (1.11):

{T,F,K} is a basis for s.a.

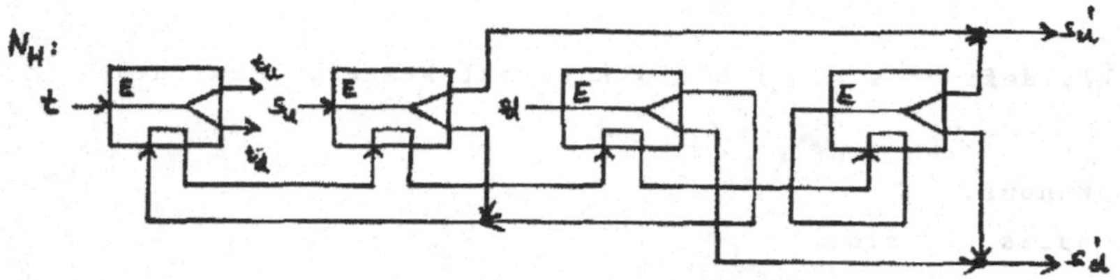

The state u(d) of H corresponds to the state of N_H in which all E-components are in the state u(d).

Theorem (1.7.):

{H,K},{E,K} are bases for s.a.

Recent proofs which take into account "size" and "time" of simulating networks can be found in [BrPrKöSchä 84].

We now turn to the "basis question", i.e. the problem to decide whether or not a given finite set of s.a. forms a basis. Unfortunately the full basis question is still open, but we know an answer in some special cases. The decision criterion is given by a"monotony"-property for s.a., which has been discovered by [Kö 76]. [Kö 76] also proved Lemma (1.9) and Theorem (1.10).

Definition (1.8):

1. A s.a. A is called monotonous iff there holds:

   $\forall s\in S_A:\forall p,q,r\in I_A^*:$

   $[(\delta_A^*(p,s) \cong s \cong \delta_A^*(qr,s) \wedge L(q)\subseteq L(p))$

   $\rightarrow L(\lambda^*(q,s))\subseteq L(\lambda^*(p,s))]$.

2. A is called antimonotonous iff A is not monotonous.

3. $\deg(A) := (|I_A|-1)\cdot(|O_A|-1)\cdot(|S_A|-1)$ is called the degree of A.

Here $\delta^*:I^*\times S\rightarrow S$ and $\lambda^*:I^*\times S\rightarrow O^*$ are the canonical extensions of δ and λ, and L(p) denotes the set of letters which occur in a word p.

Lemma (1.9.):

1. Monotony of s.a. is decidable.
2. All s.a. with degree 0 are monotonous.
3. Each basis for s.a. contains at least one antimonotonous automaton.

Theorem (1.12):

Let A be a totally defined s.a. with two internal states. Then there are equivalent:

1. A is antimonotonous.
2. {A,K,F} is a basis for s.a.

Proof:

Assume $A:(I,O,S,\delta,\lambda)$ to be antimonotonous. Then there exist $x\in I$, $p,q,r,s\in I$, $z\in S$ with:

1. r contains only letters of pxq.
2. $\delta^{*}(pxq,z)\tilde{=}z\tilde{=}\delta^{*}(rxs,z)$.
3. $\lambda(x,\delta^{*}(r,z))$ doesn't occur in $\lambda^{*}(pxq,z)$.

Therefore, A must change its state at least once while processing pxq in state z. But $|S|=2$, and so A reaches state $\delta^{*}(r,z)$ while processing pxq. Therefore without loss of generality one can assume $r=\Lambda$ (the empty word).
Set now $z':=\delta^{*}(p,z)$. Then we have $S=\{z,z'\}$. By identifying all output lines $\neq \lambda(x,z) =: y$ via K-modules, we can assume further $Y=\{y,y'\}$, and the behaviour of A is given by the following graph:

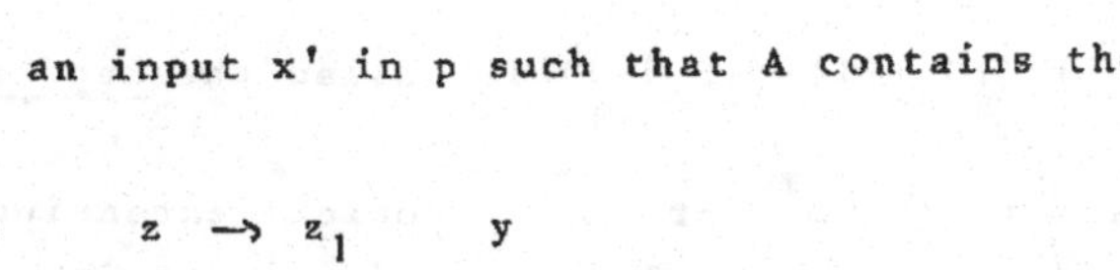

There exists an input x' in p such that A contains the following auto maton T^{*}.

T^{*} :	x	$z \rightarrow z_1$	y
	x	$z' \rightarrow z_2$	y'
	x'	$z \rightarrow z'$	y'
	x'	$z' \rightarrow z_3$	y_3

Case 1: $z_2=z$.
In this case T^{*} is antimonotonous, i.e. $\{T^{*},F,K\}$ is a basis

(Theorem 1.10), and therefore {A,F,K} is a basis, too.

<u>Case 2:</u> $z_2=z'$.

In this case there is an input x" in q such that A contains the automaton T^+:

T^+ :	x	$z \to z_1$	y	}	T^*
	x	$z' \to z_2$	y'	}	
	x'	$z \to z'$	y'	}	
	x'	$z' \to z_3$	y_3	}	
	x''	$z \to z_4$	y_4		
	x''	$z' \to z$	y'		

We may assume $x'=x''$ (otherwise replace $\xrightarrow{x'}$ by $\underset{x''}{\overset{x'}{\rightrightarrows}}\!*\!\rightarrow$).

<u>Case 2a:</u> $z_1=z'$.

In the cases $(z_3,y_3)\in\{(z,y),(z,y'),(z',y)\}$ we can apply Theorem (1.10) again to prove that {A,F,K} is a basis.
Assume now $(z_3,y_3)=(z',y')$.
The network N_T

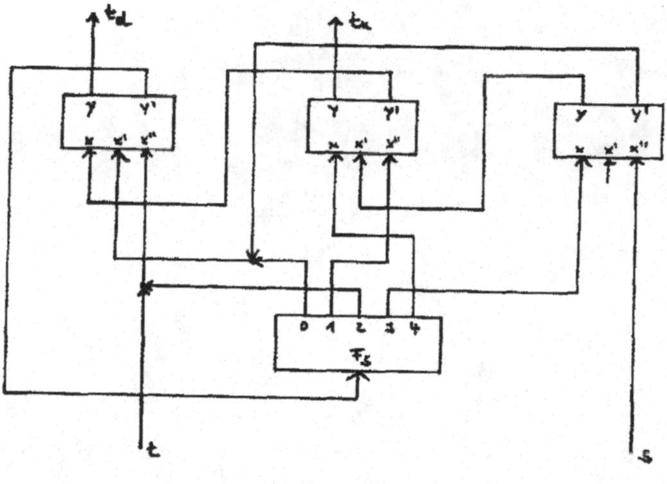

simulates T with respect to the state assignment $u \longmapsto \binom{z'z'z'}{0}$, $d \longmapsto \binom{z'z'z'}{4}$. By Lemma (1.11) we have that {A,F,K} is a basis for s.a.

<u>Case 2b:</u> $z_1 = z$.

Then 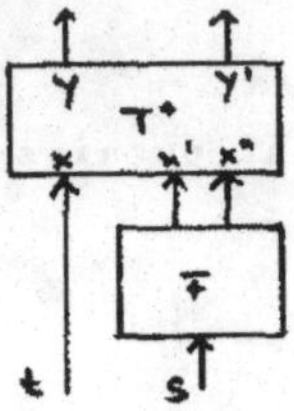 simulates D:

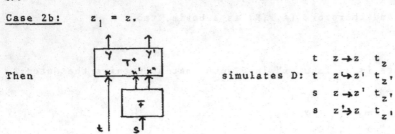

t	$z \to z$	t_z
t	$z \hookrightarrow z'$	$t_{z'}$
s	$z \to z'$	$t_{z'}$
s	$z' \to z$	$t_{z'}$

and since D is antimonotonous, {D,F,K} is a basis (Theorem 1.10), and therefore {A,F,K} is a basis, too.

It is an open problem to generalize Theorem (1.12). Especially we don't know whether

S :	t	u → d	t_u
	t	d → u	t_d
	s	u → d	t_d

or (equivalently)

S^+ :	t	u → d	t_u
	t	d → u	t_d
	t	+ → +	t_+
	s	u → d	t_d
	s	d → +	t_+
	s	+ → +	t_+

forms a basis together with F and K.

But another theorem of [Vo 80] tells us that antimonotonous automata are nearly strong enough to form a basis:

Theorem (1.13):

For each antimonotonous s.a. A there exists a monotonous s.a. B such that {A,B} is a basis for s.a.

Proof:

B will have the form B'‖F‖K. Therefore we may replace A by a network over A,F,K. Hence we can describe the antimonotony of A without loss of generality by the following graph:

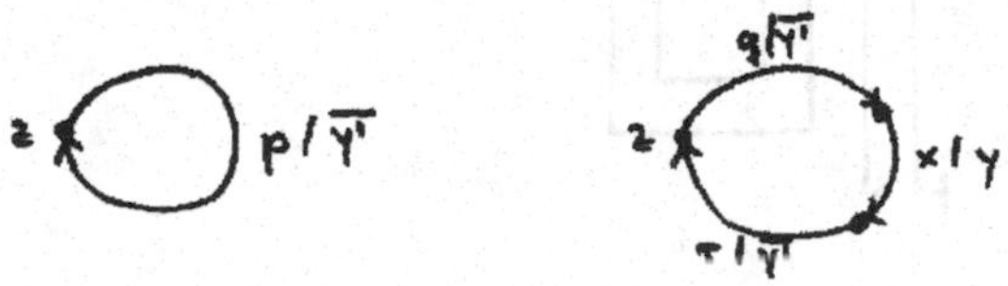

$p=i_1\ldots i_n$, $qxr=j_1\ldots j_m$, $qx=j_1\ldots j_s$, $n,s,m\geq 1$, $s<m$,
$\{j_1,\ldots,j_m\}\subseteq\{i_1,\ldots,i_n\}$. Let B' have exactly the following program lines:

1	$0 \rightarrow z_1$	i_1	
⋮			
3	$z_\nu \rightarrow z_{\nu+1}$	$i_{\nu+1}$	$(1 \leq \nu < n)$
⋮			
3	$z_n \rightarrow 1$	1	
0	$1 \rightarrow 0$	1	
0	$0 \rightarrow z'_1$	j_1	
⋮			
3	$z'_\sigma \rightarrow z'_{\sigma+1}$	$j_{\sigma+1}$	$(1 \leq \sigma < s)$
⋮			
2	$z'_s \rightarrow z'_{s+1}$	j_{s+1}	
⋮			
3	$z'_\mu \rightarrow z'_{\mu+1}$	$j_{\mu+1}$	$(\sigma < \mu < m)$
⋮			
3	$z'_m \quad 0$	0	

Because of $\{j_1, \ldots, j_m\} \subseteq \{i_1, \ldots, i_n\}$, B' is monotonous. The network N_T

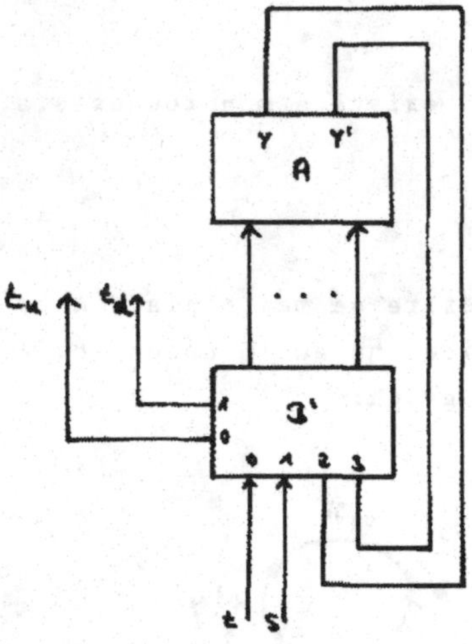

simulates T with respect to the state assignment $u \longmapsto (0,z)$, $d \longmapsto (1,z)$. Therefore the monotonous $B := B' \| F \| K$ fulfills the requirements.

2. Normal forms: the decision and the axiomatization problem

In this section we consider automata with special input and output devices that allow some special network construction principles which each can be described by a function of those special automata. Naturally, these construction principles should be universal with respect to the corresponding type of automata. In this context we investigate the expressive power of the standard-interpreted first-order-logic with variables and individual constants for automata of the specified type, function symbols for the construction principles and predicate symbols for same canonical predicates ("decision problem"). On the other hand we will investigate the "axiomatization problem", i.e. the axiomatization of the equivalence of networks built up according to the special construction principles.

The first construction principles to be investigated is concatenation of automata chains. The corresponding automata are chain automata:

Definition (2.1):

A chain automaton of type $\begin{pmatrix} k & m \\ l & n \end{pmatrix}$ is a s.a. $A=(I,O,S,\delta,\lambda)$ with

$I=\{<1<,\dots,<m<\} \cup \{>1>,\dots,>l>\}$ and

$O=\{<1<,\dots,<k<\} \cup \{>1>,\dots,>n>\}$.

$<1<,\dots,<k<$ are	the left output lines
$>1>,\dots,>l>$	the left input lines,
$<1<,\dots,<m<$	the right input lines,
$>1>,\dots,>n>$	the right output lines of A.

A chain automaton A of type $\begin{vmatrix} k & m \\ l & n \end{vmatrix}$ is represented in the following way:

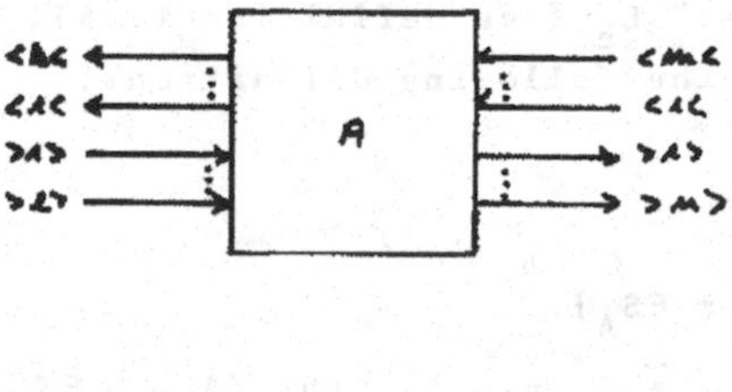

By A $\begin{pmatrix} k & m \\ l & n \end{pmatrix}$ we indicate, that A is a chain automaton of type $\begin{pmatrix} k & m \\ l & n \end{pmatrix}$.

In addition to Definition (1.6), simulation between chain automata must be compatible with the given enumeration and localization (left or right) of the input and output lines.

Definition (2.2):

Let A $\begin{pmatrix} k & m \\ 1 & n \end{pmatrix}$ and B $\begin{pmatrix} m & r \\ n & s \end{pmatrix}$ be chain automata. The concatenation AB $\begin{pmatrix} k & r \\ 1 & s \end{pmatrix}$ is the chain automaton which is uniquely determined by the following Normed Network over A and B:

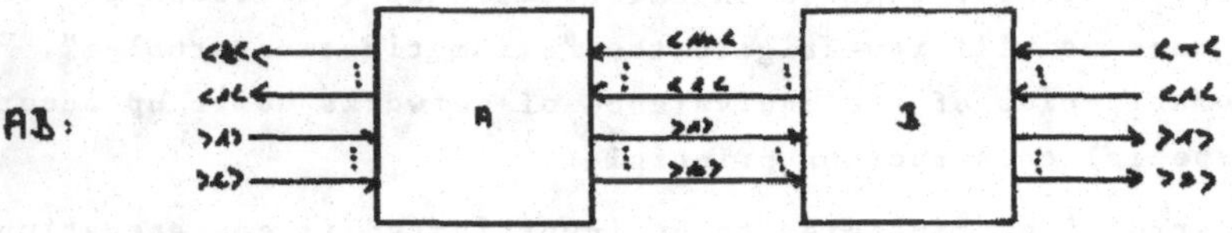

Concatenation is universal for chain automata.

Theorem (2.3):

$\exists B_o \begin{pmatrix} 2 & 2 \\ 2 & 2 \end{pmatrix}, \ldots, B_{12} \begin{pmatrix} 2 & 2 \\ 2 & 2 \end{pmatrix} : \forall \begin{pmatrix} k & m \\ 1 & n \end{pmatrix} : \exists L_{k,1} \begin{pmatrix} k & 2 \\ 1 & 2 \end{pmatrix} : \exists R_{m,n} \begin{pmatrix} 2 & m \\ 2 & n \end{pmatrix}$

$\forall A \begin{pmatrix} k & m \\ 1 & n \end{pmatrix} : \exists A_1, \ldots, A_t \in \{B_o, \ldots, B_{12}\}:$

$L_{k,1}\, A_1 \ldots A_t\, R_{m,n}$ simulates A.

A proof can be found in [KöOtt 74].

For the discussion of the decision problem we consider initial chain automata of type $\begin{pmatrix} 2 & 2 \\ 2 & 2 \end{pmatrix}$ which allow for speaking about internal states in the"concatenation-language" L_c (see definition (2.5), part 1). For the semantics of L_c we need the following definitions:

Definition (2.4):

1. $\mathbb{A} := \{(A,z_o) \mid A \begin{pmatrix} 2 & 2 \\ 2 & 2 \end{pmatrix} \wedge z_o \in S_A\}$.
2. $R(A,z_o) := \{z \in S_A \mid \exists p \in I_A^* : z \cong \delta_A^*(p,z_o)\}$ für $(A,z_o) \in \mathbb{A}$.
3. $(A,z_o) \approx (B,z_1)$

   $\longleftrightarrow \exists \varphi : R(A,z_o) \longrightarrow R(B,z_1)$ bij.$:[\varphi(z_o)=z_1 \wedge$

   $\forall x \in I_A : \forall z \in R(A,z_o)\ (\delta_A(x,z) \cong \delta_B(x,\varphi(z)) \wedge \lambda_A(x,z) \cong \lambda_B(x,\varphi(z)))]$

   for $(A,z_o), (B,z_1) \in \mathbb{A}$.

4. $(A,z_o) \sim (B,z_1) \;:\Longleftrightarrow\; \forall p \in I_A^* : \lambda_A^*(p,z_o) \simeq \lambda_B^*(p,z_1)$ für $(A,z_o),(B,z_1) \in \mathbb{A}$.

5. $(A,z_o) \subseteq (B,z_1) \;:\Longleftrightarrow\; \exists z_2 \in R(B,z_1) : (A,z_o) \approx (B,z_1)$.

In a natural way from Theorem (2.3) one can obtain a finite set $\mathbb{M} \subseteq \mathbb{A}$ which forms a basis for $\mathbb{A}$ with respect to $\approx$ via concatenation.

<u>Definition (2.5):</u>

1. L_c is the first-order-language with standard interpretation $(\mathbb{A}; (M)_{M \in \mathbb{M}}; c; \approx, \sim, \subseteq)$, where $c((A,z_o),\ (B,z_1)) := (A,z_o)(B,z_1) := (AB,(z_o,z_1))$.
2. $\mathfrak{a}_c$ is the class of all predicates over $\mathbb{A}$ which are definable in L_c. $\mathfrak{a}_c$ is called the class of <u>arithmetical predicates</u> over $\mathbb{A}$.

To justify the term "arithmetical predicate over $\mathbb{A}$"[Ott 74] has defined a natural "effective" gödelization $\gamma: A \longrightarrow \mathbb{N}$, and for the class $\mathfrak{a}_\gamma$ of the predicates over $\mathbb{A}$ which correspond to arithmetical predicates over $\mathbb{N}$ via γ he has shown:

<u>Theorem (2.6):</u>

$\mathfrak{a}_c = \mathfrak{a}_\gamma$.

[Br 83] slightly has sharpened this result:

<u>Definition (2.7):</u>

1. L_c^- is the first-order-language with standard interpretation $(\mathbb{A}; (M)_{M \in \mathbb{M}}; c; \approx, \subseteq)$.
2. $\mathfrak{a}_c^-$ is the class of predicates over $\mathbb{A}$ which are definable in L_c^-.

<u>Theorem (2.8):</u>

$\mathfrak{a}_c^- = \mathfrak{a}_\gamma$.

<u>Proof:</u>

[Ott 74] has used "counters" to represent natural numbers within $\mathbb{A}$: $(X,a_o) \subset \mathbb{A}$ is a <u>counter</u> ($\mathfrak{C}(X,a_o)$) iff $R(X,a_o) = \{a_o,\ldots,a_p\}$ and X has the following transitions:

$$\begin{array}{llll} <1< & a_i \rightarrow a_{i+1} & >1>, & \text{for } 0 \leq i < p \\ <1< & a_p \rightarrow a_o & <1< & \\ x & a_i \rightarrow a_i & x & \text{for } x \in \{<2<,>1>,>2>\} \\ & & & 0 \leq i \leq p. \end{array}$$

Naturally, a counter (X,a_o) represents the number $|S_X|-1$.

The proof of Theorem (2.8) is nearly the same as in [Ott 74] but here we must show "$\mathfrak{L} \in \mathfrak{A}_c^-$" instead of "$\mathfrak{L} \in \mathfrak{A}_c$"

We describe $\mathfrak{L}$ as in [Ott 74]:

$$\mathfrak{L}(X,a_o) \Longleftrightarrow \mathfrak{L}_o(X,a_o) \vee (\mathfrak{L}_1(X,a_o) \wedge \mathfrak{L}_2(X,a_o)), \text{ where}$$

$$\mathfrak{L}_o(X,a_o) :\Longleftrightarrow R(X,a_o) = \{a_o\} \wedge \forall x \in \{>1>,>2>,<1<,<2<\}: x\, a_o \xrightarrow[X]{} a_o$$

$$\begin{aligned}
\mathfrak{L}_1(X,a_o) :\Longleftrightarrow\ & 1) \forall a \in R(X,a_o): \forall x \in \{>1>,>2>,<2<\}: x\, a \xrightarrow[X]{} ax \\
\wedge\ & 2) \forall a \in R(X,a_o): \exists b \in R(X,a_o): (<1<a \xrightarrow[X]{} b<1< \\
& \qquad \vee <1<a \xrightarrow[X]{} b>1>) \\
\wedge\ & 3) \exists p \in \mathbb{N}_o: \exists a_1,\dots,a_p \in R(X,a_o): \\
& \quad (R(X,a_o) = \{a_o,\dots,a_p\} \wedge \forall 0 \le i < p: \delta_X(<1<,a_i) \cong a_{i+1} \\
& \qquad \wedge \delta_X(<1<,a_p) \cong a_o) \\
\wedge\ & 4) \neg\exists a \in R(X,a_o): \lambda_X^*(<1<<1<,a) \cong <1<<1< \\
\wedge\ & 5) \lambda_X(<1<,a_o) \cong >1> \\
\wedge\ & 6) \exists a \in R(X,a_o): \lambda_X(<1<,a) \cong <1< \\
\wedge\ & 7) \exists s \in \mathbb{N}_+: (s \text{ devides } |R(X,a_o)| \wedge \\
& \quad \lambda_X^*(<1<<1<\dots,a_o) \cong \underbrace{>1>\dots>1>}_{s-1}<1<\underbrace{>1>\dots>1>}_{s-1}<1<\dots
\end{aligned}$$

$$\mathfrak{L}_2(X,a_o) :\Longleftrightarrow \lambda_X^*(\underbrace{<1<\dots<1<}_{|R(X,a_o)|}) \cong \underbrace{>1>\dots>1>}_{|R(X,a_o)|-1}<1<.$$

There holds:

$$\mathfrak{L}_o(X,a_o) \Longleftrightarrow X' \approx A_1',$$

$$\begin{aligned}
\mathfrak{L}_1(X,a_o) \Longleftrightarrow\ & 1)\ \forall Z' \subseteq X': (A_2'Z'A_2' \approx A_2' \wedge A_3'Z'A_3' \approx A_3' \wedge A_4'Z'A_4' \approx A_4') \\
\wedge\ & 2)\ \forall Z' \subseteq X': (A_5'Z'A_5' \approx A_5' \vee A_6'Z'A_7' \approx A_8') \\
\wedge\ & 3)\ \forall Z' \subseteq X': X' \subseteq Z' \\
\wedge\ & 4)\ \neg\exists Z' \subseteq X': A_9'Z'A_9' \approx A_9' \\
\wedge\ & 5)\ A_6'X'A_7' \approx A_8' \\
\wedge\ & 6)\ \exists Z' \subseteq X': A_5'Z'A_5' \approx A_5' \\
\wedge\ & 7)\ A_{10}'X'A_{11}' \approx A_{12}'X'A_{13}' \quad ,
\end{aligned}$$

$$\mathfrak{L}_1(X,a_o) \Rightarrow (\mathfrak{L}_2(X,a_o) \Longleftrightarrow \neg\exists Y';Z': (\mathfrak{L}_1(Y') \wedge \mathfrak{L}_1(Z') \wedge A_{14}'Z'A_{15}'Y'A_{16}' \approx A_{17}'X'A_{16}') ,$$

where $A_1,\dots,A_{17}$ are given by the following table and $A_i' := (A_i,0)$ $(1 \le i \le 17)$, $X' := (X,a_o)$:

A_1 :
>1> 0 ⟶ 0 >1>
>2> 0 ⟶ 0 >2>
<1< 0 ⟶ 0 <1<
<2< 0 ⟶ 0 <2<

A_2 :
<2< 0 ⟶ 0 <2<

A_3 :
>1> 0 ⟶ 0 >1>

A_4 :
>2> 0 ⟶ 0 >2>

A_5 :
<1< 0 ⟶ 1 <1<

A_6 :

A_7 :
<1< 0 ⟶ 0 <1<
>1> 0 ⟶ 1 >1>

A_8 :
<1< 0 ⟶ 1 >1>

A_9 :
<1< 0 ⟶ 1 <1<
<1< 1 ⟶ 2 <1<

A_{10}:
<1< 0 ⟶ 1 >2>
<1< 1 ⟶ 1 <1<

A_{11}:
<1< 0 ⟶ 0 <1<
>1> 0 ⟶ 0 <1<
>2> 0 ⟶ 1 <1<
>1> 1 ⟶ 1 >1>
<1< 1 ⟶ 1 <1<

A_{12}:
<2< 0 ⟶ 1 >2>
<1< 1 ⟶ 1 <1<

A_{13}:
<1< 0 ⟶ 0 <2<
>2> 0 ⟶ 1 <1<
<1< 1 ⟶ 1 <1<
>1> 1 ⟶ 1 >1>

A_{14}:
<1< 0 ⟶ 0 <1<
<2< 0 ⟶ 0 <1<

A_{15}:
<1< 0 ⟶ 0 <1<
>1> 0 ⟶ 0 <2<

A_{16}:
<1< 0 ⟶ 0 <1<
>1> 0 ⟶ 0 >1>

A_{17}:
<1< 0 ⟶ <1< 0

If we replace now $A_1,\ldots,A_{17}$ by chains over $\mathbb{M}$ we obtain a description of $\mathcal{C}$ in L_c^-, and the theorem is proven.

The axiomatization problem for the equivalence of chains has been solved approximatively by [Kö 76]. He has specified 6 automata of type $\binom{3\ 3}{3\ 3}$ which can simulate each type $\binom{2\ 0}{2\ 0}$-automaton via concatenation and some simple(!) axioms and rules to axiomatize syntactically the equivalence of chains built up from the 6 basis automata.

[Kö 76] also has investigated other construction principles. His special automata are <u>acceptors</u>, i.e.s.a. A with $I_A = \{x_1,x_2\}$ and $O_A = \{y_1,y_2\}$. His special construction principles are "network functions": A network N with acceptor components $X_1,\ldots,X_n$ which itself is an acceptor represents a function F_N: Given acceptors $A_1,\ldots,A_n$ one has to replace $X_1,\ldots,X_n$ in N by $A_1,\ldots,A_n$ to receive the acceptor $F_N(A_1,\ldots,A_n)$.

<u>Definition (2.9):</u>

1. $\mathbb{B} := \{A \mid A \text{ s.a.},\ I_A = \{x_1,x_2\},\ O_A = \{y_1,y_2\}\}$.
2. For $A_1,A_2 \in \mathbb{B}$ the acceptors $F_1(A_1,A_2)$, $F_2(A_1,A_2) \in B$ are given by the following networks:

$F_1(A_1,A_2)$:

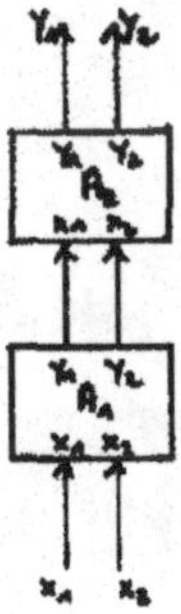

$F_2(A_1,A_2)$:

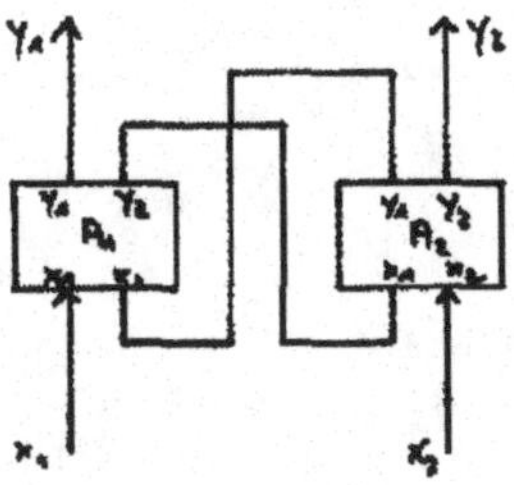

3. $D_1 := (\{x_1,x_2\}, \{y_1,y_2\}, \{u,d\}, \underset{D_1}{\longrightarrow})$,

$N := (\{x_1,x_2\}, \{y_1,y_2\}, \{0\}, \underset{N}{\longrightarrow})$ with

$$x_1\, u \underset{D_1}{\longrightarrow} u\, y_1 \qquad x_1\, 0 \underset{N}{\longrightarrow} 0\, y_2$$

$$x_1\, d \underset{D_1}{\longrightarrow} u\, y_2 \qquad x_2\, 0 \underset{N}{\longrightarrow} 0\, y_1$$

$$x_2\, u \underset{D_1}{\longrightarrow} d\, y_2$$

$$x_2\, d \underset{D_1}{\longrightarrow} u\, y_1$$

4. $\mathbb{M}\,(D_1,N;\, F_1,F_2)$ is the smallest class which contains D_1 and N and is closed under F_1 and F_2.

Theorem (2.1o):

$\forall A \in \mathbb{B}: \exists N \in \mathbb{M}$: N simulates A.

Naturally, simulation here respects the actual names of inputs and outputs. Unfortunately, the proof of theorem (2.1o) is not published and it is too long and too complicated to give it here, but it is available by xerox from the author.

We conclude this section with two open problems, namely the decision and the axiomatization problem for acceptors together with the construction principles F_1 and F_2.

3. Functionals of finite type: representation by automata

In this section we give some hints how to represent functionals of finite type by automata and thus give an alternative definition of "recursive" functionals of finite type Naturally, for type $0 \longrightarrow 0$ exactly the partial recursive functionals shall be representable. For this purpose we use initial s.a. with possibly infinite, but denumerable set of states:

Definition (3.1):

1. A recursive s.a. (r.s.a.) is a tuple $A = I_A,O_A,S_A,z_A,\delta_A,\lambda_A)$ with finite sets I_A,O_A, a finite or denumerable recursive set S_A, $z_A \in S_A$, and partial recursive functions $\delta_A: I_A \times S_A \longrightarrow S_A$ and $\lambda_A: I_A \times S_A \longrightarrow O_A$ with dom δ_A = dom λ_A.

2. $R := (\{a,s\},\{=,\neq\}, \mathbb{N}_0, 0, \delta, \lambda)$ with

$$\delta(a,n) = n+1, \ \lambda(a,n) = \neq$$

$$\delta(s,n) = n-1, \ \lambda(s,n) = \begin{cases} \neq & \text{for } n \neq 0 \\ = & \text{for } n = 0 \end{cases}$$

In this section r.s.a. are simply called automata. Now one can easily extend the definitions of networks, simulation and basis to r.s.a. and prove the following result:

<u>Lemma (3.2):</u>

$\{(H,u),(K,0),R\}$ is a basis for r.s.a.

But now let's turn to functionals of finite type.

<u>Definition (3.3):</u>

1. (Finite) types are defined inductively as usual: 0 is a type. If σ,τ are types then $\sigma \longrightarrow \tau$ is a type.
2. The classes F_τ of <u>functionals of type τ</u> are defined inductively as usual: $F_0 = \mathbb{N}_0$, $F_{\sigma \longrightarrow \tau} = \{f: F_\sigma \longrightarrow F_\tau \mid f \text{ partial function}\}$.

To relate automata to types we use the concept of "data lines":

<u>Definition (3.4.)</u>

1. A <u>data line</u> is a non empty string $d = d_1,\ldots,d_n$ of i (for input line) and o (for output line).
2. $\bar{}$ is the homomorphism of data lines with $\bar{i} = o$, $\bar{o} = i$.
3. To each type τ a data line $d(\tau)$ is associated inductively: $d(o) = i\ o\ o$, $d(\sigma \longrightarrow \tau) = \bar{\sigma}\tau$.

<u>Definition (3.5):</u>

1. An <u>automaton</u> A <u>with data line</u> $d = d_1,\ldots,d_n$ has ordered its input and output lines $I_A \dot{\cup} O_A = \{x_1,\ldots,x_n\}$ such that $x_k \neq x_l$ for $1 \leq k \neq l \leq n$, $x_k \in I_A \Longleftrightarrow d_k = i$ for $1 \leq k \leq n$.
2. An automaton of type τ is an automaton with data line $d(\tau)$.
3. i is identified with the symbols ↧ and ⤒, o is identified with the symbols ↥ and ⤓.

An automaton A, for example, with data line d = i o o (or d = $\underline{\downarrow\uparrow\uparrow}$ or d = $\overline{\downarrow\downarrow\downarrow}$ equivalently is represented by the figures

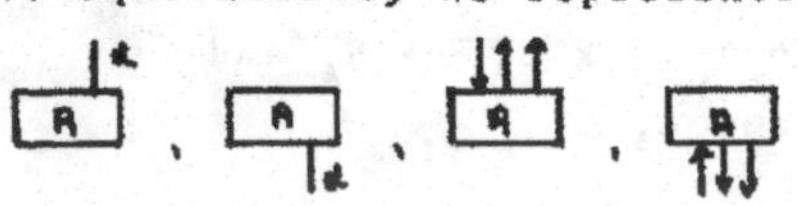

.

An automaton which "knows" some natural numbers n can give this information to its environment via a data line $\overset{x_1}{\downarrow}\overset{x_2}{\uparrow}\overset{x_3}{\uparrow}$ by answering to an input sequence x_1^{n+1} with the output sequence $x_2^n x_3$. Thus this automaton represents n.

The proposition that "an automaton A of arbitrary type τ represents a functional $f \in F_\tau$" may be made precise by the following inductive definition:

Definition (3.6):

1. An automaton A of type 0 represents a number $n \in F_o$ iff: $\lambda_A^*(x_1^{n+1}, z_A) = x_2^n x_3$ and $\delta_A^*(x_1^{n+1}, z_A) = z_A$.
2. An automaton A of type $\sigma \longrightarrow \tau$ represents a functional $f \in F_{\sigma \longrightarrow \tau}$ iff for all automata B of type σ and all $g \in F_\sigma$ which are represented by B and for the canonically defined automaton $A \bullet_\tau^\sigma B$

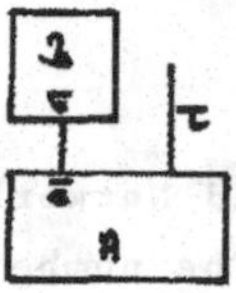

there holds:
$A \bullet_\tau^\sigma B$ represents a functional of type τ iff $f\ g\downarrow$, and if $f\ g\downarrow$ then $A \bullet_\tau^\sigma B$ represents fg.

It's not very hard to prove the following.

Lemma (3.7):

For all $f \in F_{o \longrightarrow o}$ there holds:
f is representable by an automaton of type $o \longrightarrow o$ iff f is partial recursive.

But what about functionals of higher types? The next step of investigation should be the question whether or not some special combinators like

$K \in F_{\alpha \to \beta \to \alpha}$ with $Kfg = f$,

$S \in F_{(\alpha \to \beta \to \gamma) \to (\alpha \to \beta) \to \alpha \to \gamma}$ with $Sfgh = fh(gh)$, and

$R \in F_{0 \to \alpha \to (0 \to \alpha \to \alpha) \to \alpha}$ with $R0fg = f$,
$R(n+1)fg = gn(Rnfg)$

are representable.

These questions have been investigated by [Kn 8o]. Indeed, he has represented these functionals. For this purpose he has defined his special representing automata in a somewhat restrictive manner to enable them to perform various concurrent computations respecting the condition, that these computations do not disturb one another. However, we think that we should further look for a really satisfactory solution of this problem.

4. Back to the basis question

Let's now switch back to arbitrary Normed Networks over non-specialized s.a.. In this section we refer some results about the "distance" between basis and non-basis sets of s.a.

Remember from section 1 that {H,F,K} forms a basis for s.a. but {F,K} does not. [Ott 73] has described the "distance" between {H,F,K} and {F,K} by the following theorem:

Theorem (4.1):

For each acceptor $A \in \mathbb{B}$ there exists a Normed Network $N \in \mathbf{NN}(H,F,K)$ with three H-components only that simulates A. The number of components of N_A and the number of components of N_A which are used by a signal during a simulation process are polynomially in the size of A. This result appears in a new light if one compares it with some new result of [KB 84].

Definition (4.2):

The reachability problem for a class $\mathcal{B}$ of s.a. is defined as follows:
Instance: $N \in NN(\mathcal{B})$ given by its components and the connections (feedbacks) between them, $i \in I_N, s, s' \in S_N$, $o \in O_N$.
Question: $\delta_N(i,s) \cong s' \wedge \lambda_N(i,s) \cong o$?

Theorem (4.3):

1. The reachability problem for {F,K} is decidable in polynomial time.
2. The reachability problem for {H,F,K} is PSPACE-complete.

5. Final remarks and acknowledgements

We hope to have convinced the reader that the theory of Normed Networks involves a lot of interesting combinatorial and logical problems. Especially the results about arithmetical predicates seem to justify considering Normed Networks as representatives of canonical operations over automata which are comparable with arithmetical operations of natural numbers.

Finally I would like to thank Frau Anne Brüggemann for her engaged help in the preparation of this paper and Frau Karin Grulich and Frau Petra Gehlhar for diligently typewriting the manuscript.

6. References

[Br 83] Brüggemann, A.: Personal communication

[Br Pr Rö Schä 84] Brüggemann, A./Priese, L./ Rödding, D./Schätz, R.: Modular Decomposition of Automata. In: Börger,E./ Hasenjäger,G./ Rödding, D.(eds), Proceedings of the Symposium "Rekursive Kombinatorik", Münster 1983. To appear in LNCS (1984)

[Kn 8o] Kniza,K.-P.: Automaten und rekursive Funktionale endlichen Typs, Dissertation, Inst. f. math.Logik und Grundlagenforschung, Münster 198o

[KB 84] Kleine Büning, H.: Complexity of Loop-Problems in Normed Networks. In: Börger,E./Hasenjäger,G./Rödding,D.(eds.), Proceedings of the Symposium "Rekursive Kombinatorik", Münster 1983. To appear in LNCS (1984)

[Kö Ott 74] Körber,P./Ottmann,Th.: Simulation endlicher Automaten durch Ketten aus einfachen Bausteinautomaten. EIK lo (1974)

[Kö 76] Körber,P.: Untersuchungen an sequentiellen, durch normierte Konstruktionen gewonnenen Netzwerken endlicher Automaten. Dissertation, Inst. f. math. Logik und Grundlagenforschung, Münster 1976

[Ott 73] Ottmann,Th.: Über Möglichkeiten zur Simulation endlicher Automaten durch eine Art sequentieller Netzwerke aus einfachen Bausteinen. ZmLG 19(1973)

[Ott 74] Ottmann,Th.: Arithmetische Prädikate über einem Bereich endlicher Automaten. Arch. math. Logik 16 (1974)

[Ott 78] Ottmann,Th.: Eine einfache universelle Menge endlicher Automaten. ZmLG 24 (1978)

[Rö 83] Rödding,D.: Modular Decomposition of Automata (Survey), in: Karpinski,M. (ed.) , Proceedings of the 1983 International FCT-Conference, LNCS 158

[Vo 8o] Vobl,R.: Komplexitätsuntersuchungen an Basisdarstellungen endlicher Automaten. Diplomarbeit, Inst. f. math. Logik und Grundlagenforschung, Münster (198o).

DIOPHANTINE EQUATIONS IN A FRAGMENT OF NUMBER THEORY

Ulf R. Schmerl [1])

Mathematisches Institut der Universität München

We study the following problem: Given a diophantine equation, is it possible to find out whether or not this equation can be proved impossible in the fragment Z_o of classical number theory in $0, S, +, \cdot$ and open induction?

In [1] Shepherdson constructed a counter model showing that none of the following equations can be refuted in Z_o:

$$2x+1 = 2y$$

$$(x+1)^2 = 2(y+1)^2$$

$$(x+1)^3 + (y+1)^3 = (z+1)^3 .$$

We shall use proof-theoretic methods in order to prove the following: Let $r(x_1..x_n)=s(x_1..x_n)$ be a diophantine equation in the variables $x_1..x_n$ where r and s are polynomials in these variables with coefficients in $\mathbb{N}$. Then

$\forall x_1..x_n[r(x_1..x_n)\neq s(x_1..x_n)]$ is provable in Z_o iff

there exists a natural number c such that each polynomial which can be obtained from $r(x_1..x_n)-s(x_1..x_n)\in\mathbb{Z}[x_1..x_n]$ by substitutions of $\hat{x}_i=0,1,..,c-1,x_i+c$ for the variable x_i, $\forall i$, $1\leq i\leq n$, divides some polynomial with natural number coefficients and a constant coefficient >0, i.e. $\exists c\in\mathbb{N}\forall\hat{x}_1\in\{0,1,..,c-1,x_1+c\},..,\hat{x}_n\in\{0,1,..,c-1,x_n+c\}\exists q\in\mathbb{Z}[x_1..x_n]$ $(r(\hat{x}_1..\hat{x}_n)-s(\hat{x}_1..\hat{x}_n))\cdot q\in 1+\mathbb{N}[x_1..x_n]$.

---

[1]) The author would like to thank H.Schwichtenberg and G.Kreisel for advice.

We conjecture that this property is decidable for each polynomial $p\in Z[x_1..x_n]$, i.e. that it is decidable whether or not a given diophantine equation can be refuted by Z_o.

A simple but useful criterion in order to establish the independence of diophantine equations of Z_o is the following: If in an equation r=s there is a variable x occurring in r but not in s and a variable y occurring in s but not in r (i.e. r=s has a pair of separated variables) then r≠s is not provable in Z_o. This gives a purely syntactical method of proving that the equations listed above and others like $(x+1)^n+(y+1)^n=(z+1)^n$ $(n\geq 3)$, $3x+2=y^2$, $x^2+x+1=2y$ etc. cannot be refuted in Z_o.

We also show that the given characterization of Z_o is preserved when adding the functions P (predecessor), sg, and $\overline{sg}$ (sign and co-sign).

<u>1. The equational calculus DGI.</u> We first study an equational calculus DGI; Z_o will turn out to be conservative over this calculus with respect to the provability of equations and inequations.

The system DGI in the language with 0,S,+,· is formulated as a classical sequent calculus with sequents of the form $\Gamma\vdash\Delta$, where Γ,Δ are finite, possibly empty, sets of equations. Equations of the form s=t and t=s are identified. The axioms of DGI are $r=s\vdash r=s$, axioms for equality, the defining axioms for the functions S,+,·, together with the corresponding equality axioms. The inference rules of DGI are rules for thinning, cut, and the following rules for induction:

$$\frac{\vdash r(0)=s(0) \qquad r(x)=s(x)\vdash r(Sx)=s(Sx)}{\vdash r(t)=s(t)} \quad (\mathrm{Ind}_r)$$

$$\frac{r(0)=s(0)\vdash \qquad r(Sx)=s(Sx)\vdash r(x)=s(x)}{r(t)=s(t)\vdash} \quad (\mathrm{Ind}_l)$$

where t is an arbitrary term.

The normal form F(t) of a term t is its representation as a polynomial in several variables with coefficients in the natural numbers under an arbitrary but fixed order of the monomials.

<u>1. Lemma:</u> For each term t, DGI$\vdash$ t=F(t) .
Proof: Associativity, commutativity of + and · and the ditributive laws are provable in DGI by means of the Ind-rules. Hence each term

can be brought into the polynomial form.

In the following we often identify terms with their normal forms.

<u>2. Lemma:</u> For arbitrary terms s and t: $DGI \vdash s=t$ iff $F(s)=F(t)$.
proof: From $DGI \vdash s=t$ it follows that s and t have the same value under arbitrary substitutions of natural numbers for the variables, so they represent the same polynomial, $F(s)=F(t)$. The converse follows by the preceding lemma.

If Γ is a finite and Δ an arbitrary, non empty set of equations, then $(\Gamma)|\Delta$ means that the greatest common divisor of the polynomials $r-s \in Z[x_1..x_n]$ for $r=s \in \Gamma$ divides a polynomial $u-v$ for $u=v \in \Delta$. If $\Gamma=\emptyset$ then $(\Gamma)|\Delta$ means that there exists $u=v \in \Delta$ with $u-v=0$. Let $I^c_{x_1..x_n}$ be the set $\{0,1,..,c-1,x_1+c\} \times .. \times \{0,1,..,c-1,x_n+c\}$ and let $(\hat{x}_1..\hat{x}_n)$ range over $I^c_{x_1..x_n}$. The following main lemma translates the derivability relation for DGI into a divisibility relation on polynomials.

<u>3. Lemma:</u> If the sequent $\Gamma(x_1..x_n) \vdash \Delta(x_1..x_n)$ is derivable in DGI,

then $\exists c \in N \forall d \geq c \forall (\hat{x}_1..\hat{x}_n) \in I^d_{x_1..x_n} \quad (\Gamma(\hat{x}_1..\hat{x}_n))|1+N[x_1..x_n], \Delta(\hat{x}_1..\hat{x}_n)$.

Proof: By induction on the height of the derivation.
1. Axioms: the verification is a matter of routine; example: the axiom $u_1(\bar{x})=v_1(\bar{x}), u_2(\bar{x})=v_2(\bar{x}) \vdash u_1(\bar{x}) \cdot u_2(\bar{x})=v_1(\bar{x}) \cdot v_2(\bar{x})$. For each $d \in N$,

$(\hat{x}_1..\hat{x}_n) \in I^d_{x_1..x_n}$, $p \in Z[x_1..x_n]$: If $p|u_1(\hat{\bar{x}})-v_1(\hat{\bar{x}})$ and $p|u_2(\hat{\bar{x}})-v_2(\hat{\bar{x}})$,

then $p|(u_1(\hat{\bar{x}})-v_1(\hat{\bar{x}})) \cdot u_2(\hat{\bar{x}})$, $p|(u_2(\hat{\bar{x}})-v_2(\hat{\bar{x}})) \cdot v_1(\hat{\bar{x}})$ and hence

$p|u_1(\hat{\bar{x}}) \cdot u_2(\hat{\bar{x}})-v_1(\hat{\bar{x}}) \cdot v_2(\hat{\bar{x}})$.

2. Inference rules: - the assertion is trivially preserved under thinning rules.

- cut: $$\frac{\Gamma(\bar{x}) \vdash \Delta(\bar{x}), A(\bar{x}) \quad \Pi(\bar{x}), A(\bar{x}) \vdash \Sigma(\bar{x})}{\Gamma(\bar{x}), \Pi(\bar{x}) \vdash \Delta(\bar{x}), \Sigma(\bar{x})}$$

By induction hypothesis,

$$\exists c_1 \forall d \geq c_1 \forall (\hat{\bar{x}}) \in I^d_{\bar{x}} \quad (\Gamma(\hat{\bar{x}}))|1+N[\bar{x}], \Delta(\hat{\bar{x}}), A(\hat{\bar{x}})$$

$$\exists c_2 \forall d \geq c_2 \forall (\hat{\bar{x}}) \in I^d_{\bar{x}} \quad (\Pi(\hat{\bar{x}}), A(\hat{\bar{x}}))|1+N[\bar{x}], \Sigma(\hat{\bar{x}}).$$

Let $c := \max(c_1, c_2)$, $d \geq c$, $\hat{\bar{x}} \in I^d_{\bar{x}}$.

If $(\Gamma(\hat{\hat{x}}))|1+N[\bar{x}],\Delta(\hat{\hat{x}})$ then trivially $(\Gamma(\hat{\hat{x}}),\Pi(\hat{\hat{x}}))|1+N[\bar{x}],\Delta(\hat{\hat{x}}),\Sigma(\hat{\hat{x}})$.

If $(\Gamma(\hat{\hat{x}}))|A(\hat{\hat{x}})$ then $(\Gamma(\hat{\hat{x}}),\Pi(\hat{\hat{x}}))|(\Pi(\hat{\hat{x}}),A(\hat{\hat{x}}))|1+N[\bar{x}],\Sigma(\hat{\hat{x}})$.

- Ind_r: $$\frac{\vdash r(\bar{x},0)=s(\bar{x},0) \qquad r(\bar{x},y)=s(\bar{x},y)\vdash r(\bar{x},Sy)=s(\bar{x},Sy)}{\vdash r(\bar{x},t)=s(\bar{x},t)}$$

Then $\forall n\in N \vdash r(\bar{x},n)=s(\bar{x},n) \Rightarrow \forall n\in N\ r(\bar{x},n)-s(\bar{x},n)=0 \Rightarrow$
$\forall t\ \ r(\bar{x},t)-s(\bar{x},r)=0$.

- Ind_1: $$\frac{r(\bar{x},0)=s(\bar{x},0)\vdash \qquad r(\bar{x},Sy)=s(\bar{x},Sy)\vdash r(\bar{x},y)=s(\bar{x},y)}{r(x,t)=s(x,t)\vdash}$$

By induction hypothesis, there exists c such that $\forall d\geq c\forall(\hat{x}_1..\hat{x}_n)\in I^d_{\bar{x}}$

$$r(\hat{\hat{x}},0)-s(\hat{\hat{x}},0)|1+N[\bar{x}]$$

$$r(\hat{\hat{x}},1)-s(\hat{\hat{x}},1)|1+N[\bar{x}],r(\hat{\hat{x}},0)-s(\hat{\hat{x}},0)$$
$$\vdots$$
$$r(\hat{\hat{x}},d)-s(\hat{\hat{x}},d)|1+N[\bar{x}],r(\hat{\hat{x}},d-1)-s(\hat{\hat{x}},d-1)$$

$$r(\hat{\hat{x}},y+d+1)-s(\hat{\hat{x}},y+d+1)|1+N[\bar{x}],r(\hat{\hat{x}},y+d)-s(\hat{\hat{x}},y+d),$$

hence $\forall m<d+1\ r(\hat{\hat{x}},m)-s(\hat{\hat{x}},m)|1+N[\bar{x}]$. Suppose

$r(\hat{\hat{x}},y+d+1)-s(\hat{\hat{x}},y+d+1)|r(\hat{\hat{x}},y+d)-s(\hat{\hat{x}},y+d)$. Then

$r(\hat{\hat{x}},y+d+1)-s(\hat{\hat{x}},y+d+1)=r(\hat{\hat{x}},d)-s(\hat{\hat{x}},d)|1+N[\bar{x}]$. Hence

$\forall d\geq c+1\forall(\hat{x}_1..\hat{x}_n)\in I^d_{x_1..x_ny}\ \ r(\hat{\hat{x}},\hat{y})-s(\hat{\hat{x}},\hat{y})|1+N[\bar{x},y]$.

For each term $t(\bar{x},\bar{y})$ and each $d\in N$ there exists $t'\in N[\bar{x},\bar{y}]$ such that

$$\forall(\hat{x}_1..\hat{x}_n\hat{y}_1..\hat{y}_m)\in I^d_{\bar{x},\bar{y}}\ \ t(\hat{\hat{x}},\hat{\hat{y}}) = \begin{cases}0\\1\\\vdots\\d-1\\t'+d\end{cases}$$

It follows that $\forall d\geq c+1\forall(\hat{\hat{x}},\hat{\hat{y}})\in I^d_{\bar{x}\bar{y}}\ \ r(\hat{\hat{x}},t(\hat{\hat{x}},\hat{\hat{y}}))-s(\hat{\hat{x}},t(\hat{\hat{x}},\hat{\hat{y}}))|1+N[\bar{x},\bar{y}]$.

<u>4. Lemma:</u> Let r,s,u,v be polynomials over N.
(i) If $r-s|u-v$ (divisibility in $Z[x_1..x_n]$) then there exists a poly nomial p over N such that $r=s\vdash u+p=v+p$ is provable in DGI.
(ii) If $r-s|1+N\ x_1..x_n$ then $r=s\vdash$ is provable in DGI.
Proof: (i) $r-s|u-v \Rightarrow \exists t\in Z[x_1..x_n]\ (r-s)\cdot t=u-v$. t can be written as c-d with $c,d\in N[x_1..x_n]$, i.e. $(r\cdot c+s\cdot d)-(r\cdot d+s\cdot c)=u-v$, hence $u+p=r\cdot c+s\cdot d+a$, $v+p=r\cdot d+s\cdot c+a$ for certain $a,p\in N[x_1..x_n]$. Since

$r=s \vdash r \cdot c=s \cdot c$ and $r=s \vdash r \cdot d=s \cdot d$ are provable in DGI, it follows that $r=s \vdash u+p=v+p$ is provable in DGI.

(ii) If $r-s \mid 1+N[x_1..x_n]$ then by (i) $\exists p \in N[x_1..x_n]$ such that $r=s \vdash 1+t+p=p$ with a certain $t \in N[x_1..x_n]$ is provable in DGI. Since $1+t+p=p \vdash$ is provable by Ind_1, the provability of $r=s \vdash$ in DGI follows.

5. Theorem: $r(x_1..x_n)=s(x_1..x_n) \vdash$ is provable in DGI iff

$$\exists c \in N \; \forall (\hat{x}_1..\hat{x}_n) \in I^c_{x_1..x_n} \quad r(\hat{x}_1..\hat{x}_n)-s(\hat{x}_1..\hat{x}_n) \mid 1+N[x_1..x_n] .$$

Proof: If $r(x_1..x_n)=s(x_1..x_n) \vdash$ is provable in DGI, then by lemma 3 $\exists c \forall (\hat{x}_1..\hat{x}_n) \in I^c_{x_1..x_n} \; r(\hat{x}_1..\hat{x}_n)-s(\hat{x}_1..\hat{x}_n) \mid 1+N[x_1..x_n]$. If there exists $c \in N$ such that $\forall (\hat{x}_1..\hat{x}_n) \in I^c_{x_1..x_n} \; r(\hat{x}_1..\hat{x}_n)-s(\hat{x}_1..\hat{x}_n) \mid 1+N[x_1..x_n]$ then by lemma 4 $\forall (\hat{x}_1..\hat{x}_n) \in I^c_{x_1..x_n} \; r(\hat{x}_1..\hat{x}_n)=s(\hat{x}_1..\hat{x}_n) \vdash$ is provable in DGI and the provability of $r(x_1..x_n)=s(x_1..x_n) \vdash$ follows by iterated applications of Ind_1.

If in an equation $r=s$ there is a variable x occurring in r but not in s and a variable y occurring in s but not in r, we shall say that $r=s$ has separated variables.

6. Proposition: If $r=s$ is an equation with separated variables, then $r=s \vdash$ is not provable in DGI.

Proof: Suppose $r(\bar{x},x)=s(\bar{x},y) \vdash$ were provable in DGI. Then also $r(\bar{0},x)=s(\bar{0},y) \vdash$ were provable in DGI and for some $t(x,y) \in Z[x,y]$, $c \in N$ $(r(\bar{0},x+c)-s(\bar{0},y+c)) \cdot t(x,y) \in 1+N[x,y]$. Let $r(\bar{0},x+c)=:r_1(x)$ and $s(\bar{0},y+c)=:s_1(y)$. Then $(r_1(0)-s_1(y)) \cdot t(0,y)$ and $(r_1(x)-s_1(0)) \cdot t(x,0)$ were in $1+N[x,y]$, which is impossible as can easily be seen by looking at the coefficients.

7. Corollary: The following equations are not refutable in DGI:

- Shoenfield's $D_{n,m}$ [2]: $nx+m=ny$ $(0<m<n)$
- $3x+2=y^2$, $x^2+x+1=2y$
- $(x+1)^2=2(y+1)^2$
- the instances of Fermat's equations $(x+1)^n+(y+1)^n=(z+1)^n$ $(n>2)$.

Proof: These are equations with separated variables.

2. Adding the functions "predecessor", "sign", "co-sign". We now extend the system DGI by adding new function symbols P (predecessor), sg (sign), $\overline{\mathrm{sg}}$ (co-sign), defining axioms

$$\vdash P(0)=0 \quad \vdash sg(0)=0 \quad \vdash \overline{sg}(0)=S0$$
$$\vdash P(St)=t \quad \vdash sg(St)=S0 \quad \vdash \overline{sg}(St)=0,$$

and the corresponding equality axioms for the new functions. Let DGI^+ denote the extended system. We are going to show that DGI^+ and DGI refute the same diophantine equations.

8. Lemma: (i) For each term $t(x_1..x_n)$ in DGI^+ we can find $c\in N$ such that $\forall(\hat{x}_1..\hat{x}_n)\in I^c_{x_1..x_n}$ there exists a term $t'(x_1..x_n)$ in DGI with

$$DGI^+ \vdash t(\hat{x}_1..\hat{x}_n)=t'(x_1..x_n).$$

The normal form of t' is uniquely determined.
(ii) For each term $t(x_1..x_n)$ in DGI^+ and $k\in N$ we can find $c\in N$ such that $\forall(\hat{x}_1..\hat{x}_n)\in I^c_{x_1..x_n}$

$$DGI^+ \vdash t(\hat{x}_1..\hat{x}_n) = \begin{cases} 0 \\ 1 \\ \vdots \\ k-1 \\ k+t'(x_1..x_n),\ t' \text{ in DGI} \end{cases}$$

Proof: By induction on the length of t.

9. Lemma: For each sequent $\Gamma(x_1..x_n)\vdash\Delta(x_1..x_n)$ provable in DGI^+ we can find $c\in N$ such that

$$\forall d\geq c\forall(\hat{x}_1..\hat{x}_n)\in I^d_{x_1..x_n} \quad \Gamma(\hat{x}_1..\hat{x}_n)'\vdash\Delta(\hat{x}_1..\hat{x}_n)' \text{ is provable in DGI,}$$

where $\Gamma(\hat{x}_1..\hat{x}_n)'$, $\Delta(\hat{x}_1..\hat{x}_n)'$ are obtained from $\Gamma(\hat{x}_1..\hat{x}_n)$, $\Delta(\hat{x}_1..\hat{x}_n)$ by replacing DGI^+-terms $t(\hat{x}_1..\hat{x}_n)$ by their DGI-reductions $t'(x_1..x_n)$ according to the preceding lemma.
Proof: By induction on the height of the derivation of $\Gamma\vdash\Delta$.
The assertion is trivial for the axioms and is clearly preserved under thinnings and cuts. It remains to show that the assertion is preserved under the Ind-rules

$$\frac{\vdash\!\!\!\dashv r(\bar{x},0)=s(\bar{x},0) \qquad r(\bar{x},y)=s(\bar{x},y)\vdash\!\!\!\dashv r(\bar{x},y+1)=s(\bar{x},y+1)}{\vdash\!\!\!\dashv r(\bar{x},t\bar{x})=s(\bar{x},t\bar{x})}.$$

By induction hypothesis,

$$\exists c_1\forall d\geq c_1\forall(\hat{\bar{x}},\hat{y})\in I^d_{xy} \quad \vdash\!\!\!\dashv r(\hat{\bar{x}},0)' = s(\hat{\bar{x}},0)' \quad \text{is provable in DGI}$$

$$\exists c_2\forall d\geq c_2\forall(\hat{\bar{x}},\hat{y})\in I^d_{xy} \quad r(\hat{\bar{x}},\hat{y})' = s(\hat{\bar{x}},\hat{y})'\vdash\!\!\!\dashv r(\hat{\bar{x}},\hat{y}+1)'=s(\hat{\bar{x}},\hat{y}+1)' \quad \text{is provable in DGI}$$

Let $c:=\max(c_1,c_2)$. Then for each $(\hat{x}_1..\hat{x}_n)\in I^d_{x_1..x_n}$ with $d\geq c$ there exist

DGI-terms $r'(\bar{x},y)$, $s'(\bar{x},y)$ with $r(\hat{\bar{x}},y+c+z)'=r'(\hat{\bar{x}},y+z)$, $s(\hat{\bar{x}},y+c+z)'= s'(\hat{\bar{x}},y+z)$. Hence from

$$\begin{matrix} r(\hat{\bar{x}},y+c)'=s(\hat{\bar{x}},y+c)' \;\vdash\; r(\hat{\bar{x}},y+c+1)'=s(\hat{\bar{x}},y+c+1)' \\ r'(\hat{\bar{x}},y) \quad s'(\hat{\bar{x}},y) \quad r'(\hat{\bar{x}},y+1) \quad s'(\hat{\bar{x}},y+1) \end{matrix}$$

by Ind_r, Ind_l in DGI we obtain $\vdash r'(\hat{\bar{x}},y)=s'(\hat{\bar{x}},y)$, i.e. $\vdash r(\hat{\bar{x}},y+c)'=s(\hat{\bar{x}},y+c)'$. For each $m<d$, $\vdash r(\hat{\bar{x}},m)'=s(\hat{\bar{x}},m)'$ can be obtained by transforming the induction into a sequence of cuts and substitutions. Finally, by lemma 8, we can find for each DGI^+-term $t(x_1..x_n)$ a $c\in N$ such that

$$t(\hat{x}_1..\hat{x}_n) = \begin{cases} 0 \\ 1 \\ \vdots \\ d-1 \\ d+t'(x_1..x_n), \ t' \text{ in DGI} . \end{cases}$$

<u>10. Theorem:</u> If $r(\bar{x})=s(\bar{x})\vdash$ is provable in DGI^+, where r and s are DGI-terms, then $r(\bar{x})=s(\bar{x})\vdash$ is already provable in DGI.
Proof: This is an immediate consequence of the preceding lemma.

<u>3. The system Z_o.</u> Let Z_o denote classical first order number theory in 0,S,+,· with induction restricted to quantifier-free formulae, formulated in a Gentzen-type sequent calculus. Let Z_o^+ be the system obtained when the function P is added.

Obviously Z_o is an extension of DGI, but with respect to the provability of equations and inequations (not for arbitrary sequents) Z_o is even equivalent to DGI: Since all true equations (i.e. polynomial identities) are provable in DGI, it is clear that sequents $\vdash r=s$ provable in Z_o are also provable in DGI. It remains to show that inequations provable in Z_o are also provable in DGI. We begin with two observations about Z_o^+ which are due to Shepherdson [1]. Shepherdson showed that in Z_o^+-derivations of open formulae the schema of open induction can be replaced by the following rule of open induction:

$$\frac{\vdash A(\bar{x},0) \quad A(\bar{x},y)\vdash A(\bar{x},y+1)}{\vdash A(\bar{x},\iota)} \quad A(\bar{x},y) \text{ open.}$$

In addition he gave an induction-free equivalent of Z_o^+ with induction rule which is obtained from Z_o^+ by dropping the induction rule and adding the following additional axioms:

(1) $t\neq 0 \rightarrow t=S(P(t))$

(2) $s+t=t+s$

(3) $(s+t)+u=s+(t+u)$

(4) $s+u=t+u \to s=t$

(5) $s\cdot t=t\cdot s$

(6) $(s\cdot t)\cdot u=s\cdot (t\cdot u)$

(7) $s\cdot (t+u)=s\cdot t+s\cdot u$

$$(C'_d) \quad d\cdot s=d\cdot t \to \bigvee_{i=0}^{d-1} (u+i)\cdot s=(u+i)\cdot t \qquad (d=2,3,..)$$

where s,t,u are arbitrary terms. We denote this system by T. The stronger and simpler axioms

$$(C_d) \quad d\cdot s=d\cdot t \to s=t \qquad (d=2,3,..)$$

are not provable in Z_o^+ ; this was also pointed out by Shepherdson. Let T_1^+ be the system with axioms (1)-(7) and (C_d) instead of (C'_d) and let T_1 be the restriction of T_1^+ to the language without the function P. Since T_1^+ is stronger than T each open formula provable in Z_o^+ is also provable in T_1^+. We are now going to show that T_1^+ is conservative over DGI with respect to the provability of inequations.

In a classical sequent calculus the axioms of T_1^+ can be formulated as initial sequent consisting of equations only (without propositional connectives). We shall refer to sequents $\Gamma \vdash \Delta$ as prime sequents if Γ,Δ are finite sets of equations. By cut-elimination, T_1^+-derivations of prime sequents can be transformed such that the whole derivation consists of prime sequents only, i.e. is built up by axioms and structural inference rules. Next we replace terms with the symbol P as outer function symbol via variable substitutions $x_i \to \hat{x}\in\{0,1,..,c-1,x_i+c\}$: For each term $t(x_1..x_n)$ in the language of T_1^+ we can find $c\in N$ such that $\forall(\hat{x}_1..\hat{x}_n)\in I^c_{x_1..x_n}$

$$\vdash t(\hat{x}_1..\hat{x}_n)=0 \quad \text{or} \quad \vdash t(\hat{x}_1..\hat{x}_n)=1+t'(x_1..x_n),$$

$t'(x_1..x_n)$ in the language of T_1, can be proved from the axioms of Z_o^+ (without induction) and axioms of type (2),(3),(5),(6),(7) above. A term $Pt(x_1..x_n)$ can then be replaced by its T_1-reductions 0 or $t'(\bar{x})$. In analogy to lemma 9 we have the following

<u>11. Lemma:</u> Let $\Gamma(x_1..x_n)\vdash\Delta(x_1..x_n)$ be a prime sequent provable in T_1^+. Then there exists $c\in N$ such that $\forall(\hat{x}_1..\hat{x}_n)\in I^c_{\underline{x}}$ the sequents $\Gamma(\hat{x}_1..\hat{x}_n)'\vdash\Delta(\hat{x}_1..\hat{x}_n)'$ are provable in T_1, where $\Gamma(\hat{x}_1..\hat{x}_n)'$ and $\Delta(\hat{x}_1..\hat{x}_n)'$ are obtained from $\Gamma(\hat{x}_1..\hat{x}_n)$ and $\Delta(\hat{x}_1..\hat{x}_n)$ by replacing P-terms by their T_1-reductions.

Proof: as in lemma 9; axioms of type (1) become (provable) identities, axioms of type (2)-(7) and (C_d) are replaced by P-free instances of axioms of the same type.

Finally, derivability in T_1 can be translated into a divisibility relation on polynomials in almost the same way as in lemma 3 in the case of DGI-derivations. The only difference is the fact that due to the axioms (C_d) we have now to take the greatest common <u>primitive</u> divisor $(\Gamma(\hat{x}_1..\hat{x}_n))^*$ instead of $(\Gamma(\hat{x}_1..\hat{x}_n))$.

<u>12. Lemma:</u> For each prime sequent $\Gamma(x_1..x_n) \vdash \Delta(x_1..x_n)$ provable in T_1 there exists $c \in N$ $\forall d \geq c$ $\forall (\hat{x}_1..\hat{x}_n) \in I^d_{x_1..x_n}$

$$(\Gamma(\hat{x}_1..\hat{x}_n))^* | \Delta(\hat{x}_1..\hat{x}_n), 1+N[x_1..x_n],$$

where $(\Gamma(\hat{x}_1..\hat{x}_n))^*$ is the primitive polynomial of the polynomial $(\Gamma(\hat{x}_1..\hat{x}_n))$.

Proof: As in the case of DGI-derivations, the divisibility relation is preserved under structural inference rules. For Z_o-axioms the assertion follows immediately from the proof of lemma 3 since $(\Gamma(\hat{\bar{x}}))|\Delta(\hat{\bar{x}}),1+N[\bar{x}]$ yields $(\Gamma(\hat{\bar{x}}))^*|\Delta(\hat{\bar{x}}),1+N[\bar{x}]$. For the same reason the assertion holds for axioms of type (2),(3),(5),(6), and (7) since these are provable in DGI. In the case of axioms of type (4) and (C_d) we trivially have

$$(s(\hat{\bar{x}})+u(\hat{\bar{x}})-t(\hat{\bar{x}})-u(\hat{\bar{x}}))^* | s(\hat{\bar{x}})-t(\hat{\bar{x}}) \qquad \text{and}$$

$$(d \cdot s(\hat{\bar{x}})-d \cdot t(\hat{\bar{x}}))^* = (s(\hat{\bar{x}})-t(\hat{\bar{x}}))^* | s(\hat{\bar{x}})-t(\hat{\bar{x}}) \quad .$$

<u>13. Theorem:</u> If $r=s \vdash$ is provable in T_1^+ and r, s are terms in the language of Z_o, then $r=s \vdash$ is also provable in DGI.

Proof: If $r(x_1..x_n)=s(x_1..x_n) \vdash$ is provable in T_1^+, then by lemma 11 there exists $c_o \in N$ such that $\forall (\hat{x}_1..\hat{x}_n) \in I^{c_o}_{x_1..x_n}$ $r(\hat{x}_1..\hat{x}_n)=s(\hat{x}_1..\hat{x}_n) \vdash$ is provable in T_1. Hence by lemma 12 there exists $c_1 \in N$ such that $\forall (\hat{x}_1..\hat{x}_n) \in I^{c_1}_{x_1..x_n}$ $(r(\hat{x}_1..\hat{x}_n)-s(\hat{x}_1..\hat{x}_n))^* | 1+N[x_1..x_n]$. Then clearly $r(\hat{x}_1..\hat{x}_n)-s(\hat{x}_1..\hat{x}_n) | 1+N[x_1..x_n]$ and by theorem 5 $r(x_1..x_n)=s(x_1..x_n) \vdash$ is provable in DGI.

<u>14. Corollary:</u> Z_o^+ and DGI prove the same inequations.

<u>Remark:</u> As in the case of DGI, we can again extend Z_o^+ by adding the functions sign and co-sign without changing the set of provable inequations: variable substitutions $x_i \to \hat{x}_i \in \{0,1,..,c-1,x_i+c\}$ give defi- values 0 or 1 to terms $sg(t(x_1..x_n))$ and $\overline{sg}(t(x_1..x_n))$.

References.

[1] J.C.Shepherdson, Non-standard models for fragments of number theory, Proc.Int.Symp. on Model Theory, Berkeley, 1963

[2] J.R.Shoenfield, Open sentences and the induction axiom, J.Symb. Logic 23 (1958) 7-12

GENERALIZED RULES FOR QUANTIFIERS AND THE COMPLETENESS OF THE INTUITIONISTIC OPERATORS $\&, \vee, \supset, \curlywedge, \forall, \exists$.

Peter Schroeder-Heister

Universität Konstanz

In this paper, we develop a proof-theoretic framework for the treatment of arbitrary quantifiers binding m variables in n formulas. In particular, we motivate a schema for introduction and elimination rules for such quantifiers based on a concept of 'derivation' that allows rules as assumptions which may be discharged. With respect to this schema, the system of the standard operators of intuitionistic quantifier logic turns out to be complete.

1. GENERAL INTRODUCTION

This paper, which is a sequel to [18], deals with a generalization of natural deduction systems. The calculi of natural deduction as developed by Gentzen [7] and investigated by Prawitz [13] have at least two distinctive features: Firstly they present a conceptualization of reasoning from assumptions in allowing that assumptions may be discharged in the course of a derivation, and secondly they contain a certain systematics in that the rules governing the logical signs are split up into introduction (I) and elimination (E) rules for each sign. Both aspects are especially important for intuitionistic logic. As to the first, the possibility of discharging assumptions directly admits a derivative interpretation of implication as opposed to the truth-functional one (the term 'derivative' is due to Schmidt [16]): $\alpha\supset\beta$ means that β can be derived from α as is made obvious by the $\supset$I rule

$$\frac{\begin{matrix}[\alpha]\\ \beta\end{matrix}}{\alpha\supset\beta}\ .$$

Concerning the second, the I and E rules for the intuitionistic system show a certain symmetry or duality in the sense that they can be considered inverses of each other (cf. Prawitz [13]) while in the classical case this symmetry is at least partly lost. For example, the absurdity rule

$$\frac{\curlywedge}{\alpha}$$

can be conceived as the elimination rule of the 0-place operator $\curlywedge$ and can be justified from the fact that there is no I rule for $\curlywedge$, whereas the corresponding classical rule of indirect proof

$$\frac{\begin{matrix}[\alpha\supset\curlywedge]\\ \curlywedge\end{matrix}}{\alpha}$$

cannot be so justified. Thus it is not surprizing that recent attempts at building up a proof-theoretical semantics for intuitionistic logic ('proof' here regarded not purely syntactically but in the traditional philosophical sense as the foundation of a proposition) are mainly based on systems of natural deduction (cf. Dummett [6], Prawitz [14]).

The extension of intuitionistic natural deduction which is proposed in the following, concerns both aspects. We shall first of all define notions of 'inference rule' as well as of 'derivation', according to which not only formulas but also rules themselves are allowed as assumptions which may be discharged by the application of rules. It seems quite natural to use as a hypothesis that one can pass over from derived formulas to other ones, and not only that one may start with a certain formula as a hypothesis. This can of course be achieved by formulas too; for instance, $\alpha\supset\beta$ as an assumption allows the transition from α to β by means of modus ponens. However, this is a result of our investigations which are intended to start from an intuitively plausible concept of 'derivation' which does not presuppose any specific inference rules. Derivations in a formal system can be considered derivations in a system without any basic rules, if all rules which are used are counted as assumptions. This is just the way derivations will be introduced: A derivation is defined as an arbitrary finite tree of pairs of formulas and individual variables, indicating the eigenvariables of inferences, together with a discharge function, indicating where an assumption is discharged (this is a generalization of Prawitz' notion of a discharge function in [13]). Those rules which justify such a tree as a derivation and which can be decoded from the tree are then the assumptions of this derivation, and the undischarged assumptions can be considered either assumptions on which the end-formula depends or applications of basic rules (if there are any). So the concept of a derivation is defined independently of the question of which assumptions are ad hoc, i.e. are ones on which a formula in a derivation depends, and which belong to the considered framework, i.e. are applications of basic rules - in the same way as in the usual concept of natural deduction it does not affect the intrinsic structure of a derivation (but only the set of formulas on which it depends) whether a certain formula is an axiom or an assumption. (Gentzen's system NI of natural deduction for intuitionistic logic does not contain any axiom but only proper rules, but this is a specific feature of this particular calculus and not of the type of calculus of which NI is a representative; for the classical system NK, e.g., Gentzen proposes the 'tertium non datur' as an axiom schema.)

The generalized notions of 'rule' and 'derivation' will then be used to make the often stated 'symmetry' or 'harmony' between the I and E rules for logical operators more explicit. This is done by providing a general schema for I and E rules of an arbitrary operator which captures the I and E rules for all standard intuitionistic connectives &,v, ⊃,⅄,∀,∃, and by a metalinguistic characterization in terms of derivability which justifies this general schema. For this purpose the vocabulary of our language is based on the generalized notion of a 'quantifier' or 'operator' which binds m individual variables in n formula-arguments, thus including the usual existential and universal quantifiers (m=1, n=1) and n-ary sentential connectives (m=0). The justification of this schema is based on the idea that from the conclusion of an introduction rule it should follow exactly what can be concluded from all its possible premisses (and what will be called the 'common content with respect to certain eigenvariables' of these premisses). In the case of $\exists x\alpha$, for example, all that follows from every substitution instance $\alpha[x|t]$ of α should be a consequence of $\exists x\alpha$. This is guaranteed by the $\exists$E rule

$$\frac{\exists x\alpha \quad \begin{matrix}[\alpha]\\ \beta\end{matrix}}{\beta}$$ (x not free in β nor in an assumption besides α on which β depends)

which states that whatever follows from α also follows from $\exists x\alpha$ where the eigenvariable x in a derivation from α is understood universally, i.e. representing all substitution instances of the derivation.

This idea is closely connected with the concept of rules as assumptions, because if one wants to speak of what follows from the premisses of an I rule, one must have representatives of these premisses which can serve as assumptions (if they are not simply formulas without dischargeable assumptions and eigenvariable conditions). Rules are very suitable for that purpose. For instance, the rule $\alpha \Rightarrow \beta$ ('from α you may infer β') represents the premiss (including the dischargeable assumption) of ⊃I and the rule $\Rightarrow_x \alpha$ ('for arbitrary t you may infer $\alpha[x|t]$') the premiss (including the eigenvariable condition) of ∀I. This is generalized in such a way that arbitrary finite lists of arbitrarily complex rules can be premisses of I rules and therefore assumptions of minor premisses of E rules.

This general conception of I and E rules for generalized operators suggests as a technical question whether a certain set of operators is complete with respect to this conception (in that sense of 'completeness' one uses when speaking of 'functional completeness' in classical sentential logic). It will be shown that the six standard intuitionistic operators $\&,\vee,\supset,\curlywedge,\forall,\exists$ suffice to explicitly define every other operator which falls under our general schema and that, therefore, all that can be formulated by use of rules as assumptions can be expressed with their help. This shows that rules as assumptions are superfluous once we have the standard operators at our disposal (but this is an insight for which the concept of such rules is necessary!).

2. RULES AND QUANTIFIERS

In the sentential case, as treated in [18], a rule was defined to be an arbitrary formula tree growing upwards, whose height was called its level. So the general form of a rule was

$$\frac{\frac{\Gamma_1}{\beta_1} \quad \cdots \quad \frac{\Gamma_n}{\beta_n}}{\alpha}$$

written linearly

$$\langle\Gamma_1\Rightarrow\beta_1,\ldots,\Gamma_n\Rightarrow\beta_n\rangle\Rightarrow\alpha$$

where α and the β_i are formulas, Γ_i are lists (i.e. linear graphic arrangements) of rules (where formulas are special cases of rules, viz. rules of level 1). The intended meaning of such a rule, underlying the definition of a derivation and the derivability of a formula α from a list of rules Δ, was: If, for all i ($1\leq i\leq n$), β_i has been derived from Γ_i and additional assumptions Δ_i (i.e., rules of Γ_i and Δ_i can have been used in the derivation of β_i), one may immediately infer α and consider it derived from $\Delta_1,\ldots,\Delta_n$ alone (i.e. the Γ_i may be discharged by the application of the rule).

The standard form for I and E rules for arbitrary n-ary sentential operators S, according to which, roughly speaking, the E rule allows one to establish all that is implied by the premisses of all I rules, was

$$(\text{S-I})\qquad \frac{\Phi_1(\underline{A})}{S\underline{A}} \quad \ldots \quad \frac{\Phi_m(\underline{A})}{S\underline{A}}$$

$$(\text{S-E})\qquad \frac{S\underline{A} \qquad \dfrac{\Phi_1(\underline{A})}{B} \quad \ldots \quad \dfrac{\Phi_m(\underline{A})}{B}}{B}$$

where $\underline{A}$ is a list $A_1 \ldots A_n$ of different schematic letters for formulas, B a schematic letter for formulas different from $A_1, \ldots, A_n$ and the $\Phi_i(\underline{A})$ systems of rule schemata containing at most $A_1, \ldots, A_n$ as schematic letters. The rule schemata for the standard intuitionistic connectives were

$$(\&\text{I})\quad \frac{A \quad B}{A\&B} \qquad\qquad (\&\text{E})\quad \frac{A\&B \qquad \dfrac{A \quad B}{C}}{C}$$

$$(\vee\text{I})\quad \frac{A}{A\vee B} \quad \frac{B}{A\vee B} \qquad\qquad (\vee\text{E})\quad \frac{A\vee B \qquad \dfrac{A}{C} \quad \dfrac{B}{C}}{C}$$

$$(\supset\text{I})\quad \frac{\dfrac{A}{B}}{A\supset B} \qquad\qquad (\supset\text{E})\quad \frac{A\supset B \qquad \dfrac{\dfrac{A}{B}}{C}}{C}$$

$$[\text{no } \curlywedge\text{I}] \qquad\qquad (\curlywedge\text{E})\quad \frac{\curlywedge}{A}$$

Here ⊃E, which can be shown to be equivalent to modus ponens (cf. [18], Lemma 4.4), is an example of a rule schema of level 4 (whereas usual natural deduction systems only contain rules of levels ≤3).

In order to treat quantifiers within a related framework, we must first have a concept of substitution of (individual) terms for (individual) variables at our disposal. Furthermore we must be able to express the fact that a rule holds for all substitutions of a variable by a term, as e.g. in the case of ∃I, and to express eigenvariable conditions, as e.g. in the case of ∃E. The latter include restrictions concerning the 'additional' assumptions, i.e. those assumptions on which the premisses of an application of a rule depend but which cannot be discharged by an application of that rule. These restrictions cannot be dealt with by an appropriate choice of the formulas which occur in the rule itself - at least as long as one works in a natural deduction framework where 'additional' assumptions are not made explicit in the formulation of a rule.

Our proposal is to introduce a certain kind of universal quantification into a rule. We define a variable-formula-pair (VF-pair) to be a sign $\langle \underline{x},\alpha\rangle$ consisting of a list of distinct variables $\underline{x}$ and a formula α, and define a rule of level n to be a tree of VF-pairs (growing upwards) of height n. The general schema of a rule then becomes

$$\frac{\dfrac{\Gamma_1}{\langle \underline{x}_1,\beta_1\rangle} \quad \cdots \quad \dfrac{\Gamma_n}{\langle \underline{x}_n,\beta_n\rangle}}{\langle \underline{x},\alpha\rangle} \tag{1}$$

where the Γ_i are lists of rules. If $\underline{x}$ is the empty list, we simply write α instead of $\langle \underline{x},\alpha\rangle$; so formulas are special kinds of level-1-rules. The variables $\underline{y}$ of a VF-pair $\langle \underline{y},\gamma\rangle$ are considered bound in γ and in the formulas of all VF-pairs above $\langle \underline{y},\gamma\rangle$. This is made more obvious in our linear notation of (1):

$$\langle \Gamma_1 \Rightarrow_{\underline{x}_1} \beta_1,\ldots,\Gamma_n \Rightarrow_{\underline{x}_n} \beta_n\rangle \Rightarrow_{\underline{x}} \alpha \quad .$$

A list of variables $\underline{y}$ in $\ldots\Rightarrow_{\underline{y}}\ldots$ universally quantifies $\ldots\Rightarrow\ldots$ with respect to $\underline{y}$ in a certain sense. Variants of rules, resulting by relabelling such bound variables and adding or omitting vacuous quantifications, can then be defined in the obvious way, as can the substitution $[\underline{x}|\underline{t}]$ of appropriate lists $\underline{t}$ of terms for lists $\underline{x}$ of variables in rules (for precise definitions see section 3).

The intended meaning of (1) can then be stated as follows: For any variant

$$\frac{\dfrac{\Gamma_1'}{\langle \underline{y}_1,\beta_1'\rangle} \quad \cdots \quad \dfrac{\Gamma_n'}{\langle \underline{y}_n,\beta_n'\rangle}}{\langle \underline{y},\alpha'\rangle}$$

of (1) and any appropriate list of terms $\underline{t}$: If for each i ($1\le i\le n$) $\beta_i'[\underline{y}|\underline{t}]$ has been derived from $\Gamma_i'[\underline{y}|\underline{t}]$ and additional assumptions Δ_i not containing $\underline{y}_i$ free, then one may immediately infer $\alpha'[\underline{y}|\underline{t}]$ and consider it derived from $\Delta_1,\ldots,\Delta_n$ alone (i.e. the $\Gamma_i'[\underline{y}|\underline{t}]$ may be discharged by the application of (1)).

If, for example, α is a formula containg only x as a free variable and β a formula without free variables,

$$\frac{\alpha}{\langle x,\exists x\alpha\rangle} \qquad \frac{\langle x,\alpha\rangle}{\forall x\alpha} \qquad \frac{\exists x\alpha \qquad \dfrac{\alpha}{\langle x,\beta\rangle}}{\beta} \qquad \frac{\forall x\alpha \qquad \dfrac{\langle x,\alpha\rangle}{\langle x,\beta\rangle}}{\beta}$$

are instances of ∃I, ∀I, ∃E and ∀E. The last rule is equivalent to the usual

$$\frac{\forall x\alpha}{\langle x,\alpha\rangle},$$

but falls under a standard form for E rules.

In the following section we define in detail the notion of a derivation of formulas from rules as assumptions, following the intended meaning of a rule as stated above. This theory is applied directly to a language for quantifiers in a generalized sense. Such a quantifier is considered to be an operator which gives a formula from n_1 variables and n_2 formulas, where n_1 and n_2 are natural numbers. (n_1,n_2) is then called its type. If S is a quantifier of type (n_1,n_2) and $x_1,\ldots,x_{n_1}$ are distinct variables and $\alpha_1,\ldots,\alpha_{n_2}$ formulas, then $Sx_1\ldots x_{n_1}\alpha_1\ldots\alpha_{n_2}$ is a formula, in which free occurrences of $x_1,\ldots,x_{n_1}$ in $\alpha_1,\ldots,\alpha_{n_2}$ become bound. If $n_1 = 0$, S is a sentential operator. In the following we shall simply speak of operators instead of quantifiers in the generalized sense.

Historical Remark. Our notion of a quantifier or operator of type (n_1, n_2) corresponds to the notion of a variable-binding operator of degree $(1,n_1,0,n_2)$ in Kalish/Montague [8], i.e. a variable-binding operator without terms as arguments or values. Borkowski [3] considers only quantifiers of type $(1,n_2)$. Concepts of rules of higher levels with bound variables can be found in Lorenzen [10] and Prawitz [12,15]. Both approaches differ from the one presented here in that they consider the consequence relation to be a relation between rules and allow - in our terminology - iteration of ⇒ to the right in the linear notation of a rule so that a rule does not have a tree structure. Furthermore, Lorenzen's theory is based on the concept of admissibility whereas we take the concept of a formula being derived from assumptions to be primary and not the concept of a formula being derivable if certain assumptions are derivable.

3. THE LANGUAGE. DERIVATIONS FROM RULES

When we speak of a list, we mean a (possibly empty) linear graphic arrangement of symbols which are called its members, i.e. a sequence in the graphic sense, not in the sense of an abstract mathematical entity. Its number of members is called its length. Analogously, trees are conceived as graphic objects.

As basic signs we assume to be given:

(i) Denumerably many (individual) variables (syntactical variables for them: 'x', 'y', 'z', for lists of distinct variables: '$\underline{x}$', '$\underline{y}$', '$\underline{z}$', all with and without ' and indices).

(ii) Denumerably many (individual) terms, forming a (not necessarily proper) superset of the set of variables (syntactical variables for them: 't', for lists of terms: '$\underline{t}$', both with and without ' and indices).

(iii) For each number of argument places denumerably many predicate letters (syntactical variables: 'P', with and without indices).

(iv) Finitely or denumerably many <u>operators</u>, each with an associated pair (n_1,n_2) of natural numbers as its type (syntactical variable: 'S').

<u>Atomic formulas</u> are of the form $P\underline{t}$, where the length of $\underline{t}$ is equal to the arity of P. <u>Formulas</u> are atomic formulas and signs $S\underline{x}\underline{\alpha}$ where, if S is of type (n_1,n_2), $\underline{x}$ is of length n_1 and $\underline{\alpha}$ is a list of formulas of length n_2. In the case of binary sentential operators we may write $(\alpha_1 S \alpha_2)$, where outer brackets can be omitted. <u>Parts</u> of a formula β are β itself and the parts of members of $\underline{\alpha}$ if β is $S\underline{x}\underline{\alpha}$. Syntactical variables for formulas: 'α', 'β', 'γ', for lists of formulas: '$\underline{\alpha}$', '$\underline{\beta}$', '$\underline{\gamma}$', all with and without ' and indices.

The members of $\underline{x}$ in $S\underline{x}\underline{\alpha}$ are considered binding corresponding occurrences of variables in the members of $\underline{\alpha}$. All elements of $\underline{\alpha}$ belong to the scope of $\underline{x}$. So free and bound (occurrences of) variables in formulas can be defined as usual. $\underline{x}$ is free in $\underline{\alpha}$ if for each member x of $\underline{x}$, x is free in $\underline{\alpha}$. t is free for x in α if x does not occur free in α within the scope of a variable which occurs also in t. $\underline{t}$ is free for $\underline{x}$ in α if $\underline{x}$ and $\underline{t}$ are of the same length n and for each i $(1 \leq i \leq n)$, the i-th member t_i of $\underline{t}$ is free for the i-th member x_i of $\underline{x}$ in α. $\alpha[\underline{x}|\underline{t}]$ is defined if $\underline{t}$ is free for $\underline{x}$ in α and is the result of simultaneously substituting the free occurrences of x_i in α by t_i $(1 \leq i \leq n)$ if $\underline{x}$ is $x_1 \ldots x_n$ and $\underline{t}$ is $t_1 \ldots t_n$. $\underline{\alpha}[\underline{x}|\underline{t}]$ is defined if for all members β of $\underline{\alpha}$, $\beta[\underline{x}|\underline{t}]$ is defined, and is the result of forming $\beta[\underline{x}|\underline{t}]$ for all β.

<u>Rules</u> of level n were already defined in § 2 (see (1)) as finite trees of height n of VF-pairs $\langle \underline{x},\alpha \rangle$ called their <u>elements</u>, where $\underline{x}$ binds corresponding occurrences of variables in α and in formulas above $\langle \underline{x},\alpha \rangle$, and where, if $\underline{x}$ is empty, α is identified with $\langle \underline{x},\alpha \rangle$. <u>Parts</u> of a rule ρ are ρ itself and the rules $\rho_1,\ldots,\rho_n$ if

$$\frac{\rho_1 \quad \cdots \quad \rho_n}{\langle x,\alpha \rangle}$$

is a part of ρ (i.e., the parts of ρ are ρ and all subtrees of ρ). Proper parts of ρ are parts of ρ different from ρ. Syntactical variables for rules: 'ρ', for lists of rules: 'Δ', 'Γ', all with and without ' and indices.

We say that a list of variables does not occur free in a rule if none of its members occurs free in the rule. So '$\underline{t}$ is free for $\underline{x}$ in ρ' is defined in the obvious way, and $\rho[\underline{x}|\underline{t}]$ as well. $\underline{t}$ is free for $\underline{x}$ in a list of rules Δ if $\underline{t}$ is free for $\underline{x}$ in all its members. $\Delta[\underline{x}|\underline{t}]$ is defined memberwise. When making an assertion about an $\alpha[\underline{x}|\underline{t}]$, we shall understand this assertion to be restricted to those $\underline{t}$ such that $\alpha[\underline{x}|\underline{t}]$ is defined.

A variant of ρ results from ρ by re-ordering lists of variables, omitting or adding vacuous quantifications or re-labelling the variables $\underline{x}$ of a variable-formula-pair, i.e. by once or more often replacing a part $\Delta \Rightarrow_{\underline{x}} \alpha$ of ρ by (i) $\Delta \Rightarrow_{\underline{x}_1} \alpha$ where $\underline{x}_1$ contains the same variables as $\underline{x}$ in a different order, (ii) $\Delta \Rightarrow_{\underline{x}_1} \alpha$ where $\underline{x}_1$ results form $\underline{x}$ by omitting or adding variables not occurring free in $\Delta \Rightarrow \alpha$, (iii) $\Delta[\underline{x}|\underline{y}] \Rightarrow_{\underline{y}} \alpha[\underline{x}|\underline{y}]$ provided $\underline{y}$ is free for $\underline{x}$ in $\Delta \Rightarrow \alpha$ and no member of $\underline{y}$ is free in $\Delta \Rightarrow_{\underline{y}} \alpha$.

A subrule of ρ results by arbitrarily often performing one of the following transformations, starting with ρ:
(i) Transforming a rule into a variant.
(ii) Specializing of variables $\underline{x}$ in a rule $\Delta \Rightarrow_{\underline{x}} \alpha$ to $\underline{t}$, yielding $\Delta[\underline{x}|\underline{t}] \Rightarrow \alpha[\underline{x}|\underline{t}]$.
(iii) Transforming a rule $\langle \Gamma_1 \Rightarrow_{\underline{x}_1} \beta_1, \ldots, \Gamma_n \Rightarrow_{\underline{x}_n} \beta_n \rangle \Rightarrow_{\underline{x}} \alpha$ into $\langle \Gamma'_1 \Rightarrow_{\underline{x}_1} \beta_1, \ldots, \Gamma'_n \Rightarrow_{\underline{x}_n} \beta_n \rangle \Rightarrow_{\underline{x}} \alpha$ where Γ'_i $(1 \leq i \leq n)$ results from Γ_i by omitting, duplicating, re-ordering members of Γ_i and replacing members of Γ_i by subrules of them.
If ρ' is a subrule of ρ, then, according to the intended meaning of a rule, an application of ρ' can be conceived as an application of ρ as well.

In the sentential case ([18]), a derivation was considered a pair (T,f) consisting of a finite formula tree T and a discharge function f, i.e. a function defined on the set of all formula occurrences of T such that $f(\alpha)$ is either α or a formula occurrence below α, indicating where the rule with conclusion α is discharged (it remains undischarged if $f(\alpha)$ is the lowermost formula of T). In the quantifier case we also have to make explicit the eigenvariables related to an inference step. For that purpose one could define an assignment of eigenvariables e for T to be a function which associates a list of variables $\underline{x}$ with each formula occurrence α of T besides the lowermost one, having the intended meaning that the substitution of terms $\underline{t}$ for $\underline{x}$ is blocked above α, i.e. that the inference step with α as one of its premisses need not remain valid if $\underline{x}$ is substituted by a list of terms $\underline{t}$. A derivation could then be conceived of as a triple (T,e,f) consisting of a finite formula tree T, an assignment of eigenvariables e for T and a discharge function f for T.

For technical purposes however it is easier to consider VF-pairs $\langle\underline{x},\alpha\rangle$ instead of assignments of eigenvariables to formulas and to let a discharge function operate on VF-pairs instead of formulas. So we define a *discharge function* for a finite tree T of VF-pairs to be a function f defined on the set of all elements of T such that $f(\langle\underline{x},\alpha\rangle)$ is either $\langle\underline{x},\alpha\rangle$ or a VF-pair below $\langle\underline{x},\alpha\rangle$. A *derivation* is a pair (T,f) consisting of a finite tree T of VF-pairs, in whose lowermost element the list of variables is empty, and a discharge function f for T. Obviously, this approach is equivalent to the previously sketched one using an assignment e of eigenvariables, and everything which follows could be translated into the former approach (take $\underline{x}$ in $\langle\underline{x},\alpha\rangle$ to be $e(\alpha)$ and identify $f(\langle\underline{x},\alpha\rangle)=f(\langle\underline{y},\beta\rangle)$ with $f(\alpha)=f(\beta)$). With our latter definitions, however, we can immediately take over the notions defined for rules (i.e. trees of VF-pairs): Derivations (T,f) and (T',f') are called *variants* of each other, if T and T' are variants of each other and $f(\langle\underline{x},\alpha\rangle)$ and $f'(\langle\underline{x}',\alpha'\rangle)$ are corresponding elements of T and T' whenever $\langle\underline{x},\alpha\rangle$ and $\langle\underline{x}',\alpha'\rangle$ are corresponding elements of T and T'. *Substitution* in derivations is defined as follows: If $\underline{t}$ is free for $\underline{x}$ in T, then $(T,f)[\underline{x}|\underline{t}]$ is defined to be the derivation $(T[\underline{x},\underline{t}],f')$ where $f'(\langle\underline{y},\alpha\rangle[\underline{x}|\underline{t}])= f(\langle\underline{y},\alpha\rangle)[\underline{x}|\underline{t}]$.

The rules (= the assumptions!) which are used in a derivation and thus justify its inference steps are not considered part of the derivation itself. Rather, they are assigned to it by a function g associating with each VF-pair $\langle\underline{x},\alpha\rangle$ a rule ρ in such a way that $\langle\underline{x},\alpha\rangle$ may be considered the result of an application of ρ (or of a rule of which ρ is a subrule). This coincides with the usual view that comments stating which rule is applied in a particular step belong to the metalanguage. Such a rule assignment g is determined to a great extent by the derivation (T,f): it has to be chosen in accordance with the intended meaning of a discharge function which is, that by the application of the rule one of whose premisses is $f(\langle \underline{x},\alpha\rangle)$, the application of the rule whose conclusion is $\langle \underline{x},\alpha\rangle$ is discharged; moreover with the intended meaning of the eigenvariables $\underline{x}$ of VF-pairs $\langle \underline{x},\alpha\rangle$ of T which is that they become bound in all rules which have been applied above $\langle \underline{x},\alpha\rangle$ but not yet discharged above $\langle \underline{x},\alpha\rangle$. So we define:

g is a <u>rule assignment</u> for a derivation (T,f), if g associates a rule $g(\langle \underline{x},\alpha\rangle)$ with each VF-pair $\langle \underline{x},\alpha\rangle$ of T such that:
If $\langle \underline{x},\alpha\rangle$ occurs in T as

$$\frac{\begin{matrix}\vdots & & \vdots \\ \langle \underline{x}_1,\beta_1\rangle & \cdots & \langle \underline{x}_n,\beta_n\rangle\end{matrix}}{\langle \underline{x},\alpha\rangle},$$

then

$$g(\langle \underline{x},\alpha\rangle) = \frac{\dfrac{\Gamma_1}{\langle \underline{x}_1,\beta_1\rangle} \quad \cdots \quad \dfrac{\Gamma_n}{\langle \underline{x}_n,\beta_n\rangle}}{\langle \underline{y},\alpha\rangle}$$

where for all i $(1\leq i\leq n)$, Γ_i contains all rules $g(\langle \underline{z},\gamma\rangle)$ for all $\langle \underline{z},\gamma\rangle$ such that $f(\langle \underline{z},\gamma\rangle)=\langle \underline{x}_i,\beta_i\rangle$, and where $\underline{y}$ contains all variables belonging to a VF-pair which equals or is below $\langle \underline{x},\alpha\rangle$ and is properly above $f(\langle \underline{x},\alpha\rangle)$ in T. This definition contains as a limiting case the occurrence of $\langle \underline{x},\alpha\rangle$ as a top VF-pair.

There are only finitely many rule assignments for a given derivation, since they can differ only in the order of the members of $\underline{y}$ and of the Γ_i. In particular, if g and g' are rule assignments for (T,f) and $\langle \underline{x},\alpha\rangle$ is an element of T, then $g(\langle \underline{x},\alpha\rangle)$ and $g'(\langle \underline{x},\alpha\rangle)$ are subrules of each other.

So far we have defined what a derivation looks like and how to find rules which justify the inference steps, but not on which assumptions a derivation or its lowermost formula <u>depends</u>. For that purpose one has to consider the <u>undischarged assumptions</u> of a derivation (T,f) with respect to a rule assignment g for (T,f), which, according to the intended meaning of f, are defined to be the rules assigned by g to those VF-pairs $\langle \underline{x},\alpha\rangle$ of T for which $f(\langle \underline{x},\alpha\rangle)$ is the lowermost formula of T. We define an <u>assumption system</u> for (T,f) to be a list Γ of rules such that for a given rule assignment g for (T,f) each undischarged assumption is a subrule of a member of Γ (this definition is independent of the choice of g!). Now a derivation (T,f) with α as the lowermost formula of T and Γ as an assumption system for (T,f) can be considered a derivation of α from Γ.

However, some members of Γ may be <u>basic rules</u> of a calculus, i.e. belong to the given <u>framework</u> and are not assumed <u>ad hoc</u> within a derivation, and therefore should not be counted as something on which a derived formula depends. So we define: Let a set R of rules which are distinguished as basic rules be given. Then a derivation (T,f) is a <u>derivation of α from Δ in R</u> (i.e., in the calculus having R as its set of basic rules), if α is the lowermost formula of T and there is an assumption system Γ for (T,f) such that each member of Γ belongs either to Δ or is a subrule of an element of R. This definition includes as a limiting case that R is empty, i.e. no basic rules are given. As can easily be seen, the concept of a derivation from Δ in R is decidable, if 'subrule of a rule of R' is decidable. α is <u>derivable</u> from Δ in R ('$\Delta \vdash_R \alpha$'), if there is a derivation of α from Δ in R. We write '$\vdash$' instead of '$\vdash_R$' if a statement is independent of a specific choice of basic rules or if it is obvious what is meant.

Concerning the set of basic rules R our only restriction is that basic rules contain no free variables (this will be crucial for the important lemma 3.1(iv)). When writing a basic rule in the form

$$\frac{\Delta}{\alpha}$$

we mean the rule

$$\frac{\Delta}{\langle \underline{x},\alpha\rangle}$$

where $\underline{x}$ contains all variables free in Δ or α (in a certain standard order, for the sake of uniqueness). This convention is important if we write basic rules schematically, since then different instances of a schema have different lists $\underline{x}$ of variables of that kind.

<u>Lemma 3.1</u>:

(i) If ρ_1 is a subrule of ρ_2, then $\rho_1[\underline{x}|\underline{t}]$ is a subrule of $\rho_2[\underline{x}|\underline{t}]$.

(ii) Let (T',f') be a variant of a derivation (T,f) and g be a rule assignment for (T,f). Then there is a rule assignment g' for (T',f') such that for each $\langle \underline{x}',\alpha'\rangle$ of T' which corresponds to $\langle \underline{x},\alpha\rangle$ in T, $g'(\langle \underline{x}',\alpha'\rangle)$ is a variant of $g(\langle \underline{x},\alpha\rangle)$.

(iii) Let (T,f) be a derivation. Let $\underline{t}$ be free for $\underline{x}$ in T. If g is a rule assignment for (T,f), then g' is a rule assignment for $(T,f)[\underline{x}|\underline{t}]$ where $g'(\langle\underline{y},\alpha\rangle[\underline{x}|\underline{t}])=g(\langle\underline{y},\alpha\rangle)[\underline{x}|\underline{t}]$.

(iv) If $\Gamma\vdash\alpha$, then for all $\underline{x}$, $\underline{t}$: $\Gamma[\underline{x}|\underline{t}]\vdash\alpha[\underline{x}|\underline{t}]$.

Proof: (i) - (iii) follow straightforward from our definitions. (iv) follows by use of (i) - (iii): Consider a derivation (T,f) of α from Γ, choose a variant (T',f') of (T,f) in such a way that $\underline{x}$ is free for $\underline{t}$ not only in Γ and α but also in T, and apply (ii), (iii) and (i) (note that basic rules contain no free variables).

A rule $\Delta\Rightarrow_{\underline{x}}\gamma$ is called derivable from Γ in R, if for all its variants $\Delta'\Rightarrow_{\underline{y}}\gamma'$ and for all $\underline{t}$ such that $(\Delta'\Rightarrow\gamma')[\underline{y}|\underline{t}]$ has the form $\langle\Gamma_1\Rightarrow_{\underline{x}_1}\beta_1,\dots,\Gamma_n\Rightarrow_{\underline{x}_n}\beta_n\rangle\Rightarrow\alpha$, the following holds for all $\Delta_1,\dots,\Delta_n$: If, for each i $(1\le i\le n)$: $\Gamma,\Delta_i,\Gamma_i\vdash_R\beta_i$, where no variable of $\underline{x}_i$ occurs free in Δ_i or Γ, then $\Gamma,\Delta_1,\dots,\Delta_n\vdash_R\alpha$. This definition follows the intended meaning we have given to a rule (see § 2).

We shall use $\Gamma\vdash\Delta\Rightarrow\gamma$ as an abbreviation for $\Gamma,\Delta\vdash\gamma$. $\Gamma\vdash\Delta\Rightarrow_{\underline{x}}\gamma$ expresses that for a variant $\Delta'\Rightarrow_{\underline{y}}\gamma'$ of $\Delta\Rightarrow_{\underline{x}}\gamma$ such that no variable of $\underline{y}$ is free in Γ: $\Gamma,\Delta'\vdash\gamma'$. By lemma 3.1 (iv), this then holds for any variant of this kind. Furthermore this is, again by lemma 3.1 (iv), equivalent to the statement that for all $\underline{t}$: $\Gamma,\Delta'[\underline{y}|\underline{t}]\vdash\gamma'[\underline{y}|\underline{t}]$, where $\Delta'\Rightarrow_{\underline{y}}\gamma'$ is any variant of $\Delta\Rightarrow_{\underline{x}}\gamma$ (without restriction). Our notation $\Gamma\vdash\rho$, which suggests that ρ is derivable from Γ, but is not defined in this way, will be justified by lemma 3.3 which is based on lemma 3.2.

Lemma 3.2:

(i) $\rho\vdash\rho$ (i.e. $\Delta\Rightarrow_{\underline{x}}\alpha$, $\Delta\vdash\alpha$).

(ii) If $\Delta\vdash\rho$ and $\Delta,\rho\vdash\gamma$ then $\Delta\vdash\gamma$.

Proof: See the proofs of lemmata 3.4 and 3.5 in [18]. Only a few additions are necessary which deal with bound variables.

Lemma 3.3: $\Gamma\vdash\rho$ iff ρ is derivable from Γ.

Proof: Let ρ be $\Delta\Rightarrow_{\underline{x}}\gamma$. $\Gamma\vdash\rho$ means that for all variants $\Delta'\Rightarrow_{\underline{y}}\gamma'$ of ρ and all $\underline{t}$ we have $\Gamma,\Delta'[\underline{y}|\underline{t}]\vdash\gamma'[\underline{y}|\underline{t}]$. Let this be of the form:

$$\Gamma,\Gamma_1\Rightarrow_{\underline{x}_1}\beta_1,\dots,\Gamma_n\Rightarrow_{\underline{x}_n}\beta_n\vdash\alpha \quad . \tag{2}$$

If we have for all i $(1 \le i \le m)$: $\Gamma, \Delta_i, \Gamma_i \vdash \beta_i$ where no variable of $\underline{x}_i$ occurs free in Γ or Δ_i, we obtain, since this can be written as $\Gamma, \Delta_i \vdash \Gamma_i \Rightarrow_{\underline{x}_i} \beta_i$, by (2) and lemma 3.2 (ii) (n-fold application): $\Gamma, \Delta_1, \ldots, \Delta_n \vdash \alpha$. Conversely, since by lemma 3.2 (i) it holds that for all i $(1 \le i \le n)$: $\Gamma_i \Rightarrow_{\underline{x}_i} \beta_i, \Gamma_i \vdash \beta_i$, the derivability of ρ from Γ implies (2).

By this lemma we may use $\Gamma \vdash \rho$ as an alternative formulation of the derivability of ρ from Γ. $\Delta \vdash \Gamma$ means that $\Delta \vdash \rho_i$ for all i $(1 \le i \le n)$ if Γ is the list $\rho_1 \ldots \rho_n$. As a limiting case, $\Delta \vdash \Gamma$ is considered to be true if Γ is empty. $\Delta \dashv\vdash \Gamma$ means that $\Delta \vdash \Gamma$ and $\Gamma \vdash \Delta$.

<u>Lemma 3.4</u>: Let $\alpha \dashv\vdash \beta$. Let ρ contain $\langle \underline{x}, \alpha \rangle$ as an element and let ρ' be the result of replacing this element by $\langle \underline{x}, \beta \rangle$. Then $\rho \dashv\vdash \rho'$.

<u>Proof</u>: Let ρ_1 be the part of ρ whose lowermost element is $\langle \underline{x}, \alpha \rangle$ (linearly written: $\Delta \Rightarrow_{\underline{x}} \alpha$), and let ρ_1' be $\Delta \Rightarrow_{\underline{x}} \beta$. From $\Delta \Rightarrow_{\underline{x}} \alpha, \Delta \vdash \alpha$ and $\alpha \vdash \beta$ it follows that $\Delta \Rightarrow_{\underline{x}} \alpha \vdash \Delta \Rightarrow_{\underline{x}} \beta$; analogously $\Delta \Rightarrow_{\underline{x}} \beta \vdash \Delta \Rightarrow_{\underline{x}} \alpha$. So we have $\rho_1 \dashv\vdash \rho_1'$. If ρ_1 is identical with ρ, nothing remains to be shown. Let ρ_1 occur as a proper part of a part ρ_2 of ρ of the form $\langle \ldots \rho_1 \ldots \rangle \Rightarrow_{\underline{y}} \gamma$ and let ρ_2' be $\langle \ldots \rho_1' \ldots \rangle \Rightarrow_{\underline{y}} \gamma$. Then $\rho_2, \ldots, \rho_1, \ldots \vdash \gamma$ by lemma 3.2 (i) and $\rho_2, \ldots, \rho_1', \ldots \vdash \gamma$ by lemma 3.2 (ii), i.e. $\rho_2 \vdash \rho_2'$. In the same way we obtain $\rho_2' \vdash \rho_2$. Repeated application of this procedure yields $\rho \dashv\vdash \rho'$.

4. BASIC RULES FOR OPERATORS

We shall define a standard form for schematically given basic rules (more precisely, I and E rules) for operators. For this purpose we assume to be given:

(i) Denumerably many <u>schematic letters</u> to be instantiated by <u>formulas</u> (syntactical variables for them: 'A', 'B', 'C', for lists of distinct letters: '$\underline{A}$', '$\underline{B}$', '$\underline{C}$', all with and without indices).

(ii) For each list of distinct schematic letters for formulas $\underline{A}$, denumerably many schematic letters to be instantiated by <u>variables</u> which do not occur (neither free nor bound) in a member of the list of formulas by which $\underline{A}$ is instantiated (syntactical variables for them: '$X_{\underline{A}}$', '$Y_{\underline{A}}$', '$Z_{\underline{A}}$' where 'X', 'Y', 'Z' may have an index, for lists of distinct letters: '$\underline{X}_{\underline{A}}$', '$\underline{Y}_{\underline{A}}$', '$\underline{Z}_{\underline{A}}$', for lists of distinct letters of the kind $Y_{1\underline{A}_1} \ldots Y_{n\underline{A}_n}$: '$\underline{U}$', '$\underline{V}$'). If $\underline{A}$ is empty these schematic letters can be instantiated by any variable (syntactical variables in that case: 'X', 'Y', 'Z', '$\underline{X}$', '$\underline{Y}$', '$\underline{Z}$', with and without indices).

A schematic letter $X_{\underline{A}}$ for nonempty $\underline{A}$ can be instantiated only if $\underline{A}$ is instantiated at the same time. - As for variables, we use 'x_α' or '$\underline{x}_\alpha$' in order to express that x or $\underline{x}$ does not occur in α.

Formula schemata are defined as follows: Each schematic letter for formulas is a formula schema. For all A, $\underline{U}$ and (not necessarily distinct) $X_{1\underline{A}_1},\ldots,X_{n\underline{A}_n}$, where $\underline{U}$ is of length n and all $\underline{A}_i$ ($1\le i\le n$) contain A (i.e., the $X_{i\underline{A}_i}$ must not be instantiated by variables occurring in the instance of A), $A[\underline{U}|X_{1\underline{A}_1}\ldots X_{n\underline{A}_n}]$ is a formula schema. (.[.|.] is here a sign and not an operation!) If S is an operator of type (n_1,n_2), $\underline{U}$ is of length n_1 and $F_1,\ldots,F_{n_2}$ are formula schemata, then $S\underline{U}F_1\ldots F_{n_2}$ is a formula schema. All occurrences of schematic letters of $\underline{U}$ in $S\underline{U}F_1\ldots F_{n_2}$ are called bound. A rule schema is a finite tree of pairs $\langle\underline{U},F\rangle$ where F is a formula schema. A linear notation for rule schemata is defined in the same way as for rules. Occurrences of schematic letters of $\underline{U}$ in $\langle\underline{U},F\rangle$ and above $\langle\underline{U},F\rangle$ in the considered rule schema are called bound.

The instantiation of formula/rule schemata to formulas/rules is defined as follows: Replace schematic letters for formulas A by formulas α and different schematic letters $X_{\underline{A}}$ for variables by different variables $x_{\underline{\alpha}}$ not occurring in the instance $\underline{\alpha}$ of $\underline{A}$. $\alpha[\underline{y}|z_1\ldots z_n]$, when resulting from $A[\underline{U}|X_{1\underline{A}_1}\ldots X_{n\underline{A}_n}]$ is then always defined and can be evaluated, since no z_i occurs in α because of the restriction on the $X_{i\underline{A}_i}$. A formula/rule schema is called derivable from a set of basic rules R iff all its instances are derivable from the empty list of assumptions in R.

Remark. Whereas on the level of formula/rule schemata .[.|.] is a sign, on the level of formulas/rules .[.|.] is an operation to be evaluated. So the procedure of instantiating a schema includes the evaluation of .[.|.] conceived as a metalinguistic substitution operation. This way of dealing with substitution could have been avoided by treating quantifiers not as variable-binding operators but as operators which are applied to λ-terms. Then we would have had to add rules of λ-conversion to the basic rules.

As syntactical variables we use 'F' for formula schemata, 'R' for rule schemata, 'Φ' for lists of rule schemata (all with and without indices). If $\underline{U}$ and $\underline{V}$ have no schematic letter in common, we write '$F(\underline{U},\underline{V},\underline{A})$', '$R(\underline{U},\underline{V},\underline{A})$', '$\Phi(\underline{U},\underline{V},\underline{A})$' to indicate that F, R and Φ contain no other schematic letters for variables than those of $\underline{U}$ and $\underline{V}$ and no other schematic letters for formulas than those of $\underline{A}$ (but possibly fewer). If $\underline{U}$, $\underline{V}$, $\underline{A}$

can be instantiated by $\underline{x}$, $\underline{y}$, $\underline{\alpha}$, then '$F(\underline{x},\underline{y},\underline{\alpha})$', '$R(\underline{x},\underline{y},\underline{\alpha})$', '$\Phi(\underline{x},\underline{y},\underline{\alpha})$', respectively, denote the formula, rule or list of rules which is the result of this instantiation.

We motivate our standard form for I and E rules for an operator by referring to the <u>common content</u> of lists of rules $\Gamma_1,\ldots,\Gamma_m$ with respect to a list of variables $\underline{x}$. It is defined as follows: The common content of $\Gamma_1,\ldots,\Gamma_m$ with respect to $\underline{x}$ in R is the set of all ρ such that for all $\underline{t}$ and for all i ($1\le i\le m$): $\Gamma_i[\underline{x}|\underline{t}] \vdash_R \rho$. Whereas in the propositional case the common content was the finite intersection of contents of lists of rules (see [18]), the common content with respect to a list of variables can be considered to be an infinite intersection of contents of lists of rules. The limiting case m=0 leads to intuitionistic logic since it allows one to interpret the absurdity sign.

Similar to [18], we assume that with each operator S of type (n_1,n_2) lists of rule schemata $\Phi_1(\underline{X},\underline{Y}_{\underline{A}},\underline{A}),\ldots,\Phi_m(\underline{X},\underline{Y}_{\underline{A}},\underline{A})$ ($m\ge 0$) are associated where $\underline{X}$ is of length n_1 and $\underline{A}$ of length n_2. It is required that all operators can be ordered in a sequence S_1, $S_2,\ldots$ in such a way that in the lists associated with an operator S_k at most the operators S_j for j<k occur. Furthermore, the $\Phi_i(\underline{X},\underline{Y}_{\underline{A}},\underline{A})$ ($1\le i\le m$) must fulfil the condition that letters of $\underline{Y}_{\underline{A}}$ only occur bound. This is because instances of Φ_i should not contain free variables beyond those in the corresponding instance of $S\underline{X}\underline{A}$, save variables by which $\underline{X}$ is instantiated. The reason for writing '$\underline{Y}_{\underline{A}}$' instead of '$\underline{Y}$' is that variables free in an instance of $S\underline{X}\underline{A}$ should be free in the corresponding instance of Φ_i. In other words, instances $\Phi_i(\underline{x},\underline{y},\underline{\alpha})$ and $S\underline{x}\underline{\alpha}$ of Φ_i and $S\underline{X}\underline{A}$ can, with respect to free variables, differ only in that variables of $\underline{x}$ are free in $\Phi_i(\underline{x},\underline{y},\underline{\alpha})$ but bound in $S\underline{x}\underline{\alpha}$. S is called a $\bot$-operator if m=0, i.e., if no list of rules (not even the empty list) is associated with S.

We require that the set R_S of basic rules for S be a minimal set with respect to derivability which satisfies the condition that for all $\underline{x}$, $\underline{y}_{\underline{\alpha}}$, $\underline{\alpha}$ (where $\underline{x}$ is of length n_1 and $\underline{\alpha}$ of length n_2), $S\underline{x}\underline{\alpha}$ expresses the common content of $\Phi_1(\underline{x},\underline{y}_{\underline{\alpha}},\underline{\alpha}),\ldots,\Phi_m(\underline{x},\underline{y}_{\underline{\alpha}},\underline{\alpha})$ with respect to $\underline{x}$ in R, i.e.

(*) for all $\underline{x}$, $\underline{y}_{\underline{\alpha}}$, $\underline{\alpha}$ and for all ρ: $S\underline{x}\underline{\alpha} \vdash_R \rho$ iff for all $\underline{t}$ and all i ($1\le i\le m$): $\Phi_i(\underline{x},\underline{y}_{\underline{\alpha}},\underline{\alpha})[\underline{x}|\underline{t}] \vdash_R \rho$.

By 'minimal with respect to derivability' we mean that if (*) holds for a certain R, then for each $\rho\in R_S$: $\vdash_R \rho$. Obviously, different minimal sets R and R' are interderivable in the sense that for each $\rho\in R'$: $\vdash_R \rho$, and for each $\rho \in R: \vdash_{R'} \rho$.

We shall show that R_S can be chosen as the set of all instances of I and E rule schemata of the following standard form.

$$\text{S-I} \qquad \frac{\Phi_1(\underline{X},\underline{Y}_{\underline{A}},\underline{A})}{S\underline{X}\underline{A}} \quad \cdots \quad \frac{\Phi_m(\underline{X},\underline{Y}_{\underline{A}},\underline{A})}{S\underline{X}\underline{A}}$$

Linear notation: $\Phi_i(\underline{X},\underline{Y}_{\underline{A}},\underline{A}) \Rightarrow S\underline{X}\underline{A} \qquad (1 \leq i \leq m)$

$$\text{S-E} \qquad \frac{S\underline{X}\underline{A} \qquad \dfrac{\Phi_1(\underline{X},\underline{Y}_{\underline{A}},\underline{A})}{\langle \underline{X}, B[\underline{X}|\underline{Z}_B]\rangle} \qquad \dfrac{\Phi_m(\underline{X},\underline{Y}_{\underline{A}},\underline{A})}{\langle \underline{X}, B[\underline{X}|\underline{Z}_B]\rangle}}{B[\underline{X}|\underline{Z}_B]}$$

where B is a schematic letter different from those in $\underline{A}$, and $\underline{Z}_B$ and $\underline{X}$ have no schematic letter in common, i.e., they are instantiated by different variables.
Linear notation: $\langle S\underline{X}\underline{A}, \Phi_1(\underline{X},\underline{Y}_{\underline{A}},\underline{A}) \Rightarrow_{\underline{X}} B[\underline{X}|\underline{Z}_B], \ldots, \Phi_m(\underline{X},\underline{Y}_{\underline{A}},\underline{A}) \Rightarrow_{\underline{X}} B[\underline{X}|\underline{Z}_B]\rangle$
$\Rightarrow B[\underline{X}|\underline{Z}_B]$.

One should remember that according to a convention stated in § 3 all variables are bound in instances of rule schemata of this form, so that for appropriate $\underline{z}$ containing all variables which are free in $\Phi_i(\underline{x},\underline{y}_{\underline{\alpha}},\underline{\alpha})$,

$$\frac{\Phi_i(\underline{x},\underline{y}_{\underline{\alpha}},\underline{\alpha})}{\langle \underline{z}, S\underline{x}\underline{\alpha}\rangle}$$

would be an instance of S-I.

<u>Theorem 4.1</u>: (*) holds iff S-I and S-E are derivable in R.

<u>Proof</u>: Let arbitrary $\underline{x},\underline{y}_{\underline{\alpha}},\underline{\alpha}$ be given such that $\Phi_i(\underline{x},\underline{y}_{\underline{\alpha}},\underline{\alpha})$ is defined for every i $(1 \leq i \leq m)$. Taking ρ to be $S\underline{x}\underline{\alpha}$ we obtain from (*) for all i and $\underline{t}$ $(1 \leq i \leq m)$: $\Phi_i(\underline{x},\underline{y}_{\underline{\alpha}},\alpha)[x|\underline{t}] \vdash S\underline{x}\underline{\alpha}$, in particular $\Phi_i(\underline{x},\underline{y}_{\underline{\alpha}},\underline{\alpha}) \vdash S\underline{x}\underline{\alpha}$. Thus $\Phi_i(\underline{x},\underline{y}_{\underline{\alpha}},\underline{\alpha}) \Rightarrow_{\underline{z}} S\underline{x}\underline{\alpha}$ is derivable where $\underline{z}$ contains all variables free in $\Phi_i(\underline{x},\underline{y}_{\underline{\alpha}},\underline{\alpha})$. -
By lemma 3.2 (i) we have for all i $(1 \leq i \leq m)$ if $\underline{x}$ is not free in β:
$\Phi_1(\underline{x},\underline{y}_{\underline{\alpha}},\underline{\alpha}) \Rightarrow_{\underline{x}} \beta, \ldots, \Phi_m(\underline{x},\underline{y}_{\underline{\alpha}},\underline{\alpha}) \Rightarrow_{\underline{x}} \beta, \Phi_i(\underline{x},\underline{y}_{\underline{\alpha}},\underline{\alpha}) \vdash \beta$,
therefore by lemma 3.1 (iv) for all $\underline{t}$:
$\Phi_1(\underline{x},\underline{y}_{\underline{\alpha}},\underline{\alpha}) \Rightarrow_{\underline{x}} \beta, \ldots, \Phi_m(\underline{x},\underline{y}_{\underline{\alpha}},\underline{\alpha}) \Rightarrow_{\underline{x}} \beta, \Phi_i(\underline{x},\underline{y}_{\underline{\alpha}},\underline{\alpha})[\underline{x}|\underline{t}] \vdash \beta$,

which is the same as

$\Phi_i(\underline{x},\underline{y}_{\underline{\alpha}},\underline{\alpha})[\underline{x}|\underline{t}] \vdash \langle\Phi_1(\underline{x},\underline{y}_{\underline{\alpha}},\underline{\alpha})\Rightarrow_{\underline{x}}\beta,\ldots,\Phi_m(\underline{x},\underline{y}_{\underline{\alpha}},\underline{\alpha})\Rightarrow_{\underline{x}}\beta\rangle\Rightarrow\beta$.

From (*) we obtain

$S\underline{x}\underline{\alpha} \vdash \langle\Phi_1(\underline{x},\underline{y}_{\underline{\alpha}},\underline{\alpha})\Rightarrow_{\underline{x}}\beta,\ldots,\Phi_m(\underline{x},\underline{y}_{\underline{\alpha}},\underline{\alpha})\Rightarrow_{\underline{x}}\beta\rangle\Rightarrow\beta$

which is the same as

$S\underline{x}\underline{\alpha},\Phi_1(\underline{x},\underline{y}_{\underline{\alpha}},\underline{\alpha})\Rightarrow_{\underline{x}}\beta,\ldots,\Phi_m(\underline{x},\underline{y}_{\underline{\alpha}},\underline{\alpha})\Rightarrow_{\underline{x}}\beta \vdash \beta$.

Thus $\langle S\underline{x}\underline{\alpha},\Phi_1(\underline{x},\underline{y}_{\underline{\alpha}},\underline{\alpha})\Rightarrow_{\underline{x}}\beta,\ldots,\Phi_m(\underline{x},\underline{y}_{\underline{\alpha}},\underline{\alpha})\Rightarrow_{\underline{x}}\beta\rangle\Rightarrow_{\underline{z}}\beta$

is derivable, where $\underline{z}$ contains all variables free in a $\Phi_i(\underline{x},\underline{y}_{\underline{\alpha}},\underline{\alpha})$ $(1\le i\le m)$ or β. -

Conversely, from $\Phi_i(\underline{x},\underline{y}_{\underline{\alpha}},\underline{\alpha}) \vdash S\underline{x}\underline{\alpha}$ follows $\Phi_i(\underline{x},\underline{y}_{\underline{\alpha}},\underline{\alpha})[\underline{x}|\underline{t}] \vdash S\underline{x}\underline{\alpha}$ by lemma 3.1 (iv), thus together with $S\underline{x}\underline{\alpha}\vdash\rho$: $\Phi_i(\underline{x},\underline{y}_{\underline{\alpha}},\underline{\alpha})[\underline{x}|\underline{t}]\vdash\rho$.-

Assume $\Phi_i(\underline{x},\underline{y}_{\underline{\alpha}},\underline{\alpha})[\underline{x}|\underline{t}]\vdash\rho$ for all i $(1\le i\le m)$ and all $\underline{t}$.

Choose $\underline{z}_1,\underline{z}_2$ in such a way that $\underline{x}$, $\underline{z}_1$, $\underline{z}_2$ have no variable in common, no variable of $\underline{z}_1$ occurs free in $\Phi_i(\underline{x},\underline{y}_{\underline{\alpha}},\underline{\alpha})$ for any i $(1\le i\le m)$, no variable of $\underline{z}_2$ occurs free in ρ, $\rho[\underline{x}|\underline{z}_1][\underline{z}_1|\underline{x}]$ is ρ and $\Phi_i(\underline{x},\underline{y}_{\underline{\alpha}},\underline{\alpha})[\underline{x}|\underline{z}_2][\underline{z}_2|\underline{x}]$ is $\Phi_i(\underline{x},\underline{y}_{\underline{\alpha}},\underline{\alpha})$ for all i $(1\le i\le m)$. Then we obtain $\Phi_i(\underline{x},\underline{y}_{\underline{\alpha}},\underline{\alpha})[\underline{x}|\underline{z}_2]\vdash\rho$ for all i $(1\le i\le m)$.

Therefore by two applications of theorem 3.1 (iv):

$\Phi_i(\underline{x},\underline{y}_{\underline{\alpha}},\underline{\alpha})\vdash\rho[\underline{x},\underline{z}_1]$ for all i $(1\le i\le m)$,

thus for a variant $\Gamma'\Rightarrow_{\underline{z}}\beta'$ of $\rho[\underline{x}|\underline{z}_1]$ such that $\underline{z}$ and $\underline{x}$ have no variable in common and $\underline{z}$ is not free in any $\Phi_i(\underline{x},\underline{y}_{\underline{\alpha}},\underline{\alpha})$:

$\Phi_i(\underline{x},\underline{y}_{\underline{\alpha}},\underline{\alpha}),\Gamma'\vdash\beta'$, thus

$\Gamma'\vdash\Phi_i(\underline{x},\underline{y}_{\underline{\alpha}},\underline{\alpha})\Rightarrow_{\underline{x}}\beta'$.

By S-E and lemma 3.2 (ii):

$S\underline{x}\underline{\alpha},\Gamma'\vdash\beta'$, thus

$S\underline{x}\underline{\alpha}\vdash\rho[\underline{x}|\underline{z}_1]$, thus by lemma 3.1 (iv):

$S\underline{x}\underline{\alpha}\vdash\rho$.

If one takes R_S to contain exactly the instances of S-I and S-E, then S-I and S-E are trivially derivable in R_S, so (by the theorem) R_S fulfils (*). Conversely, if R satisfies (*) then (by the theorem) S-I and S-E are derivable in R, i.e. R_S is a minimal set satisfying (*).

Basic rule schemata for the standard quantifiers $\forall,\exists$ which are of type (1,1), have the form:

$$\forall I \quad \frac{\langle X,A\rangle}{\forall XA} \qquad\qquad \forall E \quad \frac{\forall XA \qquad \dfrac{\langle X,A\rangle}{\langle X,B[X|Z_B]\rangle}}{B[X|Z_B]}$$

lin.: $\langle\Rightarrow_X A\rangle\Rightarrow\forall XA$ lin.: $\langle\forall XA,\langle\Rightarrow_X A\rangle\Rightarrow_X B[X|Z_B]\rangle\Rightarrow B[X|Z_B]$

$\exists I \quad \dfrac{A}{\exists XA}$ $\qquad\qquad\qquad$ $\exists E \quad \dfrac{\exists XA \qquad \dfrac{A}{\langle X, B[X|Z_B]\rangle}}{B[X|Z_B]}$

lin.: $\langle A\rangle \Rightarrow \exists XA$ $\qquad\qquad$ lin.: $\langle \exists XA, \langle A\rangle \Rightarrow_X B[X|Z_B]\rangle \Rightarrow B[X|Z_B]$

$\forall E$ is equivalent to the usual $\forall$ elimination rule which has the form

$$\frac{\forall XA}{A} \quad . \tag{3}$$

For letting x,α be arbitrary, β not containing x free, then

$$\cfrac{\cfrac{\overline{\forall x\alpha}}{\langle x,\alpha\rangle}}{\beta}$$

is a derivation of β for which a rule assignment is given by: $g(\forall x\alpha) = \Rightarrow_x \forall x\alpha$, $g(\langle x,\alpha\rangle) = \langle \forall x\alpha\rangle \Rightarrow \alpha$, and $g(\beta) = \langle \Rightarrow_x \alpha\rangle \Rightarrow \beta$. Since $\Rightarrow_x \forall x\alpha$ and $\langle \Rightarrow_x \alpha\rangle \Rightarrow \beta$ are subrules of $\forall x\alpha$ and $\langle \Rightarrow_x \alpha\rangle \Rightarrow_x \beta$ respectively, we have a derivation of β from $\forall x\alpha$, $\langle \Rightarrow_x \alpha\rangle \Rightarrow_x \beta$ and an instance of (3). Conversely, taking the instance $\langle \forall x\alpha, \langle \Rightarrow_x \alpha\rangle \Rightarrow_x \alpha[x|y_\alpha]\rangle \Rightarrow_{\underline{z}} \alpha[x|y_\alpha]$ of $\forall E$ where $\underline{z}$ contains all variables which are free in $\alpha[x|y_\alpha]$, we obtain $\langle \forall x\alpha, \langle \Rightarrow_x \alpha\rangle \Rightarrow \alpha[x|y_\alpha]\rangle \Rightarrow_{\underline{z}} \alpha[x|y_\alpha]$ as a variant and $\langle \forall x\alpha, \langle \Rightarrow_x \alpha\rangle \Rightarrow \alpha\rangle \Rightarrow \alpha$ as a subrule; since $\Rightarrow_x \alpha \vdash \alpha$ holds trivially, we obtain $\forall x\alpha \vdash \alpha$ by use of $\forall E$.

Examples of further operators are those whose I rule schemata have the following form, where the types of the operators are mentioned on the right:

$\langle\langle A\rangle \Rightarrow_X B\rangle \Rightarrow \Pi XAB$ (1,2)

$\langle A,B\rangle \Rightarrow \Sigma XAB$ (1,2)

$\langle \Rightarrow_Y A\rangle \Rightarrow \exists\forall XYA$ (2,1)

$\langle \Rightarrow_{Z_A} A[X_2X_3|Z_AZ_A]\rangle \Rightarrow LX_1X_2X_3A$

$\langle \Rightarrow_{Z_A} A[X_1X_3|Z_AZ_A]\rangle \Rightarrow LX_1X_2X_3A$ (3,1)

$\langle \Rightarrow_{Z_A} A[X_1X_2|Z_AZ_A]\rangle \Rightarrow LX_1X_2X_3A$

$\langle \Rightarrow_{Z_A} A[XY|Z_AZ_A]\rangle \Rightarrow DXYA$ (2,1)

$\langle\langle A\rangle \Rightarrow_{XY_A} A[X|Y_A]\rangle \Rightarrow IXA$ (1,1)

$\langle A, \langle A, A[X|Z_{AB}]\rangle \Rightarrow_{XZ_{AB}} B[Y|Z_{AB}]\rangle \Rightarrow JXYAB$ (2,2)

$\langle\langle A[X_2|Y_A], A[X_1X_2|Y_AZ_A]\rangle \Rightarrow_{X_1Y_AZ_A} A[X_2|Z_A]\rangle \Rightarrow TX_1X_2A$ (2,1)

Corresponding E rules are determined uniquely by the I rules. The ΠE rule, for instance, is

$<\Pi xAB, <<A> \Rightarrow_x B> \Rightarrow_x C[X|z_C]> \Rightarrow C[X|z_C]$.

<u>Remark</u>. At the meeting 'Konstruktive Mengenlehre und Typentheorie' (München 1980) Per Martin-Löf presented a version of his intuitionistic set theory in which the E rule for his operator Π is structurally similar to the above ΠE rule, and could be considered a rule of level 4 in our sense. He formulates it as

$$\frac{c \in \Pi(A,B) \qquad \begin{matrix}[y(x) \in B(x) \quad (x \in A)] \\ d(y) \in C(\lambda(y))\end{matrix}}{F(c,d) \in C(c)}$$

which in our notation for rules would have to be written as

$$\frac{c \in \Pi(A,B) \qquad \dfrac{\dfrac{x \in A}{<x, y(x) \in B(x)>}}{<y, d(y) \in C(\lambda(y))>}}{F(c,d) \in C(c)}$$

or linearly

$$<c \in \Pi(A,B), <<x \in A> \Rightarrow_x y(x) \in B(x)> \Rightarrow_y d(y) \in C(\lambda(y))> \Rightarrow F(c,d) \in C(c).$$

Apart from the usage of a new kind of assumption, however, this similarity only concerns certain basic ideas about the relationship between I rules and E rules, and not those specific features of Martin-Löf's system which make it behave differently from other formalizations of intuitionistic mathematics. In order to justify its rules in detail, one would need further principles, in particular about the way logical rules, i.e. rules concerning propositions, are part of a type or set theory. In the published versions of his system (e.g.[11]) Martin-Löf gives an equivalent ΠE rule which is of level 2 (for a description of Martin-Löf's theory as a formal system see [1]). - Zucker and Tragesser [20] work, when treating the completeness of the standard intuitionistic operators for quantifier logic, within the framework of Martin-Löf's theory. Their general schema for I rules (whose premisses are conceived as trees growing <u>downwards</u>) is intended to capture all kinds of operators which may be introduced in this framework. Zucker and Tragesser give, however, almost no motivation for their schema, not even an example. Similarities between their approach (when restricted to first-order logic) and the one presented here I can only suspect.

<u>Theorem 4.2</u> (Replacement theorem): Assume $\beta \dashv\vdash \beta'$. Let β occur as a part of α, and let α' be the result of replacing this part of α with β'. Then $\alpha \dashv\vdash \alpha'$. Furthermore, if $<\underline{x},\alpha>$ is an element of a member of Γ and Γ' results by replacing this element with $<\underline{x},\alpha'>$, then $\Gamma \dashv\vdash \Gamma'$.

<u>Proof</u>: From lemma 3.4 in almost the same way as in the propositional case (see [18], theorem 4.6).

<u>Theorem 4.3</u> (Relabelling of bound operator variables): $S\underline{x}\underline{\alpha} \dashv\vdash S\underline{z}(\underline{\alpha}[\underline{x}|\underline{z}])$ if $\underline{z}$ is not free in $\underline{\alpha}$.

Proof: Induction on k where S is S_k, i.e. the k-th member in the sequence of all operators. If S is S_1 and has I and E rules of the general form stated above, then the $\Phi_i(\underline{X},\underline{Y}_{\underline{A}},\underline{A})$ do not contain any operator. From the S-I rule we have

$$\Phi_i(\underline{x},\underline{y}_{\underline{\alpha}},\underline{\alpha}) \vdash S\underline{x}\underline{\alpha} \quad \text{for all } i \ (1 \leq i \leq m) \quad , \tag{4}$$

and by forming a variant of $\Phi_i(\underline{x},\underline{y}_{\underline{\alpha}},\underline{\alpha})$ and applying lemma 3.1 (iv) we obtain

$$\Phi_i(\underline{z},\underline{y}_{\underline{\alpha}[\underline{x}|\underline{z}]},\underline{\alpha}[\underline{x}|\underline{z}]) \vdash S\underline{x}\underline{\alpha} \quad \text{for all } i \ (1 \leq i \leq m) \quad . \tag{5}$$

Application of S-E yields

$$S\underline{z}(\underline{\alpha}[\underline{x}|\underline{z}]) \vdash S\underline{x}\underline{\alpha} \quad .$$

The converse follows analogously.

If S is S_k for k>1, then the situation differs only in that the $\Phi_i(\underline{X},\underline{Y}_{\underline{A}},\underline{A})$ may contain S_j for j<k. So we have, in order to pass from (4) to (5), to apply additionally the induction hypothesis and the replacement theorem 4.2.

5. THE COMPLETENESS OF THE STANDARD INTUITIONISTIC OPERATORS

We can now show that all operators given by I and E rules of our standard form are explicitly definable in terms of $\&,\vee,\supset,\curlywedge,\forall,\exists$. We assume that the calculus we are considering already contains these operators and their basic rules and that they form the first 6 members $S_1,\dots,S_6$ of the enumeration of operators. $\forall\underline{x}\alpha$ and $\exists\underline{x}\alpha$ are used as abbreviations for $\forall x_1\dots\forall x_n\alpha$ and $\exists x_1\dots\exists x_n\alpha$ if $\underline{x}$ is $x_1\dots x_n$, and similarly for schemata.

We associate with each formula schema F, rule schema R and list of rule schemata Φ a formula schema F*, R* and Φ* which contains at most $\&,\supset,\curlywedge,\forall$ as operators besides operators occurring in F, R, Φ: F* is F. R* is $(R_1^*\&\dots\&R_n^*)\supset F$ if R is $\langle R_1,\dots,R_n\rangle \Rightarrow F$. R* is $\forall\underline{X}((R_1^*\&\dots\&R_n^*)\supset F)$ if R is $\langle R_1,\dots,R_n\rangle \Rightarrow_{\underline{X}} F$. Φ^* is $R_1^*\&\dots\&R_n^*$ if Φ is the list $R_1\dots R_n$. Φ^* is $\curlywedge\supset\curlywedge$ if Φ is the empty list. In the same way we associate with each formula α, rule ρ, list of rules Δ a formula α^*, ρ^*, Δ^*.

Lemma 5.1: For all formulas α, rules ρ and lists of rules Δ: $\alpha \dashv\vdash \alpha^*$, $\rho \dashv\vdash \rho^*$, $\Delta \dashv\vdash \Delta^*$.

Proof: We consider only the case where a quantifier is involved. Assume

$$\langle\rho_1,\ldots,\rho_n\rangle \Rightarrow \alpha \dashv\vdash (\rho_1{*}\&\ldots\&\rho_n{*})\supset\alpha \quad . \tag{6}$$

Since $\langle\rho_1,\ldots,\rho_n\rangle\Rightarrow\alpha$ is a subrule of $\langle\rho_1,\ldots,\rho_n\rangle\Rightarrow_{\underline{x}}\alpha$, we have

$$\langle\rho_1,\ldots,\rho_n\rangle\Rightarrow_{\underline{x}}\alpha \vdash (\rho_1{*}\&\ldots\&\rho_n{*})\supset\alpha \quad .$$

By applying $\forall$I (possibly more than once), we obtain
$\langle\rho_1,\ldots,\rho_n\rangle\Rightarrow_{\underline{x}}\alpha\vdash \forall\underline{x}((\rho_1{*}\&\ldots\&\rho_n{*})\supset\alpha)$.
Conversely, we have by (3), which is equivalent to $\forall$E, and (6):
$\forall\underline{x}((\rho_1{*}\&\ldots\&\rho_n{*})\supset\alpha),\rho_1,\ldots,\rho_n\vdash\alpha$. Thus
$\forall\underline{x}((\rho_1{*}\&\ldots\&\rho_n{*})\supset\alpha)\vdash\langle\rho_1,\ldots,\rho_n\rangle\Rightarrow_{\underline{x}}\alpha$.

Theorem 5.2: For each S of type (n_1,n_2) there is a formula schema $F(\underline{X},\underline{Y}_{\underline{A}},\underline{A})$ containing at most $\&,\vee,\supset,\curlywedge,\forall,\exists$ as operators, where $\underline{X}$ is of length n_1, $\underline{A}$ of length n_2 and letters of $\underline{Y}_{\underline{A}}$ only occur bound, such that for all $\underline{x}$, $\underline{y}_{\underline{\alpha}}$, $\underline{\alpha}$: $S\underline{x}\underline{\alpha}\dashv\vdash F(\underline{x},\underline{y}_{\underline{\alpha}},\underline{\alpha})$.

Proof: Induction on k where S is S_k. If S is S_1, then S is one of the operators $\&,\vee,\supset,\curlywedge,\forall,\exists$; so we can take F to be $S\underline{X}\underline{A}$. Let k>0. If S is a $\bot$-operator, take F to be $\curlywedge$. Obviously $S\underline{x}\underline{\alpha}\dashv\vdash\curlywedge$. Otherwise there are lists of rule schemata $\Phi_i(\underline{X},\underline{Y}_{\underline{A}},\underline{A})$ associated with S, containing at most operators S_j for j<k. For these S_j, there are, by induction hypothesis, formula schemata $F_j(\underline{Z},\underline{Z}_{1\underline{B}},\underline{B})$ containing at most $\&,\vee,\supset,\curlywedge,\forall,\exists$ as operators such that for all $\underline{z}$, $\underline{z}_{1\underline{\beta}}$, $\underline{\beta}$
$S_j\underline{z}\underline{\beta}\dashv\vdash F_j(\underline{z},\underline{z}_{1\underline{\beta}},\underline{\beta})$,
and, if all members of $\underline{\beta}$ are members of $\underline{\alpha}$, then for all $\underline{z}$, $\underline{z}_{2\underline{\alpha}}$, $\underline{\beta}$, $S_j\underline{z}\underline{\beta}\dashv\vdash F_j(\underline{z},\underline{z}_{2\underline{\alpha}},\underline{\beta})$.
By application of the replacement theorem 4.2 we obtain lists $\Phi_i'(\underline{X},\underline{Y}_{\underline{A}},\underline{A})$ containing at most $\&,\vee,\supset,\curlywedge,\forall,\exists$ as operators such that for all $\underline{x}$, $\underline{y}_{\underline{\alpha}}$, $\underline{\alpha}$

$$\Phi_i(\underline{x},\underline{y}_{\underline{\alpha}},\underline{\alpha})\dashv\vdash\Phi_i'(\underline{x},\underline{y}_{\underline{\alpha}},\underline{\alpha}) \quad . \tag{7}$$

Now take $F(\underline{X},\underline{Y}_{\underline{A}},\underline{A})$ to be

$$\exists\underline{X}(\Phi_1'(\underline{X},\underline{Y}_{\underline{A}},\underline{A}))^* \vee \ldots \vee \exists\underline{X}(\Phi_m'(\underline{X},\underline{Y}_{\underline{A}},\underline{A}))^* \quad .$$

(This formula schema possibly contains vacuous quantifications which can be omitted.) Let $\underline{x}$, $\underline{y}_{\underline{\alpha}}$, $\underline{\alpha}$ be given. Since
$\Phi_i(\underline{x},\underline{y}_{\underline{\alpha}},\underline{\alpha})\vdash S\underline{x}\underline{\alpha}$ for all i ($1\le i\le m$), we have $(\Phi_i'(\underline{x},\underline{y}_{\underline{\alpha}},\underline{\alpha}))^*\vdash S\underline{x}\underline{\alpha}$ for all i ($1\le i\le m$) (by (7) and lemma 5.1). Thus by $\exists$E:

$\exists \underline{x}((\Phi_i^{!}(\underline{x},\underline{y}_{\underline{\alpha}},\underline{\alpha}))^* \vdash S\underline{x}\underline{\alpha}$

and by vE:

$F(\underline{x},\underline{y}_{\underline{\alpha}},\underline{\alpha}) \vdash S\underline{x}\underline{\alpha}$.

Conversely, since for all i ($1 \le i \le m$)

$\Phi_i(\underline{x},\underline{y}_{\underline{\alpha}},\underline{\alpha}) \vdash (\Phi_i^{!}(\underline{x},\underline{y}_{\underline{\alpha}},\underline{\alpha}))^*$ (by (7) and lemma 5.1)

and

$\Phi_i(\underline{x},\underline{y}_{\underline{\alpha}},\underline{\alpha}) \vdash F(\underline{x},\underline{y}_{\underline{\alpha}},\underline{\alpha})$ (by ∃I and vI) ,

and since $F(\underline{x},\underline{y}_{\underline{\alpha}},\underline{\alpha})$ does not contain $\underline{x}$ free, we obtain by S-E: $S\underline{x}\underline{\alpha} \vdash F(\underline{x},\underline{y}_{\underline{\alpha}},\underline{\alpha})$.

For example, the operators Π, Σ, ∃∀, L, D, I, J, T, for which I rules are given in § 4, are definable in terms of $\&,\vee,\supset,\curlywedge,\forall,\exists$ as follows:

ΠXAB:	$\forall X(A \supset B)$
ΣXAB:	$\exists X(A \& B)$
∃∀XYA:	$\exists X \forall Y A$
$LX_1X_2X_3A$:	$\exists X_1 \forall Z_A A[X_2X_3 \mid Z_AZ_A] \vee \exists X_2 \forall Z_A A[X_1X_3 \mid Z_AZ_A] \vee \exists X_3 \forall Z_A A[X_1X_2 \mid Z_AZ_A]$
DXYA:	$\forall Z_A A[XY \mid Z_AZ_A]$
IXA:	$\forall XY_A(A \supset A[X \mid Y_A])$
JXYAB:	$\exists XY(A \& \forall XZ_{AB}((A \& A[X \mid Z_{AB}]) \supset B[Y \mid Z_{AB}]))$
TX_1X_2A:	$\forall X_1Y_AZ_A((A[X_2 \mid Y_A] \& A[X_1X_2 \mid Y_AZ_A]) \supset A[X_2 \mid Z_A])$

J can, for instance, be used to express the relation between a one place predicate P_1 and a two place relation P_2 which holds if P_1 is satisfiable and for all x,y, if P_1x and P_1y, then P_2xy. T can be used to express the transitivity of a relation, etc.

Application of the replacement theorem 4.2 yields that for each rule ρ there is a rule ρ^+ containing at most $\&,\vee,\supset,\curlywedge,\forall,\exists$ as operators such that $\rho \dashv\vdash \rho^+$. Thus by lemma 5.1.:

$$\rho_1,\ldots,\rho_n \vdash \alpha \quad \text{iff} \quad \rho_1^{+*},\ldots,\rho_n^{+*} \vdash \alpha^{+*} ,$$

where $\rho_1^{+*},\ldots,\rho_n^{+*},\alpha^{+*}$ are formulas containing at most $\&,\vee,\supset,\curlywedge,\forall,\exists$ as operators. If we assume that

(**) every derivation of β from Δ can be transformed into a derivation of β from Δ only using rules for operators occurring in β or Δ as basic rules,

we obtain:

$$\rho_1,\ldots,\rho_n \vdash \alpha \quad \text{iff} \quad \rho_1^{+*},\ldots,\rho_n^{+*} \vdash_I \alpha^{+*} \quad ,$$

where I denotes the ordinary natural deduction system for intuitionistic logic as presented in Prawitz [13]. Furthermore, since for formulas $\beta_1,\ldots,\beta_n,\beta$ containing at most $\&,\vee,\supset,\bot,\forall,\exists$, $^{+*}$ is the identity, it holds that

$$\beta_1,\ldots,\beta_n \vdash_I \beta \quad \text{iff} \quad \beta_1,\ldots,\beta_n \vdash \beta \quad .$$

This shows that our generalized calculus can be embedded in ordinary intuitionistic logic and vice versa. (**) can be proved, since the normalization procedures and subformula principles, as given in [13], can be taken over to our generalized calculus for logical operators. This is done for the sentential case in [17], and the quantifier case does not provide additional problems in principle.

6. CONCLUDING REMARKS

The approach presented here is designed for the interpretation of intuitionistic quantifier logic. It is natural to look for a similar interpretation of classical logic. Since the classical I and E rules do not immediately fit into the characterization given in § 4 by (*) and the general form for basic rules for operators, one has to change the underlying framework for rules, derivations etc. One way is to introduce the denial $\sim\alpha$ of a sentence α besides its assertion and to construct calculi with refutation-rules, i.e. rules which govern the denials of sentences. $\sim$, as distinguished from the operator $\neg$ of negation, is here a sign which can only occur in outermost position and must therefore not be iterated. The rules of 'reductio ad absurdum', formulated by use of $\sim$ (and not of $\neg$), must then be considered fixed basic rules which are independent of the I and E rules for logical operators (for sentential logic this is carried out in [17], cf. also [9]).

Another way is to use multiple-conclusion logic (cf. [19]), i.e., to consider derivations based on rules which may have more than one conclusion. In that case one could always have 'direct' E rules like

$$\frac{\alpha\vee\beta}{\alpha \;\; \beta} \quad .$$

However, since our general form of I and E rules is based on 'indirect' E rules as present in the usual vE and ∃E rules of systems of natural deduction, it is much more promising to use the system of multiple-conclusion logic developed by Boričić [2] as a starting point. This system is, roughly speaking, a natural deduction calculus for sets of formulas (understood disjunctively) as premisses and conclusions of rules, in which e.g. the rule of vE takes the form

$$\frac{M\cup\{A\vee B\} \quad \begin{matrix}[\{A\}] \\ N\end{matrix} \quad \begin{matrix}[\{B\}] \\ N\end{matrix}}{M\cup N} \quad .$$

It can be considered an immediate natural deduction counterpart of the classical sequent calculus which differs from the intuitionistic one by allowing more than one formula in the succedent.

Instead of using a natural deduction framework one could work with sequent systems from the beginning. Such systems have the advantage of making structural assumptions explicit. This is especially useful for the treatment of modal operators. Whereas sequent calculi are not very natural for the interpretation of 'ordinary' intuitionistic or classical logic because they are 'meta-calculi' in the sense that, to explain the sequent arrow, one seems to have to refer to a derivability relation between antecedent and succedent (cf. [13]), thus presupposing something like natural deduction, this meta-perspective is just appropriate for modal logic (for, e.g., $\Box\alpha$ should express that α can be logically derived). For such a framework Došen [4,5] introduced sequents of higher levels; their exact relationship to our rules of higher levels is still to be investigated.

Concerning our method of proving the completeness of a system of operators with respect to a general form for I and E rules of operators, an application to Martin-Löf's system, more precisely, to its 'logical' part without natural numbers, well-orderings and universes (and perhaps without propositional equality too), seems promising. As stated above, this system is based on several assumptions which go beyond that which can be dealt with by the approach presented here, but it is nevertheless possible to give a general schema for I and E rules in Martin-Löf's framework, and I conjecture that the completeness of Π, Σ, + and the finite types can be established with respect to such a schema. This could explain some of the 'systematic character' of Martin-Löf's theory, i.e. its being free from stipulations which seem arbitrary, which makes it, at least from the philosophical point of view, so attractive as a foundation for logic and mathematics.

Acknowledgements. I should like to thank Edwald R. Griffor and an anonymous reviewer for helpful comments and suggestions, and Erika Fraiss for typing.

REFERENCES

[1] M. Beeson, Recursive Models for Constructive Set Theories, Annals of Mathematical Logic 23 (1982), 127-178.

[2] B. R. Boričić, Prilog teoriji intermedijalnih iskaznih logika, Dissertation, Beograd 1983.

[3] L. Borkowski, On Proper Quantifiers I, Studia Logica 8 (1958), 65-129.

[4] K. Došen, Logical Constants. An Essay in Proof Theory, Dissertation, Oxford 1980.

[5] K. Došen, Sequent-Systems for Modal Logic (forthcoming).

[6] M. Dummett, Elements of Intuitionism, Oxford 1977.

[7] G. Gentzen, Untersuchungen über das logische Schließen, Mathematische Zeitschrift 39 (1935), 176-210, 405-431.

[8] D. Kalish and R. Montague, Logic. Techniques of Formal Reasoning, New York 1964.

[9] F. v. Kutschera, Ein verallgemeinerter Widerlegungsbegriff für Gentzenkalküle, Archiv für mathematische Logik und Grundlagenforschung 12 (1969), 104-118.

[10] P. Lorenzen, Einführung in die operative Logik und Mathematik, Berlin 1955, 2nd edition 1969.

[11] P. Martin-Löf, Constructive Mathematics and Computer Programming. In: L. J. Cohen, J. Los, H. Pfeiffer and K. P. Podewski (eds.), Logic, Methodology and Philosophy of Science VI, Amsterdam 1982, 153-175.

[12] D. Prawitz, Angående konstruktiv logik och implikationsbegreppet. In: Sju filosofiska studier tillägnade Anders Wedberg, Stockholm 1963, 9-32.

[13] D. Prawitz, Natural Deduction. A Proof-Theoretical Study, Stockholm 1965.

[14] D. Prawitz, Towards a Foundation of a General Proof Theory. In: P. Suppes, L. Henkin, A. Joja, G. C. Moisil (eds.), Logic, Methodology and Philosophy of Science IV, Amsterdam 1973, 225-250.

[15] D. Prawitz, Remarks on Some Approaches to the Concept of Logical Consequence (forthcoming).

[16] H. A. Schmidt, Mathematische Gesetze der Logik I. Vorlesungen über Aussagenlogik, Berlin 1960.

[17] P. Schroeder-Heister, Untersuchungen zur regellogischen Deutung von Aussagenverknüpfungen, Dissertation, Bonn 1981.

[18] P. Schroeder-Heister, A Natural Extension of Natural Deduction (to appear in The Journal of Symbolic Logic).

[19] D. J. Shoesmith and T. J. Smiley, Multiple-Conclusion Logic, Cambridge 1978.

[20] J. I. Zucker and R. S. Tragesser, The Adequacy Problem for Inferential Logic, Journal of Philosophical Logic 7 (1978), 501-516.

ON SUBSETS OF THE SKOLEM CLASS OF EXPONENTIAL POLYNOMIALS

P.H. Slessenger
Mathematics Department
Leeds University
Leeds LS2 9JT
England

Introduction

The Skolem class $\mathcal{T}$ is defined to be the least class of functions: $N \to N$ which contains the constant function 0 and the identity function x, and which containing $f(x)$ and $g(x)$ must contain $f(x) + g(x)$, $f(x) \cdot g(x)$ and $f(x)^{g(x)}$.

$\mathcal{T}$ is well-ordered by eventual domination $f(x) \prec g(x) \leftrightarrow \exists n \in N \ \forall x > n \ f(x) < g(x)$. (Ehrenfeucht [1(1973)] using results of Richardson [7(1969)] and Kruskal [2(1960)].)

This paper presents two results concerning the class of exponential arithmetic functions. (Many of the earlier results are given, without proof, to provide a general background.) The first gives an upper bound to the points of intersection of elements of a particular sub-class. The second deals with some other sub-classes whose elements dominate initial segments of $\mathcal{T}$ which are closed variously under addition, multiplication and the raising of functions to the power "x".

Section 1, General Background

Skolem [8(1956)] defined S to be the set of functions of x that can be built from $0,1$ using the two rules of production

(1) If $f(x)$ and $g(x) \in S$ then build $f(x) + g(x)$ in S.

(2) If $f(x) \in S$ then build $x^{f(x)}$ in S.

Skolem used the means of generation of the functions in S to show firstly that for any $f(x), g(x) \in S$, one of $f(x) \prec g(x)$, $f(x) = g(x)$, $g(x) \prec f(x)$ is true, and secondly that the order type of S is ϵ_0. Skolem then went on to define $\mathcal{T}$.

Levitz [3(1975)] extended S to S' by including the rules

(3) If $f(x), g(x) \in S'$, then form $f(x)\cdot g(x)$ in S', and

(4) If $f(x) \in S'$ and $n \in N$ then form $n^{f(x)}$ in S',

with (1) and (2) above. He then uses an order preserving 1:1 map $G: S' \to \epsilon_0$ to determine the order type.

S' is, so far, the most general subset of T whose order type is known. Levitz [4(1974)] has shown the order type of T to be $\le \chi_0$.

McBeth [6(1980)] introduced the notion of height for elements of S, which has been extended to the following for T.

Definition

The height $H(f(x))$ of a function $q(x)$ in T is given by

(1) If $n \in N$ then $H(n) = 0$

(2) $H(x) = 1$

(3) $H(f(x) + g(x)) = H(f(x)\cdot g(x) = \max \{H(f(x)), H(g(x))\}$

(4) For $1 \prec f(x)$, and $x \preccurlyeq g(x)$,

$H(f(x)^{g(x)}) = \max \{H(f(x)), H(g(x))+1\}$.

Theorem 1

For $f(x), g(x) \in T$, $H(f(x)) < H(g(x)) \to f(x) \prec g(x)$.

Proof (This was produced by Reuben Gurevič (Leningrad) after he had seen the authors original, much longer, proof.)

Let $t_1(x) = x$, $t_{k+1}(x) = 2^{t_k(x)}$, so that $t_k(x)$ is a tower of k-1 2's with an x on the top. Then $H(t_k(x)) = k$, and it suffices to show that for some $n \in N$ $H(f(x)) = k$ implies

$$t_k(x) \preccurlyeq f(x) \prec t_k(x^n).$$

Proceed by induction on the construction of $f(x)$. The cases of addition and multiplication are evident.

Let $f(x) = p(x)^{q(x)}$. $H(f(x) = k+1$, $k \ge 1$.

If $H(p(x)) \ge k$ and $H(g(x)) = k$, then for some $m,n \in N$ we have:

$$t_{k+1}(x) = 2^{t_k(x)} = p(x)^{q(x)} \prec t_k(x^n)^{t_k(x^m)}$$

$$\text{and R.H.S.} = 2^{t_{k-1}(x^n)\cdot t_k(x^m)} \prec 2^{t_k(x^{(n+m)})} = t_{k+1}(x^{(n+m)})$$

and similarly for the case $H(p(x)) = k+1$, $H(q(x)) \geq k$.#

The sets S and S' are generated in such a way that anyone familiar with Cantor normal form for ordinals would expect them to have order type ϵ_0.

It seems natural to use the definition of height to restrict our rules concerning the generation of functions within a class and investigate whether we can still obtain ϵ_0.

Definition 2

Let H be the least class of functions: $N \to N$, containing the identity function x and which, containing $f(x)$ and $g(x)$ contains $f(x)\cdot g(x)$ and $f(x)^{g(x)}$ if and only if $H(f(x)) = H(g(x))$.

Theorem 3

The order type of H is ϵ_0. (Slessenger [9(1981)].)

Section 2, Upper Bounds on Intersections

All the results so far quoted have been obtained as a consequence of defining a standard form for each element of the set. That standard form has then been used to decide between $f(x) \prec g(x)$, $f(x) = g(x)$, $g(x) \prec f(x)$ for any $f(x)$, $g(x)$ in that set, with a subsequent definition of an ordinal for each $f(x)$.

Levitz [5(1983)] has, however, defined a set F, the least class of functions containing the constant and identity functions 1 and x, which is closed under addition, multiplication and base 2 exponentiation ($2^{f(x)}$) of functions, for which he has been able to prove the following.

Theorem 4

If $f(x), g(x) \in P$, then $f(x) \prec g(x) \leftrightarrow$

$f(2^{f(2)+g(2)}) < g(2^{f(2)+g(2)})$.

However, in H, because of its simpler nature we have:

Theorem 5

If $f(x), g(x) \in H$ then $f(x) \prec g(x) \leftrightarrow$

$f(f(2) + g(2)) < g(f(2) + g(2))$.

$f(2) + g(2)$ would appear to be the least practical upper bound on the intersections of $f(x)$ and $g(x)$. (The bound given by Levitz for elements of P is proved using logarithms to base 2 and so cannot be realistically lowered without a different method of proof.)

To prove the result we need a few preliminaries. (Proofs of the next three lemmas can be found in 9.)

Definition 6

$f(x) \in H$ is multiplicatively prime if and only if there are no two elements $g(x), h(x)$ of H such that $f(x) = g(x) \cdot h(x)$.

Lemma 7

If $f(x), g(x) \in H$, and $f(x)$ is multiplicatively prime, then $g(x) \prec f(x)$ implies $g(x) \cdot g(x) \prec f(x)$.

Lemma 8

If $f(x)^{g(x)}$ and $p(x)^{g(x)}$ are multiplicatively prime elements of H then

$$f(x)^{g(x)} \prec p(x)^{q(x)} \text{ iff } g(x) \prec q(x),$$
$$\text{or } g(x) = g(x) \text{ and } f(x) \prec p(x).$$

Definition 9

$H^k = \{f(x) \in H : H(f(x)) = k\}$, and let $h_k(x)$ be the least element of H^k. So e.g. $h_1(x) = x$, $h_{k+1}(x) = h_k(x)^{h_k(x)}$.

Lemma 10

If $H(f(x)) = k$ and $h_k(x)$ is the least element of H of height k, then there is no $g(x) \in H$ such that

$$f(x) \prec g(x) \prec f(x) \cdot h_k(x).$$

Lemma 11

If $H(f(x)) = k$ and $a = f(2)+h_{k+1}(2)$, then $\forall\ x > a,\ f(x) < h_{k+1}(x)$.

Proof (by induction on k)

Base step. Let $k = 1$. Then for some $m \in N$ $f(x) = x^m \prec x^x = h_2(x)$. Obviously $\forall\ x > m,\ f(x) < h_{k+1}(x)$. Hence the base step.

Induction step. Assume that for all $j < k$, if $H(f(x)) = j$, then $\forall\ x > f(2)+h_{j+1}(2),\ f(x) < h_{j+1}(x)$.

We know that for any $f(x) \in H$,

$$f(x) = \prod_{i=1}^{n} u_i(x)^{v_i(x)}.$$

We also know that $h_{j+1}(x) = h_j(x)^{h_j(x)}$. From the inductive hypothesis we have that

$$\forall\ x > \prod_{i=1}^{n} u_i(2) + h_j(2) = a, \text{ say } U(x) = \prod_{i=1}^{n} u_i(x) < h_k(x).$$

We therefore investigate a lower bound on x to ensure that

$$\sum_{i=1}^{n} v_i(x) < h_j(x).$$

Now $\sum_{i=1}^{n} v_i(2) + h_j(2) < \prod_{i=1}^{n} v_i(2) + h_j(2) = b$ say.

By the inductive hypothesis, we have $\forall\ x > b$

$$\sum_{i=1}^{n} v_i(x) < \prod_{i=1}^{n} v_i(x) = V(x) < h_j(x).$$

Now obviously $f(x) \prec U(x)^{V(x)}$ (considering what $f(x), U(x), V(x)$ are). But $\forall\ x > \max\{a,b\} = c$, say, $U(x)^{V(x)} < h_{j+1}(x)$. Therefore

$$\forall\ x > c,\ \prod_{i=1}^{n} u_i(x)^{v_i(x)} < h_{j+1}(x).$$

But by inspection $c < f(2) + h_{j+1}(2)$. Hence the result.#

We are now in a position to prove theorem 5, which states that for $F(x), g(x) \in H$, $F(x) \prec G(x)$ if and only if $\forall x > F(2) + G(2), F(x) < G(x)$. The method of proof is by induction on the height of $F(x)$ and $G(x)$.

(1) We prove the result for $H(F(x)), H(G(x)) = 1$.

(2) We assume the result for all $F(x), G(x)$ of height $\le k$.

(3) We show that $F(x) = \prod_{i=1}^{n} u_i(x)^{v_i(x)} \prec p(x)^{q(x)} = G(x)$

(where $p(x)$ is multiplicatively prime) if and only if

$$\forall x > \prod_{i=1}^{n} u_i(2)^{v_i(2)} + p(2)^{q(2)}, \quad \prod_{i=1}^{n} u_i(x)^{v_i(x)} < p(x)^{q(x)}.$$

(4) We then show 1,2 and 3 imply for any $F(x) \prec G(x) \in H^{k+1}$ that $\forall x > F(2) + G(2), F(x) < G(x)$.

This establishes the theorem.

Proof of Theorem 5

(1) If $H(F(x)) = 1 = H(G(x))$, then for some $r, s \in N, F(x) = x^r \prec x^s = G(x)$. Obviously $\forall x > r, F(x) < G(x)$, and

$$r < 2^r + 2^s.$$

(2) Assume that for all $F(x) \prec G(x) \in H^k$

$\forall x > F(2) + G(2), F(x) < G(x)$.

(3) Let $F(x) = \prod_{i=1}^{n} u_i(x)^{v(x)_i} \prec p(x)^{q(x)} = Q(x)$.

Where $p(x)$ is multiplicatively prime, and $v_1(x) \succ v_2(x) \succ \ldots \succ v_n(x)$.

Then for all $x \ge F(2) + G(2), F(x) < Q(x)$.

Proof

From the definitions of lemma 8, we have either

(a) $v_1(x) \prec q(x)$, or

(b) $v_1(x) = q(x)$ and $u_1(x) \prec p(x)$.

Suppose (a), then $v_1(x) \cdot n \prec v_1(x) \cdot h_k(x) \preceq q(x)$.

So $\forall x > v_1(2) \cdot h_k(2) + q(2), \; v_1(x) \le \dfrac{q(x)}{h_k(x)}$.

We also have, by lemma 11, that for all

$$x > \prod_{i=1}^{n} u_i(2) + h_{k+1}(2) = a_1 \quad \text{say,} \quad \prod_{i=1}^{n} u_i(x) < h_{k+1}(x).$$

We put $a = \max_i \{v_i(2) \cdot h_k(2) + q(2)\} \cup \{a_1\}$. Then $\forall\ x \geq a$

$$F(x) = \prod_{i=1}^{n} u_i(x)^{v_i(x)} < h_{k+1}(x)^{v_1(x)} \leq p(x)^{q(x)}$$

as $h_k(x) < p(x)$ for all $x > 1$ from the definition of H. We require this definition of 'a', as it is possible that $v_1(2) \cdot h_k(2) + q(2) < v_2(2) + q(2)$ etc. From the definition of a and $F(2) + G(2)$ it is obvious that $a < F(2) + G(2)$.

Now consider (b), $v_1(x) = q(x)$ and $u_1(x) \prec p(x)$. Then $\forall\ i > 1\ v_j(x) \prec q(x)$, and we may assume $\forall\ i > 1\ p(x) \prec u_i(x)$. Now $p(x)$ is, by assumption, multiplicatively prime so

$$u_1(x) \cdot h_k(x) \prec p(x), \quad \text{and therefore}$$

$$u_1(x)^{q(x)} \cdot h_k(x)^{q(x)} \prec p(x)^{q(x)}$$

$$\prod_{i=2}^{n} u_i(x)^{v_i(x)} \prec h_k(x)^{q(x)} = h_k(x)^{v_1(x)}.$$

So for $x \geq \left\{ \max_i v_i(2) + q(2), \prod_{i=2}^{n} u_i(2) + h_{k+1}(2) \right\} = a$ say

we have $\prod_{i=1}^{n} u_i(x)^{v_i(x)} < u_1(x)^{v_1(x)} \cdot h_k(x)^{v_1(x)} \prec p(x)^{q(x)}$.

(4) $f(x) \prec g(x)$ in $H \longleftrightarrow f(2)+g(2)) < g(f(2)+g(2))$

This follows from the fact that multiplicatively prime elements of H dominate initial segments of H which are closed under multiplication. So multiplying the right hand side by an element of H would only decrease the maximum intersection on N.

Therefore the result holds that for any $F(x), G(x) \in H$, $F(x) \prec G(x)$ if and only if, $\forall\ x > F(2) + G(2)$, $F(x) < G(x)$.#

Section 3. Functions that dominate closed initial segments

In showing the order type of T to be $\leq \chi_0$, Levitz defined the class of regular functions R.

Definition 12

$R = \{f(x) \in T : g(x) \in T, g(x) \prec f(x) \rightarrow g(x)^x \prec f(x)\}$.

We now show that R has the same order type as the two other subclasses, defined below.

Definition 13

$A = \{f(x) \in T : g(x) \in T, g(x) \prec f(x) \rightarrow g(x)\cdot n \prec f(x)\}$,

$W = \{f(x) \in T : g(x) \in T, g(x) \prec f(x) \rightarrow g(x)^n \prec f(x)\}$,

for $n \in N$.

W is the set of elements of T which dominate the intial segments of T closed under multiplication. A is the similarly defined set for additively closed initial segments of T. The two sets A and W would seem natural ones to study in order to determine the order type of T, and indeed are worthy of study in their own right.

Theorem 14

$p(x) \in A \leftrightarrow 2^{p(x)}, x^{p(x)} \in W$.

Proof

(1) $\leftarrow$ Obvious from the definitions of A and W.

(2) It is a consequence of the authors proof that $H(f(x)) < H(g(x)) \rightarrow f(x) \prec g(x)$ [9(1981)] that:

for any $f(x) \in T$ we can find a $g(x) \in T$ such that for some $m \in N$ either

$$2^{g(x)} \preceq f(x) \prec g^{m\cdot g(x)} \text{ or}$$

$$x^{g(x)} \preceq f(x) \prec x^{m\cdot g(x)}.$$

So, if $2^{g(x)} \prec 2^{p(x)}$, then $g(x) \prec p(x)$

$$\leftrightarrow m\cdot g(x) \prec p(x) \leftrightarrow 2^{m\cdot g(x)} \prec 2^{p(x)}.$$

So suppose $x^{g(x)} \prec 2^{p(x)}$.

Then, substituting 2^b for x and taking $\log_2$ of both sides, we have : $b \cdot g(2^b) \prec p(2^b)$.

Now, $p(x) \in A$, and cannot "mimic" a function of T multiplied by a constant and $\log_2(x)$. Hence neither can $p(2^b)$.

[This can be seen from the fact that all elements of A are finite products of elements of A. So let

$$p(x) = \prod_{i=1}^{n} p_i(x)^{v_i(x)}$$

where each $p_i(x)^{v_i(x)}$ is an element of A.

If none of the $p_i(x)$ are additive we have no problem. So assume $p_1(x) = r(x) + s(x)$. Then

$$p_1(x)^{v_1(x)} = r(x)^{v_1(x)} \cdot \left[1 + \frac{s(x)}{r(x)}\right]^{v_1(x)} .$$

Now $p_1(x)^{v_1(x)} \in A$, therefore

$$\left[1 + \frac{s(x)}{r(x)}\right]^{v_1(x)} \to \infty \quad \text{as} \quad x \to \infty .$$

From the structure of the function, it is immediate that neither it, nor a finite product of functions like it can "mimic" $\log_2(x)$ in any way.]

So $b \cdot g(2^b) \cdot m \prec p(2^b)$ i.e. $x^{m \cdot g(x)} \prec 2^{p(x)}$.

Hence $p(x) \in A \iff 2^{p(x)} \in W$. With a similar proof for $x^{p(x)}$.

Hence the result.#

Theorem 15

$x \preccurlyeq g(x) \in W \to 2^{g(x)} \in R$.

Proof

We have that $f(x) \prec g(x) \to f(x) \cdot f(x) \prec g(x)$, and thus $x \cdot f(x) \prec g(x)$. The rest follows by a similar method to the above. Note that $x^{g(x)}$ is not in R as $2^{g(x)} \prec x^{g(x)} \prec 2^{x \cdot g(x)}$.

Corollary 16

The order types of A,W and R are equal. This follows easily from the facts that $R \subset W \subset A$ and

$$f(x) \in A \to 2^{f(x)} = g(x) \in W \to 2^{g(x)} \in R.$$

Acknowledgements

The author would like to thank Dr. F.R. Drake (Leeds) and Hilbert Levitz (Florida), and acknowledges the receipt of an S.E.R.C. award.

References

[1] A. EHRENFEUCHT : Polynomial functions with exponentiation are well ordered. Algebra Universalis 3 (1973), pp. 261-262.

[2] J.B. KRUSKAL : Well-quasi-ordering, the tree theorem, and Vazsonyi's conjecture. Trans. Amer. Math. Soc. 95 (1960), pp. 210-225.

[3] H. LEVITZ : An ordered set of arithmetic functions representing the least ϵ-number. Zeitschr. f. math. Logik und Grundlagen d. Math. Bd. 21, S115-120 (1975).

[4] H. LEVITZ : An ordinal bound for the set of polynomial functions with exponentiation. Algebra Universalis 8 (1978), 233-243.

[5] H. LEVITZ : Decidability of some problems pertaining to base 2 exponential diophantine equations. To appear, Zeitschr. f. math. Logik und Grundlagen d. Math.

[6] R. McBETH : Fundamental sequences for exponential polynomials. Zeitschr. f. math. Logik und Grundlagen d. Math. Bd. 26, S115-122 (1980).

[7] D. RICHARDSON : Solution of the identity problem for integral exponential functions. Zeitschr. f. math. Logik und Grundlagen d. Math. Bd. 15, S.333-340 (1969).

[8] TH. SKOLEM : An ordered set of arithmetic functions representing the least ϵ-number. DET KONGELIGE NORSKE VIDENSKABERS SELKABS FORHANDLINGER Bind 29 (1956) Nr. 12.

[9] P.H. SLESSENGER : A height restricted generation of a set of arithmetic functions of order-type ϵ_0. To appear, Zeitschr. f. math. Logik und Grundlagen d. Math.

Effective Operators in a Topological Setting

Dieter Spreen
Lehrstuhl für Informatik I
Rheinisch-Westfälische Technische Hochschule Aachen
Büchel 29-31, D-5100 Aachen
West Germany

Paul Young[1]
Department of Computer Science, FR-35
University of Washington
Seattle, Washington 98195
U.S.A.

ABSTRACT. **It is the aim of this paper to present a uniform generalization of both the Myhill/Shepherdson and the Kreisel/Lacombe/Shoenfield theorems on effective operators. To this end we consider countable topological T_0-spaces that satisfy certain effectivity requirements which can be verified for both the set of all partial recursive functions and for the set of all total recursive functions, and we show that under appropriate further conditions all effective operators between such spaces are effectively continuous. From this general result we derive the above mentioned theorems and also the generalizations of these theorems which are due to Weihrauch, Moschovakis and Ceitin. There is a long history of interest in continuity theorems in recursive function theory, and such continuity results are basic in computer science when studying continuous partial orders introduced by Scott and by Eršov for studies of the semantics of the λ-calculus.**

[1]**Supported by NSF Research Grant MOS 7609212A, University of Washington.**

1. Introduction

If one starts to study effective operators on effectively given spaces, then one soon becomes acquainted with two theorems which each say that certain effective operators are effectively continuous, namely the theorem by Myhill/Shepherdson and the theorem by Kreisel/Lacombe/Shoenfield (cf. [7,9,13]). In the first case the domain of the effective operators is the set of all partial recursive functions, while in the second case it is the set of all total recursive functions. Although both statements are similar, there is a great difference between the topologies of the two spaces, and this difference seems to make it difficult to obtain an appropriate single generalization of the two theorems, in spite of the apparent similarity of the two proofs. With the usual topology, the set of all partial recursive functions is a separable, quasicompact and connected T_o-space which is not a T_1-space. On the other hand, the set of all total recursive functions is a separable metric space, neither compact nor connected.

Both theorems have been generalized. The natural generalization of the set of all partial functions are complete partially ordered sets (cpo), which have been independently introduced by Scott [14] and by Eršov [3,4] as a tool for studies in the semantics of the λ-calculus. If the cpo satisfies certain effectivity requirements, then it can be shown that effective operators that are defined for all computable elements of the cpo are effectively continuous (cf. [2,4,15]). Similarly, it has been shown independently by Moschovakis [10,11] and by Ceĭtin [1] that effective operators on recursively separable recursive metric spaces are effectively continuous. This generalizes the Kreisel/Lacombe/Shoenfield theorem. General versions of the Myhill/Shepherdson theorem and of the Kreisel/Lacombe/Shoenfield theorem have been investigated by Lachlan [8]. The aim of *this* paper is to obtain a *single* generalization which treats both theorems uniformly, in so far as their differing topologies permit.

We consider countable topological T_o-spaces with a countable base, and we show that effective operators between such spaces are effectively continuous provided certain effectivity requirements are satisfied. One of these conditions says that there is a procedure which generates for each point in the domain of the operator an effective sequence of a certain type which has this point as its limit. Perhaps the most important property is that of an operator having a *witness for noninclusion*, roughly: if some basic open set in the domain is *not* mapped into a given basic open set in the range, then we must be able to effectively find a witness for this, i.e. an element of the basic open set in the domain which is mapped outside the basic open set in the range. Another condition is that the indexing of the domain must be such that from the index of a recursive sequence which approximates some element of the domain we can compute an index of this element. Finally, for the indexing of the image space we assume that all indices of the elements of a basic open set can be enumerated, recursively in the domain of the indexing, and that this enumeration procedure is uniform with respect to the basic open sets. These conditions are discussed in Section 2.

In Sections 3 and 4 we study some important, standard examples of such spaces. We shall see that they easily satisfy the conditions of Section 2. In particular, in Section 3 we apply the results of Section 2 to topological spaces consisting of the computable elements of an effective cpo. The continuity theorem for *this* general class of cpo-s has been introduced by K. Weihrauch [15]. As a special case we obtain the Myhill/Shepherdson theorem.

In Section 4, we consider recursive metric spaces. Again applying the results of Section 2, we obtain the continuity theorem of Moschovakis and Ceĭtin, which in turn implies the Kreisel/Lacombe/Shoenfield theorem.

In the literature examples are known which show that effective operators are not effectively continuous in general (cf. [5,6,7,12,16,17]). In Section 5 we discuss some consequences of these examples, showing that the conditions of Section 2 are necessary.

2. Effective Spaces

In what follows, let $<,>: \omega^2 \to \omega$ be a recursive pairing function, let $P^{(n)}(R^{(n)})$ denote the set of all n-ary partial (total) recursive functions, and let W_i be the domain of the i-th partial recursive function φ_i with respect to some Gödel numbering φ. We let $\varphi_i(a)\downarrow$ mean that the computation of $\varphi_i(a)$ stops, and $\varphi_i(a)\downarrow_n$ mean that it stops within n steps. In the opposite cases we write $\varphi_i(a)\uparrow$ and $\varphi_i(a)\uparrow_n$ respectively.

Now, let (T,τ) be a countable topological T_o-space with a countable basis $\{B_j \mid j \in U\}$ of basic open sets ($U \subseteq \omega$). As is well-known, on such spaces a partial order $\leq_T$ can be defined as follows: $y \leq_T z$ iff for any open set C with $y \in C$, $z \in C$ also holds. Note that $\leq_T$ is trivial if (T, τ) is a T_1-space, i.e. $\leq_T$ coincides with the identity. Moreover, if $(y_a)_{a\in\omega}$ is a converging sequence of elements of T and $Lim\ y_a$ is the set of its limit points, then from $y \in Lim\ y_a$ and $z \leq_T y$ it follows that $z \in Lim\ y_a$.

Let $x : S \subseteq \omega \to T$ (onto) be a (partial) indexing of T. The value of x at $i \in S$ is denoted, interchangeably, by x_i or by $x(i)$. The same holds for the indexing B of the basic open sets. T is said to be *recursively separable*, if there is some recursively enumerable (r.e.) set $D \subseteq S$ such that $\{x_i \mid i \in D\}$ is dense in T, i.e., it intersects every open set. In this section we always assume that T is recursively separable, and we call $\{x_i \mid i \in D\}$ the *canonical* dense base of T.

An important property of the spaces we are going to consider is that for each point y of the space there is a sequence (y_a) of elements of the canonical dense base with $y = sup\ Lim\ y_a$. These sequences are not arbitrary, but are of a special type, and as we shall see in the applications, certain properties which we need in the proof of the main result are only valid for these special type of sequences. To this end, let Seq be a nonempty set of sequences of elements of the canonical dense base $\{x_i \mid i \in D\}$ of T. A sequence$(y_a) \in Seq$ is *recursive* if there is some function $f \in R^{(1)}$ with range $f \subseteq D$ such that $y_a = x_{f(a)}$ for all $a \in \omega$. Any Gödel number of f is called an *index* of the sequence (y_a). Then x is said to *satisfy* $(A1)$ and $(A2)$ respectively, if the following holds:

- $(A1)$. There is an r.e. set L such that for all $i \in S$ and $j \in U$, $< i, j > \in L$ iff $x_i \in B_j$. (Thus $A1$ merely says that the elements of the basic open sets are completely enumerable *relative* to the indexing of T.)
- $(A2)$. There is a function $k \in P^{(1)}$ such that, if m is an index of a converging recursive sequence $(y_a) \in Seq$ for which $sup\ Lim\ y_a$ exists, then $k(m)\downarrow$, $k(m) \in S$ and $x_{k(m)} = sup\ Lim\ y_a$. (Thus, if $A2$ holds, we can pass effectively from effective converging sequences of base elements to the sups of their Lim sets. Note that if T is a T_1-space, then k is often the identity.)

Finally, the space T is called *effective* if there exist functions $p, q \in P^{(1)}$ such that for $i \in S$, (i) $p(i)$ is a Gödel number of a function $g \in R^{(1)}$ such that range $g \subseteq U$ and for all $a \in \omega$, $x_i \in B_{g(a)}$ and $B_{g(a+1)} \subseteq B_{g(a)}$, and (ii) $q(i)$ is an index of a recursive sequence $(y_c) \in Seq$ with $x_i = sup\ Lim\ y_c$. (Thus, a space is effective if, given any element, x_i, of the space, we can pass effectively to a recursive sequence (y_c) of elements of its canonical dense base the sup of whose limit points is x_i. Furthermore, we can find a potentially, decreasing sequence of basic open sets $(B_{g(a)})$ each containing x_i.) Obviously, for interesting applications, we will want the basic open sets $(B_{g(a)})$ to "approach" x_i in consort with the sequence (y_c). The necessary conditions will be guaranteed when we explain below what we mean by an effective operator to *have a witness for noninclusion*. For now we content ourselves with the comment that Conditions $(A1)$, $(A2)$, and the notion of an *effective* space all seem to us to be entirely natural (if minimal) conditions for any notion of effective topologies on T_o-spaces.

Now let (T', r') be a second countable topological T_o-space with countable base $\{B'_j \mid j \in U'\}$ of basic open sets $(U' \subseteq \omega)$, and let $x' : S' \subseteq \omega \to T'$ (onto) be a (partial) indexing of T'. Moreover, let $F : T \to T'$. Then, F is *effective*, if there is a function $t \in P^{(1)}$ such that $t(i)\downarrow$, $t(i) \in S'$ and $Fx_i = x'_{t(i)}$, for all $i \in S$. Furthermore, we say that F is *effectively continuous* if there is a function $h \in P^{(2)}$ such that, if $Fx_i \in B'_j$, then $h(i,j) \in U$, $x_i \in B_{h(i,j)}$ and $F(B_{h(i,j)}) \subseteq B'_j$. Let T be effective. Then F has a *witness for noninclusion*, if there exist functions $s \in P^{(2)}$ and $r \in P^{(3)}$ such that for $i \in S$, $j \in U'$ and $a \in \omega$ with $Fx_i \in B'_j$ and $F(B(\varphi_{p(i)}(a))) \not\subseteq B'_j$ the following hold: (i) $s(i,j)\downarrow$, $s(i,j) \in U'$ and $Fx_i \in B'_{s(i,j)} \subseteq B'_j$, (ii) $r(i,a,j)\downarrow$, $r(i,a,j) \in D$ and $Fx_{r(i,a,j)} \notin B'_{s(i,j)}$ and (iii) if $y_c = x(\varphi_{q(i)}(c))$, for $c < a$, and $y_c = x_{r(i,a,j)}$, otherwise, then $(y_c) \in Seq$.

To understand how this relates to continuity, suppose that the canonical sequence of open sets for x_i, $(B(\varphi_{p(i)}(a)))$, somehow converges to x_i. Suppose that $F(x_i) \in B'_j$ but that the neighborhood $B(\varphi_{p(i)}(a))$ of x_i does not map into the neighborhood B'_j, of $F(x_i)$. Then we can effectively find a (possibly) smaller neighborhood $B'_{s(i,j)}$ with $F(x_i) \in B'_{s(i,j)} \subseteq B'_j$, and a point $x_{r(i,a,j)}$ which, under F, maps, not necessarily outside of B'_j, but at least outside of $B'_{s(i,j)}$. Furthermore, if we modify x_i's canonical sequence $(x(\varphi_{q(i)}(c)))$ of members of the recursively dense base to look like $x(\varphi_{q(i)}(c))$ for $c < a$ but to "stick" at $x_{r(i,a,j)}$ for $c \geq a$, then the resulting sequence (y_c) will "look" like it is going to converge to something which maps inside the neighborhood $B'_{s(i,j)}$. But once it "hits" the canonical neighborhood $B(\varphi_{p(i)}(a))$, of x_i which does not map inside $B'_{s(i,j)}$, (y_c) "stops" on an element $x_{r(i,a,j)}$ which does *not* map inside of $B'_{s(i,j)}$.

We next prove our general continuity theorem: *Every effective* operator having a witness for noninclusion and *mapping an effective T_o-space* in which effective sequences of elements from the canonical base can always be found for every element of the space and in which one can effectively pass from a convergent sequence of elements from the canonical base to (a name for) the sup Lim of the sequence *to a* T_o*-space* in which the elements of the basic open sets are completely enumerable *must* be effectively continuous:

THEOREM 2.1. Let T be effective, and let x and x' respectively satisfy $(A2)$ and $(A1)$. Moreover, let $F: T \to T'$ be effective, and let F have a witness for noninclusion. Then F must be effectively continuous.

PROOF. Let $L \subseteq \omega$, $k, t \in P^{(1)}$ and $p, q \in R^{(1)}$ respectively be as in $(A1)$, $(A2)$ with the definitions of F and T being effective. Moreover, let $v \in R^{(1)}$ with $W_{v(j)} = \{i \mid\ < i, j > \in L\}$ and let (s, r) be a witness for noninclusion for F. Set

$$\hat{g}(n,i,j) = \mu m \ : \ \varphi_{v(s(i,j))}(t(k(n)))\downarrow_m$$

and let $f \in R^{(3)}$ be such that

$$\varphi_{f(n,i,j)}(c) = \begin{cases} \varphi_{q(i)}(c), & if\ c < \hat{g}(n,i,j) \\ r(i, \hat{g}(n,i,j), j), & otherwise. \end{cases}$$

By the recursion theorem there is then a function $d \in R^{(2)}$ with

$$\varphi_{f(d(i,j),i,j)} = \varphi_{d(i,j)}.$$

Let $g(i,j) = \hat{g}(d(i,j),i,j)$, and suppose that $g(i,j)\uparrow$ for some $i \in S$ and $j \in U'$ with $Fx_i \in B'_j$. Then $d(i,j)$ is an index of the sequence $(y_c) \in Seq$ with $y_c = x(\varphi_{q(i)}(c))$ ($c \in \omega$). Since $x_i = sup\ Lim\ y_c$, it follows by $(A2)$ that $k(d(i,j))\downarrow$ and $x_i = x_{kd(i,j)}$. Moreover since $kd(i,j) \in S$, we furthermore obtain that $t(kd(i,j))\downarrow$ and $Fx_i = x'_{tkd(i,j)}$. Thus $x'_{tkd(i,j)} \in B'_{s(i,j)}$, which implies that $g(i,j)\downarrow$. This contradicts our assumption. Therefore $g(i,j)\downarrow$ for all $i \in S$ and $j \in U'$ with $Fx_i \in B'_j$.

Assume next that $F(B(\varphi_{p(i)}(g(i,j)))) \not\subseteq B'_j$, for some $i \in S$ and $j \in U'$ with $Fx_i \in B'_j$. Then $r(i, g(i,j), j)\downarrow$ and

$$Fx_{r(i,g(i,j),j)} \notin B'_{s(i,j)}. \tag{1}$$

Moreover, it follows for the sequence (y_c) with $y_c = x(\varphi_{d(i,j)}(c))$ that $(y_c) \in Seq$ and $sup\ Lim\ y_c = x_{r(i,g(i,j),j)}$. As above we obtain from this that $tkd(i,j)\downarrow$ and

$Fx_{r(i,g(i,j)),j)} = x'_{tkd(i,j)}$. Because of (1) it now follows that $\varphi_{v(s(i,j))}(tkd(i,j)) \uparrow$. Hence $g(i,j)\uparrow$, contradicting what we have shown above. Thus $F(B(\varphi_{p(i)}(g(i,j)))) \subseteq B'_j$ for all $i \in S$ and $j \in U'$ with $Fx_i \in B'_j$. Set $h(i,j) = \varphi_{p(i)}(g(i,j))$. Then h satisfies the condition for F being effectively continuous.

The requirements of this theorem seem to us to be very natural, and in practice it seems always easy to verify that the range and domain satisfy conditions A1 and A2. We shall discuss the condition for noninclusion more as we go along. We shall see that it too holds in many natural situations, but we shall also see that the theorem is false if the condition that F has a witness for noninclusion is not satisfied.

In the next two sections we study two important types of spaces. As we shall see, these spaces are effective, and in interesting cases effective mappings between these spaces must always have a witness for noninclusion. First we consider spaces that consist of the computable elements of an effective cpo, and then we consider recursive metric spaces.

3. The CPO Case

Let $(Q, \sqsubseteq)$ be a partial order. A subset $M \subseteq Q$ is *directed*, if $M \neq \emptyset$ and for all $y_1, y_2 \in M$ there is some $u \in M$ with $y_1, y_2 \sqsubseteq u$. $Q = (Q, \sqsubseteq, \bot)$ is a *complete partial order* (cpo), if $\bot \in Q$ is the smallest element of $(Q, \sqsubseteq)$ and for each directed subset of Q there exists its supremum. On Q a binary relation $\ll$ is defined as follows: $y_1 \ll y_2$ iff for directed subsets $M \subseteq Q$ the relation $y_2 \sqsubseteq sup\ M$ always implies the existence of a $u \in M$ with $y_1 \sqsubseteq u$. Note that $\ll$ is transitive.

A subset $Z \subseteq Q$ is a *basis* of the cpo Q, if for any $y \in Q$ the set $Z_y = \{z \in Z \mid z \ll y\}$ is directed and $y = sup\ Z_y$. If the cpo Q has a basis, then it is said to be *continuous*. For such cpo-s it is shown in [15,Lemma 2,3] that for $y_1, y_2, y_3 \in Q$.

LEMMA 3.1. (i) $y_1 \ll y_2 \Rightarrow y_1 \sqsubseteq y_2$.

(ii) $y_1 \ll y_2 \sqsubseteq y_3 \Rightarrow y_1 \ll y_3$.

(iii) $y_1 \ll y_2 \Rightarrow \exists z \in Z\ y_1 \ll z \ll y_2$.

Moreover, for continuous cpo-s a canonical topology τ_* can be defined (cf. [14]): A subset $X \subseteq Q$ is open if $(O1)$ and $(O2)$ hold:

$(O1)$ $\forall u, y \in Q\ (u \in X \wedge u \sqsubseteq y \Rightarrow y \in X)$

$(O2)\ \forall u \in X\ \exists y \in X\ y \ll u.$

The topology thus defined is called the *Scott topology* of Q. If Z is a basis of Q, then $\{0_z \mid z \in Z\}$ with $0_z = \{y \in Q \mid z \ll y\}$ is a base for this topology. As follows from Lemma 3.1, the partial order $\leq_Q$ induced by τ_s is identical with $\sqsubseteq$. Furthermore, the supremum of a monotonically increasing sequence of elements of Z is a limit point of this sequence. Hence, each such sequence (y_a) converges and *sup Lim* $y_a = sup(y_a)$. In this section we let Seq be the set of all such sequences.

Let Z be a basis of Q. If there exists an indexing $e : w \rightarrow Z$ (onto) of Z such that $\{<i,j>\mid e_i \ll e_j\}$ is r.e., then $Q = (Q, \sqsubseteq, \perp, Z, e)$ is called an *effective* cpo. An element $y \in Q$ is said to be *computable*, if $\{i \mid e_i \ll y\}$ $[= e^{-1}(Z_y)]$ is r.e. Let Q_c denote the set of all computable elements of Q, and let ρ_s be the relativization of τ_s to Q_c. Then (Q_c, ρ_s) is a countable T_o-space with basic open sets $B_j = O_{e(j)} \cap Q_c$ $(j \in \omega)$. Moreover, Z is dense in Q_c.

An indexing $x\colon \omega \rightarrow Q_c$ (onto) of the computable cpo elements is called *admissible* if it satisfies the following axioms $(CA1)$ and $(CA2)$:

$(CA1)$ $\{<i,j>\mid e_i \ll x_j\}$ is r.e.

$(CA2)$ There is a function $d \in R^{(1)}$ with $x_{d(i)} = sup\ e(W_i)$, for all indices $i \in \omega$ such that $e(W_i)$ is directed.

As it is shown in [15, Satz 5] such indexings exist. Moreover, it is easy to verify that $(A1)$ holds for any indexing x of Q_c which satisfies $(CA1)$. Since there is some $g \in R^{(1)}$ with $e = x \circ g$, if x satisfies $(CA2)$, it follows that x also satisfies $(A2)$ in this case. Because Z is dense in Q_c, we furthermore obtain that Q_c is recursively separable.

Our major theorem of this section guarantees that under rather general conditions, *all* effective operators on cpo-s are monotone and *must* have a witness for noninclusion. It thus follows from Theorem 2.1 that all such operators are effectively continuous.

THEOREM 3.2. Let $(Q, \sqsubseteq, \perp, Z, e)$ be an effective cpo and (T', τ') be a countable T_o-space with countable base $\{B'_j \mid j \in U'\}$. Moreover, let x be an admissible indexing of Q_c, and let $x' : S' \subseteq \omega \rightarrow T'$ (onto) be a (partial) indexing of T' which satisfies $(A1)$. Finally, let $F\colon Q_c \rightarrow T'$ be effective. Then the following statements hold:

(i) Q_c is effective.

(ii) F is montone with respect to $\sqsubseteq$ and $\leq_{T'}$.

(iii) F has a witness for noninclusion.

PROOF. We first show (ii). To this end let $x_i \sqsubseteq x_j$ and suppose that $Fx_i \not\leq_{T'} Fx_j$. Then there is some basic open set C with $Fx_i \in C$ but $Fx_j \notin C$. Now, let $f \in R^{(1)}$ with $\varphi_{f(n)}(a) = i$ if $\varphi_n(n) \uparrow_a$, and $\varphi_{f(n)}(a) = j$, otherwise. Then $\{x(\varphi_{f(n)}(a)) \mid a \in \omega\}$ is directed. As it is shown in [15, Lemma 11], there is some $v \in R^{(1)}$ such that $x_{v(m)} = sup\{x(\varphi_m(a)) \mid a \in \omega\}$, if $\{x(\varphi_m(a)) \mid a \in \omega\}$ is directed. Then $x_{vf(n)} = x_i$ if $n \notin K = \{c \mid \varphi_c(c) \downarrow\}$, and $x_{vf(n)} = x_j$ otherwise. Since x' satisfies $(A1)$, there is some r.e. set E with $i \in E$ iff $x'_i \in C$, for all $i \in S'$. Let $t \in P^{(1)}$ be as in the definition of F being effective. Then we have that $tvf(n) \in E$ iff $x'_{tvf(n)} \in C$. Because $x'_{tvf(n)} = Fx_{vf(n)} \in C$ iff $n \notin K$, we obtain that $tvf(n) \in E$ iff $n \notin K$. This contradicts the fact that $\omega \setminus K$ is not r.e.

Now, we prove (i). As it is shown in [15], there is a function $p \in R^{(1)}$ such that for all $i \in \omega$, $\varphi_{p(i)} \in R^{(1)}$, $e(\varphi_{p(i)}(a)) \ll e(\varphi_{p(i)}(a+1))$ $(a \in \omega)$, and $x_i = sup(e(\varphi_{p(i)}(a)))$. Moreover, there is some $g \in R^{(1)}$ with $e = x \circ g$. Let $q \in R^{(1)}$ with $\varphi_{q(i)} = g \circ \varphi_{p(i)}$. Then p and q satisfy the requirements for Q_c being effective.

For the proof of (iii) define $s \in R^{(2)}$ and $r \in R^{(3)}$ by $s(i,j) = j$ and $r(i,a,j) = \varphi_{q(i)}(a)$. Then (s,r) is a witness for noninclusion for F. In order to see this, assume that $F(B(\varphi_{p(i)}(a))) \not\subseteq B'_j$. Then $Fx(\varphi_{q(i)}(a)) = Fe(\varphi_{p(i)}(a)) \notin B'_j$, by (ii).

With Theorem 2.1 we thus have

THEOREM 3.3. Let Q be an effective cpo and let (T', τ') be a countable T_o-space with countable base. Moreover, let Q_c and T' respectively have an admissible indexing and an indexing satisfying $(A1)$. Then every effective operator $F: Q_c \to T'$ is effectively continuous.

If (T', τ') also consists of the computable elements of an effective cpo, then we obtain the generalization of the Myhill/Shepherdson theorem to effective cpo-s mentioned earlier (cf. [15]). To see how this theorem follows for the partial recursive functions, let PF be the set of all partial functions from ω into ω, let FIN be the subset of the functions with finite domain, and let $\eta : \omega \to FIN$ (onto) be a canonical indexing of FIN. Moreover, let div be the nowhere defined function, and for $f, g \in P^{(1)}$, define $f \sqsubseteq g$, to mean that g is an extension of f. Then $(PF, \sqsubseteq, div, FIN, \eta)$ is an effective cpo, each partial recursive function is a computable element of this cpo, and every Gödel numbering is admissible (cf. [15]). Hence, we obtain (cf. [9])

COROLLARY 3.4 (Myhill/Shepherdson). Every effective operator on the partial recursive functions is effectively continuous.

4. The Metric Case

Let $\mathbb{R}$ denote the set of all real numbers, and let ν be some canonical indexing of the rational numbers. Then a real number z is said to be *computable* if there is a function $f \in R^{(1)}$ such that for all $m, n \in \omega$ with $m \leq n$, the inequality $| \nu_{f(m)} - \nu_{f(n)} | < 2^{-m}$ holds and $z = lim\ \nu_{f(m)}$. Any Gödel number of the function f is called an *index* of z. This defines a partial indexing γ of the set $\mathbb{R}_c$ of all computable real numbers.

Now, let (M, δ) be a countable metric space with range $\delta \subseteq \mathbb{R}_c$, and let $x: S \subseteq \omega \rightarrow M$ (onto) be a (partial) indexing of M. Then (M, x, δ) is said to be a *recursive* metric space, if the distance funtion δ is effective. As is well-known, there is a canonical Hausdorff topology on M. The collection of sets $B_{<i,m>} = \{y \in M \mid \delta(x_i, y) < 2^{-m}\}$ $(i \in S, m \in \omega)$ is a base of this topology. Since there is an r.e. set A with $< i, j > \in A$ iff $\gamma_i < \gamma_j$, for i, j in the domain of γ (cf. [10, Lemma 5]), it follows that this choice of the basic open sets and their indexing B makes x satisfy $(A1)$ (cf. [11, Lemma 1]). Moreover, it follows that there is some r.e. set L' with $< j, < i, m >> \in L'$ iff $\delta(x_i, x_j) > 2^{-m}$, for all $i, j \in S$ and $m \in \omega$.

Let (M, x, δ) be recursively separable and let $g \in R^{(1)}$ be such that $\{x_{g(n)} \mid n \in \omega\}$ is dense in M. Then the set of all $B_{<g(i),m>}(i, m \in \omega)$ is also a base of the topology on M. In the case that (M, x, δ) is recursively separable, we shall always use this base and the numbering $H_{<i,m>} = B_{<g(i),m>}$. With respect to this base, the indexing x also satisfies $(A1)$. In the remainder of this section we let Seq be the set of all sequences (y_m) with $y_m \in \{x_{g(c)} \mid c \in \omega\}$ and $\delta(y_m, y_n) < 2^{-m}$, for all $m, n \in \omega$ with $m \leq n$.

Our first theorem of this section guarantees that any effective operator *from* a recursively separable recursive metric space in which we can pass effectively from converging sequences of base elements to their limit points *to* a recursive metric space *must* have a witness for noninclusion. Again it follows from Theorem 2.1 that *all* effective operators from such a recursive metric space to a recursive metric space *must* be effectively continuous.

THEOREM 4.1 Let (M, x, δ) be recursively separable with x satisfying $(A2)$. Moreover, let (M', x', δ') be a recursive metric space, and let $F : M \to M'$ be effective. Then M is effective and F has a witness for noninclusion.

PROOF. Let $g \in R^{(1)}$ be such that $\{x_{g(n)} \mid n \in \omega\}$ is dense in M. Then it follows from above that there is some r.e. set E such that for $i \in S$ and $j, m \in \omega$, $< i, j, m > \in E$ iff $\delta(x_i, x_{g(j)}) < 2^{-m}$. Let some enumeration of E be fixed, and for $i, a \in \omega$ let $< \hat{i}, \hat{j}, \hat{m} >$ be the first number in this enumeration with $\hat{i} = i$ and $\hat{m} = a + 2$. Set $v(i, a) = g(\hat{j})$ and $f(i, a) = < \hat{j}, a >$. Then $v, f \in P^{(2)}$. If $i \in S$ then it follows by the density of $\{x_{g(n)} \mid n \in \omega\}$ in M that $v(i, a)\downarrow$ and $f(i, a)\downarrow$, for all $a \in \omega$. Moreover, it follows from the triangle inequality that $x_i \in H_{f(i,a)}$ and $H_{f(i,a+1)} \subseteq H_{f(i,a)}$. Finally, we have that $x_{v(i,a)} \in H_{f(i,a)}$. Hence $(x_{v(i,a)})_{a\in\omega} \in Seq$ and $x_i = lim\ x_{v(i,a)}$. Now, let $p, q \in R^{(1)}$ be such that $\varphi_{p(i)}(a) = f(i, a)$ and $\varphi_{q(i)}(a) = v(i, a)$. Then p and q satisfy the requirements for M being effective.

In order to see that F has a witness for noninclusion, let $t \in P^{(1)}$ be as in the definition of F being effective. Then there is some r.e. set E' such that for $i \in S, j \in S'$ and $b, m \in \omega$, $< i, j, b, m > \in E'$ iff $\delta'(x'_j, x'_{t(i)}) + 2^{-b} < 2^{-m}$. For $i, j, m \in \omega$ let $< \hat{i}, \hat{j}, \hat{b}, \hat{m} >$ be the first element in some fixed enumeration of E' with $\hat{i} = i$, $\hat{j} = j$ and $\hat{m} = m$, and define $s(i, < j, m >) = < t(i), \hat{b} >$. Then $s \in P^{(2)}$. Moreover, if $i \in S$, $j \in S'$ and $m \in \omega$ with $Fx_i \in B'_{<j,m>}$, then $s(i, < j, m >)\downarrow$, $Fx_i \in B'_{s(i,<j,m>)}$ and $\overline{B'_{s(i,<j,m>)}} \subseteq B'_{<j,m>}$ where $\overline{B'_{<a,c>}} = \{y \in M' \mid \delta'(x'_a, y) \leq 2^{-c}\}$.

Now, assume that $F(H_{f(i,a)}) \not\subseteq B'_{<j,m>}$, for some $a \in \omega$. Then $F(H_{f(i,a)}) \not\subseteq \overline{B'_{s(i,<j,m>)}}$. As has already been noted, there are r.e. sets L and L' such that $< b, c > \in L$ iff $x_b \in H_c$, for $b \in S$ and $c \in \omega$. Furthermore, $< b, c > \in L'$ iff $x'_b \notin \overline{B'_c}$, for $b \in S'$ and c in the domain of B'. Then the set

$$A = \{< i, j, a, m, n > \mid < g(n), f(i, a) > \in L \quad \wedge \quad < tg(n), s(i, < j, m >) > \in L'\}$$

is also r.e. For $i, j, a, m \in \omega$ let $< \hat{i}, \hat{j}, \hat{a}, \hat{m}, \hat{n} >$ be the first element in some fixed enumeration of A with $\hat{i} = i$, $\hat{j} = j$, $\hat{a} = a$ and $\hat{m} = m$. Define $r(i, a, < j, m >) = g(\hat{n})$. Then $r \in P^{(3)}$. Moreover, if $i \in S$, $j \in S'$ and $a, m \in \omega$ are such that $Fx_i \in B'_{<j,m>}$ but $F(H_{f(i,a)}) \not\subseteq B'_{<j,m>}$, and if there is some $y \in H_{f(i,a)} \cap \{x_{g(n)} \mid n \in \omega\}$ with $Fy \notin$

$\overline{B'_{s(i,<j,m>)}}$, then it follows that $r(i,a,<j,m>)\downarrow$ and $Fx_{r(i,a,<j,m>)} \notin \overline{B'_{s(i,<j,m>)}}$. Since furthermore $x_{r(i,a,<j,m>)} \in H_{f(i,a)}$, we also obtain that if $y_c = x_{v(i,c)}$ for $c < a$ and $y_c = x_{r(i,a,<j,m>)}$ otherwise, then $(y_c) \in Seq$.

Thus, it remains to show that $F(H_{f(i,a)}) \not\subseteq \overline{B'_{s(i,<j,m>)}}$ implies the existence of some $y \in H_{f(i,a)} \cap \{x_{g(n)} \mid n \in \omega\}$ with $Fy \notin \overline{B'_{s(i,<j,m>)}}$. To this end let $z \in H_{f(i,a)}$ be such that $Fz \notin \overline{B'_{s(i,<j,m>)}}$. Then, in the same way as above, a recursive sequence $(y_c) \in Seq$ can be constructed such that $y_c \in H_{f(i,a)} \cap \{x_{g(n)} \mid n \in \omega\}$ $(c \in \omega)$ and $lim\, y_c = z$. Let $\bar{b}$ be an index for this sequence and let $k \in P^{(1)}$ be as in condition $(A2)$. Moreover, let $W_{b'} = \{c \mid < c, s(i,<j,m>) > \in L'\}$. By the recursion theorem there is then some $b \in \omega$ with

$$\varphi_b(c) = \begin{cases} \varphi_{\bar{b}}(c), & if\ \varphi_{b'}tk(b)\uparrow_c \\ \varphi_{\bar{b}}(\mu\bar{c} : \varphi_{b'}tk(b)\downarrow_{\bar{c}}), & otherwise. \end{cases}$$

It is now easy to see that the assumption that $\varphi_{b'}tk(b)\uparrow$ leads to a contradiction. Thus $\varphi_{b'}tk(b)\downarrow$, which means that $F\, lim\, x(\varphi_b(c)) \notin \overline{B'_{s(i,<j,m>)}}$. Furthermore it follows for $\hat{c} = \mu\bar{c} : \varphi_{b'}tk(b)\downarrow_{\bar{c}}$ that $\varphi_b(c) = \varphi_{\bar{b}}(\hat{c})$ for all $c \geq \hat{c}$. Hence $x_{k(b)} = lim\, x(\varphi_b(c)) = x(\varphi_{\bar{b}}(\hat{c})) = y_{\hat{c}}$. Thus there is some $y \in H_{f(i,a)} \cap \{x_{g(n)} \mid n \in \omega\}$ with $Fy \notin \overline{B'_{s(i,<j,m>)}}$, namely $y = y_{\hat{c}}$. This shows that (s,r) is a witness for noninclusion for F.

From Theorem 2.1 we now have immediately (cf. [1,10,11]):

THEOREM 4.2 (Moschovakis, Ceǐtin). Let (M,x,δ) be recursively separable with x satisfying $(A2)$, and let (M',x',δ') be a recursive metric space. Then every effective operator $F : M \to M'$ is effectively continuous.

As is well-known, $R^{(1)}$ with the Baire metric

$$\delta(f,g) = \begin{cases} 0, & if\ f = g \\ 2^{-\mu a: f(a) \neq g(a)}, & otherwise \end{cases}$$

is a metric space. It is easily shown that it is also a recursive metric space. Moreover, $(R^{(1)},\varphi,\delta)$ is recursively separable: The functions that are eventually zero form an enumerable dense subset. Finally, the Gödel numbering φ satisfies $(A2)$. In order to see this we need only note that the limit of a recursive sequence $(\varphi_{f(a)})$ with $\delta(\varphi_{f(m)},\varphi_{f(n)}) < 2^{-m}$, for all m,n with $m \leq n$, is the function $g \in R^{(1)}$ with

$g(i) = \varphi_{f(i)}(i)$. Since in addition ω with the metric $\delta'(a,b) = O$, if $a = b$, and $\delta'(a,b) = 1$, otherwise, is a recursive metric space, we thus obtain (cf. [7,13])

COROLLARY 4.3 (Kreisel/Lacombe/Shoenfield).

(i) Every effective operator $F: R^{(1)} \to R^{(1)}$ is effectively continuous.

(ii) Every effective functional $F: R^{(1)} \to \omega$ is effectively continuous.

Concluding Remarks

In the preceding sections we have shown that under some rather general conditions effective operators between countable topological T_0-spaces must be effectively continuous. Moreover, we have shown that these conditions are always satisfied either (i) if the domain is generated by an effective cpo, or (ii) if the domain is a recursively separable recursive metric space and the range is a recursive metric space.

From the literature, examples are known which show that in general effective operators are not effectively continuous (cf. [5,6,7,12,16,17]). One of these examples, namely that of Friedberg [5,13], is an effective map $G: R^{(1)} \to P^{(1)}$. Hence, it is an effective map from a recursively separable recursive metric space into the set of all computable elements of an effective cpo. This shows that in the case not covered by Theorems 3.3 and 4.2 an effective continuity result does not hold. Since $R^{(1)}$ is effective and the Gödel numbering φ fulfills the corresponding requirements in Theorem 2.1, it follows that G has no witness for noninclusion and thus that Theorem 2.1 is false without this condition.

Acknowledgements

Thanks are due to Professor Dana Scott for discussions which helped the first author see the results of this paper in a different light. Both authors wish to thank their colleague Gisela Schäfer for useful discussions, interesting ideas about cpo-s, and patience with early presentations of false proofs.

References

[1] Ceǐtin, G.S.: Algorithmic operators in constructive metric spaces. *Trudy Mat. Inst. Steklov* 67, 295-361 (1962); English transl., *Amer. Math. Soc. Transl.* (2) 64, 1-80 (1967).

[2] Egli, H., Constable, R.L.: Computability concepts for programming language semantics. *Theoret. Comp. Sci.* 2, 133-145 (1976).

[3] Eršov, Ju.L.: Computable functionals of finite types. *Algebra i Logika* 11, 367-437 (1972); English transl., *Algebra and Logic* 11, 203-242 (1972).

[4] ———: Model C of partial continuous functionals. Logic Colloquium 76 (Gandy, R., Hyland, M., eds.), 455-467. Amsterdam: North-Holland (1977).

[5] Friedberg, R.: Un contre-exemple relatif aux fonctionelles récursives. Compt. Rend. Acad. Sci. Paris 247, 852-854 (1958).

[6] Helm, J.: On effectively computable operators. *Zeitschr. f. Math. Logik Grundl. d. Math.* 17, 231-244 (1971).

[7] Kreisel, G., Lacombe, D., Shoenfield, J.: Partial recursive functionals and effective operations. Constructivity in Mathematics (Heyting, A., ed.), 290-297. Amsterdam: North-Holland (1959).

[8] Lachlan, A.: Effective operations in a general setting. *J. Symbolic Logic* 29, 163-178 (1964).

[9] Myhill, J., Shepherdson, J.C.: Effective operations on partial recursive functions. *Zeitschr. f. math. Logik Grundl. d. Math.* 1, 310-317 (1955).

[10] Moschovakis, Y.N.: Recursive analysis. Ph.D. Thesis, Univ. of Wisconsin, Madison, Wis. (1963).

[11] ———: Recursive metric spaces. *Fund. Math.* 55, 215-238 (1964).

[12] Pour-El, M.B.: A comparison of five "computable" operators. *Zeitschr. f. math. Logik Grundl. d. Math.* 6, 325-340 (1960).

[13] Rogers, H., Jr.: *Theory of Recursive Functions and Effective Computability.* New York: McGraw-Hill (1967).

[14] Scott, D.: Outline of a mathematical theory of computation. Techn. Monograph PRG-2, Oxford Univ. Comp. Lab. (1970).

[15] Weihrauch, K., Deil, Th.: Berechenbarkeit auf cpo-s. Schriften zur Angew. Math. u. Informatik Nr. 63, RWTH Aachen (1980).

[16] Young, P.: An effective operator, continuous but not partial recursive. *Proc. Amer. Math. Soc.* 19, 103-108 (1968).

[17] Young, P., Collins W.: Discontinuities of provably correct operators on the provably recursive real numbers. *J. Symbolic Logic* 48, 913-920 (1983).

An axiomatization of the apartness fragment of the theory DLO^+ of dense linear order

T. Uesu

Department of Mathematics

Kyushu University, Fukuoka

In [S], Smorynski presented the following problem:

Axiomatize the equality and apartness fragments of the theory DLO^+ of dense linear order.

The purpose of this paper is to give an axiomatization of the apartness fragment of DLO^+. An axiomatization of the equality fragment will be given in [U2].

The theory DLO^+ is the intuitionistic theory of dense linear order with the following axioms:

$$\neg x<x,$$

$$x<y \wedge y<z \supset x<z,$$

$$x<y \supset x<z \vee z<y,$$

$$\exists x \exists y\ x<y,$$

$$\exists z((x<y \supset x<z<y) \wedge (y<x \supset y<z<x) \wedge (x=y \supset x=z=y)).$$

In DLO^+, the apartness relation is defined in terms of the order relation:

$$x\#y \Longleftrightarrow x<y \vee y<x,$$

and the equality relation is defined in terms of the apartness relation:

$$x=y \Leftrightarrow \neg x\#y.$$

We give the intuitionistic theory AP^+ of apartness over which DLO^+ is conservative. It is proof-theoretically shown that DLO^+ is conservative over AP^+. Developing the method of the proof, we can proof-theoretically prove axiomatization results, which were semi-model-theoretically proved in [S], and present a method of axiomatizing fragments of theories. That will be precisely presented in [U2].

In [S], it is mentioned that a plausible guess for the apartness fragment of DLO^+ is given by the system

$$AP_\omega + \forall x_1 \forall x_2 \exists x_3 \cdots \exists x_{n+2} Exp_{n+2}(x_1, x_2, x_3, \ldots, x_{n+2}) \quad (n \geq 1),$$

where AP_ω is the intuitionistic theory given by the following axioms:

$$\neg\, x\#x,$$

$$x\#y \supset y\#x,$$

$$x\#y \supset x\#z \vee z\#y,$$

$$\exists x_1 \cdots \exists x_n \bigwedge_{1\leq i<j\leq n} x_i\#x_j \quad (n\geq 2),$$

$Exp_{n+2}(x_1, x_2, x_3, \ldots, x_{n+2})$ is the formula

$$\bigwedge_{1\leq i<j\leq n+2} (x_i\#x_j \supset \bigwedge_{1\leq k<\ell\leq n+2} x_k\#x_\ell)$$

and Exp stands for "explosive"; however, AP^+ is a proper extension of the system. We prove this fact by giving a Kripke model of the system but not of AP^+.

1. Preliminary

We consider formal theories which are Gentzen's LK-type: A <u>theory</u> over a language consists of initial sequents and inference figures of the language.

If a sequent $\Gamma \to \Delta$ is provable in a theory T, then we write

$$\vdash_{T} \Gamma \to \Delta.$$

If a sequent $\Gamma \to \Delta$ is provable in the theory which is obtained from a theory T by adding all the instances of sequents in a set G of sequents as initial sequents, then we write

$$G \vdash_{T} \Gamma \to \Delta.$$

For a sequent $A_1,\dots,A_m \to B_1,\dots,B_n$, a formula <u>corresponding to</u> the sequent is the formula $A_1 \wedge \cdots \wedge A_m \supset B_1 \vee \cdots \vee B_n$ if $m \neq 0$ and $n \neq 0$; the formula $B_1 \vee \cdots \vee B_n$ if $m=0$ and $n \neq 0$; the formula $\neg(A_1 \wedge \cdots \wedge A_m)$ if $m \neq 0$ and $n=0$; the false sentence $\curlywedge$ if $m=n=0$.

For each n-tuple $(x_1,\dots,x_n)$ of pairwise distinct variables and for each n-tuple $(t_1,\dots,t_n)$ of (not necessarily pairwise distinct) terms, the figure

$$\begin{pmatrix} x_1 & \cdots & x_n \\ t_1 & & t_n \end{pmatrix}$$

is called a <u>substitution</u> for $x_1,\dots,x_n$ (cf. [Ul] p.334 or [R] p.31). If a substitution θ has the form $\begin{pmatrix} x_1 & \cdots & x_n \\ t_1 & & t_n \end{pmatrix}$, then, for each formula A, $A\theta$ denotes the formula which results from replacing all free occurrences of $x_1,\ldots,x_n$ in A by $t_1,\ldots,t_n$, respectively. We agree that whenever $A\theta$ appears, $t_1,\ldots,t_n$ are free at the free occurrences of $x_1,\ldots,x_n$, respectively, in A, as usual. If Σ is a sequence of formulas $A_1,\ldots,A_n$, then $\Sigma\theta$ denotes the sequence $A_1\theta,\ldots,A_n\theta$.

Inference figures according to the following schema are called <u>substitutions</u>:

$$\frac{A_1,\ldots,A_m \rightarrow B_1,\ldots,B_n}{A_1\theta,\ldots,A_m\theta \rightarrow B_1\theta,\ldots,B_n\theta},$$

where θ is a substitution.

The theory RSL_L over a language L is the theory which has no initial sequent and whose inference figures are interchanges, contractions, cuts and substitutions.

A <u>clause</u> is a sequent whose formulas are prime.

For a set G of clauses the set

$$\{\Gamma \rightarrow \Delta \mid G \vdash_{RSL_L} \Gamma \rightarrow \Delta\}$$

is called the <u>closure</u> of G for RSL_L.

The name "RSL_L" is connected with "resolution" in [R], and the term "clause" is due to [R] also.

For a set S of inference figures of a language L, LK+S denotes the theory over L obtained from LK over L by adding all inference figures in S, and LK+S-Cut denotes the theory obtained from LK+S by deleting cuts. Similarly reading LJ in place of LK.

2. Theory AP^+

Let # and < be binary predicate symbols, and B a ternary predicate symbol.

Let Σ be a finite sequence of prime formulas containing the predicate symbol B only, T a finite sequence of prime formulas containing the predicate symbol <, and $A_{\Sigma,T}$ the set of assignments α of natural numbers to variables such that for B(x,y,z) in Σ $\alpha(x) < \alpha(z) < \alpha(y)$, $\alpha(y)<\alpha(z)<\alpha(x)$ or $\alpha(x)=\alpha(z)=\alpha(y)$; and for v<w in T $\alpha(v)<\alpha(w)$.

A sequent is a <u>section</u> of the sequences Σ and T if

(1) each formula in it is a prime formula with the predicate symbol #,

(2) it is valid for each assignment in $A_{\Sigma,T}$, where the interpretation of a#b by an assignment α is $\alpha(a)\neq\alpha(b)$,

(3) each variable in it occurs in Σ or T, and

(4) no formula occurs twice in the antecedent or in the succedent.

Note that from the conditions (3) and (4), the number of sections of the sequences Σ and T is finite.

Let $\Phi^0(\Sigma;T)$ be the conjunction of formulas corresponding to sections of Σ and T, and let $\Phi^{k+1}(\Sigma;T)$ be the formula

$$\forall x\forall y\exists z\forall v\forall w(v\#w \supset \Phi^k(\Sigma,B(x,z,y);T,v<w) \vee \Phi^k(\Sigma,B(x,z,y);T,w<v)),$$

where none of the variables x,y,z,v,w occurs in Σ or T. If Σ and T are empty, then we denote $\Phi^k(\Sigma;T)$ by Φ^k. Φ^0 is the true sentence.

Let AP^+ be the intuitionistic theory with the following axioms:

$$\exists x \exists y\ x \# y,\ \Phi_1,\ \Phi_2,\ \ldots$$

Theorem. The theory DLO^+ is conservative over the theory AP^+, i.e. AP^+ axiomatizes the apartness part of DLO^+.

3. Proof of Theorem

Let S_0 be the set of inference figures of the language $\{\#,<,B\}$ of the form

$$\frac{B(x,z,y),\ \Gamma \rightarrow \Delta}{\Gamma \rightarrow \Delta},$$

where the variable z differs from x and y, and does not occur free in the lower sequent $\Gamma \rightarrow \Delta$.

Let S_1 be the set of inference figures of the language $\{\#,<,A\}$ of the form

$$\frac{\Gamma \rightarrow \Delta, v\#w \qquad v<w, \Pi \rightarrow \Lambda \qquad w<v, \Pi \rightarrow \Lambda}{\Gamma, \Pi \rightarrow \Delta, \Lambda}.$$

Along the line of Gentzen's proof of the Hauptsatz, we can establish the following lemma:

Lemma 1. Let G be a set of clauses of the language $\{\#,<,B\}$, and $\bar{G}$ the closure of G for $RSL_{\{\#,<,B\}}$. Then $G \vdash_{LK+S_1} \Pi \rightarrow \Lambda$ implies $\bar{G} \vdash_{LK+S_1\text{-Cut}} \Pi \rightarrow \Lambda$.

Similarly reading LJ and $S_0 \cup S_1$ in place of LK and S_1, respectively.

The above lemma is a cut elimination theorem for systems with new initial sequents and new inference figures. For systems with new initial sequents a similar theorem was given in [A].

A more general from, which covers the above lemma and Akaboshi's one in [A], will be given in [U2].

Let G_0 be the set of the following sequents:

$$B(x,z,y), x<y \rightarrow x<z \quad ;$$
$$B(x,z,y), x<y \rightarrow z<y \quad ;$$
$$B(x,z,y), x<z \rightarrow z<y \quad ;$$
$$B(x,z,y), z<y \rightarrow x<z \quad ;$$
$$B(x,z,y) \rightarrow B(y,z,x) \quad ;$$
$$x<x \rightarrow \quad ;$$
$$x<y, y<z \rightarrow x<z \quad ;$$
$$x<y \rightarrow x\#y \quad ;$$
$$y<x \rightarrow x\#y \quad .$$

Let $DLO^+(B)$ be the theory obtained from LJ over $\{\#,<,B\}$ by adding all the instances of sequents in G_0 and the sequents $\rightarrow \exists x \exists y\ x\#y$ and $\rightarrow \forall x \forall y \forall z(x\#y \supset x\#z \vee z\#y)$ as initial sequents, and all the inference figures in $S_0 \cup S_1$.

Then we have the following lemma.

Lemma 2. $DLO^+(B)$ is conservative over DLO^+.

Proof. It is obvious that $DLO^+(B)$ is an extension of DLO^+.

If a sequent $\Gamma \rightarrow \Delta$ is provable in $DLO^+(B)$, then the sequent which is obtained from $\Gamma \rightarrow \Delta$ by replacing all the occurrences of $B(x,z,y)$ by the formula

$$(x<y \supset x<z<y) \wedge (y<x \supset y<z<x) \wedge (x=y \supset x=z=y)$$

is provable in DLO^+.

Therefore $DLO^{+}(B)$ is conservative over DLO^{+}.

Lemma 3. Let Σ be a finite sequence of prime formulas with the predicate symbol B, and T a finite sequence of prime formulas with the predicate symbol <. If a sequent $\Gamma \rightarrow \Delta$ of the language {#} has no logical symbol and each variable in $\Gamma \rightarrow \Delta$ occurs in Σ or T, and if no formula occurs twice in Γ nor in Δ, then the following conditions are equivalent:

(1) $\Gamma \rightarrow \Delta$ is a section of Σ and T.

(2) $\overline{G}_0 \vdash_{\overline{LJ+S_1\text{-Cut}}} \forall x \forall y \forall z(x\#y \supset x\#z \vee z\#y), \Sigma, T, \Gamma \rightarrow \bigvee\!\!\!\bigvee \Delta$, where $\overline{G}_0$ is the closure of G_0 for $RSL_{\{\#,<,B\}}$.

Proof. Let $\tilde{\Sigma}$ be the sequence which results from replacing each formula $B(x,z,y)$ in Σ by the formula

$$x<z<y \vee y<z<x \vee x=z=y.$$

Let G_1 be the set of the following sequents:

$$x<x \rightarrow \;;\quad x<y, y<z \rightarrow x<z \;;\quad x<y \rightarrow x<z, z<y \;;$$

$$x<y \rightarrow x\#y \;;\quad y<x \rightarrow x\#y \;;\quad x\#y \rightarrow x<y, y<x.$$

Consider the following three conditions:

(3) $G_1 \vdash_{\overline{LK}} \tilde{\Sigma}, T, \Gamma \rightarrow \Delta$.

(4) $G_0 \vdash_{\overline{LK+S_1}} \forall x \forall y \forall z(x\#y \supset x\#z \vee z\#y), \Sigma, T, \Gamma \rightarrow \Delta$.

(5) $\overline{G}_0 \vdash_{\overline{LK+S_1\text{-Cut}}} \forall x \forall y \forall z(x\#y \supset x\#z \vee z\#y), \Sigma, T, \Gamma \rightarrow \Delta$.

Then it is clear that (1) is equivalent to (3). Since

$$G_0 \vdash_{\overline{LK+S_1}} B(x,z,y) \rightarrow x<z<y \vee y<z<x \vee x=z=y,$$

(3) is equivalent to (4). By Lemma 1, we see that (4) is equivalent to (5). By induction on the number of inference

figures in the proof figure, we can show that if there is a proof figure of $\Pi \to \Lambda$ from $\overline{G_0}$ in LK+S_1-Cut without inference figures for $\neg$-right, $\supset$-right, nor $\forall$-right, then $\Pi \to \bigvee\Lambda$ is provable from $\overline{G_0}$ in LJ+S_1-Cut. Since a proof figure of the sequent

$$\forall x\forall y\forall z(x\#y \supset x\#z \vee z\#y), \Sigma, T, \Gamma \to \Lambda$$

from $\overline{G_0}$ in LK+S_1-Cut does not contain inference figures for $\neg$-right, $\supset$-right nor $\forall$-right, (5) is equivalent to (2).

Lemma 4. (1) $\vdash_{\overline{LJ}} \Phi^k(\Sigma';T') \to \Phi^k(\Sigma;T)$,

if each formula in Σ occurs in Σ' and each formula in T occurs in T'.

(2) For each natural number j greater than k

$$\vdash_{\overline{LJ}} \exists x\exists y\ x\#y,\ \Phi^j(\Sigma;T) \to \Phi^k(\Sigma;T).$$

(3) For each substitution θ

$$\vdash_{\overline{LJ}} \forall x\ x=x, \Phi^k(\Sigma;T)\theta \to \Phi^k(\Sigma\theta;T\theta).$$

Proof. (1) By induction on k.

(2) It is obvious from the definition of $\Phi^{k+1}(\Sigma;T)$ and (1).

(3) Let θ be a substitution of the form

$$\begin{pmatrix} x_{1,1} & \cdots & x_{1,n_1} & x_{2,1} & \cdots & x_{2,n_2} & \cdots\cdots & x_{m,1} & \cdots & x_{m,n_m} \\ y_1 & & y_1 & y_2 & & y_2 & & y_m & & y_m \end{pmatrix},$$

where $x_{1,1},\dots,x_{1,n_1},x_{2,1},\dots,x_{2,n_2},\dots,x_{m,1},\dots,x_{m,n_m}$ are all the variables occurring in Σ or T, and $y_1,\dots,y_m$ are pairwise distinct. Let θ^- be the substitution

$$\begin{pmatrix} y_1 & y_2 & & y_m \\ & & \cdots & \\ x_{1,1} & x_{2,1} & & x_{m,1} \end{pmatrix},$$

and let Π be the sequence of formulas $x_{1,1}\#x_{1,2},\dots,x_{1,1}\#x_{1,n_1}$, $x_{2,1}\#x_{2,2},\dots,x_{2,1}\#x_{2,n_2},\dots,x_{m,1}\#x_{m,2},\dots,x_{m,1}\#x_{m,n_m}$. Then, for each section $\Gamma \to \Delta$ of $\Sigma\theta$ and $T\theta$, the sequent $\Gamma\theta^- \to \Delta\theta^-,\Pi$ is a section of Σ and T, and $\Gamma\theta^-\theta \to \Delta\theta^-\theta,\Pi\theta$ is the sequent

$$\Gamma \to \Delta, y_1\#y_1,\dots,y_1\#y_1, y_2\#y_2,\dots,y_2\#y_2,\dots,y_m\#y_m,\dots,y_m\#y_m.$$

Therefore

$$\vdash_{LJ} \forall x\ x=x, F\theta, \Gamma \to \Delta$$

for each section $\Gamma \to \Delta$ of $\Sigma\theta$ and $T\theta$, where F is the formula corresponding to the section $\Gamma\theta^- \to \Delta\theta^-,\Pi$ of Σ and T, and so

$$\vdash_{LJ} \forall x\ x=x, \Phi^0(\Sigma;T)\theta \to \Phi^0(\Sigma\theta;T\theta).$$

Thus we can show the result by induction on k.

Lemma 5. Let Σ be a finite sequence of prime formulas with the predicate symbol B, T a finite sequence of prime formulas with the predicate symbol $<$, and $\Gamma \to \Delta$ a sequent of the language $\{\#\}$. Assume

$$\overline{G}_0 \vdash_{LJ+S_0 \cup S_1\text{-Cut}} \Sigma, T, \Gamma \to \Delta.$$

Then there is a natural number k such that

$$\vdash_{LJ} \Phi^{k}(\Sigma;T), \forall x\ x{=}x, \exists x\exists y\ x\#y, \Gamma \to \Delta.$$

Proof. Let Π be a finite sequence of formulas in Σ, T or Γ, and let Λ be a subsequence of Δ, i.e. Δ or empty. Let P be a proof figure of the sequent $\Pi \to \Lambda$ from $\bar{G}_0$ in LJ+$S_0 \cup S_1$-Cut. By induction on the number n of inference figures in P, we show that there is a number k such that

$$\vdash_{LJ} \Phi^{k}(\Sigma;T), \forall x\ x{=}x, \exists x\exists y\ x\#y, \Gamma \to \Delta.$$

Suppose n=0. Then, by Lemma 3, Lemma 4 (1) and the definition of $\Phi^{0}(\Sigma;T)$,

$$\vdash_{LJ} \Phi^{0}(\Sigma;T), \Gamma \to \Delta.$$

Suppose n>0. We distinguish the following cases.

(i) The last inference figure in P is an inference figure in S_0 of the form

$$\frac{B(a,c,b), \Pi \to \Lambda}{\Pi \to \Lambda}.$$

In this case, by the induction hypothesis there is a number k such that

$$\vdash_{LJ} \Phi^{k-1}(\Sigma, B(a,c,b);T), \forall x\ x{=}x, \exists x\exists y\ x\#y, \Gamma \to \Delta.$$

By the condition on the variable c,

$$\vdash_{LJ} \exists z \Phi^{k-1}(\Sigma, B(a,z,b);T), \forall x\ x{=}x, \exists x\exists y\ x\#y, \Gamma \to \Delta.$$

On the other hand, by Lemma 4(1)

$$\vdash_{\overline{LJ}} \Phi^k(\Sigma;T), \exists x \exists y\ x\#y \to \exists z \Phi^{k-1}(\Sigma, B(x,z,y);T)\begin{pmatrix} x & y \\ a & b \end{pmatrix},$$

and by Lemma 4(3)

$$\vdash_{\overline{LJ}} \exists z \Phi^{k-1}(\Sigma, B(x,z,y);T)\begin{pmatrix} x & y \\ a & b \end{pmatrix}, \forall x\ x=x \to \exists z \Phi^{k-1}(\Sigma, B(a,z,b);T),$$

where x and y do not occur in Σ nor T. Therefore

$$\vdash_{\overline{LJ}} \Phi^k(\Sigma;T), \forall x\ x=x, \exists x \exists y\ x\#y, \Gamma \to \Delta.$$

(ii) The last inference figure in P is an inference figure in S_1 of the form

$$\frac{\Pi_1 \to a\#b \quad a<b, \Pi_2 \to \Lambda \quad b<a, \Pi_2 \to \Lambda}{\Pi_1, \Pi_2 \to \Lambda}.$$

In this case, by the induction hypothesis there are numbers k_1, k_2 and k_3 such that

$$\vdash_{\overline{LJ}} \Phi^{k_1}(\Sigma;T), \forall x\ x=x, \exists x \exists y\ x\#y, \Gamma \to a\#b,$$

$$\vdash_{\overline{LJ}} \Phi^{k_2}(\Sigma;T, a<b), \forall x\ x=x, \exists x \exists y\ x\#y, \Gamma \to \Delta,$$

and

$$\vdash_{\overline{LJ}} \Phi^{k_3}(\Sigma;T, b<a), \forall x\ x=x, \exists x \exists y\ x\#y, \Gamma \to \Delta.$$

By Lemma 4, for every number k greater than k_1, k_2 and k_3,

$$\vdash_{\overline{LJ}} \Phi^k(\Sigma;T), \forall x\ x=x, \exists x \exists y\ x\#y \to \Phi^{k_1}(\Sigma;T)$$

and

$$\vdash_{\overline{LJ}} \Phi^k(\Sigma;T), \forall x\ x=x, \exists x \exists y\ x\#y,\ a\#b \to \Phi^{k_2}(\Sigma, T, a<b) \vee \Phi^{k_3}(\Sigma;T, b<a).$$

Therefore there is a number k such that

$$\vdash_{LJ} \Phi^k(\Sigma;T), \forall x\ x=x,\ \exists x \exists y\ x\#y,\ \Gamma \to \Delta.$$

(iii) The other cases. These are clear by the induction--hypothesis.

This completes the proof.

Lemma 6. For each natural number k $\vdash_{DLO^+} \Phi^k$.

Proof. By induction on k, we can show that for each finite sequence Σ of prime formulas with the predicate symbol B and for each finite sequence T of prime formulas with the predicate symbol <

$$\vdash_{DLO^+(B)} \Sigma, T \to \Phi^k(\Sigma;T).$$

Therefore, by Lemma 2, $\vdash_{DLO^+} \Phi^k$ for each natural number k.

This completes the proof.

Proof of Theorem. By Lemma 6, DLO^+ is an extension of AP^+. Let $\Gamma \to \Delta$ be a sequent of the language {#} such that $\vdash_{DLO^+} \Gamma \to \Delta$. By Lemma 2, $\vdash_{DLO^+(B)} \Gamma \to \Delta$. Therefore

$$G_0 \vdash_{LJ+S_0 \cup S_1} \exists x \exists y\ x\#y, \forall x \forall y \forall z (x\#y \supset x\#z \vee z\#y), \Gamma \to \Delta,$$

and so, by Lemma 1,

$$\overline{G}_0 \vdash_{LJ+S_0 \cup S_1 - Cut} \exists x \exists y\ x\#y,\ \forall x \forall y \forall z (x\#y \supset x\#z \vee z\#y),\ \Gamma \to \Delta.$$

Therefore, by Lemma 5,

$$\vdash_{LJ} \Phi^k, \forall x\ x=x, \exists x \exists y\ x\#y,\ \forall x \forall y \forall z (x\#y \supset x\#z \vee z\#y),\ \Gamma \to \Delta$$

for some k. On the other hand

$$\vdash_{LJ} \Phi^1 \to \forall x\ x=x \wedge \forall x \forall y \forall z (x\#y \supset x\#z \vee z\#y).$$

Thus $\vdash_{AP^+} \Gamma \to \Delta$.

Therefore DLO^+ is conservative over AP^+.

This completes the proof of Theorem.

4. Φ^k does not imply Φ^{k+1}

Let AP^0_ω be the intuitionistic theory with the following axioms:

$$\neg\, x\#x,$$
$$x\#y \supset y\#x,$$
$$x\#y \supset x\#z \vee z\#y,$$
$$\exists x \exists y\ x\#y,$$
$$\forall x_1 \forall x_2 \exists x_3 \cdots \exists x_{n+2} Exp_{n+2}(x_1, x_2, x_3, \ldots, x_{n+2}) \quad (n=1,2,\ldots),$$

where $Exp_{n+2}(x_1,x_2,x_3,\ldots,x_{n+2})$ is the formula

$$\bigwedge_{1 \le i < j \le n+2} (x_i \# x_j \supset \bigwedge_{1 \le k < \ell \le n+2} x_k \# x_\ell).$$

In [S], C. Smorynski made a guess: AP^0_ω is the apartness fragment of DLO^+.

The following proposition gives a counter example to it.

Proposition. For each k, not $\vdash_{AP^0_\omega} \Phi^k \to \Phi^{k+1}$.

Proof. Let k be a positive integer.

Let $S^0_0, S^0_1, \ldots; S^1_0, S^1_1, \ldots; \ldots; S^{k+1}_0, S^{k+1}_1, \ldots;$ and T be mutually disjoint countable sets, and let $r_0, r_1, \ldots, r_k$ and r_{k+1} be mutually distinct objects which are contained in none of the above sets.

We set

$$R = \{r_0, r_1, \ldots, r_{k+1}\},$$
$$S^0 = \bigcup_{n\in\omega} S_n^0,$$
$$S^1 = \bigcup_{n\in\omega} S_n^1,$$
$$\vdots$$
$$S^{k+1} = \bigcup_{n\in\omega} S_n^{k+1},$$

and

$$S = S^0 \cup S^1 \cup \cdots \cup S^{k+1}.$$

Let $\underline{K}^{(k)}$ be the Kripke model $(K^{(k)}, <, D, \Vdash)$, where

(i) $K^{(k)}$ is the set which consists of 0 and ordered k+2-tuples of natural numbers;

(ii) $0<\mu$ for μ in $K^{(k)}-\{0\}$;

(iii) $D_\mu = R\cup S\cup T$ for μ in $K^{(k)}$;

(iv) $0\Vdash p\# q$ for p,q such that $p\neq q$ and $\{p,q\}\not\subseteq R\cup S$;

(v) $\mu\Vdash p\# q$ for p,q and μ such that $\mu=(\mu_0,\mu_1,\ldots,\mu_{k+1})\in K^{(k)}$, $p\neq q$, $\{p,q\}\not\subseteq\{r_0\}\cup S^0_{\mu_0}$, $\{p,q\}\not\subseteq\{r_1\}\cup S^1_{\mu_1}$, ..., and $\{p,q\}\not\subseteq\{r_{k+1}\}\cup S^{k+1}_{\mu_{k+1}}$.

$\underline{K}^{(k)}$ is clearly a model of AP^0_ω.

First we show that $\underline{K}^{(k)}\not\Vdash\phi^{k+1}$. Suppose

$$\underline{K}^{(k)}\Vdash\phi^0(B(r_0,s_1,r_1), B(r_0,s_2,r_2), \ldots, B(r_0,s_{k+1},r_{k+1}); \varepsilon),$$

where $s_1, s_2, \ldots, s_{k+1}\in R\cup S\cup T$ and ε is the empty sequence of formulas. Then

$$\underline{K}^{(k)}\Vdash Exp_3(r_0,s_1,r_1)\wedge Exp_3(r_0,s_2,r_2)\wedge\cdots\wedge Exp_3(r_0,s_{k+1},r_{k+1}).$$

Since, for each j in $\{1,2,\ldots,k+1\}$, $\underline{K}^{(k)} \Vdash \mathrm{Exp}_3(r_0,s_j,r_j)$ implies

$$s_j \notin \{r_0,r_j\} \cup S^0 \cup S^j \cup T,$$

there are mutually distinct numbers $j_1,\ldots,j_i$ in $\{1,2,\ldots,k+1\}$ such that

$$s_{j_1} \in \{r_{j_2}\} \cup S^{j_2},\ s_{j_2} \in \{r_{j_3}\} \cup S^{j_3},\ldots,s_{j_{i-1}} \in \{r_{j_i}\} \cup S^{j_i} \text{ and}$$

$$s_{j_i} \in \{r_{j_1}\} \cup S^{j_1}.$$

Therefore there are mutually distinct numbers $j_1,\ldots,j_i$ and an element μ in $K^{(k)}-\{0\}$ such that

$$\mu \Vdash r_0 \# r_{j_1} \wedge r_0 \# r_{j_2} \wedge \cdots \wedge r_0 \# r_{j_i} \wedge \neg s_{j_1} \# r_{j_2} \wedge \neg s_{j_2} \# r_{j_3} \wedge \cdots \wedge \neg s_{j_{i-1}} \# r_{j_i}$$
$$\wedge \neg s_{j_i} \# r_{j_1}.$$

On the other hand, the sequent

$$x_{j_1} \# y_{j_1}, x_{j_2} \# y_{j_2}, \ldots, x_{j_i} \# y_{j_i} \rightarrow z_{j_1} \# y_{j_2}, z_{j_2} \# y_{j_3}, \ldots, z_{j_{i-1}} \# y_{j_i},$$
$$z_{j_i} \# y_{j_1}, x_{j_1} \# x_{j_2}, x_{j_1} \# x_{j_3}, \ldots, x_{j_1} \# x_{j_i}$$

is a section of $B(x_1,z_1,y_1),\ldots,B(x_{k+1},z_{k+1},y_{k+1})$ and ε, and so

$$\mu \Vdash r_0 \# r_{j_1}, r_0 \# r_{j_2}, \ldots, r_0 \# r_{j_i} \rightarrow s_{j_1} \# r_{j_2}, s_{j_2} \# r_{j_3}, \ldots,$$
$$s_{j_{i-1}} \# r_{j_i}, s_{j_i} \# r_{j_1}.$$

It is a contradiction. Therefore the formula

$$\forall x_1 \forall y_1 \exists z_1 \forall x_2 \forall y_2 \exists z_2 \cdots \forall x_{k+1} \forall y_{k+1} \exists z_{k+1} \Phi^0(B(x_1,z_1,y_1), B(x_2,z_2,y_2),\ldots,$$
$$B(x_{k+1},z_{k+1},y_{k+1});\varepsilon)$$

is not valid in $\underline{K}^{(k)}$. On the other hand this formula is provable from Φ^{k+1} and $\exists x \exists y\ x\#y$ in LJ.

Thus $\underline{K}^{(k)} \not\Vdash \Phi^{k+1}$.

To show that $\underline{K}^{(k)} \Vdash \Phi^k$, we need to use the following two lemmas.

Lemma 7. Let $LK_=$ be the classical theory of equality and $LK_=^{\omega}$ the theory obtained from $LK_=$ by adding the axioms

$$\exists x_1 \exists x_2 \cdots \exists x_n \bigwedge_{1 \leq i < j \leq n} x_i \# x_j \quad (n=2,3,\ldots),$$

where we consider the formula $x\#y$ as the abbreviation of $\neg x=y$. Then

(1) $\vdash_{LK_=} \Phi^0(\Sigma;T), u\#v \rightarrow \Phi^0(\Sigma;T,u<v), \Phi^0(\Sigma;T,v<u)$,

(2) $\vdash_{LK_=^{\omega}} \Phi^0(\Sigma;T) \rightarrow \forall x \forall y \exists z \Phi^0(\Sigma, B(x,z,y);T)$,

and

(3) for each number i

$$\vdash_{LK_=^{\omega}} \Phi^0(\Sigma,T) \rightarrow \Phi^i(\Sigma;T).$$

Lemma 8. Let G_k be the set of all ordered 3k-tuples $(h_1, i_1, j_1, \ldots, h_k, i_k, j_k)$ of numbers in k+2 such that for some order $\prec$ on k+2, and for each n in $\{1,\ldots,k\}$, $h_n \prec i_n \prec j_n$, $j_n \prec i_n \prec h_n$ or $h_n = i_n = j_n$. Then there is a k-tuple $(f_1, f_2, \ldots, f_k)$ of functions such that

$$(h_1, f_1(h_1,j_1), j_1, h_2, f_2(h_1,j_1,h_2,j_2), j_2, \ldots, h_k, f_k(h_1,j_1,h_2,j_2,\ldots, h_k,j_k), j_k) \in G_k$$

for each 2k-tuple $(h_1,j_1,h_2,j_2,\ldots,h_k,j_k)$ of elements in k+2.

We call such a k-tuple $(f_1,f_2,\ldots f_k)$ of functions as in the above lemma a <u>winning strategy</u> for G_k.

Proof of Lemma 7. (1) Let α be an assignment of natural numbers to variables satisfying $\Phi^0(\Sigma;T)\wedge u\#v$. Suppose $\alpha(u)<\alpha(v)$. Let Γ be the sequence of all the formulas $x\#y$ such that $\alpha(x)\neq\alpha(y)$, and x and y occur in $\Sigma,T,u<v$. Then $\Gamma\rightarrow$ is not a section of Σ and $T,u<v$. By the discussion in Proof of Lemma 3, we have $G_1 \not\vdash_{LK} \tilde{\Sigma},T,u<v,\Gamma\rightarrow$. Therefore there is an assignment β in $A_{\Sigma,T,u<v}$ satisfying $\bigwedge\Gamma$. Such β satisfies $\Phi^0(\Sigma;T,u<v)$, and $\beta(x)\neq\beta(y)$ if and only if $\alpha(x)\neq\alpha(y)$ for each pair x,y of variables in $\Sigma,T,u<v$. Therefore α satisfies $\Phi^0(\Sigma;T,u<v)$.

(2) Let α be an assignment satisfying $\Phi^0(\Sigma;T)$, and let x,y and z be pairwise distinct variables which occur in neither Σ nor T. Let β be an assignment such that

$$\beta(u)=\alpha(u) \quad \text{for all u but z;}$$

$$\beta(z)=\alpha(x) \quad \text{if } \alpha(x)=\alpha(y);$$

$$\beta(z)\neq\alpha(u) \quad \text{for all u in } \Sigma,T,x,y \text{ if } \alpha(x)\neq\alpha(y).$$

Then β satisfies $\Phi^0(\Sigma;T)$ and $B(x,z,y)$. By the same way as in the proof of (1), we can show that β satisfies $\Phi^0(\Sigma,B(x,z,y);T)$. Therefore α satisfies $\forall x\,\forall y\,\exists z\Phi^0(\Sigma,B(x,z,y);T)$.

(3) By (1) and (2),

$$\vdash_{LK^\omega_=} \Phi^0(\Sigma;T) \rightarrow \forall x\forall y\exists z\forall u\forall v(u\#v\supset\Phi^0(\Sigma,B(x,z,y);T,u<v)$$

$$\vee\,\Phi^0(\Sigma,B(x,z,y);T,v<u)).$$

Therefore, by induction on i, we have

$$\vdash_{LK^{\omega}_{=}} \Phi^0(\Sigma;T) \to \Phi^i(\Sigma;T).$$

A proof of Lemma 8 is given in the next section.

We are now to prove that $K^{(k)} \Vdash \Phi^k$.

Let $\prec^{\#}$ be a partial order on $R \cup S \cup T$ such that for each s in $R \cup S$, $s \prec^{\#} t$ if and only if $t \in T$, and the restriction of $\prec^{\#}$ to T is a dense linear order on T without end points. Let (p_1,q_1), $(s_1,t_1),(p_2,q_2),(s_2,t_2),\ldots,(p_m,q_m),(s_m,t_m)$ be a sequence of ordered pairs of elements in $R \cup S \cup T$ such that $s_1 \prec^{\#} t_1, s_2 \prec^{\#} t_2, \ldots,$ $s_m \prec^{\#} t_m$ and $m \leq k$. Let $(h_1,j_1),(h_2,j_2),\ldots,(h_m,j_m)$ be the sequence of ordered pairs of elements in k+2 such that, for each n in $\{1,2,\ldots,m\}$, $p_n \in \{r_{h_n}\} \cup S^{h_n}$ and $q_n \in \{r_{j_n}\} \cup S^{j_n}$; or $\{p_n,q_n\} \cap T \neq \phi$ and $(h_n,j_n)=(0,0)$. Let $(f_1,f_2,\ldots,f_k)$ be a winning strategy for G_k and let

$$i_1 = f_1(h_1,j_1),$$
$$i_2 = f_2(h_1,j_1,h_2,j_2),$$
$$\vdots$$
$$i_m = f_m(h_1,j_1,h_2,j_2,\ldots,h_m,j_m).$$

Choose a sequence $u_1,u_2,\ldots,u_m$ of elements in $R \cup S \cup T$ so that for each n in $\{1,2,\ldots,m\}$

(i) if $p_n=q_n$, then $u_n=p_n$;

(ii) if $p_n \neq q_n$ and $\{p_n,q_n\} \cap T \neq \phi$, then $p_n \prec^{\#} u_n \prec^{\#} q_n$ or $q_n \prec^{\#} u_n \prec^{\#} p_n$, and

$u_n \in T - \{p_1,q_1,s_1,t_1,u_1,p_2,q_2,s_2,t_2,u_2,\ldots,$

$p_{n-1},q_{n-1},s_{n-1},t_{n-1},u_{n-1}\};$

(iii) if $p_n \neq q_n$ and $\{p_n,q_n\} \subset \{r_j\} \cup S^j_m$, then

$u_n \in S^j_m - \{p_1,q_1,s_1,t_1,u_1,p_2,q_2,s_2,t_2,u_2,\ldots,$

$p_{n-1},q_{n-1},s_{n-1},t_{n-1},u_{n-1},p_n,q_n\};$

(iv) if $p_n \in S^j_m$, $q_n \in S^j_{m'}$, and $m \neq m'$, then

$u_n \in S^j - S^j_m \cup S^j_{m'} \cup \{p_1,q_1,s_1,t_1,u_1,p_2,q_2,s_2,u_2,\ldots,$

$p_{n-1},q_{n-1},s_{n-1},t_{n-1},u_{n-1}\},$

(v) if $h_n \neq j_n$, then

$u_n \in S^{i_n} - \{p_1q_1,s_1,t_1,u_1,p_2,q_2,s_2,t_2,u_2,\ldots,p_{n-1},$

$q_{n-1},s_{n-1},t_{n-1},u_{n-1}\}.$

Let μ be an element in $K^{(k)}-\{0\}$ of the form $(\mu_0,\mu_1,\ldots,\mu_{k+1})$. We define an equivalence relation $\sim$ on the set $\{p_1,q_1,s_1,t_1,u_1,$ $p_2,q_2,s_2,t_2,u_2,\ldots,p_m,q_m,s_m,t_m,u_m\}$ by

$p \sim q$ if and only if $p=q$ or $\{p,q\} \subset \{r_n\} \cup S^n_{\mu_n}$ for some n in k+2.

Let $|p|$ be the equivalence class to which p belongs under this relation $\sim$. Let $\prec$ be an order on k+2 such that, for each n in $\{1,2,\ldots,m\}$, $h_n \prec i_n \prec j_n$, $j_n \prec i_n \prec h_n$ or $h_n=i_n=j_n$, and let $\prec^*$ be the partial order on the set of the equivalence classes such that

$|p| \prec^* |q|$ if and only if $p \prec^{\#} q$, or $p \in \{r_i\} \cup S^i$ and $q \in \{r_j\} \cup S^j$ for some i,j in k+2 with $i \prec j$.

Then we can extend $\prec^*$ to a total order on the set of the equivalence classes so that, for each n in {1,2,...,m}, unless $|p_n|=|u_n|=|q_n|$,

$$|p_n|\prec^*|u_n|\prec^*|q_n| \quad \text{or} \quad |q_n|\prec^*|u_n|\prec^*|p_n|.$$

Therefore

$$\mu\Vdash\phi^0(B(p_1,u_1,q_1),\ldots,B(p_m,u_m,q_m);s_1<t_1,\ldots,s_m<t_m).$$

Hence, by Lemma 4(1) and Lemma 7, for each number i

$$\mu\Vdash s\#t \rightarrow \phi^i(B(p_1,u_1,q_1),\ldots,B(p_m,u_m,q_m);s_1<t_1,\ldots,s_{m-1}<t_{m-1},s<t),$$

$$\phi^i(B(p_1,u_1,q_1),\ldots,B(p_m,u_m,q_m);s_1<t_1,\ldots,s_{m-1}<t_{m-1},t<s).$$

Moreover, if m=k, then

$$0\Vdash\phi^0(B(p_1,u_1,q_1),\ldots,B(p_m,u_m,q_m);s_1<t_1,\ldots,s_m<t_m).$$

Therefore, by induction on k-m,

$$\underline{K}^{(k)}\Vdash\phi^{k-m}(B(p_1,u_1,q_1),\ldots,B(p_m,u_m,q_m);s_1<t_1,\ldots,s_m<t_m).$$

Especially, $\underline{K}^{(k)}\Vdash\phi^k$.

This completes the proof.

5. Proof of Lemma 8

Consider the following condition (C) for a family of ordered sets whose domains are subsets of k+2:

(C) The intersection of any two distinct ordered sets in it is the singleton {0}, and the union of the domains of all the ordered sets in it is k+2.

Let F be a family of ordered sets satisfying the condition (C), and assume that F has two or more ordered sets. Then, for each pair h,j in k+2, we can choose a quadruple (i,K,L,M) so that $K \in F$; $L \in F$; M is an ordered set which is an extension of K and L, or an extension of K and the dual ordered set of L; h,i and j are elements in M; $h <_M i <_M j$, $j <_M i <_M h$ or h=i=j, where $<_M$ is the order of M; and $(F-\{K,L\}) \cup \{M\}$ satisfies the condition (C). To the chosen quadruple (i,K,L,M), we denote the ordered pair $(i,(F-\{K,L\}) \cup \{M\})$ by G(h,j,F). Now let

$$F_0 = \{(\{0,1\},<), (\{0,2\},<), \ldots, (\{0,k+1\},<)\}.$$

For a sequence $h_1, j_1, h_2, j_2, \ldots, h_k, j_k$ of numbers less than k+2, let

$$(i_1, F_1) = G(h_1, j_1, F_0),$$

$$(i_2, F_2) = G(h_2, j_2, F_1),$$

$$\vdots$$

$$(i_k, F_k) = G(h_k, j_k, F_{k-1}).$$

Then $i_1, i_2, \ldots, i_k, F_1, F_2, \ldots, F_k$ are defined and each i_m is determined by $h_1, j_1, h_2, j_2, \ldots, h_m, j_m$ for m less than k+1. Thus there is a winning strategy for G_k.

This completes the proof of Lemma 8.

The essential idea of the above proof is due to Kenichi Hibino. The author proved Lemma 8 independently of K. Hibino; however, the author's original proof is rather more complicated than Hibino's one.

References

[A] J. Akaboshi, On elimination theorems (in Japanese), Master thesis, Kyushu University (1981).

[R] J.A. Robinson, A machine-oriented logic based on the resolution principle, J. ACM 12 (1965), 23-41.

[S] C. Smorynski, On axiomatizing fragments, J. Symb. Logic 42 (1977), 530-544.

[U1] T. Uesu, Intuitionistic theories and toposes, Springer Lecture Notes in Math., 891 (1981), pp. 323-358.

[U2] T. Uesu, A method of axiomatizing fragments of intuitionistic theories, in preparation.

Tadahiro Uesu
Department of Mathematics
Faculty of Science
Kyushu University
Fukuoka, Postal No. 812, Japan

Vol. 952: Combinatorial Mathematics IX. Proceedings, 1981. Edited by E. Billington, S. Oates-Williams, and A.P. Street. XI, 443 pages. 1982.

Vol. 953: Iterative Solution of Nonlinear Systems of Equations. Proceedings, 1982. Edited by R. Ansorge, Th. Meis, and W. Törnig. VII, 202 pages. 1982.

Vol. 954: S.G. Pandit, S.G. Deo, Differential Systems Involving Impulses. VII, 102 pages. 1982.

Vol. 955: G. Gierz, Bundles of Topological Vector Spaces and Their Duality. IV, 296 pages. 1982.

Vol. 956: Group Actions and Vector Fields. Proceedings, 1981. Edited by J.B. Carrell. V, 144 pages. 1982.

Vol. 957: Differential Equations. Proceedings, 1981. Edited by D.G. de Figueiredo. VIII, 301 pages. 1982.

Vol. 958: F.R. Beyl, J. Tappe, Group Extensions, Representations, and the Schur Multiplicator. IV, 278 pages. 1982.

Vol. 959: Géométrie Algébrique Réelle et Formes Quadratiques, Proceedings, 1981. Edité par J.-L. Colliot-Thélène, M. Coste, L. Mahé, et M.-F. Roy. X, 458 pages. 1982.

Vol. 960: Multigrid Methods. Proceedings, 1981. Edited by W. Hackbusch and U. Trottenberg. VII, 652 pages. 1982.

Vol. 961: Algebraic Geometry. Proceedings, 1981. Edited by J.M. Aroca, R. Buchweitz, M. Giusti, and M. Merle. X, 500 pages. 1982.

Vol. 962: Category Theory. Proceedings, 1981. Edited by K.H. Kamps D. Pumplün, and W. Tholen, XV, 322 pages. 1982.

Vol. 963: R. Nottrot, Optimal Processes on Manifolds. VI, 124 pages. 1982.

Vol. 964: Ordinary and Partial Differential Equations. Proceedings, 1982. Edited by W.N. Everitt and B.D. Sleeman. XVIII, 726 pages. 1982.

Vol. 965: Topics in Numerical Analysis. Proceedings, 1981. Edited by P.R. Turner. IX, 202 pages. 1982.

Vol. 966: Algebraic K-Theory. Proceedings, 1980. Part I. Edited by R.K. Dennis. VIII, 407 pages. 1982.

Vol. 967: Algebraic K-Theory. Proceedings, 1980. Part II. VIII, 409 pages. 1982.

Vol. 968: Numerical Integration of Differential Equations and Large Linear Systems. Proceedings, 1980. Edited by J. Hinze. VI, 412 pages. 1982.

Vol. 969: Combinatorial Theory. Proceedings, 1982. Edited by D. Jungnickel and K. Vedder. V, 326 pages. 1982.

Vol. 970: Twistor Geometry and Non-Linear Systems. Proceedings, 1980. Edited by H.-D. Doebner and T.D. Palev. V, 216 pages. 1982.

Vol. 971: Kleinian Groups and Related Topics. Proceedings, 1981. Edited by D.M. Gallo and R.M. Porter. V, 117 pages. 1983.

Vol. 972: Nonlinear Filtering and Stochastic Control. Proceedings, 1981. Edited by S.K. Mitter and A. Moro. VIII, 297 pages. 1983.

Vol. 973: Matrix Pencils. Proceedings, 1982. Edited by B. Kågström and A. Ruhe. XI, 293 pages. 1983.

Vol. 974: A. Draux, Polynômes Orthogonaux Formels – Applications. VI, 625 pages. 1983.

Vol. 975: Radical Banach Algebras and Automatic Continuity. Proceedings, 1981. Edited by J.M. Bachar, W.G. Bade, P.C. Curtis Jr., H.G. Dales and M.P. Thomas. VIII, 470 pages. 1983.

Vol. 976: X. Fernique, P.W. Millar, D.W. Stroock, M. Weber, Ecole d'Eté de Probabilités de Saint-Flour XI – 1981. Edited by P.L. Hennequin. XI, 465 pages. 1983.

Vol. 977: T. Parthasarathy, On Global Univalence Theorems. VIII, 106 pages. 1983.

Vol. 978: J. Ławrynowicz, J. Krzyż, Quasiconformal Mappings in the Plane. VI, 177 pages. 1983.

Vol. 979: Mathematical Theories of Optimization. Proceedings, 1981. Edited by J.P. Cecconi and T. Zolezzi. V, 268 pages. 1983.

Vol. 980: L. Breen. Fonctions thêta et théorème du cube. XIII, 115 pages. 1983.

Vol. 981: Value Distribution Theory. Proceedings, 1981. Edited by I. Laine and S. Rickman. VIII, 245 pages. 1983.

Vol. 982: Stability Problems for Stochastic Models. Proceedings, 1982. Edited by V. V. Kalashnikov and V. M. Zolotarev. XVII, 295 pages. 1983.

Vol. 983: Nonstandard Analysis-Recent Developments. Edited by A.E. Hurd. V, 213 pages. 1983.

Vol. 984: A. Bove, J. E. Lewis, C. Parenti, Propagation of Singularities for Fuchsian Operators. IV, 161 pages. 1983.

Vol. 985: Asymptotic Analysis II. Edited by F. Verhulst. VI, 497 pages. 1983.

Vol. 986: Séminaire de Probabilités XVII 1981/82. Proceedings. Edited by J. Azéma and M. Yor. V, 512 pages. 1983.

Vol. 987: C. J. Bushnell, A. Fröhlich, Gauss Sums and p-adic Division Algebras. XI, 187 pages. 1983.

Vol. 988: J. Schwermer, Kohomologie arithmetisch definierter Gruppen und Eisensteinreihen. III, 170 pages. 1983.

Vol. 989: A.B. Mingarelli, Volterra-Stieltjes Integral Equations and Generalized Ordinary Differential Expressions. XIV, 318 pages. 1983.

Vol. 990: Probability in Banach Spaces IV. Proceedings, 1982. Edited by A. Beck and K. Jacobs. V, 234 pages. 1983.

Vol. 991: Banach Space Theory and its Applications. Proceedings, 1981. Edited by A. Pietsch, N. Popa and I. Singer. X, 302 pages. 1983.

Vol. 992: Harmonic Analysis, Proceedings, 1982. Edited by G. Mauceri, F. Ricci and G. Weiss. X, 449 pages. 1983.

Vol. 993: R.D. Bourgin, Geometric Aspects of Convex Sets with the Radon-Nikodým Property. XII, 474 pages. 1983.

Vol. 994: J.-L. Journé, Calderón-Zygmund Operators, Pseudo-Differential Operators and the Cauchy Integral of Calderón. VI, 129 pages. 1983.

Vol. 995: Banach Spaces, Harmonic Analysis, and Probability Theory. Proceedings, 1980–1981. Edited by R.C. Blei and S.J. Sidney. V, 173 pages. 1983.

Vol. 996: Invariant Theory. Proceedings, 1982. Edited by F. Gherardelli. V, 159 pages. 1983.

Vol. 997: Algebraic Geometry – Open Problems. Edited by C. Ciliberto, F. Ghione and F. Orecchia. VIII, 411 pages. 1983.

Vol. 998: Recent Developments in the Algebraic, Analytical, and Topological Theory of Semigroups. Proceedings, 1981. Edited by K.H. Hofmann, H. Jürgensen and H. J. Weinert. VI, 486 pages. 1983.

Vol. 999: C. Preston, Iterates of Maps on an Interval. VII, 205 pages. 1983.

Vol. 1000: H. Hopf, Differential Geometry in the Large. VII, 184 pages. 1983.

Vol. 1001: D.A. Hejhal, The Selberg Trace Formula for PSL(2, ℝ). Volume 2. VIII, 806 pages. 1983.

Vol. 1002: A. Edrei, E.B. Saff, R.S. Varga, Zeros of Sections of Power Series. VIII, 115 pages. 1983.

Vol. 1003: J. Schmets, Spaces of Vector-Valued Continuous Functions. VI, 117 pages. 1983.

Vol. 1004: Universal Algebra and Lattice Theory. Proceedings, 1982. Edited by R.S. Freese and O.C. Garcia. VI, 308 pages. 1983.

Vol. 1005: Numerical Methods. Proceedings, 1982. Edited by V. Pereyra and A. Reinoza. V, 296 pages. 1983.

Vol. 1006: Abelian Group Theory. Proceedings, 1982/83. Edited by R. Göbel, L. Lady and A. Mader. XVI, 771 pages. 1983.

Vol. 1007: Geometric Dynamics. Proceedings, 1981. Edited by J. Palis Jr. IX, 827 pages. 1983.

Vol. 1008: Algebraic Geometry. Proceedings, 1981. Edited by J. Dolgachev. V, 138 pages. 1983.

Vol. 1009: T.A. Chapman, Controlled Simple Homotopy Theory and Applications. III, 94 pages. 1983.

Vol. 1010: J.-E. Dies, Chaînes de Markov sur les permutations. IX, 226 pages. 1983.

Vol. 1011: J.M. Sigal. Scattering Theory for Many-Body Quantum Mechanical Systems. IV, 132 pages. 1983.

Vol. 1012: S. Kantorovitz, Spectral Theory of Banach Space Operators. V, 179 pages. 1983.

Vol. 1013: Complex Analysis – Fifth Romanian-Finnish Seminar. Part 1. Proceedings, 1981. Edited by C. Andreian Cazacu, N. Boboc, M. Jurchescu and I. Suciu. XX, 393 pages. 1983.

Vol. 1014: Complex Analysis – Fifth Romanian-Finnish Seminar. Part 2. Proceedings, 1981. Edited by C. Andreian Cazacu, N. Boboc, M. Jurchescu and I. Suciu. XX, 334 pages. 1983.

Vol. 1015: Équations différentielles et systèmes de Pfaff dans le champ complexe – II. Seminar. Edited by R. Gérard et J.P. Ramis. V, 411 pages. 1983.

Vol. 1016: Algebraic Geometry. Proceedings, 1982. Edited by M. Raynaud and T. Shioda. VIII, 528 pages. 1983.

Vol. 1017: Equadiff 82. Proceedings, 1982. Edited by H.W. Knobloch and K. Schmitt. XXIII, 666 pages. 1983.

Vol. 1018: Graph Theory, Łagów 1981. Proceedings, 1981. Edited by M. Borowiecki, J.W. Kennedy and M.M. Sysło. X, 289 pages. 1983.

Vol. 1019: Cabal Seminar 79–81. Proceedings, 1979–81. Edited by A.S. Kechris, D.A. Martin and Y.N. Moschovakis. V, 284 pages. 1983.

Vol. 1020: Non Commutative Harmonic Analysis and Lie Groups. Proceedings, 1982. Edited by J. Carmona and M. Vergne. V, 187 pages. 1983.

Vol. 1021: Probability Theory and Mathematical Statistics. Proceedings, 1982. Edited by K. Itô and J.V. Prokhorov. VIII, 747 pages. 1983.

Vol. 1022: G. Gentili, S. Salamon and J.-P. Vigué. Geometry Seminar "Luigi Bianchi", 1982. Edited by E. Vesentini. VI, 177 pages. 1983.

Vol. 1023: S. McAdam, Asymptotic Prime Divisors. IX, 118 pages. 1983.

Vol. 1024: Lie Group Representations I. Proceedings, 1982–1983. Edited by R. Herb, R. Lipsman and J. Rosenberg. IX, 369 pages. 1983.

Vol. 1025: D. Tanré, Homotopie Rationnelle: Modèles de Chen, Quillen, Sullivan. X, 211 pages. 1983.

Vol. 1026: W. Plesken, Group Rings of Finite Groups Over p-adic Integers. V, 151 pages. 1983.

Vol. 1027: M. Hasumi, Hardy Classes on Infinitely Connected Riemann Surfaces. XII, 280 pages. 1983.

Vol. 1028: Séminaire d'Analyse P. Lelong – P. Dolbeault – H. Skoda. Années 1981/1983. Edité par P. Lelong, P. Dolbeault et H. Skoda. VIII, 328 pages. 1983.

Vol. 1029: Séminaire d'Algèbre Paul Dubreil et Marie-Paule Malliavin. Proceedings, 1982. Edité par M.-P. Malliavin. V, 339 pages. 1983.

Vol. 1030: U. Christian, Selberg's Zeta-, L-, and Eisensteinseries. XII, 196 pages. 1983.

Vol. 1031: Dynamics and Processes. Proceedings, 1981. Edited by Ph. Blanchard and L. Streit. IX, 213 pages. 1983.

Vol. 1032: Ordinary Differential Equations and Operators. Proceedings, 1982. Edited by W.N. Everitt and R.T. Lewis. XV, 521 pages. 1983.

Vol. 1033: Measure Theory and its Applications. Proceedings, 1982. Edited by J.M. Belley, J. Dubois and P. Morales. XV, 317 pages. 1983.

Vol. 1034: J. Musielak, Orlicz Spaces and Modular Spaces. V, 222 pages. 1983.

Vol. 1035: The Mathematics and Physics of Disordered Media. Proceedings, 1983. Edited by B.D. Hughes and B.W. Ninham. VII, 432 pages. 1983.

Vol. 1036: Combinatorial Mathematics X. Proceedings, 1982. Edited by L.R.A. Casse. XI, 419 pages. 1983.

Vol. 1037: Non-linear Partial Differential Operators and Quantization Procedures. Proceedings, 1981. Edited by S.I. Andersson and H.-D. Doebner. VII, 334 pages. 1983.

Vol. 1038: F. Borceux, G. Van den Bossche, Algebra in a Localic Topos with Applications to Ring Theory. IX, 240 pages. 1983.

Vol. 1039: Analytic Functions, Błażejewko 1982. Proceedings. Edited by J. Ławrynowicz. X, 494 pages. 1983

Vol. 1040: A. Good, Local Analysis of Selberg's Trace Formula. III, 128 pages. 1983.

Vol. 1041: Lie Group Representations II. Proceedings 1982–1983. Edited by R. Herb, S. Kudla, R. Lipsman and J. Rosenberg. IX, 340 pages. 1984.

Vol. 1042: A. Gut, K.D. Schmidt, Amarts and Set Function Processes. III, 258 pages. 1983.

Vol. 1043: Linear and Complex Analysis Problem Book. Edited by V.P. Havin, S.V. Hruščëv and N.K. Nikol'skii. XVIII, 721 pages. 1984.

Vol. 1044: E. Gekeler, Discretization Methods for Stable Initial Value Problems. VIII, 201 pages. 1984.

Vol. 1045: Differential Geometry. Proceedings, 1982. Edited by A.M. Naveira. VIII, 194 pages. 1984.

Vol. 1046: Algebraic K-Theory, Number Theory, Geometry and Analysis. Proceedings, 1982. Edited by A. Bak. IX, 464 pages. 1984.

Vol. 1047: Fluid Dynamics. Seminar, 1982. Edited by H. Beirão da Veiga. VII, 193 pages. 1984.

Vol. 1048: Kinetic Theories and the Boltzmann Equation. Seminar, 1981. Edited by C. Cercignani. VII, 248 pages. 1984.

Vol. 1049: B. Iochum, Cônes autopolaires et algèbres de Jordan. VI, 247 pages. 1984.

Vol. 1050: A. Prestel, P. Roquette, Formally p-adic Fields. V, 167 pages. 1984.

Vol. 1051: Algebraic Topology, Aarhus 1982. Proceedings. Edited by I. Madsen and B. Oliver. X, 665 pages. 1984.

Vol. 1052: Number Theory. Seminar, 1982. Edited by D.V. Chudnovsky, G.V. Chudnovsky, H. Cohn and M.B. Nathanson. V, 309 pages. 1984.

Vol. 1053: P. Hilton, Nilpotente Gruppen und nilpotente Räume. V, 221 pages. 1984.

Vol. 1054: V. Thomée, Galerkin Finite Element Methods for Parabolic Problems. VII, 237 pages. 1984.

Vol. 1055: Quantum Probability and Applications to the Quantum Theory of Irreversible Processes. Proceedings, 1982. Edited by L. Accardi, A. Frigerio and V. Gorini. VI, 411 pages. 1984.

Vol. 1056: Algebraic Geometry. Bucharest 1982. Proceedings, 1982. Edited by L. Bădescu and D. Popescu. VII, 380 pages. 1984.

Vol. 1057: Bifurcation Theory and Applications. Seminar, 1983. Edited by L. Salvadori. VII, 233 pages. 1984.

Vol. 1058: B. Aulbach, Continuous and Discrete Dynamics near Manifolds of Equilibria. IX, 142 pages. 1984.

Vol. 1059: Séminaire de Probabilités XVIII, 1982/83. Proceedings. Edité par J. Azéma et M. Yor. IV, 518 pages. 1984.

Vol. 1060: Topology. Proceedings, 1982. Edited by L.D. Faddeev and A.A. Mal'cev. VI, 389 pages. 1984.

Vol. 1061: Séminaire de Théorie du Potentiel. Paris, No. 7. Proceedings. Directeurs: M. Brelot, G. Choquet et J. Deny. Rédacteurs: F. Hirsch et G. Mokobodzki. IV, 281 pages. 1984.